Lecture Notes in Mathematics

Edited by A. Dold and B. Eckmann

788

Topology Symposium Siegen 1979

Proceedings of a Symposium Held at the University of Siegen, June 14–19, 1979

Edited by
U. Koschorke and W. D. Neumann

Springer-Verlag
Berlin Heidelberg New York 1980

Editors

Ulrich Koschorke
Mathematik V
Universitat-Gesamthochschule Siegen
Holderlinstr 3
5900 Siegen
Federal Republic of Germany

Walter D Neumann
Department of Mathematics
University of Maryland
College Park, MD 20742
USA

AMS Subject Classifications (1980) 55-02, 55 P xx, 55 Q xx, 55 R xx, 55 S xx, 55 T xx, 57-02, 57 M 25, 57 R xx, 57 S xx

ISBN 3-540-09968-9 Springer-Verlag Berlin Heidelberg New York
ISBN 0-387-09968-9 Springer-Verlag New York Heidelberg Berlin

Library of Congress Cataloging in Publication Data Topology Symposium, University of Siegen, 1979 Topology Symposium, Siegen 1979 (Lecture notes in mathematics, 788) 1 Topology--Congresses I Koschorke, Ulrich, 1941- II Neumann, Walter D III Title IV Series Lecture notes in mathematics (Berlin), 788 QA3 L28 no 788 [QA611 A1] 510s [514] 80-12366

Printed in Germany

Printing and binding Beltz Offsetdruck, Hemsbach/Bergstr
2141/3140-543210

P R E F A C E

These are the proceedings of an international topology symposium which was held at the Gesamthochschule (Comprehensive University) of Siegen, June 14 - 19, 1979. There was a rich program of plenary lectures, special sessions and mathematical films. Some of these activities are reflected in this collection of research papers and lecture notes.

We would like to thank everyone who contributed to the success of the symposium. In particular, we are most grateful to J.F. Adams, W. Browder, J. Cerf, M. Kervaire and T. Petrie for their help in planning the scientific program. Also, there was very dedicated assistance from many members of the Gesamthochschule, especially from the young topologists Bernd Lübcke, Willi Meier and Christof Olk.

The symposium was made possible by generous grants from the Deutsche Forschungsgemeinschaft and from the Minister für Wissenschaft und Forschung von Nordrhein-Westfalen.

Siegen, October, 1979

Ulrich Koschorke   Walter Neumann

CONTENTS

Homotopy theory and algebraic topology

PARTICIPANTS

H. Abels (Bielefeld)
J. F. Adams (Cambridge)
J. Aguadé (Barcelona)
J. Anselment (Berlin)
M. Aubry (Nice)
A. Bahri (Oxford)
A. Bak (Bielefeld)
J. Barge (Orsay)
M. Barratt (Chicago)
H. Baues (Bonn)
E. Bayer (Genève)
J. Berrick (London)
Th. Bier (Gottingen)
S. Blank (Boston)
M. Breuer (Marburg)
W. Browder (Oxford)
E. H. Brown (Waltham)
S. R. Bullett (London)
D. Burghelea (New Brunswick)
J. Cerf (Orsay)
Z. Čerin (Zagreb)
H. Chaltin (Meise)
D. S. Chess (Princeton)
F. W. Clarke (Swansea)
R. Cohen (Chicago)
F. Connolly (Notre Dame)
M. Crabb (Bonn)
U. Deibnek (Bochum)
D. Y. Didet (Nantes)
K. H. Dovermann (Chicago)
P. J. Eccles (Manchester)
R. D. Edwards (Paris)
D. Erle (Dortmund)
A. Flanchec (Nantes)
S. Ferry (Princeton)
J. C. Gómez-Larrañaga (Cambridge)
D. H. Gottlieb (W. Lafayette)
I. Hambleton (Princeton)
H. Hauschild (Gottingen)
C. Hausmann (Princeton)
F. Hegenbarth (Dortmund)
W. C. Hsiang (Princeton)
J. Huebschmann (Heidelberg)
D. Husemoller (Haverford)
W. Jaco (Princeton)
S. Jackowski (Warszawa)
D. James (New Brunswick)
I. M. James (Oxford)
J. D. S. Jones (Oxford)
F. N. Kamber (Urbana)
M. Karoubi (Paris)
L. H. Kauffman (Chicago)
M. Kervaire (Geneva)
K. H. Knapp (Bonn)
D. Koll (Bonn)
W. Kohler (Gießen)
U. Koschorke (Siegen)
A. Kosinski (Princeton)
C. Kosniowski (Newcastle)
M. Kreck (Mainz)

I. Kupka (Dijon)
K. Lamotke (Koln)
W. Lellmann (Wuppertal)
P. W. H. Lemmens (Utrecht)
N. Levitt (New Brunswick)
Ch. Leytem (Cambridge)
A. Liulevicius (Chicago)
G. Loibel (Sao Carlos)
P. Loffler (Gottingen)
B. Lubcke (Siegen)
I. Madsen (Aarhus)
M. Mahowald (Evanston)
K. H. Mayer (Dortmund)
W. Meier (Siegen)
W. Meyer (Bures-sur-Yvette)
H. Minami (Osaka)
G. Mislin (Zurich)
B. Morin (Strassbourg)
H. J. Munkholm (Odense)
W. Neumann (Maryland)
A. Nofech (Beer-Sheva)
Ch. Okonek (Gottingen)
R. Oliver (Cambridge, Mass.)
C. Olk (Siegen)
E. Ossa (Wuppertal)
E. P. Peterson (Cambridge, Mass.)
T. Petrie (New Brunswick)
S. Priddy (Evanston)
A Prontê (Nantes)
V. Puppe (Konstanz)
R. Randell (Ann Arbor)
M. Raußen (Gottingen)
N. Ray (Manchester)
E. Rees (Oxford)
N. Rogler (Munchen)
W. Rouse (Oxford)
B. I. Sanderson (Coventry)
A. Scharf (Bonn)
H. Scheerer (Berlin)
H. Schulte-Crooneburg (Wuppertal)
R. Schultz (W. Lafayette)
C. Schupp (Munchen)
L. Schwartz (Orsay)
R. Schwanzl (Osnabruck)
L. Siebenmann (Orsay)
S. Sigrist (Neuchâtel)
W. Singhoff (Koln)
T. Skjelbred (Oslo)
J. Smith (Honolulu)
R. E. Staffeldt (Princeton)
K. Steffen (Dusseldorf)
H. Steinlein (Munchen)
A. Stieglitz (Bochum)
E. Stone (East Lansing)
S. Stolz (Bonn)
N. Stoltzfuß (Genêve)
R. Stocker (Bochum)
U. Suter (Neuchâtel)
Y. Suwa-Bier (Gottingen)
R. M. Switzer (Gottingen)
A. Szucs (Szeged)
Ch. Thomas (London)
A. Van de Ven (Leiden)
P. Vogel (Nantes)
W. Vogell (Princeton)
E. Vogt (Berlin)
R. Vogt (Osnabruck)
R. Wagener (Cambridge)
F. Waldhausen (Bielefeld)
G. Wassermann (Regensburg)
C. Weber (Genêve)
D. Whitgift (Cambridge)
R. Wiegmann (Osnabruck)
C. Wissemann-Hartmann (Bochum)
R. Wood (Manchester)
L. M. Woodward (Durham)
B. Zimmermann (Munchen)
P. Zschauer (Dortmund)

PROJECTIVE SPACE IMMERSIONS,

BILINEAR MAPS AND STABLE HOMOTOPY GROUPS OF SPHERES

A.J. Berrick

(Imperial College, London University)

The first two sections of this article are based on my talk at Siegen, much of which was in turn based on [10], [11]. The results in §3 were obtained after the symposium (and indeed were motivated by it).

The final section is a tabulation of known (to me) immersions and embeddings of (real) projective spaces. Aside from a modification forced by §3 below, the tables are those I circulated during the symposium. Although they exclude most negative results, which defy simple presentation (but see [6], [12], [14]), they do provide a revealing comparison with the previously published tables [18], [25] of a decade ago. However their inclusion here is not for their historical charm but in hope that they may prove of some value to other practitioners of the art.

1. Three classical problems and their (non-classical) interdependence

The three problems are

A Determine for which n,t there exists an immersion (or embedding) of P^n in $\mathbb{R}^{2n-t}$.

B Determine for which m,n,t there exists a non-singular bilinear map $f: \mathbb{R}^{m+1} \times \mathbb{R}^{n+1} \to \mathbb{R}^{m+n+1-t}$.

C Provide a geometric description of elements of π_*^S, the stable homotopy groups of spheres.

Clearly A and B are well-posed (once one knows that f non-singular - a term we in future suppress - means $f^{-1}(0) = \mathbb{R}^{m+1} \vee \mathbb{R}^{n+1}$). C allows for greater variety of interpretation, as witness the symposium contributions of Eccles, Jones, Löffler, Ossa et al. (The philosophical point is that it would be nice - and even useful - to give some geometric substance to such elements which in many cases are only known as the survivors of some purely algebraic spectral sequence to the bitter end.) Certainly, some elements of π_t^S are bilinearly representable, which is to say obtainable from a bilinear map f by the Hopf construction H, where

$$Hf: S^{m+n+1} \to S^{m+n+1-t}$$

$$(x \cos\theta,\ y \sin\theta) \mapsto (\cos 2\theta,\ \sin 2\theta . f(x,y)/\|f(x,y)\|),$$

$$x \in S^m,\ y \in S^n,\quad -\tfrac{\pi}{2} < \theta \le \pi/2.$$

K.Y. Lam [27] and L. Smith [39] observed that the Kahn-Priddy theorem gives valuable information about the image of H. In particular,

(1.1) THEOREM [27]: If $\alpha \in \pi_*^S$, then 2α is bilinearly representable.

To this encouraging result Al-Sabti and Bier soon replied

(1.2) THEOREM [5]: There exist elements of π_*^S which are not bilinearly representable.

An example, $\bar{\kappa} \in \pi_{20}^S$, is discussed further below. Much earlier H. Hopf himself, and then Ginsburg, had related B to A. (Recall that a map f with domain $X \times X$ is symmetric if invariant with respect to the interchange involution, i.e. $f(y,x) = f(x,y)$.)

(1.3) THEOREM [23]: If there exists a symmetric bilinear map $f: \mathbb{R}^{n+1} \times \mathbb{R}^{n+1} \to \mathbb{R}^{2n+1-t}$, then P^n embeds in $\mathbb{R}^{2n-t}$.

(1.4) THEOREM [16]: If there exists a bilinear map $\mathbb{R}^{n+1} \times \mathbb{R}^{n+1} \to \mathbb{R}^{2n+1-t}$ $(t < n)$, then P^n immerses in $\mathbb{R}^{2n-t}$.

(1.3) has the only truly constructive proof of all the results here: restrict f to the diagonal and normalise to obtain an embedding in S^{2n-t}. For progress in the converse direction one has to consider a topological version of B:-

B' Determine for which m,n,t there exists a biskew map $g: S^m \times S^n \to S^{m+n-t}$.

Of course biskew means $g(-x,y) = g(x,-y) = -g(x,y)$. Evidently the normalisation of any bilinear map is biskew, and so H may be regarded as a construction on biskew maps. Although an example exists of (m,n,t) (=(12,27,8)[19]) for which there is a biskew map which is not the restriction/normalisation of a bilinear one, the loss of information through this specialisation of the problem is not felt to be large. In particular, after Hurwicz-Radon and Adams one knows that for $m = t$ the problems B and B' are equivalent, as they also are for $m = n < 19$ [1]. Moreover, if one adopts a thoroughly stable point of view then

K.Y. Lam has shown that the Hopf construction fails to distinguish between $\underset{\sim}{B}$ and $\underset{\sim}{B}'$. Specifically,

(1.4) THEOREM [27]: <u>If $\alpha \in \pi_*^S$ may be obtained by the Hopf construction on some biskew map, then α is bilinearly representable.</u>

The proof of (1.4) uses Weierstrass approximation and so allows no control over the dimensions involved. In particular the condition $m = n$, which we impose from now on, has to be sacrificed in obtaining the bilinear map. However, the topological gain in restricting attention to biskew maps is enormous. For there is a precise fit between problems $\underset{\sim}{A}$ and $\underset{\sim}{B}'$ $(m = n)$.

(1.5) THEOREM [10]: <u>For $2t \leq n-4$ there is a one-to-one correspondence between biskew maps $S^n \times S^n \to S^{2n-t}$ and immersions $P^n \to \mathbb{R}^{2n-t}$ and, by restriction, between symmetric biskew maps and embeddings, with biskew homotopy classes corresponding to regular homotopy classes and symmetric biskew homotopy classes corresponding to isotopy classes.</u>

The numerical condition of (1.5) is needed only when $n \leq 15$ [3], [10], but is deemed satisfied for the remainder of this section. The correspondence consists of the composition of two functions $\pi^*, \wp$ which we now describe. Given an n-manifold M, let $\mathrm{Imm}(M, \mathbb{R}^{2n-t})$, $\mathrm{Emb}(M, \mathbb{R}^{2n-t})$ denote respectively the set of (regular homotopy classes of) immersions, embeddings of M in $\mathbb{R}^{2n-t}$. Then the double covering projection $\pi: S^n \to P^n$ induces $\pi^*: \mathrm{Imm}(P^n, \mathbb{R}^{2n-t}) \to \mathrm{Imm}(S^n, \mathbb{R}^{2n-t})$. Meanwhile $\wp$ is defined through the induced tangent bundle monomorphism $a_*: \tau S^n \to (2n-t)\varepsilon$ of an immersion $a: S^n \to \mathbb{R}^{2n-t}$ (ε being the trivial line bundle). If one adds a trivial line bundle so as to trivialise the tangent bundle and then restricts to sphere bundles the resulting map

$$S(a_* \oplus 1) : S^n \times S^n \to S^n \times S^{2n-t}$$

projects to the latter fibre to give $\varphi_a = pr_2 \circ S(a_* \oplus 1)$. This map is right-skew (indeed it comes from a right-linear map) and further biskew if $a \in \mathrm{Im}\ \pi^*$.

Not only does $a \mapsto \varphi_{\pi^* a} = \varphi \circ \pi^*(a)$ give the correspondence of Theorem 1.5, but composition with the Hopf construction H yields the following links between problems $\underset{\sim}{A}$ and $\underset{\sim}{C}$.

(1.6) THEOREM [11]: <u>Let</u> $t \in \mathbb{N}$.

(i) <u>There exist infinitely many values of</u> n <u>for which</u>

$$H \circ \varphi\ \mathrm{Imm}(S^n, \mathbb{R}^{2n-t}) = \pi_t^S.$$

(ii) $$H \circ \varphi\ \mathrm{Emb}(S^n, \mathbb{R}^{2n-t}) = 0 .$$

(iii) <u>There exist infinitely many values of</u> n <u>for which</u>

$$2\pi_t^S \subset H \circ \varphi \circ \pi^*\ \mathrm{Imm}(P^n, \mathbb{R}^{2n-t}).$$

(iv) <u>There exists</u> t' <u>such that</u>

$$H \circ \varphi \circ \pi^*\ \mathrm{Imm}(P^n, \mathbb{R}^{2n-t'}) \neq \pi_{t'}^S .$$

(v) <u>If</u> t <u>is even, then</u>

$$2.H \circ \varphi \circ \pi^*\ \mathrm{Emb}(P^n, \mathbb{R}^{2n-t}) = 0.$$

(1.7) EXAMPLE: Consider $\bar{\kappa}$, the generator of the 20-stem, of order 8. It may be represented by an immersion of S^n in $\mathbb{R}^{2n-20}$ which can neither be regular homotopic to an embedding nor factor through P^n. On the other hand, $2\bar{\kappa}$ can be represented by some P^n immersed in $\mathbb{R}^{2n-20}$, however again this immersion cannot be regular homotopic to an embedding.

2. Inductive methods for projective space immersions

The proof of (1.6) depends crucially on the commutative diagram [9]

$$
\begin{array}{ccccc}
 & & \pi_{2n+1}(S^{2n+1-t}) & & \\
 & H\circ\varphi \nearrow & & \nwarrow J & \\
\mathrm{Imm}(S^n, \mathbb{R}^{2n-t}) & & \xrightarrow{\ \ \phi\ \ } & & \pi_n(V_{2n-t+1,n+1}) \\
 & \Omega \searrow & & \nearrow j_* & \\
 & & \pi_n(V_{2n-t,n}) & &
\end{array}
$$

in which J is the generalised J-homomorphism, $j: V_{2n-t,n} \hookrightarrow V_{2n-t+1,n+1}$, Ω is the Smale invariant isomorphism [38], and, given an immersion $a: S^n \to \mathbb{R}^{2n-t}$, $\phi_a \in \pi_n(V_{2n-t+1,n+1})$ is related to the right-skew map $\varphi_a: S^n \times S^n \to S^{2n-t}$ by exponential correspondence. (Recall that φ_a is actually induced from a right-linear map $pr_2\circ(a_* \oplus 1): S^n \times \mathbb{R}^{n+1} \to \mathbb{R}^{2n-t+1}$.) If we begin instead with an immersion $a: P^n \to \mathbb{R}^{2n-t}$, $\phi_{a\circ\pi}$ is called the <u>Smale invariant</u> of a [9]; the corresponding $\varphi_{a\circ\pi}$ is biskew. When a is an embedding then by (1.5) $\varphi_{a\circ\pi}$ is symmetric, otherwise its twisting $\varphi_{a\circ\pi} \circ T$ by the interchange involution T on $S^n \times S^n$ may, by (1.5), represent a different regular homotopy class of immersions, namely those <u>adjoint</u> to a. In any event the element of $\pi_n(V_{2n-t+1,n+1})$ corresponding to $\varphi_{a\circ\pi} \circ T$ is called the <u>adjoint Smale invariant</u> of a. Thus an embedding is self-adjoint and its Smale invariants coincide. In general these two (regular homotopy) invariants carry much (sometimes all) information concerning whether or not a extends to an immersion of P^{n+1}.

This comes about in the following way. Corresponding to the decomposition of S^{n+1} into a copy of $S^n \times I$ and two copies of B^{n+1} according as its last coordinate lies in $[-\frac{1}{3}, \frac{1}{3}]$ or $[-1, -\frac{1}{3}]$, $[\frac{1}{3}, 1]$, there is a decomposition of P^{n+1} as $P^{n+1} = BH_n \cup B^{n+1}$, where H_n is the canonical line bundle over P^n.

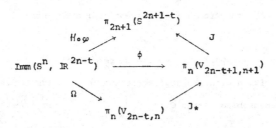

(2.1) LEMMA [10]: <u>An immersion of P^n in $\mathbb{R}^{2n-t}$ extends to an immersion of BH_n in $\mathbb{R}^{2n-t}$ if and only if its adjoint Smale invariant is zero.</u>

This is the key lemma. Note that it involves an extension of the original immersion itself and not merely an extension up to regular homotopy. This carries the bonus that if we start with an embedding, then a neighbourhood of the zero-section of BH_n is also embedded. Thus

(2.2) COROLLARY [10]: <u>An embedding of P^n in $\mathbb{R}^{2n-t}$ extends to an embedding of BH_n in $\mathbb{R}^{2n-t}$ if and only if its Smale invariant is zero.</u>

It now remains to extend the embedding over $BH_n \cup B^{n+1}$. This can cost a dimension (and when n+1 is a power of 2 must cost a dimension) if one wishes to avoid creating double points. In other words, the following theorem is best possible in general.

(2.3) THEOREM [10]: <u>An embedding of P^n in $\mathbb{R}^{2n-t}$ with trivial Smale invariant extends to an immersion of P^{n+1} in $\mathbb{R}^{2n-t}$ and embedding of P^{n+1} in $\mathbb{R}^{2n-t+1}$.</u>

Theorem 3.1 below is an application of this result. Predictably, relaxation of conditions on the original immersion weakens the conclusion about its extension.

(2.4) THEOREM [10]: <u>An immersion of P^n in $\mathbb{R}^{2n-t}$ with trivial Smale and adjoint Smale invariants extends to an immersion of P^{n+1} in $\mathbb{R}^{2n-t+1}$.</u>

3. Embeddings and immersions of even-dimensional spaces

As an application of the methods discussed in the previous section we establish general embedding and immersion results for even-dimensional projective spaces.

Inevitably we need a few number-theoretic functions, even for statements of results. Discussion of their properties is deferred until after (3.2), while an early glance at (3.1), (3.6) may help to motivate their digestion by the reader. It is also helpful to adopt two notations for the value of functions β,γ: $\{0,1,2,3,\ldots\} \to \{0,1,2,3,\ldots\}$ at n, namely β_n if β is bounded above, and $\gamma(n)$ if γ is not so bounded. Suppose

$$n = \sum_{i=0}^{\tau(n)} u_i 2^i , \qquad u_i \in \{0,1\} \qquad u_{\tau(n)} = 1.$$

Then define

$$\kappa_n = u_0 + (u_1+u_2)^2 - \max(u_0,u_1,u_2),$$

$$\nu(n) = \max\{i \geq 0 | u_0 = \ldots = u_{i-1} = 0\},$$

$$\alpha(n) = \sum_{i\geq 0} u_i ,$$

$$\rho(n) = u_1 + \sum_{i\geq 3} u_i 2^{i-3} = u_1 + [\tfrac{n}{8}] ,$$

$$\phi(n) = \min(\alpha(n),\rho(n)),$$

$$\theta(n) = \min(\alpha(n) + \kappa_n, \rho(n)).$$

The main result of this section applies our inductive techniques to the odd-dimensional embeddings e_S and immersions i_M of Steer [41], [8] and Milgram [33] respectively.

(3.1) THEOREM: Let $n \geq 32$ be even. Then

$$e_S: P^{n-1} \to \mathbb{R}^{2n-1-\alpha(n-1)-\kappa_{n-1}}$$

extends to an immersion

$$P^n \to \mathbb{R}^{2n-\alpha(n)-\kappa_n}$$

and embedding

$$P^n \to \mathbb{R}^{2n+1-\alpha(n)-\kappa_n} .$$

Apart from the case $n \equiv 0 \pmod 8$, when an improvement of one dimension is effected, the Euclidean space dimensions are those obtained by restriction of the embedding/immersion of P^{n+1} established in [41], [33]. (For odd m, Milgram's $k(m) = \kappa_m$.) Because of the geometric nature of our induction technique, we are able to avoid getting immersed (sic) in the algebraic machinations of [33], [41], beyond checking the following (by means of [33] §3(4)).

(3.2) LEMMA: Milgram's immersions i_M are regular homotopic under restriction (of projective spaces) and inclusion (of Euclidean spaces).

Given this, it is easy enough to see that $i_M: P^{n-1} \to \mathbb{R}^{2n-\alpha(n)-\kappa_n}$ has trivial Smale invariant ((3.7) below); the crucial lemma is that e_S does too. To decide in which Euclidean space $i_M(P^{n-1})$ and $e_S(P^{n-1})$ have the same Smale invariant we need to know something about the function θ. First note that ρ satisfies the iterative condition:-

(3.3) For $n \geq 8$,

$$\rho(n) = \rho(2^{\tau(n)} - 1) + \rho(n - 2^{\tau(n)}).$$

Further,

(3.4) For m odd,

$$\rho(m) - \alpha(m) \leq 0 \qquad \text{if} \quad m < 32,$$

$$0 \leq \rho(m) - \alpha(m) \leq \kappa_m \qquad \text{if} \quad 32 < m < 64, \quad m \equiv 7 \pmod 8,$$

$$\rho(m) - \alpha(m) \geq \kappa_m \qquad \text{otherwise}.$$

From (3.3) and (3.4) it is simple enough to verify the following valuable observation, which says that θ is a good function for the purposes of Lemma 3.6. (Indeed, thanks to (3.3) there is none better.)

(3.5) LEMMA: Let $m > 8$ be odd. Then

$$\theta(m) \leq \theta(2^{\tau(m)} - 1) + \phi(m - 2^{\tau(m)}).$$

(3.6) LEMMA: Let m be odd. Then e_S and i_M: $P^m \to \mathbb{R}^{2m+1-\theta(m)}$ have the same Smale invariants.

Proof. By induction on m. For $m = 1,5$ $\theta(m) = 0$ and the Smale invariants lie in the trivial group $\pi_m(V_{2m+1,m})$. For $m = 3,7$ $\theta(m) = 1$ and, after [10] (3.4), the Smale invariants are the non-trivial element of $\pi_m(V_{2m,m}) \approx \mathbb{Z}/2$. If $m \geq 8$, then $e_S \circ \pi\colon S^m \to \mathbb{R}^{2m+1-\alpha(m)-\kappa_m}$ restricts on $S^{2^\tau-1} \times S^{m-2^\tau} \subset S^m$ (τ being $\tau(m)$ and $\pi\colon S^m \to P^m$ the double covering projection) to the product $e_1 \circ \pi \times e_2 \circ \pi$, where (up to regular homotopy) e_1, e_2 are embeddings

$$e_1\colon P^{2^\tau-1} \to \mathbb{R}^{2(2^\tau-1)-\tau-3},$$

$$e_2\colon P^{m-2^\tau} \to \mathbb{R}^{2(m-2^\tau)-\alpha(m-2^\tau)+1}.$$

Likewise $i_M \circ \pi$ restricts to $i_1 \circ \pi \times i_2 \circ \pi$ in $\mathbb{R}^{2(2^\tau-1)-\tau-4} \times \mathbb{R}^{2(m-2^\tau)-\alpha(m-2^\tau)}$. By inductive hypothesis (and Lemma 3.2), $e_1 \circ \pi$ and $i_1 \circ \pi$ are regular homotopic in $\mathbb{R}^{2(2^\tau-1)+1-\theta(2^\tau-1)}$, as are $e_2 \circ \pi$ and $i_2 \circ \pi$ in $\mathbb{R}^{2(m-2^\tau)+1-\phi(m-2^\tau)}$. It follows from Lemma 3.5 that the product immersions $e_1 \circ \pi \times e_2 \circ \pi$ and $i_1 \circ \pi \times i_2 \circ \pi$ are regular homotopic in $\mathbb{R}^{2m-\theta(m)}$. Hence $e_S \circ \pi$ and $i_M \circ \pi$ are regular homotopic in $\mathbb{R}^{2m+1-\theta(m)}$ when restricted to $S^{2^\tau-1} \times S^{m-2^\tau} \times I$ (the $\mathbb{Z}/2 \times \mathbb{Z}/2$-cover of the disc bundle $pr_1^*(H) \otimes pr_2^*(H)$ over $P^{2^\tau-1} \times P^{m-2^\tau}$).

To complete the proof we must consider the obstruction to extending a regular homotopy (of $\mathbb{Z}/2 \times \mathbb{Z}/2$-invariant immersions) from $S^{2^\tau-1} \times S^{m-2^\tau} \times I$ to $(S^{2^\tau-1} \times B^{m-2^\tau+1} \times 0) \cup (S^{2^\tau-1} \times S^{m-2^\tau} \times I)$. After [22] (5.9) this is equivalent to the obstruction to a linear homotopy between certain pairs of vector bundle monomorphisms, viz. a difference element of

$$\mathrm{Mono}(\tau(S^{2^\tau-1} \times B^{m-2^\tau+1}),\ (S^{2^\tau-1} \times B^{m-2^\tau+1}) \times \mathbb{R}^{2m+1-\theta(m)})$$

$$= \mathrm{Mono}(pr_1^*\tau(S^{2^\tau-1}) \times \mathbb{R}^{m-2^\tau+1},\ pr_1^*(S^{2^\tau-1} \times \mathbb{R}^{m+2^\tau-\theta(m)}) \times \mathbb{R}^{m-2^\tau+1})$$

$$\cong \mathrm{Mono}(\tau(S^{2^\tau-1}),\ S^{2^\tau-1} \times \mathbb{R}^{m+2^\tau-\theta(m)})$$

by stability and contractibility of $B^{m-2^\tau+1}$. This last difference element is that between two Smale invariants (of projective space immersions) in $\pi_{2^\tau-1}(V_{2^\tau+m-\theta(m),2^\tau-1})$. Since $m-\theta(m) \geq 2^\tau-2$ the obstruction vanishes (with occasional help from [10] (3.4)). Similarly, the obstruction to extending the regular homotopy over $(S^{2^\tau-1} \times S^{m-2^\tau} \times I) \cup (B^{2^\tau} \times S^{m-2^\tau} \times 1)$ also vanishes, whence $e_S \circ \pi$ and $i_M \circ \pi$ are regular homotopic over

$$S^m = (S^{2^\tau-1} \times B^{m-2^\tau+1} \times 0) \cup (S^{2^\tau-1} \times S^{m-2^\tau} \times I) \cup (B^{2^\tau} \times S^{m-2^\tau} \times 1),$$

which is to say, e_S and i_M have the same Smale invariant.

An immediate consequence of this lemma is that the Smale and adjoint Smale invariants of $i_M: P^m \to \mathbb{R}^{2m+1-\theta(m)}$ are equal. However, more than this is true. For the bilinear map [33] Corollary 7 used to construct i_M has the additional property ([33] Theorem 6(i)) that the interchange involution on $\mathbb{R}^{m+1} \times \mathbb{R}^{m+1}$ commutes with a sign change on all components of the range $\mathbb{R}^{2m-\alpha(m)-\kappa_m+1}$ of i_M other than the first. Thus the induced axial map $P^m \times P^m \to P^{2m-\alpha(m)-\kappa_m}$ is symmaxial provided $\alpha(m) + \kappa_m$ is even; otherwise its inclusion into $P^{2m-\alpha(m)-\kappa_m+1}$ is symmaxial. Equivalently $i_M: P^m \to \mathbb{R}^{2m-\alpha(m)-\kappa_m+\varepsilon}$ is self-adjoint where $\varepsilon = 0$ or 1 according as $\alpha(m) + \kappa_m \equiv 0$ or $1 \pmod 2$.

The theorem now follows immediately from Theorem 2.3, Lemma 3.6 and the following lemma.

(3.7) LEMMA: <u>Let</u> n <u>be even</u>. <u>Then</u> $i_M: P^{n-1} \to \mathbb{R}^{2n-\alpha(n)-\kappa_n}$ <u>has trivial Smale invariant</u>.

This is in turn a consequence of the following much more general result. It seems worth providing a proof here as the literature contains several statements of the assertion - all incorrect!

(3.8) PROPOSITION: <u>Milgram's system of non-singular bilinear maps includes</u>

$$\mathbb{R}^{n+1} \times \mathbb{R}^{m+1} \to \mathbb{R}^{n+m+1-\alpha(n)+\alpha(n-m)-(2^d-d-1)},$$

<u>where</u> $n \geq m$ <u>and</u> $d = \min(3, \nu(n+1), \nu(m+1))$.

<u>Proof</u>. Let F be the division algebra of dimension 2^d over $\mathbb{R}$, and choose $k > \tau(n)$. Milgram constructs (cf. [26] (7.1))

$$F^{2^k} \times F^{2^k} \to \mathbb{R}^{2^{k+d+1}-k-2^d}.$$

The inclusions

$$\mathbb{R}^{n+1} = F_{2^k - \frac{n+1}{2^d}+1} \oplus \cdots \oplus F_{2^k} \subset F_1 \oplus \cdots \oplus F_{2^k}$$

$$\mathbb{R}^{m+1} = F_{2^k - \frac{n+1}{2^d}+1} \oplus \cdots \oplus F_{2^k - \frac{n-m}{2^d}} \subset F_1 \oplus \cdots \oplus F_{2^k}$$

induce a restriction to $\mathbb{R}^{n+1} \times \mathbb{R}^{m+1}$ whose first

$$2^d.2(2^k - \frac{n+1}{2^d}) - \alpha(2(2^k - \frac{n+1}{2^d}))$$

and last

$$2^d.\frac{n-m}{2^d} - \alpha(\frac{n-m}{2^d})$$

components vanish, after [33] Lemma 11 and the correct version ([41] Lemma 4.2(i)) of [33] Lemma 9. (In other words, in this last reference replace $V^{i-m+2,k}$ by $V^{m,k}$. This allows the induction when $2^{k-1} < m$. On the other hand, application of the original version leads eventually, via (3.9), (3.10), to a contradiction.)

Here is a construction which suggests an easy recapture of Steer's embedding of P^n in $\mathbb{R}^{2n-\alpha(n)-\kappa_n+1}$, by combining with (3.8) for $p = n-m-1$ as the mod 8 residue of n, and using the $m \equiv 7 \pmod 8$ case as inductive hypothesis. (However, an analogue of (3.6) is needed for iteration of the 7 (mod 8) case.)

(3.9) PROPOSITION: Suppose an embedding of $(q+1)H_p$ in $\mathbb{R}^M$ restricts to an embedding of P^p in $\mathbb{R}^{M-(2q+1-N)}$. If P^q immerses in $\mathbb{R}^N$ then P^{p+q+1} immerses in $\mathbb{R}^{M+N}$, embeds in $\mathbb{R}^{M+N+1}$, provided $M+N-2 > p+q+\max(p,q)$.

<u>Proof</u>. We have an embedding of P^p in $\mathbb{R}^{M-(2q+1-N)}$ with normal bundle ν_1 such that $(q+1)H_p \subset \nu_1 \oplus (2q+1-N)\varepsilon$. By [22] and stability any embedding of P^q in $\mathbb{R}^{2q+1}$ has $(2q+1-N)\varepsilon$ as a subbundle of its normal bundle ν_2. Thus $pr_2^*(q+1)H_p$ is a subbundle of the normal bundle $pr_1^*\nu_1 \oplus pr_2^*\nu_2$ of the product embedding $P^p \times P^q \to \mathbb{R}^{M-(2q+1-N)} \times \mathbb{R}^{2q+1}$. So the disc bundle $pr_1^* B((q+1)H_p)$ embeds in $\mathbb{R}^{M+N}$. After [21] the connectivity of the embedding restricted to $pr_1^*S((q+1)H_p) = pr_2^*S((p+1)H_q)$ ensures that it extends from (a neighbourhood of) the sphere bundle to an embedding over $pr_2^*B((p+1)H_q)$ (cf. [41] Proposition 3.1). The union of these two embeddings is then an immersion of

$$P^{p+q+1} = pr_1^*B((q+1)H_p) \cup pr_2^*B((p+1)H_q)$$

in $\mathbb{R}^{M+N}$ which regularly deforms to an embedding in $\mathbb{R}^{M+N+1}$.

This result considerably strengthens [31] Theorem 1.2. Unsurprisingly, there is also a biskew-map version. (In this we employ the join identifications $[w_1,0,w_2] = w_1$, $[w_1,1,w_2] = w_2$.)

(3.10) PROPOSITION: <u>Let</u> $f'\colon S^p \times S^p \to S^P$, $g'\colon S^q \times S^q \to S^Q$ <u>be symmetric biskew maps which extend respectively to biskew</u>

$$f\colon S^{p+q+1} \times S^p \to S^{P+r}$$

$(r \geq 0)$ <u>and to</u>

$$g\colon S^q \times (S^q * S^{r-1}) \to S^Q$$

<u>with</u>

$$g(-x,[-z_1,u,z_2]) = g(x,[z_1,u,z_2]) = -g(x,[-z_1,u,-z_2]) \quad x,z_1 \in S^q,\ z_2 \in S^{r-1},\ u \in I.$$

<u>Then there is a symmetric biskew map</u>

$$h: S^{p+q+1} \times S^{p+q+1} \to S^{P+Q+1}.$$

<u>Proof</u>. Given $x = [x_1, s, x_2]$, $y = [y_1, t, y_2] \in S^p * S^q$ with $s \geq t$, if $t < 1$ let

$$f([x_1, \frac{s-t}{1-t}, x_2], y_1) = [w, u, z] \in S^P * S^{r-1},$$

so that $u = 0$ if $s = t$ or $r = 0$.

Then define

$$h(x,y) = \begin{cases} [w, \max(2t-1, u), g(x_2, [y_2, \max(1-2t, 0), z])] & t < 1, \\ g(x_2, y_2) & t = 1, \end{cases}$$

and $h(y,x) = h(x,y)$.

4. Summary of immersion and embedding results

The number t listed below for a given n is the largest known value of t for which P^n immerses ($\subseteq$) or embeds ($\subset$) in $\mathbb{R}^{2n-t}$. An asterisk indicates that this value is also known to be best possible, a cross that if $8 \leq n \pmod{16} \leq 11$ then this value can be improved to t+1. Where the preceding bound or reference continues to apply the relevant square is left blank.

α(n)	1	2	3	4	5	6	7	8	9	10-13	14	15	16	≥17
n≡0 (mod 8)														
⊆	1*	9					11+		13		α			
Refs	[44]	[36]					[14]		[15]		[26]			
⊂	0*	6						α-1						
Refs	[43]	[42]						(3.1)						
n≡1 (mod 8)														
⊆	0*	3*	4*	8		11	13	13+	15			α		
Refs		[38]	[2]*	[36]		[14]			[15]			[33]		
⊂	0*	1*	3	6			α-1							
Refs		[28]*	[29]	[42]			[41]							
n≡2 (mod 8)														
⊆	1*	4*	6*	7	8	13	15+						α	
Refs	[44]	[38]		[14]	[13]	[14]							[33]	
		[7]*	[2]*											
⊂	0*	3*	4	6			α-1							
Refs	[23]	[37]		[42]			[41]-[8]							
		[30]*-[28]*												
α(n)	1	2	3	4	5	6	7	8	9	10-13	14	15	16	≥17

$\alpha(n)$	1	2	3	4	5	6	7	8	9	10-13	14	15	16	≥17
$n\equiv 3 \pmod 8$														
$\subseteq$		2*	6*	8*	9	10	15	17	17^+					$\alpha+1$
Refs		[34]	[38]-[2]*		[14]	[13]	[14]							[33]
$\subset$		1*	3	6	7		α							
Refs		[23]	[37]	[42]	[10]		[41]-[8]							
$n\equiv 4 \pmod 8$														
$\subseteq$	1*	6*	7	9	10		11	13			α			
Refs		[35]	[36]	[10]		[14]		[15]			[33]			
		[17]*												
$\subset$	0*	3	6				$\alpha-1$							
Refs		[29]	[42]				[41]-[8]							
$n\equiv 5 \pmod 8$														
$\subseteq$		3*	4*	8		12		13	15		$\alpha+1$			
Refs		[22]	[38]	[36]		[10]	[14]		[15]		[33]			
			[2]*											
$\subset$		1*	3*	6			7	α						
Refs		[23]	[29]	[42]		[37]		[41]-[8]						
		[32]*	[4]*											
$n\equiv 6 \pmod 8$														
$\subseteq$		5*	6*	8	9	14		15	16		$\alpha+3$			
Refs		[22]	[38]	[10]	[14]	[10]	[14]				[33]			
			[2]*											
$\subset$		2	5*	7		$\alpha+2$								
Refs		[37]	[4]*	[10]		[41]-[8]								
$\alpha(n)$	1	2	3	4	5	6	7	8	9	10-13	14	15	16	≥17

$\alpha(n)$	1	2	3	4	5	6	7	8	9	10-13	14	15	16	≥ 17
$n \equiv 7 \pmod 8$, $n \neq 63$														
$\subseteq$			6*	8*	9	14		16	17	18	$\alpha+4$			
Refs			[22]	[39]	[14]						[33]			
				[2]*										
$\subset$			4	7*	8	$\alpha+3$								
Refs			[37]	[4]*		[41]-[8]								
$n=2^{\alpha}-1 \geq 15$														
$\subseteq$														
$\alpha \equiv 0 \pmod 4$				2α*										
$\alpha \equiv 1,2 \pmod 4$				$2\alpha-1$*										
$\alpha \equiv 3 \pmod 4$				$2\alpha+1$*										
Refs				[24]-[20]*										
$\alpha(n)$	1	2	3	4	5	6	7	8	9	10-13	14	15	16	≥ 17

REFERENCES

1. J. ADEM, 'Some immersions associated with bilinear maps', Bol. Soc. Mat. Mexicana (2) 13 (1968), 95-104.

2. J. ADEM and S. GITLER, 'Secondary characteristic classes and the immersion problem', Bol. Soc. Mat. Mexicana (2) 8(1963), 53-78.

3. J. ADEM, S. GITLER and I.M. JAMES, 'On axial maps of a certain type', Bol. Soc. Mat. Mexicana (2) 17 (1972), 59-62.

4. J. ADEM, S. GITLER and M. MAHOWALD, 'Embedding and immersion of projective spaces', Bol. Soc. Mat. Mexicana (2) 10 (1965), 84-88.

5. G. AL-SABTI and T. BIER, 'Elements in the stable homotopy groups of spheres which are not bilinearly representable', Bull. London Math. Soc. 10 (1978), 197-200.

6. L. ASTEY and D.M. DAVIS, 'Nonimmersions of RP implied by BP', to appear.

7. P. BAUM and W. BROWDER, 'The cohomology of quotients of classical groups', Topology 3 (1965), 305-336.

8. A.J. BERRICK, Oxford Univ. D.Phil. thesis, Oxford (1973).

9. ____________, 'Induction on symmetric axial maps and embeddings of projective spaces', Proc. Amer. Math. Soc. 60 (1976), 276-278.

10. ____________, 'The Smale invariants of an immersed projective space', Math. Proc. Camb. Phil. Soc. 86 (1979), 401-412.

11. ____________, 'Consequences of the Kahn-Priddy theorem in homotopy and geometry', to appear.

12. D.M. DAVIS, 'Connective coverings of BO and immersions of projective spaces', Pacific J. Math. 76 (1978), 33-41.

13. D.M. DAVIS and M.E. MAHOWALD, 'The geometric dimension of some vector bundles over projective spaces', Trans. Amer. Math. Soc. 205 (1975), 295-315.

14. __________, 'The immersion conjecture for $RP^{8\ell+7}$ is false', Trans. Amer. Math. Soc. 236 (1978), 361-383.

15. __________, 'Immersions of complex projective spaces and the generalized vector field problem', Proc. London Math. Soc. (3) 35(1977), 333-344.

16. M. GINSBURG, 'Some immersions of projective spaces in Euclidean space', Topology 2 (1963), 69-71.

17. S. GITLER, 'The projective Stiefel manifolds II Applications', Topology 7 (1968), 47-53.

18. ________, 'Immersion and embedding of manifolds', Proc. Sympos. Pure Math. 22, A.M.S. Providence R.I. (1971), 87-96.

19. S. GITLER and K.Y. LAM, 'The generalized vector field problem and bilinear maps', Bol. Soc. Mat. Mexicana (2) 14 (1969), 65-69.

20. S. GITLER and M. MAHOWALD, 'Some immersions of real projective spaces', Bol. Soc. Mat. Mexicana (2) 14 (1969), 9-21.

21. A. HAEFLIGER, 'Plongements différentiables des variétés dans variétés', Comment. Math. Helv. 36 (1962), 47-82.

22. M.W. HIRSCH, 'Immersions of manifolds', Trans. Amer. Math. Soc. 93 (1959), 242-276.

23. H. HOPF, 'Systeme symmetrischer Bilinearformen und euklidische Modelle der projectiven Räume', Vjschr. naturf. Ges. Zurich 85 (1940), 165-177.

24. I.M. JAMES, 'On the immersion problem for real projective spaces', Bull. Amer. Math. Soc. 69 (1963), 231-238.

25. __________, 'Two problems studied by Heinz Hopf', Lecture Notes in Math. 279 Springer-Verlag, Berlin (1972), 134-174.

26. K.Y. LAM, 'Construction of some non-singular bilinear maps', Bol. Soc. Mat. Mexicana (2) 13 (1968), 88-94.

27. ________, 'Non-singular bilinear maps and stable homotopy classes of spheres', Math. Proc. Camb. Phil. Soc. 82 (1977), 419-425.

28. J. LEVINE, 'Imbedding and immersion of real projective spaces', Proc. Amer. Math. Soc. 14 (1963), 801-803.

29. M.E. MAHOWALD, 'On obstruction theory in orientable fibre bundles', Trans. Amer. Math. Soc. 110 (1964), 315-349.

30. ____________, 'On the embeddability of the real projective spaces', Proc. Amer. Math. Soc. 13 (1962), 763-764.

31. M. MAHOWALD and R.J. MILGRAM, 'Embedding real projective spaces', Annals of Math. (2) 87 (1968), 411-422.

32. W.S. MASSEY, 'On the imbeddability of the real projective spaces in Euclidean space', Pacific J. Math. 9 (1959), 783-789.

33. R.J. MILGRAM, 'Immersing projective spaces', Annals of Math. (2) 85 (1967), 473-482.

34. J. MILNOR, 'On the immersion of n-manifolds in (n+1)-space', Comment. Math. Helv. 30 (1956), 275-290.

35. F. NUSSBAUM, 'Obstruction theory of possibly non-orientable fibrations', Northwestern Univ. Ph.D. thesis, Evanston Ill. (1970).

36. D. RANDALL, 'Some immersion theorems for projective spaces', Trans. Amer. Math. Soc. 147 (1970), 135-151.

37. E. REES, 'Embeddings of real projective spaces', Topology 10 (1971), 309-312.

38. B.J. SANDERSON, 'Immersions and embeddings of projective spaces', Proc. London Math. Soc. (3) 14 (1964), 137-153.

39. S. SMALE, 'The classification of immersions of spheres in euclidean spaces', Annals of Math. (2) 69 (1959), 327-344.

40. L. SMITH, 'Nonsingular bilinear forms, generalized J homomorphisms, and the homotopy of spheres I', Indiana Univ. Math. J. 27 (1978), 697-737.

41. B. STEER, 'On the embedding of projective spaces in Euclidean space',

Proc. London Math. Soc. (3) 31 (1970), 489-501.

42. E. THOMAS, 'Embedding manifolds in Euclidean space', Osaka J. Math. 13 (1976), 163-186.

43. H. WHITNEY, 'The self-intersections of a smooth n-manifold in 2n-space', Annals of Math. (2) 45 (1944), 220-246.

44. __________, 'The singularities of a smooth n-manifold in (2n-1)-space', Annals of Math. (2) 45 (1944), 247-293.

Multiple points of codimension one immersions

Peter John Eccles

§1. The problem.

Given a self-transverse immersion $i: M \looparrowright \mathbb{R}^{n+1}$ of a connected n-dimensiona (closed, compact, smooth) manifold in Euclidean space, a point of $\mathbb{R}^{n+1}$ is an r-fold intersection point of i if it is the image of r distinct points of M. Let $\theta(i)$ be the number of $(n+1)$-fold intersection points of i (finite since they form a compact 0-dimensional manifold). Two problems spring to mind.

1.1 Problems. Determine the possible values of $\theta(i)$

(i) given the manifold M,

(ii) given the dimension n.

The problems may be illustrated by and completely solved for low dimensions where manifolds are completely classified. There is no interest in the case $n = 0$ and from now on *I will take* n *to be strictly positive.*

For $n = 1$ M must be a circle and $\theta(i)$ may take any value.

1.2 Diagram.

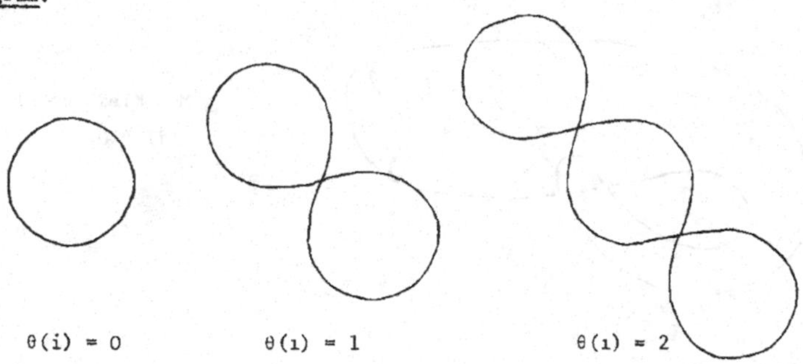

$\theta(i) = 0$ $\theta(i) = 1$ $\theta(i) = 2$

For $n = 2$ if M is a 2-sphere an immersion with two triple points may be constructed as follows. Put three mutually intersecting 2-spheres in general position and then attach two handles to form their connected sum, an immersed 2-sphere (Diagram 1.3). By taking the connected sum of sufficiently many copies of this immersion any even number can be obtained for $\theta(i)$ when M

1.3 Diagram.

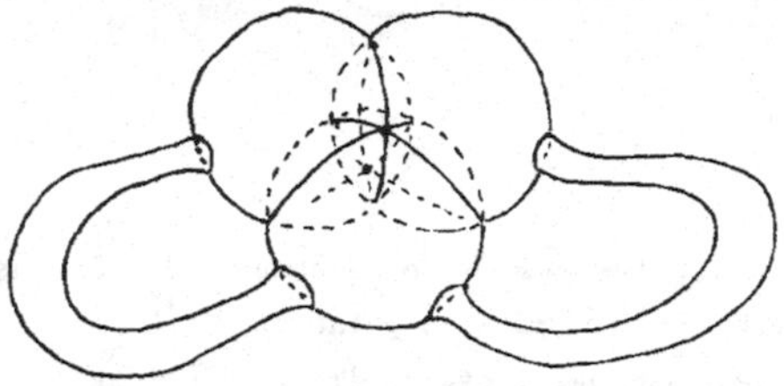

$M = S^2$

$\theta(i) = 2$

is a 2-sphere. Moreover any orientable surface can be embedded in $\mathbb{R}^3$ and so by taking the connected sum of an embedding and an appropriate immersion of a 2-sphere any even number can be obtained for $\theta(i)$ when M is an orientable surface. This also applies to non-orientable surfaces of even genus since they can be immersed in $\mathbb{R}^3$ with no triple points (by taking the connected sum of the appropriate number of copies of the usual immersion of the Klein bottle in Diagram 1.4).

1.4 Diagram.

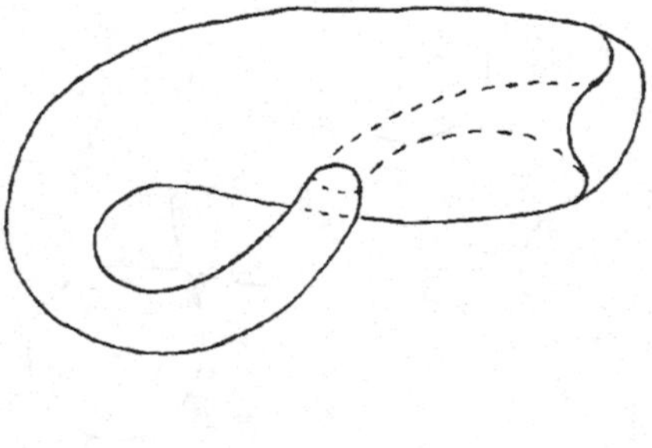

M = Klein bottle

$\theta(i) = 0$

On the other hand, the immersion of the projective plane constructed by W. Boy in his Göttingen thesis in 1901 has a single triple point. The following construction of this immersion is taken from unpublished lecture notes of R.M.W. Wood based on the description in [18;pp.150ff.] which in turn is based on Boy's original description [5] (see also [11;pp.280ff.] and [22]). Recall that the projective plane may be obtained by identifying the boundary circles of a disc and a Möbius band. The idea of the construction is to immerse a Möbius band in $\mathbb{R}^3$ so that the boundary circle can be spanned by a disc without introducing further intersection points. This is done by first immersing a Mobius band as in Diagram 1.5 and then drawing out three flat 'pods' as in Diagram 1.6.

1.5 Diagram.

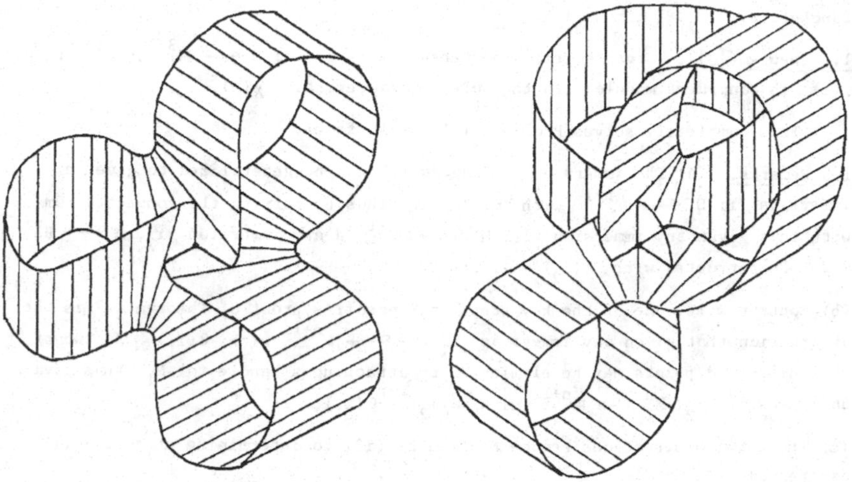

1.6 Diagram.

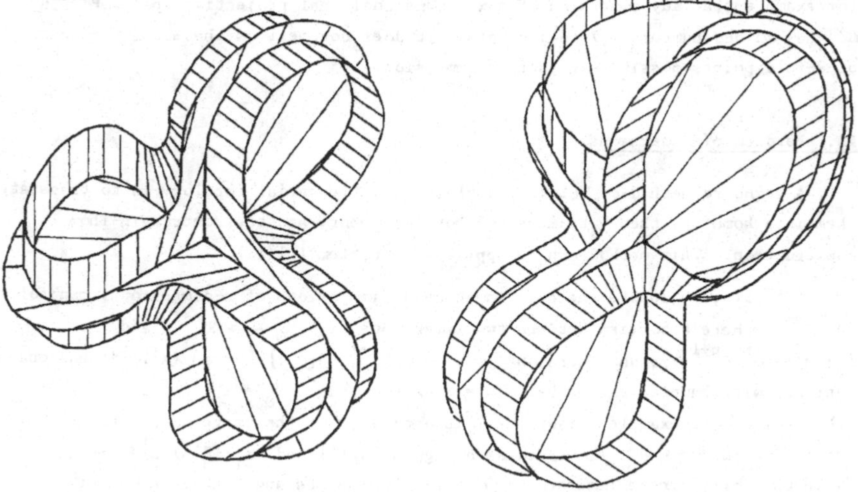

Boy's surface is now obtained by sewing a disc onto the boundary of this immersed Möbius band which can be done without introducing further intersection points. It is clear that there is a single triple point.

Finally, by taking the connected sum of this immersion with immersions previously described any odd number can be obtained for $\theta(1)$ when M is a surface of odd genus.

In fact these are the only possibilities by the following theorem of T.F. Banchoff's.

1.7 Theorem([3]). Given a self-transverse immersion $i: M \looparrowright \mathbb{R}^3$ of a surface, $\theta(i)$ is congruent modulo 2 to the Euler characteristic $\chi(M)$.

This completely solves Problems 1.1 for surfaces.

1.8 Remarks. (a) The immersion of Diagram 1.3 can be generalized to give an immersion $i: S^n \looparrowright \mathbb{R}^{n+1}$ with $\theta(i) = 2$. Thus, by forming the connected sum with this, given any immersion $i_1: M^n \looparrowright \mathbb{R}^{n+1}$ a new immersion $i_2. M \looparrowright \mathbb{R}^{n+1}$ may be constructed with $\theta(i_2) = \theta(i_1) + 2$.

(b) Contrariwise, one of the key steps in Banchoff's proof of Theorem 1.7 is the observation that given any immersion $i_1: M_1^n \looparrowright \mathbb{R}^{n+1}$ with $\theta(i_1) \geqq 2$ two of the (n+1)-fold points may be eliminated by attaching a handle to M_1 thus giving an immersion $i_2: M_2^n \looparrowright \mathbb{R}^{n+1}$ with $\theta(i_2) = \theta(i_1) - 2$.

(c) These two observations reduce Problem 1.1(ii) to determining whether $\theta(i)$ can be odd.

(d) B. Hill-Tout has recently generalized Theorem 1.7 to all even n. This implies for example that any immersion of six dimensional real projective space P^6 in $\mathbb{R}^7$ has an odd number of 7-fold points. It does not tell us the minimum number of 7-fold points possible in such an immersion.

§2. Bordism of immersions.

A standard method of solving problems in differential topology is to translate them into homotopy theory by means of bordism theory and the Pontrjagin-Thom construction. This method can be applied to Problem 1.1(ii).

Let $I(n,1)$ denote the bordism group of immersions of n-dimensional manifolds in $\mathbb{R}^{n+1}$. Here a bordism between two immersions $i_0: M_0 \looparrowright \mathbb{R}^{n+1}$ and $i_1: M_1 \looparrowright \mathbb{R}^{n+1}$ is an immersion $j: W \looparrowright \mathbb{R}^{n+1} \times [0,1]$ of an (n+1)-dimensional manifold with boundary such that $\partial W = M_0 \cup M_1$ and $j|M_0 = i_0 \times \{0\}$, $j|M_1 = i_1 \times \{1\}$. For example, Diagram 2.1 is a picture of the image of a bordism between $i_0: S^1 \looparrowright \mathbb{R}^2$ (immersed as a figure eight) and $i_1: S^1 \cup S^1 \looparrowright \mathbb{R}^2$ (with the first circle immersed with two double points and the second as i_0).

In the usual way bordism defines an equivalence relation and we may form the group $I(n,1)$ of bordism classes. Addition of bordism classes is given by the union of representatives and the inverse of a class may be obtained by reflecting a representative in a hyperplane.

2.1 Diagram.

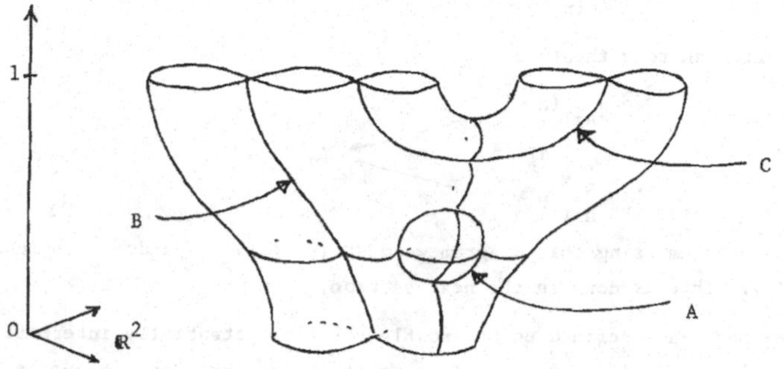

2.2 Proposition. Assigning to a self-transverse immersion $i: M^n \looparrowright \mathbb{R}^{n+1}$ the number (modulo 2) of its $(n+1)$-fold intersection points defines a homomorphism

$$\theta: I(n,1) \longrightarrow \mathbb{Z}_2.$$

This is proved by observing that any immersion is bordant to a self-transverse one and that any two bordant self-transverse immersions are bordant via a self-transverse immersion $j: W \looparrowright \mathbb{R}^{n+1} \times [0,1]$. Now the $(n+1)$-fold intersection set of j consists of immersed 1-manifolds, i.e. immersed circles (e.g. A in Diagram 2.1) and immersed closed intervals which pair off $(n+1)$-fold points in the image of the boundary (e.g. B and C in Diagram 2.1).

Alternatively we can consider oriented manifolds and define a homomorphism

$$\theta_o: SI(n,1) \longrightarrow \mathbb{Z}_2$$

on the bordism group of immersions of oriented n-dimensional manifolds in $\mathbb{R}^{n+1}$ by counting (modulo 2) the number of $(n+1)$-fold intersection points.

2.3 Problem. For which values of n is $\theta = 0$ or $\theta_o = 0$?

Of course $\theta = 0$ implies that $\theta_o = 0$. Also, in view of Remark 1.8(d), this problem (for θ) is equivalent to Problem 1.1(ii).

Having now expressed our problem in terms of bordism theory we can translate it into homotopy theory. R. Wells [26] has shown that replacing embeddings by immersions in bordism theory corresponds to replacing homotopy groups of Thom complexes by their stable counterparts. In particular, $I(n,1)$ is isomorphic to $\pi^S_{n+1}(P^\infty)$ where P^∞ is infinite dimensional real projective space (i.e. $MO(1)$) and $SI(n,1)$ is isomorphic to $\pi^S_{n+1}(S^1) \cong \pi^S_n$, the stable n-stem ($S^1 = MSO(1)$).

So the geometrical diagram

$$\begin{array}{ccc} SI(n,1) & \xrightarrow{\theta_o} & \\ \text{forget orientation} \downarrow & & \mathbb{Z}_2 \\ I(n,1) & \xrightarrow{\theta} & \end{array}$$

translates into homotopy theory as

$$\begin{array}{ccc} \pi_n^S \cong \pi_{n+1}^S(S^1) & \xrightarrow{\theta_o} & \\ \downarrow i_* & & \mathbb{Z}_2 \\ \pi_{n+1}^S(P^\infty) & \xrightarrow{\theta} & \end{array}$$

To study the problem using this diagram we must identify θ and θ_o homotopy theoretically. This is done in the next section.

Any new invariant defined on the stable stems is potentially interesting and so having defined one, like θ_o, one fears that it is simply an old one in disguise. M.H. Freedman [10] showed that $\theta_o \neq 0$ for $n = 3$ and conjectured (in this pessimistic vein) that $\theta_o \neq 0$ if and only if $n = 1, 3, 7$ ("θ_o is the stable Hopf invariant"). In support of this U. Koschorke showed that $\theta_o = 0$ for $n \leq 19$ (and $n \neq 1, 3, 7$) and for n even [15] and that $\theta_o | \mathrm{im} J = 0$ ($n \neq 1, 3, 7$) [16]. Geometrically, $\mathrm{im} J$ is the subgroup of $SI(n,1)$ arising from immersions of S^n. His arguments are differential topological but information about the structure of π_*^S and $\mathrm{im} J$ known from homotopy theory is required.

§3. The identification of θ and θ_o.

As previously remarked, to complete the translation of our problem into homotopy theory we must reformulate the homomorphisms θ and θ_o in that framework. This requires an understanding of Wells' result identifying the bordism groups with the stable homotopy groups which he proved indirectly via the classical result for embeddings using Hirsch theory [12]. Fortunately, Koschorke and B. Sanderson have recently given a beautiful direct proof [17] of this result which tells us what we need. For simplicity I will restrict my attention in the discussion which follows to the oriented case simply stating the result in the non-oriented case at the end of the section.

For any pointed space X write QX for the direct limit $\lim_{\to n} \Omega^n \Sigma^n X$ where Σ is the reduced suspension functor and Ω is the loop space functor. Then $\pi_i^S(X)$ is naturally isomorphic to $\pi_i(QX)$. Thus Wells' result tells us that QS^1 is the "Thom complex" for the bordism groups of codimension one oriented immersions. Koschorke and Sanderson exhibit the configuration space model for QS^1 as a Thom complex for the bordism groups. So let us recall this model (see [21] or [24] for more details).

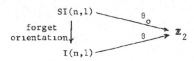

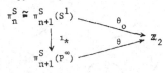

Write $\tilde{\Gamma}_r X$ for the space

$$\{ \{ (v_1,x_1),\ldots\ldots,(v_r,x_r) \} \mid v_i \in \mathbb{R}^\infty,\ x_i \in X,\ i \neq j \Rightarrow v_i \neq v_j \}$$

topologized in the obvious way as a quotient space of a subspace of $(\mathbb{R}^\infty \times X)^r$. This is the space of unordered r-tuples of distinct points in $\mathbb{R}^\infty$ each point labelled by a point of X. An equivalence relation is defined on $\bigcup_{r\geq 0} \tilde{\Gamma}_r X$ by ignoring points in $\mathbb{R}^\infty$ whose label is the base point $*$ of X. More formally it is generated by

$$\{ (v_1,x_1),\ldots\ldots,(v_{r-1},x_{r-1}),(v_r,*) \} \sim \{ (v_1,x_1),\ldots\ldots,(v_{r-1},x_{r-1}) \}.$$

Write ΓX for the space obtained by factoring out by this relation and $\Gamma_r X$ for the image of $\tilde{\Gamma}_r X$ in ΓX. It is now a well-established result that, for (reasonable) connected spaces X, ΓX is homotopy equivalent to QX.

Koschorke and Sanderson have shown how an immersion $M^n \looparrowright \mathbb{R}^{n+1}$ of an oriented manifold gives rise to a map $S^{n+1} \longrightarrow \Gamma S^1$, well-defined up to homotopy, by a generalization of the Pontrjagin-Thom construction. This induces a homomorphism

$$\Pi\colon SI(n,1) \longrightarrow \pi_{n+1}(\Gamma S^1)$$

which they prove to be an isomorphism by exhibiting an inverse. Of course I am here discussing a very special and simple case of their construction.

We simply need the construction of Π. Suppose we are given an oriented immersion $M^n \looparrowright \mathbb{R}^{n+1}$. Then

(i) extend this to an immersion of a tubular neighbourhood

$$i\colon M \times \mathbb{R} \looparrowright \mathbb{R}^{n+1}$$

using the orientation,

(ii) choose any embedding

$$f\colon M \longrightarrow \mathbb{R}^\infty.$$

Now define

$$S^{n+1} \cong (\mathbb{R}^{n+1})^* \longrightarrow \Gamma(\mathbb{R}^*) \cong \Gamma S^1$$

by mapping a point which lies in precisely r fibres of the tubular neighbourhood, say $i(m_1,t_1) = \ldots\ldots = i(m_r,t_r)$, to the point

$$\{ (f(m_1),t_1),\ldots\ldots,(f(m_r),t_r) \}$$

of the configuration space. It can now be checked that up to homotopy this map is independent of the choices of i and f and depends only on the oriented bordism class of the original immersion. Thus Π is defined and is fairly clearly a homomorphism.

It should be noted that if the original immersion is self-transverse then i can be taken to be self-transverse so that the above construction gives a map

$$S^{n+1} \longrightarrow \Gamma_{n+1}S^1.$$

In fact this induces an isomorphism

$$\Pi_{(n+1)}: SI_{(n+1)}(n,1) \longrightarrow \pi_{n+1}(\Gamma_{n+1}S^1)$$

where $SI_{(n+1)}(n,1)$ is the bordism group of oriented immersions with no more than (n+1)-fold intersection points (with the same restriction on bording immersions). By the remarks following Proposition 2.2 the obvious map $SI_{(n+1)}(n,1) \longrightarrow SI(n,1)$ is an epimorphism.

We can now look to reformulating θ_o in this context. Given a self-transverse oriented immersion $M^n \looparrowright \mathbb{R}^{n+1}$, the (n+1)-fold intersection set is a finite set of points in $\mathbb{R}^{n+1}$ each of which is provided with an unordered (n+1)-frame (one vector being provided by the normal vector to each sheet of M which passes through the point). Denote the bordism group of such objects by $E_{\widetilde{fr}}(0,n+1)$. It is clear that, since an unordered frame does not provide an orientation, this bordism group is isomorphic to $\mathbb{Z}_2$ an isomorphism being given by the number (modulo 2) of points. Thus the correct bordism formulation of the self-intersection invariant is as a map

$$\theta_o: SI(n,1) \longrightarrow E_{\widetilde{fr}}(0,n+1).$$

To interpret this homotopically write $D_{n+1}S^1$ for the quotient space $\Gamma_{n+1}S^1/\Gamma_n S^1$ and $\xi: \Gamma_{n+1}S^1 \longrightarrow D_{n+1}S^1$ for the quotient map. Then a construction similar to that above gives a Pontrjagin-Thom isomorphism

$$\Pi_{\widetilde{fr}}: E_{\widetilde{fr}}(0,n+1) \longrightarrow \pi_{n+1}(D_{n+1}S^1)$$

so that the following diagram commutes.

$$\begin{array}{ccc}
SI(n,1) & \xrightarrow[\cong]{\Pi} & \pi_{n+1}(\Gamma S^1) \\
\uparrow & & \uparrow \\
SI_{(n+1)}(n,1) & \xrightarrow[\cong]{\Pi_{(n+1)}} & \pi_{n+1}(\Gamma_{n+1}S^1) \\
\downarrow & & \downarrow \xi_* \\
\mathbb{Z}_2 \cong E_{\widetilde{fr}}(0,n+1) & \xrightarrow[\cong]{\Pi_{\widetilde{fr}}} & \pi_{n+1}(D_{n+1}S^1)
\end{array}
\qquad \theta_o: SI(n,1) \to \mathbb{Z}_2 \cong E_{\widetilde{fr}}(0,n+1)$$

This shows that θ_o corresponds to a stable Hopf invariant $h^S_{n+1}: \pi^S_{n+1}(S^1) \longrightarrow \pi^S_{n+1}(D_{n+1}S^1) \cong \pi_{n+1}(D_{n+1}S^1)$ (see [4] and [25]). It implies that $\theta_o \neq 0$ if and only if $\xi_* \neq 0$ and so the geometric problem is completely reduced to homotopy theory.

Now apply the Hurewicz homomorphism.

$$\begin{array}{ccc}
\pi_{n+1}(\Gamma_{n+1}S^1) & \xrightarrow{h} & H_{n+1}(\Gamma_{n+1}S^1) \\
\downarrow \xi_* & & \downarrow \xi_* \\
\mathbb{Z}_2 \cong \pi_{n+1}(D_{n+1}S^1) & \xrightarrow[\cong]{h} & H_{n+1}(D_{n+1}S^1)
\end{array}$$

Here and throughout H_* denotes homology with $\mathbb{Z}_2$-coefficients. The bottom map is an isomorphism by the Hurewicz isomorphism theorem. Thus we have proved

<u>3.1 Proposition.</u> $\theta_o \neq 0$ if and only if there is a spherical class $a \in H_{n+1}(\Gamma_{n+1}S^1)$ such that $\xi_*(a) \neq 0$.

In the non-oriented case a similar argument gives

<u>3.2 Proposition.</u> $\theta \neq 0$ if and only if there is a spherical class $a \in H_{n+1}(\Gamma_{n+1}P^\infty)$ such that $\xi_*(a) \neq 0 \in H_{n+1}(D_{n+1}P^\infty) \cong \mathbb{Z}_2$.

§4. $\xi_*: H_*(\Gamma_{n+1}X) \longrightarrow H_*(D_{n+1}X)$.

The original geometric Problem 1.1(ii) has now been reduced to an equivalent problem in homology theory. To study this problem it is necessary to recall the structure of the homology groups involved.

S Araki and T. Kudo have defined [2] certain operations

$$Q^i: H_n(QX) \longrightarrow H_{n+i}(QX)$$

which are zero for $i < n$ and equal to the Pontrjagin square for $i = n$. Here X may be any pointed space. These operations can be used to describe the homology of QX (with $\mathbb{Z}_2$-coefficients) in terms of the homology of X. E. Dyer and R.K. Lashof have shown [7] that, if X is connected and $\{x_\alpha\}$ is a $\mathbb{Z}_2$-basis of homogeneous elements for $\tilde{H}_*(X)$, then $H_*(QX)$ is a polynomial ring on generators

$$Q^{i_1}Q^{i_2}\cdots\cdots Q^{i_r}x_\alpha$$

where $(i_1, i_2, \ldots\ldots, i_r)$ is a sequence of integers which is admissible (i.e. $i_j \leq 2i_{j+1}$ for $1 \leq j < r$) and of excess (i.e. $i_1 - i_2 - \cdots\cdots - i_r$) greater than the dimension of x_α. I refer to the set of monomials in these generators as the <u>Dyer-Lashof basis</u> for $H_*(QX)$.

Following G. Nishida [23] define a <u>height function</u> h on this basis by $h(x_\alpha) = 1$, $h(Q^i\xi) = 2h(\xi)$ and $h(\xi\eta) = h(\xi) + h(\eta)$. Then the grading of homology by height corresponds to the filtration of $\Gamma X \simeq QX$ by the subspaces $\Gamma_r X$. More precisely, the homomorphism

$$\iota_*: H_*(\Gamma_r X) \longrightarrow H_*(\Gamma X) \cong H_*(QX)$$

is a monomorphism onto the subgroup spanned by the basis elements of height at most r and so $H_*(\Gamma_r X)$ may be identified with that subgroup. Furthermore

$$\xi_*: H_*(\Gamma_r X) \longrightarrow H_*(D_r X)$$

may be thought of as a projection map onto the subgroup spanned by the basis elements of height precisely r.

§5. The oriented case.

We are now in a position to completely solve Problem 1.1(ii) for orientable manifolds. Let $a_1 \in H_1(S^1)$ be the generator. Then there is a single basis element in $H_{n+1}(\Gamma_{n+1}S^1)$ of height n+1, namely a_1^{n+1}. This means that

$$\xi_*: H_{n+1}(\Gamma_{n+1}S^1) \longrightarrow H_{n+1}(D_{n+1}S^1) \cong \mathbf{Z}_2$$

maps a_1^{n+1} to the generator and all other basis elements to 0.

To summarize the argument so far, suppose that $\alpha \in SI(n,1) \cong \pi_{n+1}(\Gamma S^1) \cong \pi_{n+1}(QS^1)$ pulls back to $\bar{\alpha} \in \pi_{n+1}(\Gamma_{n+1}S^1)$ (which is always possible for dimensional reasons). Then

$$\theta_o(\alpha) \neq 0 \Leftrightarrow \xi_*(\bar{\alpha}) \neq 0 \in \pi_{n+1}(D_{n+1}S^1)$$

$\Leftrightarrow \xi_* h(\bar{\alpha}) = h\xi_*(\bar{\alpha}) \neq 0 \in H_{n+1}(D_{n+1}S^1)$ (Proposition 3.1)

$\Leftrightarrow h(\bar{\alpha}) \in H_{n+1}(\Gamma_{n+1}S^1)$ involves a_1^{n+1} (when written in terms of the Dyer-Lashof basis)

$\Leftrightarrow h(\alpha) \in H_{n+1}(QS^1)$ involves a_1^{n+1}.

Let us first look at the Hopf invariant 1 dimensions. In I. Madsen's thesis [19] we find

$$h(\eta) = a_1^2 \ (=Q^1a_1) \in H_2(QS^1),$$

$$h(\nu) = Q^3a_1 + a_1^4 \in H_4(QS^1),$$

$$h(\sigma) = Q^7a_1 + (Q^3a_1)^2 \in H_8(QS^1).$$

Here η, ν, σ are generators of the appropriate homotopy groups. This tells us that $\theta_o(\eta) = 1$, $\theta_o(\nu) = 1$ (which we knew) but that $\theta_o(\sigma) = 0$ disproving Freedman's conjecture.

Turning to other dimensions, it has been conjectured by Madsen that all elements there have zero Hurewicz image in $H_*(QS^1)$. The truth of this would immediately imply that $\theta_o = 0$. Failing this, it is possible to prove that for $n \neq 1, 3$ and $\alpha \in \pi_{n+1}(QS^1)$, $h(\alpha)$ does not involve a_1^{n+1} and so $\theta_o(\alpha) = 0$. Thus $\theta_o = 0$ for $n \neq 1$ or 3.

This is done by first observing that a spherical class in $H_{n+1}(QS^1)$ is necessarily in the image of the suspension homomorphism

$$\sigma: H_n(QS^0) \longrightarrow H_{n+1}(QS^1).$$

This tells us that if $\theta_o \neq 0$ then n+1 is a power of 2. Then we observe that a spherical class $a \in H_n(QS^0)$ is necessarily primitive (with respect to the cup coproduct) and has $Sq_*^j a = 0$ for $j > 0$ where Sq_*^j is the homology dual of the Steenrod cohomology operation Sq^j. The details may be found in [8]. Substantial use is made of Madsen's description [19] of the primitives in $H_*(QS^0)$.

Restating the conclusion in terms of the original geometric Problem 1.1(ii) gives

<u>5.1 Theorem.</u> A self-transverse immersion $M^n \looparrowright \mathbb{R}^{n+1}$ of a closed compact smooth orientable n-manifold can have an odd number of (n+1)-fold intersection points if and only if $n = (0,)\ 1$ or 3.

It should be observed that the classical Hopf invariant 1 result [1] lies much deeper than this. An element $\alpha \in \pi_{n+1}(QS^1)$ has Hopf invariant 1 if and only if $h(\alpha)$ involves $Q^n a_1$, for this means that the adjoint $\tilde{\alpha} \in \pi_{2n}(QS^n)$, which pulls back to $\pi_{2n}(\Omega S^{n+1})$ for dimensional reasons, has $h(\tilde{\alpha}) = a_n^2$ and so has James-Hopf invariant $(\pi_{2n}(\Omega S^{n+1}) \longrightarrow \pi_{2n}(\Omega S^{2n+1}) \cong \mathbb{Z})$ an odd integer. Thus in this case the problem is to decide whether there is a spherical class in $H_{n+1}(QS^1)$ involving $Q^n a_1$. The elementary methods used to prove the above theorem reduce us to the case where $n+1$ is a power of 2 and to deciding whether

$$Q^{2^i-1}a_1 + (Q^{2^{i-1}-1}a_1)^2 \in H_{2^i}(QS^1)$$

is spherical. They are not strong enough to settle this.

<u>§6. The non-oriented case.</u>

Let $a_i \in H_i(P^\infty) \cong \mathbb{Z}_2$ be the generator, $i \geq 1$. Suppose that $\alpha \in I(n,1) \cong \pi_{n+1}(QP^\infty)$. Proposition 3.2 together with §4 imply that $\theta(\alpha) \neq 0$ if and only if $h(\alpha) \in H_{n+1}(QP^\infty)$ involves a_1^{n+1} when written in terms of the Dyer-Lashof basis.

Specific calculation gives

$$h(\hat{\eta}\circ\eta) = a_1^2 \in H_2(QP^\infty),$$
$$h(\Diamond) = a_1^3 + a_1a_2 + a_3 + Q^2a_1 \in H_3(QP^\infty),$$
$$h(\hat{\eta}\circ\nu) = a_1^4 + Q^3a_1 \in H_4(QP^\infty),$$
$$h(\Diamond\circ\nu) = a_1^6 + a_1^2a_2^2 + a_3^2 + (Q^2a_1)^2 \in H_6(QP^\infty),$$
$$h(\hat{\theta}) = a_1^7 + a_1^5a_2 + a_1^4a_3 + a_1^4Q^2a_1 + a_1^3a_4 + a_1^2a_2a_3 + a_1^2a_5 + a_1^2Q^3a_2 +$$
$$a_1a_2^3 + a_1a_3^2 + a_1a_6 + a_2^2a_3 + a_2^2Q^2a_1 + a_2a_5 + a_3a_4 + a_7 + Q^4a_3 \in H_7(QP^\infty),$$

where $\hat{\alpha} \in \pi_1(QP^\infty)$ denotes an element such that $\lambda_*(\hat{\alpha}) = \alpha \in \pi_1(QS^0)$ (λ defined below). Thus θ takes the non-zero value when applied to each of the elements in this list. Here $\hat{\eta}\circ\eta$ and $\hat{\eta}\circ\nu$ are simply the images of the elements $\eta \in \pi_2(QS^1)$ and $\nu \in \pi_4(QS^1)$ already considered and represented by immersions of spheres. The other elements may only be represented by immersing non-orientable manifolds. In fact $\Diamond$ is represented by Boy's surface (modulo 2), $\Diamond\circ\nu$ by an immersion of $S^3 \times P^2$ in $\mathbb{R}^6$ and $\hat{\theta}$ by an immersion of P^6 in $\mathbb{R}^7$ (see

Remark 1.8(d).

Turning from these examples to the general problem a complete solution has not been obtained and this is work in progress. General results have been obtained in certain cases ($n \not\equiv 3$, modulo 4) and they show that the problem lies deeper than in the oriented case. Details will appear in [9].

Case 1: $n \equiv 0$, modulo 2.

Suppose that $\alpha \in \pi_{n+1}(QP^{\infty})$ pulls back to $\bar{\alpha} \in \pi_{n+1}(QP^{n+1})$ (which is always possible for dimensional reasons). Then

$\theta(\alpha) \neq 0 \Leftrightarrow h(\alpha) \in H_{n+1}(QP^{\infty})$ involves a_1^{n+1}

$\Leftrightarrow h(\alpha)$ and so $h(\bar{\alpha})$ involve a_{n+1} (using $n+1$ odd and the primitivity of $h(\alpha)$)

$\Leftrightarrow \bar{\alpha}: S^{n+1} \longrightarrow QP^{n+1}$ is a stable reduction of P^{n+1}.

Such a map exists if and only if $n+1 = (1,)\ 3$ or 7 [1].

This proves the following geometrical result.

6.1 Theorem. If n is even, a self-transverse immersion $M^n \looparrowright \mathbb{R}^{n+1}$ of a closed compact smooth n-manifold can have an odd number of (n+1)-fold intersection points if and only if $n = (0,)\ 2$ or 6.

Thus this result is equivalent to the Hopf invariant 1 result. The stable reducibility of projective spaces can be related to the formulation of the Hopf invariant 1 problem in terms of cohomology operations by using the reflection map $P^{\infty} \longrightarrow QS^0$. This is defined, up to homotopy, by composing the map $P^{\infty} \longrightarrow \bar{S}O$ given by reflection in the orthogonal hyperplane with $J_0: \bar{S}O \longrightarrow QS^0$ given by $J_0(x) = J(x)*1$. Here J denotes the stable J-homomorphism, $*$ denotes the track sum and 1 is any map of degree one. Its adjoint is a stable map

$$\lambda: P^{\infty} \longrightarrow S^0$$

which induces an epimorphism of the 2-component of stable homotopy groups in positive dimensions [14]. In the mapping cone of λ, $Sq^i(\iota_0) \neq 0$ for all $i \geq 2$.

Returning to the above argument, let $\tilde{\alpha} \in \pi^S_{n+1}(P^{\infty})$ be the stable adjoint of $\alpha \in \pi_{n+1}(QP^{\infty})$. Then

$h(\alpha) \in H_{n+1}(QP^{\infty})$ involves a_{n+1}

$\Leftrightarrow h(\tilde{\alpha}) = a_{n+1} \in H_{n+1}(P^{\infty})$

$\Leftrightarrow \lambda_*(\tilde{\alpha}) \in \pi^S_{n+1}(S^0)$ is detected by Sq^{n+2}

$\Leftrightarrow \lambda_*(\tilde{\alpha})$ has Steenrod-Hopf invariant 1 (by definition).

Case 2: $n \equiv 1$, modulo 4.

Suppose that $n+1 = 2m$ and $\alpha \in \pi_{2m}(QP^{\infty})$ with stable adjoint $\tilde{\alpha} \in \pi^S_{2m}(P^{\infty})$. Then

$$\theta(\alpha) \neq 0 \Leftrightarrow h(\alpha) \in H_{2m}(QP^\infty) \text{ involves } a_1^{2m}$$
$$\Leftrightarrow h(\alpha) \text{ involves } a_m^2 = Q^m a_m \text{ (using } m \text{ odd and primitivity)}$$
$$\Leftrightarrow \tilde{\alpha} \text{ is detected by } Sq^{m+1} \text{ on the cohomology dual of } a_m$$
$$\Leftrightarrow \lambda_*(\tilde{\alpha}) \in \pi_{2m}^S(S^0) \text{ is detected by the secondary cohomology}$$
operation based on $Sq^{m+1}Sq^{m+1}$ ([13]).

Such a map can exist only if $m+1$ is a power of 2 [1] say $m+1 = 2^j$ when the secondary operation is usually denoted $\phi_{j,j}$. It does exist for $j = 1$ (α = ηoη), $j = 2$ (α = νoν), $j = 3$ (α = σoσ), $j = 4$ (see [20]) and $j = 5$ (M.G. Barratt and M.E. Mahowald, unpublished).

Geometrically this result reads

6.2 Theorem. If $n \equiv 1$, modulo 4, a self-transverse immersion $M^n \looparrowright \mathbb{R}^{n+1}$ can have an odd number of $(n+1)$-fold intersection points only when $n+3$ is a power of 2, say 2^{j+1}. In that case such an immersion exists if and only if there is an element of π_{n+1}^S detected by the secondary operation $\phi_{j,j}$.

Since the existence of elements detected by $\phi_{j,j}$ has a geometric interpretation in terms of the Kervaire invariant of a framed manifold [6] it may be of interest to set out the direct correspondence between the two geometrical interpretations. This involves giving a geometrical interpretation to the map λ which has been done by Koschorke [15] and by B. Gray as follows.

Suppose that $i: M^n \looparrowright \mathbb{R}^{n+1}$ is an immersion with a normal bundle $\nu(i)$. An immersion in $\mathbb{R}^{n+2}$ of the circle bundle N of $\nu(i) \oplus \underline{1}$ may be obtained by placing a figure eight in each normal plane of the immersion

$$M \overset{i}{\looparrowright} \mathbb{R}^{n+1} \cong \mathbb{R}^{n+1} \times \{0\} \subset \mathbb{R}^{n+2}.$$

This immersion provides a trivialization of $\tau(N) \oplus \underline{1}$ and therefore represents an element of π_{n+1}^S. Conversely, a right inverse for λ (modulo odd primes) is obtained by taking the double point set of an oriented (self-transverse) immersion $N^{n+1} \looparrowright \mathbb{R}^{n+2}$ which gives an immersion $M^n \looparrowright \mathbb{R}^{n+2}$ with a normal vector field (taking the average of the unordered 2-frame induced on M by the orientation of N) and so, by Hirsch theory [12], an immersion $i: M^n \looparrowright \mathbb{R}^{n+1}$.

Theorem 6.2 and the main result of [6] tell us that, if $n \equiv 1$, modulo 4, $\theta(i)$ is odd if and only if N, with the given trivialization of its stable tangent bundle, has Kervaire invariant 1. Similarly, if n is even, the proof of Theorem 6.1 tells us that $\theta(i)$ is odd if and only if N has Hopf invariant 1 and this means that when

$$N \looparrowright \mathbb{R}^{n+2} \cong \mathbb{R}^{n+2} \times \{0\} \subset \mathbb{R}^{2n+2}$$

is made self-transverse it has an odd number of double points.

§7. Further remarks on immersions of surfaces in $\mathbb{R}^3$.

In this paper I have discussed the simplest and most striking of a series of invariants which may be defined on immersions. Given a self-transverse immersion $M^n \looparrowright \mathbb{R}^{n+k}$ with some structure on its normal bundle, the (r+1)-fold intersection set is the image of an immersion $L^{n-rk} \looparrowright \mathbb{R}^{n+k}$ with a corresponding structure on its normal bundle. This is formulated in [8] where it is related to the work of Koschorke and Sanderson's. The next simplest of these invariants is the double point set of a surface immersed in $\mathbb{R}^3$.

The bordism group of surfaces immersed in $\mathbb{R}^3$, I(2,1) is isomorphic to $\pi_3^S(P^\infty) \cong \mathbb{Z}_8$ generated by $\hat{\upsilon}$. The triple point invariant θ which takes its values in a group isomorphic to $\mathbb{Z}_2$ tells us that Boy's surface represents a generator. But it cannot detect elements of order less than 8 in the bordism group. For this the double point invariant is required.

In the commutative diagram

$$\begin{array}{ccc} \mathbb{Z}_8 \cong I(2,1) \cong \pi_3^S(P^\infty) & \longrightarrow & \pi_3^S(D_2P^\infty) \\ \uparrow & & \uparrow \\ \mathbb{Z}_2 \cong SI(2,1) \cong \pi_3^S(S^1) & \longrightarrow & \pi_3^S(D_2S^1) \end{array}$$

the horizontal maps correspond to the double point invariant (the lower map in the oriented case) and the vertical maps (induced by $S^1 \cong P^1 \subset P^\infty$) correspond to forgetting the orientation. Both of the horizontal maps are isomorphisms and the vertical maps are monomorphisms.

The group $\pi_3^S(D_2S^1)$ corresponds to the bordism group of immersed 1-manifolds with an unordered normal 2-frame (cf. §3). A generator is represented by an embedded circle with the usual Hopf framing, for $\mathbb{Z}_2 \cong \pi_3^S(S^2) \longrightarrow \pi_3^S(D_2S^1)$, induced by $S^2 \cong \Sigma P^1 \subset \Sigma P^\infty \cong D_2S^1$, is an **isomorphism** (it corresponds to forgetting the order of the frame).

The group $\pi_3^S(D_2P^\infty)$ corresponds to the bordism group of immersed 1-manifolds with an unordered normal projective 2-frame. (A projective n-frame of R^n is an n-tuple of elements in P^{n-1} such that any lift to S^{n-1} provides a basis for R^n.) D_2P^∞ is stably homotopy equivalent to a CW-complex whose 4-skeleton is

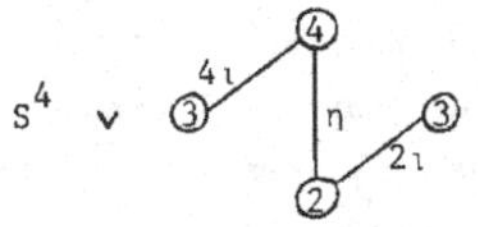

and so $\pi_3^S(D_2P^\infty) \cong \mathbb{Z}_8$ with a generator ι_3 so that $4\iota_3 = \iota_2 \circ \eta$. Geometrically a generator is represented by an embedded circle whose unordered projective 2-frame rotates through $\frac{1}{2}\pi$ with repect to a trivial framing on passing once round the circle. This may be seen by observing that four times this element is represented by the same construction with $\frac{1}{2}\pi$ replaced by 2π; this corresponds to the Hopf framing.

If the double point invariant is evaluated on Boy's surface this generator is obtained (up to bordism). This demonstrates that the double point invariant is an isomorphism. If the invariant is evaluated on the usual immersion of the Klein bottle (Diagram 1.4) a circle with a trivial framing is obtained. This shows that this immersion represents the trivial element in the bordism group, i.e. it is the boundary of an immersion of a 3-manifold with boundary in $\mathbb{R}^4$. To obtain an immersion of the Klein bottle representing an element of order 4 take a cylinder immersed so that each circular cross-section is immersed as a figure eight

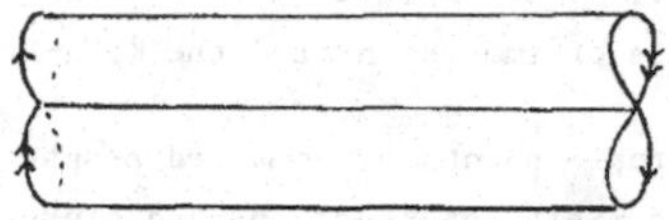

and identify the ends after twisting through an angle π. Evaluating the double point invariant on this immersion gives a circle with an unordered projective 2-frame rotating through π on passing once round the circle. This represents twice a generator.

Acknowledgements.

I am grateful to Ulrich Koschorke for discussions on his joint work with Brian Sanderson (§3) and for sending to Manchester a copy of [15] which aroused my interest in this problem, to Reg Wood for many hours spent discussing the problem (I learnt most of §§1,2 from him), and to Barry Hill-Tout for two letters giving details of his work (1.8(d)).

References.

1. J.F.Adams, On the non-existence of elements of Hopf invariant one, Ann. of Math., (2) 72 (1960), 20-104.

2. S.Araki and T.Kudo, Topology of H_n-spaces and H-squaring operations, Mem. Fac. of Sci., Kyusyu Univ., Ser.A, 10 (1956), 85-120.

3. T.F.Banchoff, Triple points and surgery of immersed surfaces, Proc. Amer. Math. Soc., 46 (1974), 407-413.

4. M.G.Barratt and P.J.Eccles, Γ^+-structures III: The stable structure of $\Omega^\infty\Sigma^\infty A$, Topology, 13 (1974), 199-207.

5. W.Boy, Über die Curvatura integra und die Topologie geschlossener Flächen, Math. Ann., 57 (1903), 151-184.

6. W.Browder, The Kervaire invariant of framed manifolds and its generalization, Ann. of Math., (2) 90 (1969), 157-186.

7. E.Dyer and R.K.Lashof, Homology of iterated loop spaces, Amer. J. Math., 84 (1962), 35-88.

8. P.J.Eccles, Multiple points of codimension one immersions of oriented manifolds, Math. Proc. Cambridge Philos. Soc., to appear.

9. P.J.Eccles, Multiple points of codimension one immersions of non-oriented manifolds, in preparation.

10. M.H.Freedman, Quadruple points of 3-manifolds in S^4, Comment. Math. Helv., 53 (1978), 385-394.

11. D.Hilbert und S.Cohn-Vossen, Anschauliche Geometrie, Springer, Berlin, 1932.

12. M.W.Hirsch, Immersions of manifolds, Trans. Amer. Math. Soc., 93 (1959), 242-276.

13. J.D.S.Jones, Thesis, University of Oxford, 1976.

14. D.S.Kahn and S.B.Priddy, The transfer and stable homotopy theory, Math. Proc. Cambridge Philos. Soc., 83 (1978), 103-111.

15. U.Koschorke, Multiple points of immersions and the Kahn-Priddy theorem, Math. Z., to appear.

16. U.Koschorke, On the (n+1)-tuple points of immersed n-spheres, to appear.

17. U.Koschorke and B Sanderson, Self intersections and higher hopf invariants, Topology, 17 (1978), 283-290.

18. W Lietzmann, Anschauliche Topologie, Oldenbourg, München, 1955.

19. I.Madsen, Thesis, University of Chicago, 1970.

20. M.E.Mahowald and M.C.Tangora, Some differentials in the Adams spectral sequence, Topology, 6 (1967), 349-369.

21. J.P.May, The geometry of iterated loop spaces, Lecture Notes in Math., 271, Springer-Verlag, Berlin, 1972.

22. B.Morin et J.-P Petit, Problématique du retournement de la sphère, C. R. Acad. Sc. Paris, 287 (1978), 767-770.

23. G.Nishida, The nilpotency of elements of the stable homotopy groups of spheres, J. Math. Soc. Japan, 25 (1973), 707-732.

24. G.B.Segal, Configuration spaces and iterated loop spaces, Invent. Math., 21 (1973), 213-221.

25. V.P.Snaith, A stable decomposition of $\Omega^n S^n X$, J. London Math. Soc., (2) 7 (1974), 577-583.

26. R.Wells, Cobordism groups of immersions, Topology, 5 (1966), 281-294.

Department of Mathematics,
The University,
Manchester, England.
M13 9PL

h-cobordisms and twisted doubles

Jean-Claude Hausmann

Let W be a compact CAT-manifold with non-empty boundary ∂W (CAT = DIFF or PL). Let $\beta : \partial W \to \partial W$ be a CAT-homeomorphism. Define the twisted double $TD(W,\beta)$ as

$$TD(W,\beta) = W \cup_\beta W$$

which is a closed CAT-manifold. A special class of twisted doubles is constituated by the open books : if V^{n-1} is a compact CAT-manifold with non-empty boundary and if $\alpha : V \to V$ is a CAT-homeomorphism equals to the identity on a neighborhood of ∂V, the open book with page V and monodromy α can be defined as $OB(V,\alpha) = TD(V \times I, \beta_\alpha)$ where $\beta_\alpha : \partial(V \times I) \to \partial(V \times I)$ is defined by:

$$\beta_\alpha(x,t) = \begin{cases} (x,t) & \text{if } t \neq 1 \\ (\alpha(x),1) & \text{for } t = 1. \end{cases}$$

This definition coincides with the usual one: $OB(V,\alpha) = M(\alpha) \cup \partial V \times D^2$ (union over $\partial M(\alpha) = \partial V \times S^1$), where $M(\alpha)$ is the mapping torus of α (see [Ha 1, (2.2)].

In this conference, we showed how open book decompositions are transformed by an h-cobordism. (Recall that a cobordism (R, M, N) is an h-cobordism if both inclusions $M \subset R$ and $N \subset R$ are homotopy equivalences.) More precisely, let $\sigma \in Wh(\partial V)$, where the latter denotes the Whitehead group of ∂V [Co] endowed with the involution $\sigma \to \bar{\sigma}$ defined in [Mi, p. 373 and 398]. A σ-enlargement of V is a manifold $V' = V \cup_{\partial V} U$, where $(U,\partial V,\partial V')$ is an h-cobordism with $\tau^0(U,\partial V) = \sigma$. ($\tau^0(f)$ is the Whitehead torsion of the homotopy equivalence $f : X \to Y$ measured in $Wh(X)$; one denotes $\tau(f) = f_*(\tau^0(f)) \in Wh(Y)$; if $i : X \subset Y$ is an inclusion which is a homotopy equivalence, one denotes $\tau^0(Y,X) = \tau^0(i)$ and $\tau(Y,X) = \tau(i)$; $i_{X,Y} : Wh(X) \to Wh(Y)$ denotes the homomorphism induced by i.)

For any σ-enlargement V_σ of V, one defines $e(V_\sigma)(\alpha) : V_\sigma \to V_\sigma$ by :

$$e(V_\sigma)(\alpha) = \begin{cases} \alpha \text{ on } V \\ \text{id elsewhere.} \end{cases}$$

<u>Theorem 1</u> Let $M^n = OB(V,\alpha)$ with $n \geq 5$. Let (R,M,N) be an h-cobordism with $\tau^0(R,M) = i_{V,M}(\sigma)$ for $\sigma \in Wh(\partial V)$. Then, for any σ-enlargement V_σ of V, the manifold N has an open book decomposition $N \cong OB(V_\sigma, e(V_\sigma)(\alpha))$.

This theorem is proved in [Ha] together with results on inertial h-cobordisms for open books also presented in this conference. The aim of this note is to show how Theorem 1 can (or cannot) be generalized to twisted doubles.

Let $W^n_\sigma = W \cup_{\partial W} U_\sigma$ be a σ-enlargement of the CAT-manifold W ($n \geq 6$). Let $\beta : \partial W \to \partial W$ be a CAT-homeomorphism such that $\beta_*(\sigma) = \sigma$. As $n \geq 6$, the h-cobordisms from ∂W are classified by their Whitehead torsion and thus β can be extended to a CAT-homeomorphism $\beta_U : U_\sigma \to U_\sigma$. Write $\bar{e}(\partial W_\sigma)(\beta) : \partial W_\sigma \to \partial W_\sigma$ for the restriction of β_U to ∂W_σ.

<u>Theorem 2</u> Let $M^n = TD(W,\beta)$, with $n \geq 6$. Let $\sigma \in Wh(\partial W)$ such that $\beta_*(\sigma) = \sigma$, and let (R,M,N) be an h-cobordism with $\tau^0(R,M) = i_{\partial W,M}(\sigma)$. Then, for any σ-enlargement W_σ of W, the manifold N has a twisted double decomposition $N \cong TD(W_\sigma, \bar{e}(\partial W_\sigma)(\beta))$.

One shows easily that Theorem 2 implies Theorem 1 (for $n \geq 6$), which gives a new proof of the latter. The condition $\beta_*(\sigma) = \sigma$ is essential, as shown by Theorem 3 below. Indeed, suppose that W has a handle decomposition with handles of index $\leq r$. Then any enlargement of W will have a handle decomposition with handles of index $\leq \max\{3,r\}$. (An h-cobordism from ∂W can be constructed by adding handles of index 2 and 3 to $\partial W \times I$, see [Ke]. Let $L_{p,q}$ denote as usual [Co,§27] the 3-dimensional lens space of type (p,q), with fundamental group cyclic of order p. Thus, the manifold $W = L_{p,q} \times D^m$ admits a handle decomposition with handles of index ≤ 3. If $m \geq 3$, one has $\pi_1(\partial W) \cong \pi_1(W) \cong \pi_1(TD(W,\beta))$ for any self-homeomorphism β of ∂W. Therefore, any h-cobordism from $TD(W,\beta)$ satisfies all the hypotheses of Theorem 2 except possibly the condition on β_*. However, one has:

<u>Theorem 3</u> Let $W = L_{5,1} \times D^{2k}$, with $2k \geq 6$. There exists a CAT-homeomorphism $\beta : \partial W \to \partial W$ with the following property: for any h-cobordism $(R, TD(W,\beta), N)$ whose torsion $\tau(R,TD(W,\beta))$ is not divisible by 2, the manifold N has no twisted double decomposition $N \cong TD(T,\gamma)$ with T admitting a handle decomposition with handles of index $\leq k$.

<u>Remark:</u> As N^{2k+3} is an odd dimensional manifold, it follows from [La 2] that $N \cong OB(V,\alpha) \cong TD(V \times I,\beta_\alpha)$ with $V \times I$ admitting a handle decomposition with handles of index $\leq k + 1$.

2. Proof of Theorem 2

Let us introduce more precise notations for the definition of $TD(W,\beta)$. Let W_i, $i = 1$ or 2, be two copies of W, and consider the homorphism β as a map from $\partial W_2 \to \partial W_1$. Thus

$$M = TD(W,\beta) = W_1 \amalg W_2 / \{\beta(x_2) = x_1, \text{ for all } x_1 \in W_1\}.$$

Let $W_\sigma = W \cup U$ be a σ-enlargement of W and denote by $(W_1)_\sigma = W_1 \cup U_1$ the copy of W_σ considered as a σ-enlargement of W_1. Construct an h-cobordism (R,M,N) as follows:

$$R = M \times [0,1] \cup_{W_1} (W_1)_\sigma \times [1,2]$$

where the union is made over the manifold W_1 included in each component as follows:

$$W_1 \subset W_1 \times 1 \subset M \times 1 \subset M \times [0,1]$$

$$W_1 \subset (W_1)_\sigma \times 1 \subset (W_1)_\sigma \times [1,2]$$

By excision [Co, (20.3)], the Whitehead torsion $\tau^0(R,M)$ is equal to

$$i_{W_1,M}\tau^0((W_1)_\sigma \times [1,2], W_1 \times 1) = i_{\partial W,M}\tau^0(U,\partial W) = i_{\partial W,M}(\sigma).$$

Now, the manifold N can be described as follows:

$$\begin{aligned} N &= \left[\partial((W_1)_\sigma \times [1,2]) - \operatorname{int}(W_1 \times 1)\right] \cup W_2 \times 1 \\ &= \left[(W_1)_\sigma \cup_{\partial (W_1)_\sigma} U_1\right] \amalg W_2 / \{\beta(x_2) = x_1 \in U_1, \text{ for all } x_2 \in W_2\} \\ &= \left[(W_1)_\sigma \cup U_1\right] \amalg (W_2)_\sigma / \{\beta_U(x_2) = x_1,\ x_1 \in U_1\} \\ &= (W_1)_\sigma \amalg (W_2)_\sigma / \{\bar{e}(\partial W_\sigma)(\beta)(x_2) = x_1,\ x_1 \in \partial(W_1)_\sigma\} \\ &= TD(W_\sigma, \bar{e}(\partial W_\sigma)(\beta)). \end{aligned}$$

This proves Theorem 2.

<u>Remark</u> When $n \geq 6$, Theorem 2 implies Theorem 1. Indeed, one has $OB(V_\sigma, e(V_\sigma)(\alpha)) = TD(V_\sigma \times I, \beta_{e(V_\sigma)(\alpha)})$. One can identify $V_\sigma \times I$ with $X = V_\sigma \times [-\varepsilon, 1+\varepsilon]$, for $\varepsilon > 0$, and the latter is an $i_{\partial V, \partial(V \times I)}(\sigma)$-enlargement of $V \times I$. Under this identification, one gets $\overline{e}(\partial X)(\beta_\alpha) = \beta_{e(V_\sigma)(\alpha)}$.

3. Proof of Theorem 3

We need first to determine the set $E(L_{5,1})$ of torsions $\tau(f) \in Wh(L_{5,1})$ of self-homotopy equivalences $f : L_{5,1} \to L_{5,1}$. In [Ha, (6.1)], we proved that $E(L_{5,1})$ contains 2 elements.

The fundamental group of $L_{5,1}$ is the cyclic group C_5 of order 5 with generator t. Recall that $Wh(L_{5,1}) = \mathbb{Z}$ with a generator u represented by the unit $-1 + t + t^{-1}$ of ZC_5 [Mi, Ex.6.6].

(3.1) <u>Lemma</u> $E(L_{5,1}) = \{0, u\}$

<u>Proof:</u> By [Ha, proof of (6.1)], it suffices to prove that $\tau(f) = u$, where f is the self-homotopy equivalence of $L_{5,1}$ inducing the homomorphism $t \to t^2$ on $\pi_1(L_{5,1})$. Consider f as a map from L to L', where L and L' are copies of $L_{5,1}$. Let M(f) denote the mapping cylinder of f. The universal covering of a space X is denoted by $\check{X}$. One has the following short exact sequence of cellular chain complexes:

$$0 \to C_*(\check{L}) \xrightarrow{\ i\ } C_*(\check{M}(f)) \xrightarrow{\ q\ } C_*(\check{M}(f); \check{L}) \to 0 \tag{1}$$

Each of these complexes is a complex of $\mathbb{Z}C_5$-modules, since C_5 is the covering transformation group of every space under consideration. The maps i and q will be $\mathbb{Z}C_5$-homomorphisms provided that $C_*(\tilde{M}(f))$ and $C_*(\tilde{M}(f);\tilde{L})$ are endowed with the $\mathbb{Z}C_5$-structure where t acts by the multiplication by $i_*(t)$ in $C_*(\tilde{M}(f))$ and by $q_* i_*(t)$ in $C_*(\tilde{M}(f);\tilde{L})$.

Let $\Sigma = 1 + t + t^2 + t^3 + t^4 \in \mathbb{Z}C_5$. If Y is a $\mathbb{Q}C_5$-module, define $NY = Y/\Sigma\cdot Y$, as in [Mi, §12]. From (1), one gets the short exact sequence of complexes:

$$0 \to N(C_*(\tilde{L}) \otimes \mathbb{Q}) \xrightarrow{i} N(C_*(\tilde{M}(f)) \otimes \mathbb{Q}) \xrightarrow{q} N(C_*(\tilde{M}(f), \tilde{L}) \otimes \mathbb{Q}) \to 0 \tag{2}$$

These three complexes are acyclic [Mi, p.405]. Therefore, their Milnor's special torsion is defined [Mi, §12]. We denote by $\Delta(L)$, $\Delta(M(f))$ and $\Delta(M(f);L)$ these torsions which are elements of $U(N)/\pm C_5$, where N is the algebra $\mathbb{Q}C_5/\mathbb{Q}\Sigma$ and $U(N)$ its group of units. As in [Mi, Theorem 3.1], one proves that

$$\Delta(L) \,.\, i_*^{-1}\, q_*^{-1}(\Delta(M(f);L)) = i_*^{-1}(\Delta(M(f))) \tag{3}$$

Let $j : M(f) \to L'$ be the natural retraction, which is a simple homotopy equivalence. One has $j_* \circ i_* = f_*$, $j_*(\Delta M(f)) = \Delta(L')$ and $j_* \circ q_*^{-1}(\Delta(M(f);L))$ is the image $\tau_N(f)$ of $\tau(f)$ in $U(N)/\pm C_5$. Hence, one gets from (3) the following formula:

$$\tau_N(f) = \Delta(L_{5,1}) \,.\, [f_*(\Delta(L_{5,1}))]^{-1} \tag{4}$$

which generalizes [Mi, Lemma 12.5].

For an element $x \in \mathbb{Q}C_5$ which is a unit modulo $\mathbb{Q}\cdot\Sigma$, denote by $[x]$ its class in $U(N)/\pm C_5$. By [Mi, p. 406], one has

$\Delta(L_{5,1}) = [t - 1]^2$. Thus, Formula (4) implies:

$\tau_N(f) = [t - 1]^2\, [t^2 - 1]^{-2} = [t + 1]^{-2}$

As $(t+1)^2 = (-t)(-1+t^2+t^3) + \Sigma$ in QC_5, one has

$$\tau_N(f)^{-1} = [\tau(f)]^{-1} = [-1+t^2+t^3] = [-1+t+t^{-1}]^{-1}. \qquad (5)$$

Observe that the map $x \to [x]$ induces an injective homomorphism $U(QC_5)/\pm C_5 \to U(N)/\pm C_5$. Indeed, if $[x] = 1$, then $x = \pm t^k + m\Sigma$. The sum of all coefficients of x is then $1 + 5m$ which implies that $m = 0$ if x is a unit. This together with Formula (5) shows that $\tau(f) = -1 + t + t^{-1} = u$.

(3.2) Lemma Let $W = L_{5,1} \times D^{2k-1} \times I$ $(k \geq 3)$ and let $W_0 = L_{5,1} \times D^{2k-1} \times 0 \subset \partial W$. There exists a CAT-homeomorphism $\beta : \partial W \to \partial W$ such that $\beta(W_0) \subset \mathrm{int} W_0$ and $f = \beta | W_0 : W_0 \to W_0$ is a homotopy equivalence with $\tau(f) = u$.

Proof: By Lemma (3.1), there exists a homotopy equivalence $f : L_{5,1} \to L_{5,1}$ with $\tau(f) = u$. By general position, one can realize f up to homotopy by an embedding $f : L_{5,1} \to W_0$. As W_0 is parallelizable, any regular neighborhood of $f(L_{5,1})$ is CAT-homeomorphic to W_0 [Wa 1, Proposition 5]. Hence one gets a CAT-embedding $F ; W_0 \to \mathrm{int} W_0$ with $\tau(F) = u$. We shall extend F to a CAT-homeomorphism $\beta : \partial W \to \partial W$.

Observe that W is an h-cobordism from $F(W_0)$ to $V = \partial W - \mathrm{int} F(W_0)$ with $\tau(W, F(W_0)) = u$. Choose a retraction by deformation from W onto $F(W_0)$ and call $r : V \to F(W_0)$ its restriction to V. As $u = \bar{u}$, the duality formula [Mi, p.394] implies that r is a simple homotopy equivalence (inducing a homeomorphism on the boundaries). Therefore, the pair (V,r) defines a class in $\mathcal{S}^s_{CAT}(W_0 \text{ rel } \partial W_0)$, the set of s-cobordism classes of simple homotopy CAT-structures on $(W_0 \text{ rel } \partial W_0)$ [Wa 2, Chapter 10, p. 102]. As in [LA 1, § 2], the class of (V,r) is the image of u under the composed homomorphism:

$$\{\sigma \in Wh(W) \mid \sigma = (-1)^{2k+2}\bar{\sigma}\} \twoheadrightarrow H^{2k+2}(Z_2, Wh(W)) \xrightarrow{\delta} L^s_{2k+1}(C_5) \xrightarrow{\alpha} \mathcal{S}^s_{CAT}(W_0 \text{ rel } W_0)$$

where $L^s_{2k+1}(C_5)$ is the Wall surgery obstruction group for simple homotopy equivalences [Wa 2, Chapter 6], α is the action of $L_{2k+1}(C_5)$ on (W_0, id) [Wa 2, p. 107-108] and δ is the boundary homomorphism of the Rothenberg exact sequence [Sh]. As $L_{2k+1}(C_5) = 0$ [Bk], the pair (V,r) is s-cobordant to (W_0, id), that is f is homotopic relative to the boundary to a CAT-homeomorphism. Thus F can be extended to a CAT-self-homeomorphism β of ∂W with the required properties.

Proof of Theorem 3. We take β as in Lemma 3.2, and consider it as a map from $\partial W_2 \to \partial W_1$, where W_i are copies of W. Let (R,M,N) be an h-cobordism, with $M = TD(W,\beta)$. As $\tau^0(R,M)$ is in the image of $i_{W_1,M}$, the h-cobordism R is homeomorphic to $W_2 \times I \cup R_0$ where (R_0, W_1, L) is an h-cobordism. Write $\sigma = \tau^R(R_0, W_1)$. (For $X \subset Y$ a homotopy equivalence between subspaces X and Y or R, we denote $\tau^R(Y,X) = i_{Y,R}(\tau(Y,X)) \in Wh(R)$.)

As in Lemma 3.2, we consider W_1 as an h-cobordism from $\beta((W_2)_0)$, where $(W_2)_0 = L_{5,1} \times D^{2k-1} \times 0 \subset \partial(L_{5,1} \times D^{2k-1} \times I) = \partial W_2$. The torsion $\tau^R(W_1, \beta((W_2)_0))$ is equal to the generator $v = i_{W_1,R}(u)$ of $Wh(R) \cong \mathbb{Z}$ (by Lemma 3.2). The duality formula together with the fact the $v = \bar{v}$ implies that

$$\tau^R(R_0, L) = -\overline{\tau^R(R_0, W_1)} = -\bar{v} = -v$$

which implies that $\tau^R(L, \beta((W_2)_0) \times 1) = v + 2\sigma$. (1)

Now suppose that

$$N = TD(T,\gamma) = T_1 \amalg T_2 / \{\beta(x_2) = x_1 \in \partial T_1, \text{ for all } x_2 \in \partial T_2\}$$

with T having a handles decomposition with handles of index $\leq k$. By general position, one has $T = T_0 \times I$ which gives $T_i = (T_i)_0 \times I$ for the corresponding copies. By general position and unicity of regular neighborhood in N, the embedding $W_2 \subset N$ can be deformed by an ambiant isotopy so that:

$$(W_2)_0 \subset \operatorname{int}(T_2)_0)$$

$$T_2 = (W_2)_0 \times I \subset U \times I \text{ where } U = (T_2)_0 - \operatorname{int}(W_2)_0.$$

As both inclusions $T_2 \subset N$ and $W_2 \subset N$ are $(k+2)$-connected, the inclusion $(W_2)_0 \subset (T_2)_0$ is a homotopy equivalence. As $\pi_1(\partial T_0) \cong \pi_1(T_0)$ and $\pi_1(\partial W_0) \cong \pi_1(W_0)$, the cobordism $(U, \partial(W_2)_0, \partial(T_2)_0)$ is an h-cobordism and $(T_2)_0$ is an η_0-enlargement of $(W_2)_0$ for some $\eta_0 \in Wh(\partial W_2)_0)$. Write η for $\iota_{\partial(W_2)_0,R}(\eta_0)$.

We are now in a situation where we can apply the results of [Ha, §4]. This gives the following two statements:

i) The inclusion $\gamma((T_2)_0) \subset T_1$ is a homotopy equivalence ([Ha, Lemma (4.1)]; this can also be seen with the argument of connectivity and dimension used above.)

ii) $\tau^R(T_1, \gamma((T_2)_0)) = \tau^R(W_1, \beta((W_2)_0) - 2\eta$ [Ha, main formula (4.3)].

Combining (i) amd (ii), one obtains $\tau^R(T_1, \gamma((T_2)_0) = v + 2\sigma - 2\eta$.

As T_1 and T_2 are copies of T, the torsion $v + 2\sigma - 2\eta$ belongs to $\iota_{T_1,R}(E(T))$.

The set E(T) can be computed using the fact that any self-homotopy equivalence f of T is homotopic to a composite map $\iota \circ g \circ r$, where:

$\iota : W \to T$ is the inclusion corresponding to $(W_2)_0 \subset (T_2)_0$; thus, one has

$$\tau^0(\iota) = \iota_{\partial W, W}(\eta_0).$$

$g : W \to W$ is a homotopy equivalence.

$r : W \to T$ is a homotopy inverse for ι.

Hence $\tau(f) = \tau(\iota) + \iota_*(\tau(g)) + \iota_* g_*(\tau(r))$.

Accordingly [Ha, proof of (6.1)] there are now two possibilities:

a) $g_* = id$, in which case $\tau(g) = 0$ and thus $\tau(f) = 0$, because $\tau(i) = -i_*(\tau(r))$.

b) $g_*(x) = -x$ and $\tau(g) \neq 0$. It follows from Lemma (3.1) that $\tau(g) = u$ and thus $\tau(f) = u + 2\tau(i)$.

This proves that $i_{T_1,R}(E(T)) = \{0, v + 2\eta\}$. The equation $v + 2\partial - 2\eta = 0$ is impossible since v is a generator of $Wh(R) \cong Z$. The other possibility is $v + 2\partial - 2\eta = v + 2\eta$ which implies $\sigma = 2\eta$. Theorem 3 is thus proved.

References

[Bk] Bak. A. Odd dimension surgery groups of odd torsion groups vanish. Topology 14 (1975), 367-374.

[Co] Cohen M. M. A course in simple homotopy theory. Springer Graduate Texts in Math. 10 (1973).

[Ha] Hausmann J-Cl. Open books and h-cobordisms. To appear.

[Ke] Kervaire M. Le théorème de Barden-Mazur-Stallings, Comm. Math. Helv. 40 (1965), 31-42.

[La 1] Lawson T. Inertial h-cobordisms with finite fundamental group. Proc. AMS 44 (1974), 492-496.

[La 2] Lawson T. Open book decompositions for odd dimensional manifolds. Topology 17 (1978), 189-192.

[Mi] Milnor J. Whitehead torsion. Bull. AMS 72 (1966), 358-428.

[Sh] Shaneson J. Wall's surgery obstruction groups for Z×G. Annals of Math. 90 (1969), 226-234.

[Wa 1] Wall C.T.C. Classification problems in topology IV (thickenings) Topology 5 (1966), 73-94.

[WA 2] Wall C.T.C. Surgery on compact manifolds. Academic Press 1970.

University of Warwick
Coventry CV4 7AL, England

University of Geneva,
Switzerland
(current address)

Homotopy invariants of foliations

by

S. Hurder and F. W. Kamber*)

1. In this note we propose to study the homotopy groups of $B\Gamma_G^q$, the classifying space of G-foliated microbundles [H1]. A foliation F on a manifold X is a G-foliation if it is defined by local submersions into a q-dimensional model manifold B, such that the local transition functions preserve a G-structure on B. With respect to an adapted connection ω, the Chern-Weil homomorphism defines a map $h(\omega):I(G)_\ell \longrightarrow A^\bullet(X)$, where $I(G)_\ell$ is the ring of invariant polynomials on G modulo the ideal of elements of degree $> 2\ell$. The index ℓ depends on G and whether the G-structure is integrable; it can always be taken $\leq q$ [B].

For any commutative DG-algebra A, we let $\varphi:M_A \longrightarrow A$ be a minimal model [S]. Let $\overline{M}_A$ denote the augmentation ideal; the quotient $\pi^*(A) \overset{\text{def}}{=} \overline{M}_A/\overline{M}_A{}^2$ is called the dual homotopy of A. For any (semi-simplicial) manifold X, we set $\pi^*(X) = \pi^*(A^\bullet(X))$ [D], [S], where $A^\bullet(X)$ is the deRham algebra in the sense of Sullivan-Dupont. If X is 1-connected and of finite rational type, then there is a natural isomorphism $\pi^*(X) \cong \mathrm{Hom}(\pi_*(X),\mathbb{R})$. For a DG-algebra A, whose cohomology is of finite type, we define

$$\pi_*(A) = \mathrm{Hom}_{\mathbb{R}}(\pi^*(A),\mathbb{R}).$$

Let X be a G-foliated manifold. The following result is proved in [Hu 1]

1.1. THEOREM. *The Chern-Weil homomorphism induces a map* $h^\#:\pi^*(I(G)_\ell) \longrightarrow \pi^*(X)$ *which depends only on the concordance class of the foliation.*

If X is of finite rational type, the map $h^\#$ induces by transposition a mapping

(1.2) $$h_\#:\Pi(X) \longrightarrow \Pi(I(G)_\ell),$$

*) Partially supported by NSF grant MCS79-00256

where $\Pi(X) = s^{-1}(\pi_*(X)\otimes\mathbb{R})$ and $\Pi(A) = s^{-1}\pi_*(A)$ are the desuspended graded homotopy groups. It is known [B-L] that the functor Π has values in the category of graded Lie algebras. As $h(\omega):I(G)_\ell \longrightarrow A^\bullet(X)$ is a homomorphism of DG-algebras, $h_\#$ is a homomorphism of graded Lie algebras. This construction extends to G-microbundles in an obvious way and therefore defines a homomorphism of Lie algebras

$$\tilde{h}_\#:\Pi(B\Gamma^q_G) \longrightarrow \Pi(I(G)_\ell).$$

If $f:X \longrightarrow B\Gamma^q_G$ denotes the classifying map of the G-foliation F on X, the diagram

$$(1.3)\qquad \begin{array}{ccc} \Pi(X) & \xrightarrow{h_\#} & \Pi(I(G)_\ell) \\ & {\scriptstyle f_\#}\searrow \quad \nearrow{\scriptstyle \tilde{h}_\#} & \\ & \Pi(B\Gamma^q_G) & \end{array}$$

is commutative. It is the purpose of this note to determine the Lie algebra structure of $\Pi(I(G)_\ell)$ (Section 2) and to detect elements in the image of $\tilde{h}_\#$ via appropriate choices of (X,f) (Section 4). In 3 we study the relationship of $h_\#$ with the characteristic homomorphism Δ_* for G-foliations [K-T 1], [K-T 2].

2. *The structure of* $\Pi(I(G)_\ell)$

In this section we determine the structure of the graded Lie algebra $\Pi(I(G)_\ell)$. Let G be a reductive Lie group. In order to simplify the following discussion, we will assume that G is connected in which case $I(G) \cong \mathbb{R}[c_1,\dots,c_r]$ is a polynomial algebra generated by the characteristic classes c_j of even degree. As before, we denote by

$$I_\ell = I(G)_\ell = \mathbb{R}[c_1,\dots,c_t]/(\phi(c_1,\dots,c_t)\,|\deg\phi > 2\ell)$$

the truncated polynomial algebra, where $c_1,\dots,c_t$ denote the generators of degree $\leq 2\ell$.

Let $A_\ell = \Lambda P_{(2\ell)} \otimes I_\ell$ be the DG-algebra introduced in section 3. The inclusion $0 \longrightarrow I_\ell \longrightarrow A_\ell$ dualizes to give an epimorphism of DG-coalgebras, $A_\ell^* \xrightarrow{j^*} I_\ell^* \longrightarrow 0$. Applying Quillen's L construction [Q], [B-L], we get an exact sequence of free DG-Lie algebras

$$(2.1) \qquad 0 \longrightarrow \ker j^* \longrightarrow L(A_\ell^*) \longrightarrow L(I_\ell^*) \longrightarrow 0,$$

where $L(C) \equiv \mathbb{L}(s^{-1}\bar{C})$ is the free DG-Lie algebra generated by a suspended reduced DG-coalgebra C [B-L], [N-M]. Passing to cohomology we get an exact sequence

$$(2.2) \qquad \begin{array}{ccccc} H_*(L(A_\ell^*)) & \longrightarrow & H_*(L(I_\ell^*)) & \xrightarrow{\delta} & H_{*-1}(\ker j^*) \\ \| \wr & & \| \wr & & \\ \Pi(A_\ell) & \xrightarrow{j_\#} & \Pi(I_\ell) & & \end{array}$$

2.3. THEOREM. *There is an extension of graded Lie algebras*

$$0 \longrightarrow \Pi(A_\ell) \xrightarrow{j_\#} \Pi(I_\ell) \longrightarrow P^*_{(2\ell)} \longrightarrow 0,$$

where $P^*_{(2\ell)}$ *is an abelian Lie algebra and* $\Pi(A_\ell) \cong L(H^\cdot(A_\ell)^*)$ *is a free Lie algebra.*

We remark that the Lie algebra structure of the extension $\Pi(I_\ell)$ is uniquely determined by the induced representation of $P^*_{(2\ell)}$ in the Lie algebra of outer derivations of $\Pi(A_\ell)$. This follows from the fact that the free Lie algebra $\Pi(A_\ell)$ has trivial center and from the general theory of extensions of Lie algebras (compare [Ho] for the ungraded case). The proof of this theorem, culminating in the determination of this induced action, will occupy the rest of this section.

First note that by taking a Λ-minimal model of the KS - extension $0 \longrightarrow I_\ell \longrightarrow A_\ell \longrightarrow \Lambda P_{(2\ell)} \longrightarrow 0$ [Ha], there is a long exact dual homotopy sequence with injective coboundary ∂^* and therefore a short exact sequence

$$(2.4) \qquad 0 \longrightarrow \partial^* P_{(2\ell)} \longrightarrow \pi^*(I_\ell) \longrightarrow \pi^*(A_\ell) \longrightarrow 0.$$

Dualizing this sequence gives the exact sequence of the theorem. The elements of $P^*_{(2\ell)}$ all have odd degree, so as a Lie algebra this must be abelian. We want to analyze how the elements in $P^*_{(2\ell)}$ act on the image of $\Pi(A_\ell)$, and this will show that $\Pi(A_\ell)$ is an ideal in $\Pi(I_\ell)$.

The algebra A_ℓ admits a subalgebra $Z_\ell \subset A_\ell$ with trivial differential and products, which induces an isomorphism in cohomology $Z_\ell \xrightarrow{\cong} H^*(A_\ell)$ [K-T3]. Therefore A_ℓ is biformal and we have isomorphisms

$$L(Z^*_\ell) \cong H_*(L(Z^*_\ell)) \xrightarrow{\cong} H_*(L(A^*_\ell)) \cong \Pi(A_\ell). \tag{2.5}$$

It follows that $\Pi(A_\ell)$ is a free graded Lie algebra generated by $s^{-1}Z^*_\ell$.

The algebra Z_ℓ and the isomorphism $Z_\ell \cong H(A_\ell)$ have been described in [K-T1], [K-T3]. We use here a slightly different notation, which is more convenient in the present context. For ordered sequences $I = (i_1 < \cdots < i_s)$, $J = (j_1 \leq \cdots \leq j_m)$, the symbol (I/J) is called *admissible* if it satisfies

$$\deg c_J \leq 2\ell, \quad c_J = \prod_{\alpha=1}^{m} c_{j_\alpha}, \tag{2.6}$$

$$\deg c_{i_1} c_J > 2\ell; \tag{2.7}$$

$$i_1 \leq j_1. \tag{2.8}$$

For an admissible symbol (I/J), the cochain

$$z_{(I/J)} = y_I \otimes c_J = y_{i_1} \wedge \cdots \wedge y_{i_s} \otimes c_{j_1} \cdots c_{j_m} \in A_\ell$$

is clearly a cocycle and the product of any two such cocycles $= 0$. The algebra Z_ℓ is then given by the linear space spanned by 1 and the cocycles $z_{(I/J)}$ for (I/J) admissible.

Let $I = \mathbb{R}[c_1,\ldots,c_t]$; the canonical quotient map $I \longrightarrow I_\ell$ dualizes to an inclusion $L(I^*_\ell) \subset L(I^*)$. Since I^* has a canonical Hopf algebra structure and trivial differential, we find easily that

$$H_*(L(I^*)) \cong P^*_{(2\ell)} = \operatorname{span}\{Y_1,\ldots,Y_t\},$$

where $Y_j = s^{-1}c_j^*$, $j = 1,\ldots,t$. Hence all the cycles in $L(I^*)$ of degree $\geq 2\ell$ are boundaries. Thus all the cycles in $L(I_\ell^*)$ of degree $\geq 2\ell$ are boundaries in $L(I^*)$. With this observation in mind, we produce explicitly a set of cycles in $L(I_\ell^*)$, which will generate $H(L(I_\ell^*))$.

For any monomial c_K, $K = (k_1 \leq \cdots \leq k_m)$ in I, we set $Y_K = s^{-1}c_K^* \in I^*$. The diagonal Δ in I^* is given by

$$\Delta(c_K^*) = c_K^* \otimes 1 + 1 \otimes c_K^* + \frac{1}{2} \sum_{(\alpha,\beta)} (c_\alpha^* \otimes c_\beta^* + c_\beta^* \otimes c_\alpha^*), \tag{2.9}$$

where (α,β) runs over all ordered proper partitions of the set $\{k_1,\ldots,k_m\}$. By definition, the differential d_L of $L(I^*)$ is determined by the formula

$$d_L Y_K = \frac{-1}{2} \sum_{(\alpha,\beta)} [Y_\alpha, Y_\beta] \in L(I^*). \tag{2.10}$$

For an admissible symbol (I/J) with $I = (i)$, it follows that $d_L Y_{(i,J)} \in L(I_\ell^*)$ is a cycle of degree $\geq 2\ell$, and we define

$$s^{-1}u_{(i/J)}^* = -d_L Y_{(i,J)} = \frac{1}{2} \sum_{(\alpha,\beta)} [Y_\alpha, Y_\beta] \tag{2.11}$$

If (I/J) is an arbitrary admissible symbol, we set

$$s^{-1}u_{(I/J)}^* = \mathrm{ad}(Y_{i_s}) \circ \cdots \circ \mathrm{ad}(Y_{i_2}) s^{-1}u_{(i_1/J)}^*, \quad s > 1, \tag{2.12}$$

where $\mathrm{ad}(Y) = [Y,-]$ denotes the adjoint representation. Clearly the Y_j and the $s^{-1}u_{(I/J)}^*$ are cycles in $L(I_\ell^*)$. Their corresponding homology classes in $H_*(L(I_\ell^*))$ are denoted by the same symbol. Observe that the elements $s^{-1}u_{(i/J)}^*$ correspond exactly to a minimal set of relations $c_i c_J \sim 0$ for the quotient algebra I_ℓ. For $z_{(I/J)} \in Z_\ell$, denote by $z_{(I/J)}^*$ the corresponding dual basis element of Z_ℓ^*.

2.13. LEMMA. *The injective homomorphism* $j_\#: \Pi(A_\ell) \longrightarrow \Pi(I_\ell)$ *of Theorem* 2.3 *is induced by the homomorphism of free* DG-*Lie algebras* $L(Z_\ell^*) \longrightarrow L(I_\ell^*)$, *which is determined on the generators by* $s^{-1}z_{(I/J)}^* \longrightarrow s^{-1}u_{(I/J)}^*$ *for* (I/J) *admissible.*

It follows that the homology classes $s^{-1}u^{*}_{(I/J)}$ generate a free subalgebra in $L(I^{*}_{\ell})$. The formulas in the following Proposition have to be understood with the convention: Whenever a symbol (I'/J') is not admissible, the term in which the symbol occurs must be replaced by 0.

2.14. PROPOSITION. Let (I/J) be admissible and $k < j$ in $\{1,\dots,t\}$. Then the following formulas hold in $H_{\bullet}(L(I^{*}_{\ell}))$:

(2.15) $\mathrm{ad}(Y_k)(Y_j) = s^{-1}u^{*}_{(k/j)}$;

(2.16) $\mathrm{ad}(Y_k)s^{-1}u^{*}_{(I/J)} = s^{-1}u^{*}_{(i_1\cdots i_s\, k|J)}$, for $k > i_s$;

(2.17) $$\mathrm{ad}(Y_k)s^{-1}u^{*}_{(I/J)} = \sum_{\beta=\alpha}^{s}(-1)^{s-\beta}\mathrm{ad}(Y_{i_s})\circ\cdots\circ\mathrm{ad}(Y_{i_{\beta+1}})[s^{-1}u^{*}_{(k/i_\beta)}, s^{-1}u^{*}_{(i_1\dots i_{\beta-1}|J)}]$$
$$+ (-1)^{s-\alpha+1}s^{-1}u^{*}_{(i_1\cdots i_{\alpha-1}k i_\alpha\cdots i_s|J)},\quad \text{for } i_{\alpha-1} < k < i_\alpha,\ 1 < \alpha \le s;$$

(2.18) $$\mathrm{ad}(Y_k)s^{-1}u^{*}_{(I/J)} = \sum_{\beta=\alpha}^{s}(-1)^{s-\beta}\mathrm{ad}(Y_{i_s})\circ\cdots\circ\mathrm{ad}(Y_{i_{\beta+1}})[s^{-1}u^{*}_{(k/i_\beta)}, s^{-1}u^{*}_{(i_1\dots i_{\beta-1}|J)}]$$
$$+ (-1)^{s+1}\mathrm{ad}(Y_{i_s})\circ\cdots\circ\mathrm{ad}(Y_{i_2})\circ\mathrm{ad}(Y_k)s^{-1}u^{*}_{(i_1|J)},\quad \text{for } k < i_1;$$

(2.19) $$\mathrm{ad}(Y_k)s^{-1}u^{*}_{(i/J)} = -\sum_{j_\beta = i}^{t} s^{-1}u^{*}_{(k,j_\beta/j_0\cdots\hat{j}_\beta\cdots j_m)},\quad \text{for } k < j_0 = i.$$

Together with corresponding formulas for $k = i_\alpha$, $\alpha = 1,\dots,s$, (2.15) to (2.19) completely determine the Lie algebra extension in Theorem 2.3 and hence the structure of $\Pi(I_\ell) = H(L(I^{*}_{\ell}))$. They also show that the subalgebra $L(Z^{*}_{\ell}) \subset H(L(I^{*}_{\ell}))$ is an ideal. We denote by $D(Y_k)$ the derivation on $L(Z^{*}_{\ell})$ induced by $\mathrm{ad}(Y_k)$, $k = 1,\dots,t$. (2.15) implies

(2.20) $$[D(Y_k),D(Y_j)] = D(Y_k)\circ D(Y_j) + D(Y_j)\circ D(Y_k) = \mathrm{ad}(s^{-1}z^{*}_{(k|j)}),\quad k < j.$$

D therefore induces a representation $\overline{D}$ of the abelian Lie algebra $P^{*}_{(2\ell)}$ in $\mathrm{Der}(L(Z^{*}_{\ell}))/\mathrm{IntDer}$. This is the representation canonically associated to the extension (2.3).

As an example, we describe $\Pi(I_2)$ for $I = I(\underline{gl}(2)) = \mathbb{R}[c_1,c_2]$. We have $I_2 = [c_1,c_2]/(c_1^3,c_1c_2,c_2^2)$. The Lie algebra $\Pi(I_2) \cong H(L(I^*))$ is generated by $Y_j = s^{-1}c_j^*$, $j = 1, 2$ and $s^{-1}u^*_{(1/11)} = [Y_1,Y_{(1,1)}]$, $dY_{(1,1)} = \frac{1}{2}[Y_1,Y_1]$. The free subalgebra $L(Z_2^*) \subset H(L(I_2^*))$ is generated by $s^{-1}u^*_{(1,11)}$ and the elements $s^{-1}u^*_{(1/2)} = [Y_1,Y_2]$, $s^{-1}u^*_{(2/2)} = \frac{1}{2}[Y_2,Y_2]$, $s^{-1}u^*_{(12/11)} = [Y_2,[Y_1,Y_{(1,1)}]]$ and $s^{-1}u^*_{(12/2)} = [Y_2,[Y_1,Y_2]]$. The non-zero brackets in Prop. 2.14 are given by

$$[Y_1,s^{-1}u^*_{(2/2)}] = - s^{-1}u^*_{(12/2)},$$

$$[Y_1,s^{-1}u^*_{(12/11)}] = = [s^{-1}u^*_{(1/2)},s^{-1}u^*_{(1/11)}],$$

$$[Y_1,s^{-1}u^*_{(12/2)}] = [s^{-1}u^*_{(1/2)},s^{-1}u^*_{(1/2)}] = 0 ,$$

$$[Y_2,s^{-1}u^*_{(12/11)}] = [s^{-1}u^*_{(2/2)},s^{-1}u^*_{(1/11)}]$$

and

$$[Y_2,s^{-1}u^*_{(12/2)}] = [s^{-1}u^*_{(2/2)}, s^{-1}u^*_{(1/2)}].$$

Theorem 2.3. has the following consequence

2.21. THEOREM. *The minimal algebra* $M(I_\ell)$ *appears as an extension of* DG-*algebras*

(2.22)
$$\begin{array}{ccccccccc} 0 & \longrightarrow & Id(c_1,\dots,c_t) & \longrightarrow & M(I_\ell) & \xrightarrow{M_1} & M(A_\ell) & \longrightarrow & 0 \\ & & \| & & \wr\| & & \wr\| & & \\ 0 & \longrightarrow & Id(c_1,\dots,c_t) & \longrightarrow & C^\bullet(\Pi(I_\ell)) & \longrightarrow & C^\bullet(L(Z_\ell^*)) & \longrightarrow & 0, \end{array}$$

where $C^\bullet(L) \equiv S^\bullet(sL)$ *denotes the cochain complex of a graded Lie algebra. As* A_ℓ *is biformal, the isomorphism* $M(A_\ell) \cong C^\bullet(L(Z_\ell^*))$ *preserves differentials. For* I_ℓ, *which is only formal, the cochain differential* d_C *describes only the quadratic terms of* d_M.

The 1-cochains $u_{(I/J)} \in M(I_\ell)$, dual to the basis elements $u^*_{(I/J)}$ described earlier, are mapped to the cocycles $z_{(I/J)} \in Z_\ell \subset A_\ell$. The obvious relation in $M(I_\ell)$

$$d_M u_{(i/J)} = c_i c_J \tag{2.23}$$

shows that d_M has non-quadratic terms (for c_J decomposable); hence I_ℓ is not coformal for $\ell > 1$. The minimal algebras of the form $C(L(Z_\ell^*))$ have the homotopy type of a finite wedge of spheres, and their study goes back to P. J. Hilton (compare e.g. [H1], [H2]). By contrast the minimal algebra $M(I_\ell)$ appears to be quite complicated as far as the differential is concerned. Details of these constructions will appear elsewhere [K-T4] .

3. *The relationship of* $h^\#$ *with* Δ_*

Let X have a G-foliation; we assume that the G-frame bundle $F(Q)$ of the normal bundle admits an H-reduction, where $H \subseteq G$ is closed and (G,H) is a reductive, CS-pair [K-T2]. Let $P \subseteq \Lambda \underline{g}^*$ be the space of primitives: $P \cong \operatorname{span}\{Y_1,\ldots,Y_r\}$ where Y_j is the cohomology suspension of c_j. Let $P = \hat{P} \oplus \tilde{P}$ be a Samelson decomposition. Denote the image of the transgression mapping by $V = \tau_g P = \hat{V} \oplus \tilde{V} \subseteq I(\underline{g})$, so that $\operatorname{Ideal}(\hat{V}) = \ker(i^* \cdot I(\underline{g}) \longrightarrow I(\underline{h}))$. We denote by $V^{(2\ell)}$, resp. $V_{(2\ell)}$ the subspace generated by the elements of degree $> 2\ell$, resp. $\leq 2\ell$. Similarly we decompose $P = P^{(2\ell)} \oplus P_{(2\ell)}$.

The complex A_ℓ used in section 2 is defined by $A_\ell = \Lambda P_{(2\ell)} \otimes I(\underline{g})_\ell$, with differential defined by the transgression. A similar relative complex is defined by $\hat{A}_\ell = \Lambda \hat{P}_{(2\ell)} \otimes I(\underline{g})_\ell$. The relative Weil algebra of the pair then has cohomology [K-T 3]

$$\begin{aligned} H^*(W(\underline{g},\underline{h})_\ell) &\cong H^*(\Lambda P \otimes I(\underline{g})_\ell \otimes I(\underline{h})) \\ &\cong \Lambda \hat{P}^{(2\ell)} \otimes H^*(\hat{A}_{(2\ell)}) \underset{I(\underline{g})}{\otimes} I(\underline{h}). \end{aligned}$$

The Chern-Weil theory gives a characteristic homomorphism for G-foliations [K-T,2]

$$\Delta : W(\underline{g},\underline{h})_\ell \xrightarrow{k} A^\cdot(F(Q)/H) \xrightarrow{s^*} A^\cdot(X),$$

giving a commutative diagram of minimal models

$$
(3.1)\qquad
\begin{array}{ccc}
M(W(\underline{\underline{g}},\underline{\underline{h}})_\ell) & \xrightarrow{Mk} & M(F(Q)/H) \\
\Big\uparrow {\scriptstyle MJ} & \searrow^{M\Delta} & \Big\downarrow {\scriptstyle Ms^*} \\
M(I(G)_\ell) & \xrightarrow{Mh} & M(X)
\end{array}
$$

We deduce two results from (3.1): First, there is a relationship between $h^\#$ and Δ_*, given by [Hul]

3.2. THEOREM. *The diagram*

$$
(3.3)\qquad
\begin{array}{ccccc}
\pi^*(I(G)_\ell) & & \xrightarrow{h^\#} & & \pi^*(X) \\
\Big\uparrow {\scriptstyle \zeta} & & & & \Big\uparrow {\scriptstyle \mathcal{H}^*} \\
H^*(\hat{A}_\ell) & & \xrightarrow{\Delta_*} & & H^*(X) \\
& \searrow & & \nearrow_{\Delta_*} & \\
& & H^*(W(\underline{\underline{g}},\underline{\underline{h}})_\ell) & &
\end{array}
$$

naturally commutes, where $\mathcal{H}^*$ *is the dual Hurewicz map and* ζ *is the inclusion mapping* ${}^{z}(I/J) \longrightarrow {}^{u}(I/J)$.

This result gives a new method for showing the non-triviality of Δ_*: a class which is non-zero in the image of $h^\# \circ \zeta$ is mapped to a non-zero class by Δ_*. Conversely, the non-triviality of Δ_* for a given X can be used to show $h^\#$ is non-trivial, if the map $\mathcal{H}^*$ is known. Section 4 will indicate what can be shown using these techniques.

Let $F\Gamma^q_G$ be the classifying space of trivialized, G-foliated microbundles; let $K \subseteq G$ be a maximal compact group. By the functoriality of $h_\#$ and $\Delta_\#$, dualizing (3.1) gives

3.4. THEOREM. *Let* $f: X \longrightarrow B\Gamma^q_G$ *classify a* G-*foliation on* X . *In a natural way, there are defined maps so that the diagram commutes:*

(3.5)
$$\begin{array}{ccc}
\Pi(F\Gamma^q_G) & \xrightarrow{\tilde{\Delta}_\#} & \Pi(W(\underline{g},\underline{e})_\ell) \\
\downarrow & & \downarrow \\
\Pi(B\Gamma^q_G) & \xrightarrow{\tilde{\Delta}_\#} & \Pi(W(\underline{g},\underline{k})_\ell) \\
{\scriptstyle f_\#}\uparrow & \searrow^{\tilde{h}_\#} & \downarrow{\scriptstyle j_\#} \\
\Pi(X) & \xrightarrow[h_\#]{} & \Pi(I(G)_\ell)
\end{array}$$

The cokernel of $j_\#$ is $\tilde{P}^*$, (3.5) forces $\tilde{h}_\#$ to have cokernel $\supseteq \tilde{P}^*$. The obvious question is whether equality holds: Does image $\tilde{h}_\#$ = image $j_\#$?

4. *The homotopy of* $B\Gamma^q_G$

For the three standard types of G-foliations, we indicate the extent to which $\tilde{h}_\#$ is known.

Let $G = G\ell(q,\mathbb{R})$. Mather and Thurston [T] have shown that $\nu: B\Gamma^q \longrightarrow BO(q)$ is $(q+2)$ connected. Therefore

(4.1) $\tilde{h}_\#$ maps onto Y_{2j} for $4j \leq q + 2$.

(4.2) $\pi_m(B\Gamma^q) \otimes Q \longrightarrow H_m(B\Gamma^q;Q)$ is an isomorphism (resp. onto) for $m \leq 2q + 2$ (resp. $m = 2q + 3$).

By (4.1) we see that $\tilde{h}_\#$ is onto $s^{-1}u^*_{(2m,2m)}$, for $q = 4m - 2 > 3$ or $q = 4m - 1$. Theorem 3.2 implies $\Delta_*(Y_{2m}c_{2m}) \neq 0$ in $H^{8m-1}(F\Gamma^q)$. This is a rigid class for q even. Other Whitehead products are similarly in the image of $\tilde{h}_\#$ [Hu1].

Many more results follow from the Theorems of Heitsch [He] or Fuks [F] on the variability of the classes in the image of Δ_*.

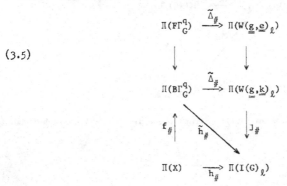

Using (4.2) we conclude there is a surjection of $\pi_{2q+1}(B\Gamma^q) \longrightarrow \mathbb{R}^d$, for some $d > 0$. For example, by Fuks we have $\Pi_{2q}(B\Gamma^q) \xrightarrow{\tilde{h}_\#} \Pi_{2q}(I_q)$ is onto. The homotopy of $B\Gamma^q$ therefore maps onto a rather large Lie subalgebra of $\Pi(I_q)$. The 2q connectivity of ν would imply $\tilde{h}_\#$ is almost onto, within the restrictions of (3.5).

When $G = G\ell(n,\mathbb{C})$, the classes $Y_1,\ldots,Y_s$ are in the image of $\tilde{h}_\#:\Pi(B\Gamma_n^{\mathbb{C}}) \longrightarrow \Pi(I_n)$, where $s = [\sqrt{n}]$. Coupled with the Theorem of Baum and Bott [B-B], this shows $\Pi(B\Gamma_n^{\mathbb{C}}) \longrightarrow \Pi(I_n)$ is onto a much larger subalgebra than originally considered in [H3] Further details are in [Hu1].

When $G = SO(q)$ the map $\tilde{h}_\#:\Pi(BR\Gamma^q) \longrightarrow \Pi(I_{q'})$, $q' = [q/2]$, is onto, and complete variation occurs [Hu2]. In this case the Lie algebra $\Pi(I_{q'})$ is injected into $\Pi(BR\Gamma^q)$. The variability of the classes implies there are uncountably many distinct ways of choosing a section $\Pi(I_q) \longrightarrow \Pi(BR\Gamma^q)$.

Note: The example $\pi(I_2)$ was first calculated by R. Hain, using K.T. Chen's method of power series connections (Bull. AMS, 83 (1977), 831-879).

References

[B-L]. H. J. Baues and J. M. Lemaire, *Minimal models in homotopy theory*, Math. Ann. 225, 219-242 (1977).

[B-B]: P. Baum and R. Bott, *Singularities of holomorphic foliations*, J. Diff. Geom. 7, 279-342 (1972).

[B]: R. Bott, *On a topological obstruction to integrability*, Proc, Symp. Pure Math., AMS vol. 16, 127-131 (1970).

[D]: J. L. Dupont, *Simplicial de Rham cohomology and characteristic classes of flat bundles*, Topology 15, 233-245 (1976).

[F]: D. B. Fuks, *Non-trivialité des classes caractéristiques de* g*-structures*, C. R. Acad. Sci. Paris 284, 1017-1019 and 1105-1107 (1977).

[H1]: A. Haefliger, *Feuilletages sur les variétés ouvertes*, Topology 9, 183-194 (1970).

[H2]: ___________, *Sur la cohomologie de l'algèbra de Lie des champs de vecteurs*, Ann. Sci. École Norm. Sup. 9, 503-532 (1976).

[H3]: ___________, *Whitehead products and differential forms*, Proceedings-Rio de Janeiro, LNM vol. 652, 13-24 (1978).

[Ha]: S. Halperin, *Lectures on Minimal Models*, Université de Lille I (1978).

[He]: J. Heitsch, *Independent variation of secondary classes*, Annals of Math. 108, 421-460 (1978).

[Ho]: G. Hochschild, *Lie algebra kernels and cohomology*, Amer. J. Math 76, 698-716 (1954).

[Hu1]: S. Hurder, *Dual homotopy invariants of* G*-foliations*, University of Illinois preprint (1979).

[Hu2]: ___________, *Some remarks on the homotopy and cohomology of* $B R\Gamma^q$, University of Illinois preprint (1979).

[K-T1]: F. Kamber and Ph. Tondeur, *Foliated bundles and characteristic classes*, LNM vol. 493 (1975).

[K-T2]: ___________, G*-foliations and their characteristic classes*, Bulletin AMS 84, 1086-1124 (1978).

[K-T3]: ___________, *Characteristic classes and Koszul complexes*, Proc. Symp. Pure Math. vol 32, AMS, 159-166 (1978).

[K-T4]: ___________, *The rational homotopy type of truncated relative Weil algebras*, to appear.

[McL]: S. Mac Lane, *Homology*, Springer-Verlag (1963).

[N-M]: J. Neisendorfer and T. Miller, *Formal and coformal spaces*, Illinois Journal of Math. 22, 565-579 (1978).

[Q]: D. Quillen, *Rational homotopy theory*, Ann. Math. 90, 205-295 (1969).

[S]: D. Sullivan, *Infinitesimal computations in topology*, IHES 47, 269-331 (1977).

[T]: W. Thurston, *Foliations and groups of diffeomorphisms*, Bulletin AMS 80, 304-307 (1974).

Department of Mathematics
University of Illinois
Urbana, IL 61801

USA

On Ends of Groups and Johannson's Deformation Theorem

William Jaco [1]

I give a new proof of a theorem of K. Johannson [Jo_2] describing homotopy equivalences between compact, irreducible, sufficiently-large 3-manifolds (Haken-manifolds) in the case of non-empty boundary. The motivation for the proof given here is after the ideas of A. Swarup (Lemma 1.2 and Lemma 1.4) Since one of the purposes of this presentation is to make the proof (and the understanding) of this theorem (Theorem 3.1) more accessible, I have included a lot of detail. A proof with all the details appears in Chapter X of [J].

In §1 I present the ideas due to Swarup and prove Johannson's Theorem in the special case that there are no essential, embedded annuli. The analysis of homotopy equivalences between Haken-manifolds with incompressible boundary, in the general case, is described by using the characteristic Seifert pair of a Haken-manifold. I give the notation and basic concepts involved with the characteristic Seifert pair in §2. Complete details of the existence, uniqueness and properties of the characteristic Seifert pair may be found in [J-S], [Jo_2] or [J].

---

1. Research partially supported by NSF Grant MCS 78 06116 and a grant from the Institute for Advanced Study.

§1. A SPECIAL CASE: NO ESSENTIAL ANNULI

There are many examples of "exotic" homotopy equivalences between Haken-manifolds (see [W_2] or [J]). However, the phenomenon of these examples is quite limited The results of this section are after the work of Waldhausen [W_1] and Johannson [Jo_1, Jo_2]; and they show that no "exotic" homotopy equivalences exist in the absence of essential annuli. The proof I give of Theorem 1.6 follows from the ideas of A. Swarup (Lemma 1.2 and Lemma 1.4) reported to me by P. Scott.

1.1 THEOREM: Let M and M' be Haken manifolds and suppose that $f:(M, \partial M) \to (M', \partial M')$ is a map. If f is a homotopy equivalence, then f can be deformed to a homeomorphism. Furthermore, if $f|\partial M: \partial M \to \partial M'$ is a homeomorphism, then the deformation may be taken to be constant on ∂M.

Let $\tilde{X}$ be a polyhedron If C is a compact subpolyhedron of $\tilde{X}$, then $\tilde{X} - C$ has at most a finite number of components; some of which have noncompact closure, if $\tilde{X}$ is not itself compact. Let $e(\tilde{X},C)$ denote the number of components of $\tilde{X}-C$ that have noncompact closure. The number of ends of $\tilde{X}$, written $e(\tilde{X})$, is defined to be $e(\tilde{X}) = \sup\{e(\tilde{X},C): C$ is a compact subpolyhedron of $\tilde{X}\}$, if this number exists; otherwise, $e(\tilde{X})$ is infinite.

Now, I want to use the above idea to define an invariant of a pair (G,H), where G is a finitely generated group and H is a subgroup of G. Recall that if G is a finitely generated group, then there exists a polyhedron X with finite one-skeleton such that $\pi_1(X) \approx G$. So, suppose that G is a finitely generated group and H is a subgroup of G. Let X be a polyhedron with finite one-skeleton so that $\pi_1(X) \approx G$; and let $\tilde{X}$ be the covering space of X corresponding to the conjugacy class of H in G. Define $e(G,H) = e(\tilde{X})$ to be the <u>number of ends of the pair</u> (G,H). Then $e(G,H)$ depends only on G and the cosets of H in G.

If M is a 3-manifold, a subgroup H of $\pi_1(M)$ is <u>peripheral</u> if there exists a component B of ∂M such that H is conjugate into $\mathrm{Im}\,(\pi_1(B) \to \pi_1(M))$.

1.2. LEMMA: <u>Let</u> M <u>be a Haken-manifold with incompressible boundary</u>. <u>Set</u> $G = \pi_1(M)$ <u>and let</u> H <u>be a subgroup of</u> G <u>isomorphic to the fundamental group of a closed, orientable surface</u>. <u>If</u> $e(G,H) = 1$, <u>then</u> H <u>is peripheral</u>.

Proof: Let $\tilde{M}(H)$ denote the covering space of M corresponding to the conjugacy class of H in $G = \pi_1(M)$. Let $p\colon \tilde{M}(H) \to M$ denote the covering projection.

Let F be a closed, orientable surface with $\pi_1(F) \approx H$. There exists a map $f\colon F \to M$ such that the induced homomorphism $f_*\colon \pi_1(F) \to \pi_1(M)$ is an isomorphism onto H. Hence, f lifts to $\tilde{f}\colon F \to \tilde{M}(H)$. Since $H \approx \pi_1(F)$ is neither infinite cyclic nor a nontrivial free product, it follows that there exists a compact submanifold N in $\tilde{M}(H)$ such that the inclusion induced homomorphism of $\pi_1(N)$ to $\pi_1(\tilde{M}(H))$ is an isomorphism. Now, by hypothesis $e(\tilde{M}(H)) = 1$; so $\tilde{M}(H) - N$ has precisely one unbounded component. I may assume that $\tilde{M}(H) - N$ is connected (and unbounded). However, $\pi_1(N)$ being isomorphic to the fundamental group of a closed surface implies that N is an I-bundle; and the orientability implies that N is a product I-bundle. Hence, one component $\tilde{B}$ of ∂N is contained in $\partial\tilde{M}(H)$ and the inclusion $\pi_1(\tilde{B}) \to \pi_1(\tilde{M}(H))$ is a homotopy equivalence. Set $B = p(\tilde{B})$ Then B is a component of ∂M and H is conjugate in $\pi_1(M)$ into $\mathrm{Im}(\pi_1(B) \to \pi_1(H))$; i e., H is peripheral.

1.3. REMARK: Notice that if M is a compact 3-manifold, $G = \pi_1(M)$ and H is a finitely-generated subgroup of G, which has infinite index in G, then $e(G,H)$ finite implies that ∂M is incompressible. In particular, in the preceding lemma I did not need to assume that ∂M is incompressible; it is necessarily incompressible by the assumptions on H.

1.4. PROPOSITION: Let M be a Haken-manifold with incompressible boundary. Set $G = \pi_1(M)$ and let H be a subgroup of G isomorphic to the fundamental group of a closed, orientable surface. If M contains no essential, embedded annuli and H is peripheral, then $e(G,H) = 1$.

Proof: Since H is peripheral, there is a component B of ∂M such that H is conjugate into $\mathrm{Im}(\pi_1(B) \to \pi_1(M))$ Furthermore, since H is a closed surface group, the conjugacy class of H in $\mathrm{Im}(\pi_1(B) \to \pi_1(M))$ has finite index Let $\widetilde{M}(B)$ denote the covering space of M corresponding to the conjugacy class of $\mathrm{Im}(\pi_1(B) \to \pi_1(M))$; and let $\widetilde{M}(H)$ denote the covering space of M corresponding to the conjugacy class of H in $\pi_1(M)$. The manifold $\widetilde{M}(B)$ admits a manifold compactification. Since $\widetilde{M}(H)$ is a finite sheeted covering of $\widetilde{M}(B)$, the manifold $\widetilde{M}(H)$ admits a manifold compactification. What this implies is that if $p\colon \widetilde{M}(H) \to M$ is the covering projection, then there is a component $\widetilde{B}$ of $p^{-1}(B)$ such that $p|\widetilde{B}\colon \widetilde{B} \to B$ is a finite sheeted covering and $\widetilde{M}(H)$ is homeomorphic to $\widetilde{B} \times I$ minus a closed subset Z of $\widetilde{B} \times \{1\}$ via a homeomorphism taking $\widetilde{B}$ to $\widetilde{B} \times \{0\}$. Since M has no essential annuli, it follows that each component of $p^{-1}(\partial M)$ except $\widetilde{B}$ is simply connected; hence, Z is connected. So $\widetilde{M}(H)$ has only one end. This completes the proof.

1.5. REMARK: Let M be a Haken-manifold and suppose that ∂M is incompressible. Then there is an algebraic characterization of when the manifold M has essential, embedded annuli; namely, the manifold M has an essential, embedded annulus iff $\pi_1(M)$ splits as a nontrivial free product with amalgamation along the infinite cyclic group or as an HNN over the infinte cyclic group (see Lemma 2.5).

1.6. THEOREM (Johannson [Jo_2]): Let M and M' be Haken-manifolds with incompressible boundary. Suppose that $f: M \to M'$ is a homotopy equivalence. If M does not contain an essential, embedded annulus, then there is a homotopy $f_t: M \to M'$ such that $f_0 = f$ and $f_1: M \to M'$ is a homeomorphism.

Proof: By Theorem 1.1 it is sufficient to prove that f can be deformed to a boundary preserving map $f': (M, \partial M) \to (M', \partial M')$.

Let B be a component of ∂M By Lemma 1.4, $e(\pi_1(M), \pi_1(B)) = 1$. Since f_* is an isomorphism, $e(\pi_1(M'), f_*(\pi_1(B))) = 1$. Hence, by Lemma 1.2 the subgroup $f_*(\pi_1(B))$ is peripheral in $\pi_1(M')$. The desired deformation of f can now be established, since for each component B of ∂M the map $f|B$ is homotopic to a map taking B into $\partial M'$.

§2 CHARACTERISTIC SEIFERT PAIR.

A pair (X,Y) is a <u>polyhedral pair</u> if X is a polyhedron and Y is a subpolyhedron of X. A polyhedral pair (X,Y) is <u>connected</u> if X is connected. A <u>component</u> of a polyhedral pair (X,Y) is a polyhedral pair (X',Y') where X' is a component of X and $Y' = Y \cap X'$. The polyhedral pair (X',Y') is <u>contained in</u> the polyhedral pair (X,Y), written $(X',Y') \subset (X,Y)$, if X' is a subpolyhedron of X and Y' is a subpolyhedron of Y.

An n-<u>manifold pair</u> is a polyhedral pair (M,T) where M is an n-manifold and T is an (n-1)-manifold contained in ∂M. A 3-manifold pair (S,F) is an <u>I-pair</u> if there exists a homeomorphism h of S onto the total space of an I-bundle over a compact 2-manifold, not necessarily orientable, such that $h(F)$ is the total space of the corresponding ∂I-bundle. A 3-manifold pair (S,F) is an <u>S^1-pair</u> if there exists a homeomorphism h of S onto the total space of a Seifert fibered 3-manifold such that $h(F)$ is a saturated subset in some Seifert fibration. A 3-manifold pair (S,F) is a <u>Seifert pair</u> if each component is either an I-pair or an S^1-pair.

Let M be a 3-manifold. The 3-manifold pair $(\Sigma,\Phi) \subset (M, \partial M)$ is <u>well-embedded</u> (<u>in</u> M) if $\Sigma \cap \partial M = \Phi$ and $\mathrm{Fr}\Sigma$ is incompressible. The 3-manifold pair $(\Sigma,\Phi) \subset (M,\partial M)$ is <u>perfectly-embedded</u> (in M) if (Σ,Φ) is well-embedded, each component of $\mathrm{Fr}\Sigma$ is essential (i.e., if C is a component of $\mathrm{Fr}\Sigma$, then the inclusion map of the pair $(C,\partial C)$ into $(M,\partial M)$ is not homotopic (as a map of pairs) to a map taking C into ∂M) and no component (σ,φ) of (Σ,Φ)

can be homotoped (as a pair) into $(\Sigma-\sigma, \Phi-\varphi)$.

Two well-embedded 3-manifold pairs $(\Sigma', \Phi') \subset (M, \partial M)$ and $(\Sigma, \Phi) \subset (M, \partial M)$ are equivalent if there is a homeomorphism $J: M \to M$, isotopic to the identity on M such that $J(\Sigma') = \Sigma$ and $J(\Phi') = \Phi$. The well-embedded pair $(\Sigma', \Phi') \subset (M, \partial M)$ is "less than or equal to" the well-embedded pair $(\Sigma, \Phi) \subset (M, \partial M)$, written $(\Sigma', \Phi') \leq (\Sigma, \Phi)$, if there is a homeomorphism $J: M \to M$ isotopic to the identity on M such that $J(\Sigma') \subset \mathrm{Int}_M(\Sigma)$ and $J(\Phi') \subset \mathrm{Int}_{\partial M}(\Phi)$.

2.1 THEOREM ([J-S], [Jo_1], [J]): Let M be a Haken-manifold that is closed or has incompressible boundary. Then there exists a unique (up to ambient isotopy of M), maximal, perfectly-embedded Seifert pair in $(M, \partial M)$.

This unique, maximal, perfectly-embedded Seifert pair is called the characteristic Seifert pair for M.

A map $f: (X,Y) \to (M, \partial M)$ is essential if f is not homotopic (as a map of pairs) to a map taking X into ∂M. If in addition the map $f_*: \pi_1(X) \to \pi_1(M)$ is an injection, then f is nondegenerate.

2.2. THEOREM: Let M be a Haken-manifold that is closed or has incompressible boundary Let $(\Sigma, \Phi) \subset (M, \partial M)$ be a Seifert pair. The following are equivalent:

(i) (Σ, Φ) is the characteristic Seifert pair of M.

(ii) (Σ, Φ) is perfectly-embedded and any nondegenerate map from a Seifert pair (S,F), which is distinct from $(D^2 \times I, D^2 \times \partial I)$, $(D^2 \times S^1, \emptyset)$, $(S^2 \times S^1, \emptyset)$ or $(S^3, \emptyset)$, into $(M, \partial M)$ is homotopic (as a map of pairs) to a map from (S,F) into $(M, \partial M)$ with the image of S in Σ and the image of F in Φ.

2.3 COROLLARY. Let M be a Haken-manifold that is closed or has incompressible boundary. Let $(\Sigma, \Phi) \subset (M, \partial M)$ be the characteristic Seifert pair for M. Then any nondegenerate map from either $(S^1 \times I, S^1 \times \partial I)$ or $(S^1 \times S^1, \emptyset)$ into $(M, \partial M)$ is homotopic (as a map of pairs) to a map (from either $(S^1 \times I, S^1 \times \partial I)$ or $(S^1 \times S^1, \emptyset)$ into $(M, \partial M)$, resepctively) with its image contained in (Σ, Φ).

The condition of Corollary 2.3 does not characterize the characteristic Seifert pair of M (see Examples IX.21 (g) and (h) of [J]).

The 3-manifold pair (X,Y) is simple if every incompressible annulus or torus W in X with $\partial W \subset \text{Int } Y$ is either parallel into Y or parallel into $\overline{(\partial X - Y)}$.

2.4. LEMMA: Let M be a Haken-manifold that is closed or has incompressible boundary. Let (Σ, Φ) be the characteristic Seifert pair for M. Let X be a component of $\overline{(M - \Sigma)}$ and set $Y = X \cap \partial M$. The 3-manifold pair (X,Y) is simple.

If M is a Haken-manifold that is closed or has incompressible boundary and (Σ, Φ) is the characteristic Seifert pair for M, then a component (σ,φ) of (Σ,Φ) is a Seifert factor of M; and if X is a component of $(M-\Sigma)$ and $Y = X \cap \partial M$, then (X,Y) is a simple factor of M. Both the Seifert factors of M and the simple factors of M are uniquely determined up to ambient isotopy of M. I do point out a common misunderstanding about the simple factors. It is often believed that a simple factor has no essential, embedded annuli Quite often to the contrary, a simple factor may have essential, embedded annuli.

If M is a Haken-manifold with incompressible boundary, the peripheral characteristic Seifert pair of M is the collection of Seifert factors of M that meet ∂M. The peripheral characteristic Seifert pair of M is unique up to ambient isotopy of M

2.5 LEMMA If M is a Haken-manifold with incompressible boundary, then the following are equivalent·

(i) M does not contain an essential, embedded annulus

(ii) the peripheral characteristic Seifert pair of M is empty

(iii) $\pi_1(M)$ does not split as a nontrivial free product $A \underset{C}{*} B$ or an HNN group $A \underset{C}{*}$ where C is a cyclic group

§3. THE GENERAL CASE

3.1 THEOREM (Johannson [Jo]). Let M and M′ be Haken-manifolds with incompressible boundary. Let (Λ,Ψ) and (Λ',Ψ') denote the peripheral characteristic Seifert pairs for M and M′, respectively. Suppose that $f: M \to M'$ is a homotopy equivalence. Then there is a homotopy $f_t: M \to M'$ such that

(i) $f_0 = f$

(ii) $f_1 | \overline{(M-\Lambda)}: \overline{(M-\Lambda)} \to \overline{(M'-\Lambda')}$ is a homeomorphism, and

(iii) $f_1 | \Lambda: \Lambda \to \Lambda'$ is a homotopy equivalence.

PROOF: I shall first make some observations and notational conventions The proof is carried out by establishing a number of assertions. These assertions can be used as "The outline" of the proof.

3 2. OBSERVATIONS AND NOTATION: Let (Σ, Φ) (respectively, (Σ', Φ')) be the characteristic Seifert pair of M (respectively, M').

(1) Suppose that either $M = \Sigma$ or $M' = \Sigma'$. Say $M = \Sigma$. If $M = \Sigma$ is a Seifert fibered manifold, then it can be easily argued that M' is a Seifert fibered manifold and so $M' = \Sigma'$. If $M = \Sigma$ is an I-bundle, then M' is an I-bundle and so, again, $M' = \Sigma'$

(2) Suppose that either $\Lambda = \emptyset$ or $\Lambda' = \emptyset$. Say $\Lambda = \emptyset$. Then by Remark 2.5, it follows that $\Lambda' = \emptyset$. In this case Theorem 3.1 follows from Theorem 1 6.

(3) The Seifert factors of M (respectively, M') partition into three classes.

(i) The Seifert factors that have fundamental group isomorphic to $\mathbb{Z}$ are tubes; I denote this collection by $\mathfrak{T}$ (respectively, $\mathfrak{T}'$).

(ii) The Seifert factos that are S^1-pairs and not tubes; I denote this collection by $\mathfrak{S}$ (respectively, $\mathfrak{S}'$).

(iii) Seifert factors that are I-pairs and not tubes; I denote this collection by $\mathcal{B}$ (respectively, $\mathcal{B}'$).

(Since $M \neq \Sigma$ ($M' \neq \Sigma'$) and $\Lambda \neq \emptyset$ ($\Lambda' \neq \emptyset$), it follows that $\mathcal{B}(\mathcal{B}')$ is the collection of I-pairs of (Σ, Φ) $((\Sigma', \Phi'))$ that are not S^1-pairs.)

The next observation is very important and is employed throughout the proof of Theorem 3.1.

(4) _If_ (X,Y) _is a simple factors of_ $M(M')$, _where_ $Y = X \cap \partial M$ $(Y = X \cap \partial M')$, _then either there is no component of_ Y _that is an annulus or_ Y _is the union of precisely two annuli and there is a pair isomorphism from_ (X,Y) _to the product I-pair_ $(S^1 \times I \times I, S^1 \times I \times \partial I)$.

Outline of proof of (4): Suppose that y is a component of Y and y is an annulus. Let $\partial_0 y$ and $\partial_1 y$ denote the components of ∂y. Let y_0 and y_1 be the components of FrX (y_0 and y_1 are annuli) such that $\partial_i y \subset y_i$, $i = 0$ or 1. There are two possibilities to consider. Either $y_0 = y_1$ or $y_0 \neq y_1$. In the case that $y_0 = y_1$, a contradiction to Theorem 2.2 occurs. In the case that $y_0 \neq y_1$, again there is a contradiction to Theorem 2.2 or the pair (X,Y) is pair isomorphic to $(S^1 \times I \times I, S^1 \times I \times \partial I)$. So, to establish (4), I have used that (Σ, Φ) (respectively, (Σ', Φ')) is the characteristic Seifert pair of M (respectively, M').

(5) The simple factors of M (respectively, M') partition into three classes.

(i) The simple factors of M (respectively, M′) that have fundamental group isomorphic to $\mathbb{Z}$ are <u>simple</u> <u>tubes</u>, i.e., if $(Q, Q \cap \partial M)$ is a simple factor of M and $\pi_1(Q) \approx \mathbb{Z}$, then the pair $(Q, Q \cap \partial M)$ is pair isomorphic to the product I-pair $(S^1 \times I \times I, S^1 \times I \times \partial I)$ Furthermore, each simple tube meets precisely two components of (Σ, Φ); one is an I-pair and <u>not</u> an S^1-pair, while the other is an S^1-pair and <u>not</u> an I-pair

To see this, observe that Q is homeomorphic to a solid torus; hence, $Q \cap \partial M \neq \emptyset$ and each component of $Q \cap \partial M$ must be an annulus. The conclusion follows from Observation (4) above. I denote this collection by $\mathcal{Q}$ (respectively, $\mathcal{Q}'$)

(ii) The simple factors of M (respectively, M′) that have fundamental group isomorphic to $\mathbb{Z} \times \mathbb{Z}$ are <u>simple</u> <u>shells</u>; i.e., if $(P, P \cap \partial M)$ is a simple factor of M and $\pi_1(P) \approx \mathbb{Z} \times \mathbb{Z}$, then the pair $(P, P \cap \partial M)$ is pair isomorphic to the product I-pair $(S^1 \times S^1 \times I, \emptyset)$ Furthermore, each simple shell meets only components of (Σ, Φ) that are S^1-pairs and not I-pairs. (I am assuming, that $\partial M \neq \emptyset$, so, M is not a torus bundle over S^1)

To see this, observe that P is homeomorphic to $S^1 \times S^1 \times I$. Hence, $P \cap \partial M$ is either an annulus or a torus. The former cannot happen by Observation (4) above. The latter cannot happen since FrP is essential I denote this collection by $\mathcal{P}$ (respectively, $\mathcal{P}'$).

(iii) The simple factors of M (respectively, M′) that have <u>non-abelian</u> <u>fundamental</u> group. I denote this collection by $\mathcal{N}$ (respectively, $\mathcal{N}'$).

The first thing to prove is that *there is a homotopy* $f_t\colon M \to M'$ *such that*

(i) $f_0 = f$,

(ii) $f_1|(M-\Sigma)\colon (M-\Sigma) \to (M'-\Sigma')$ *is a homeomorphism, and*

(iii) $f_1|\Sigma\colon \Sigma \to \Sigma'$ *is a homotopy equivalence.*

This is done by establishing a number of assertions. As each assertion is established, each of the preceding assertions remains true; if an assertion is made in terms of f, then it is implicitly understood that the same is true about f′ (with the appropriate use of notation, where f′ is the homotopy inverse of f (and vice-versa); and to avoid too much notation, after each deformation, I continue to call the deformed map f (respectively, f′)

ASSERTION 1. *There is a deformation such that each component of* $f^{-1}(\mathrm{Fr}\Sigma')$ *is an essential, incompressible, embedded annulus or torus and the number of components of* $f^{-1}(\mathrm{Fr}\Sigma')$ *that are annuli is as small as possible.*

ASSERTION 2. *It follows that if* y *is the closure of a component of* $\partial M - f^{-1}(\mathrm{Fr}\Sigma')$ *and* y *is an annulus, then the map* $f|y\colon (y, \partial y) \to (M', \mathrm{Fr}\Sigma')$ *is essential, and therefore, nondegenerate.*

If a component y of $\partial M - f^{-1}(\mathrm{Fr}\Sigma')$ is an annulus and $f|y\colon (y, \partial y) \to (M', \mathrm{Fr}\Sigma')$ is not essential. Then there is a deformation of f decreasing the number of annuli in the preimage of $\mathrm{Fr}\Sigma'$. This is a contradiction to the choice of f in Assertion 1.

ASSERTION 3. *There is a deformation such that if* X *is a component of*

$f^{-1}(\Sigma')$ _and_ $Y = X \cap \partial M$, _then the pair_ (X,Y) _is a perfectly embedded Seifert pair_. (In fact, if (σ',φ') is a component of (Σ',Φ') and X is a component of $f^{-1}(\sigma')$, then for (σ',φ') an I-pair and not and S^1-pair, the pair (X,Y) is a perfectly-embedded I-pair and not an S^1-pair and for (σ',φ') an S^1-pair, the pair (X,Y) is an S^1-pair.)

If (σ',φ') is a component of (Σ',Φ') and X is a component of $f^{-1}(\sigma')$, set $Y = X \cap \partial M$ There are two cases to consider.

Suppose (σ',φ') is an I-pair and not an S^1-pair. Let y be any component of Y. Using the I-bundle structure of σ', it follows from Assertion 2, that $(f|y)_*(\pi_1(y))$ has finite index in $\pi_1(\sigma')$, and therefore, $\pi_1(y)$ has finite index in $\pi_1(X)$. So, either X is an I-bundle with y a component of the corresponding ∂I-bundle or $\pi_1(X) \approx \mathbb{Z}$. However, the latter situation cannot happen, since $(f|y)_*(\pi_1(y))$ is of finite index in $\pi_1(\sigma')$, which is not a cyclic group (of course, here is where I use that the pair (σ',φ') is not an S^1-pair). It follows that (X,Y) is an I-pair and not an S^1-pair.

Suppose (σ',φ') is an S^1-pair. If $\pi_1(\sigma')$ is abelian, then $\pi_1(X)$ is abelian and X admits a Seifert fibration. In any case, this is the first step in proving that (X,Y) is an S^1-pair. So, consider the case that $\pi_1(\sigma')$ is not abelian.

Now, the map $f'|\sigma'\colon \sigma' \to M$ has the property that $(f'|\sigma')_*$ is an injection. From the previous paragraph, it follows that $f'|\sigma'\colon (\sigma',\varphi) \to (M,\partial M)$ is a nondegenerate map. So, by Theorem 2.2, the map $f'|\sigma'$ is homotopic to map $g.(\sigma',\phi) \to (M,\partial M)$ such that $g(\sigma')\subset\Sigma$. Say $g(\sigma') \subset \sigma$, where (σ,φ) is a component of (Σ,Φ). Since f' is the homotopy inverse of f, it follows that X homotopes into σ (however, the pair (X,Y) need not homotop, as a pair into (σ,φ)) Let $\tilde{M}_\sigma$ be the covering space of M corresponding to $\pi_1(\sigma)$. The manifold $\tilde{M}_\sigma$ compactifies to a Siefert fibered manifold. Since X homotopes into σ, X lifts to $\tilde{M}_\sigma$. From this I conclude that X admits a Seifert fibration. So, in any case X admits a Seifert fibration; and therefore, it remains to show that Y is saturated in some Seifert fibration of X. The only problem in doing this is when X is homeomorphic to $S^1 \times S^1 \times I$. To obtain the desired conclusion in this case, it may be necessary to deform the map f. The details appear in [J]; and it follows (X,Y) is an S^1-pair.

ASSERTION 4. <u>There is a deformation such that</u> $f(\overline{(M-\Sigma)}) \subset \overline{(M'-\Sigma')}$.

It follows from Assertion 3 that if X is a component of $f^{-1}(\Sigma')$ and $Y = X \cap \partial M$, then (X,Y) is perfectly-embedded Seifert pair. By Theorem 2.2, there is an ambient isotopy of M taking (X,Y) into (Σ,Φ). So, there is an ambient isotopy of M taking $f^{-1}(\Sigma')$ into Σ. The conclusion of Assertion 4 follows.

It is easier to express the next three assertions if I introduce some terminology. Let $f: M \to M'$ be a homotopy equivalence between the 3-manifolds M and M'. Suppose that F' is a two-sided manifold embedded in M'. The homotopy equivalence f <u>splits at</u> F' if

(i) f is transverse on F' (set $F = f^{-1}(F')$),

(ii) $f|F: F \to F'$ is a homotopy equivalence and

(iii) $f|M-F: M-F \to M'-F'$ is a homotopy equivalence.

ASSERTION 5. There is a deformation such that f splits at Fr $\mathfrak{n}'$.

Recall that $\mathfrak{n}$ (respectively, $\mathfrak{n}'$) is the collection of non-abelian simple factors of M (respectively, M').

Let N be a component of $\mathfrak{n}$. By Assertion 4, there is a component N' of $\overline{(M'-\Sigma')}$ such that $f(N) \subset N'$. Since N is a nonabelain simple factor of M and $f|N$ induces an injection on $\pi_1(N)$, the group $\pi_1(N')$ is nonabelian. Hence $N' \in \mathfrak{n}'$. Similarly, there is a component $\tilde{N}$ of $\mathfrak{n}$ such that $f'(N') \subset \tilde{N}$. It follows that N deforms into $\tilde{N}$. Since $\pi_1(N)$ is nonabelian, it can be proved that $\tilde{N} = N$.

If C is a component of $f^{-1}(N')$ and $C \cap N = \emptyset$, then, as above, it follows that C is either a tube $(C \approx D^2 \times S^1)$ or a shell $(C \approx S^1 \times S^1 \times I)$. Set $D = C \cap \partial M$. It follows from Assertion 2 and Observation 3.2 (4) that $D = \emptyset$ and C is a simple shell; i.e. $(C,D) \approx (S^1 \times S^1 \times I, \emptyset)$.

Now, $f|C: (C,\partial C) \to (N', \text{Fr}N')$ and $(f|C)$ induces an injection on $\pi_1(C)$. Since $\pi_1(N')$ is nonabelian and N' is not homeomorphic to the twisted I-bundle over the Klein bottle (N' is a component $(M' - \Sigma')$ where (Σ', Φ') is the characteristic Seifert pair), the map $f|C$ is not essential. Therefore, there is a deformation such that $f^{-1}(N')$ has precisely one component, which contains N. Using similar arguments to the preceding, it follows that there is a deformation such that $f^{-1}(N') = N$(and $f'^{-1}(N) = N'$).

The fact that f splits at $Fr\,\eta'$ is established using covering space arguments.

ASSERTION 6. There is a deformation such that f splits at $Fr(\eta' \cup \rho')$.

Recall that ρ (respectively ρ') is the collection of simple factors of M (respectively, M') that have $\mathbb{Z} + \mathbb{Z}$ fundamental group; hence, each component of ρ (respectively, ρ') is a simple shell.

Let P be a component of ρ. By Assertion 4 $f(P)$ is contained in a component of $\overline{(M'-\Sigma')}$. After, Assertion 5, the only possibility is that $f(P)$ is contained in a component of ρ' or $\mathfrak{D}'$, where $\mathfrak{D}'$ is the collection of simple factors of M that have infinite cyclic fundamental group. So, the only possibility is that there exists a component P' of ρ' and $f(P) \subset P'$. Similarly, there exists a component $\tilde{P}$ of ρ and $f'(P') \subset \tilde{P}$. Now, the argument in this case is very similar to the argument used in Assertion 5 to prove that $\tilde{P} = P$; and so, there is a deformation such that f splits at $\rho' \cup \eta'$.

ASSERTION 7. There is a deformation such that f splits at $Fr\Sigma'$.

Recall that $\mathfrak{D}$ (respectively, $\mathfrak{D}'$) is the collection of simple factors of M (respectively, M') that have infinite cyclic fundamental group; hence, by Observation 4.1 (5) part (i), a component of $\mathfrak{D}$ (respectively, $\mathfrak{D}'$) is a simple tube.

To show that f splits at $Fr\Sigma'$, it is sufficient (after Assertions 5 and 6) to show that f splits at $Fr\,\mathfrak{D}'$. The argument here is not parallel to the arguments given in Assertions 5 and 6. However, there is a good deal of control on the situation at this point.

First, there is one more observation to make, I shall label it in sequence with the earlier observations (Observations 3.2 (1) through (5)).

(6) *If* $\tilde{\tilde{Q}}$ *is a component of* $f^{-1}(\mathfrak{Q}')$, y *is a component of* $\tilde{\tilde{Q}} \cap \partial M$ *and* A_1 *and* A_2 *are distinct components of* $\mathrm{Fr}\tilde{\tilde{Q}}$ *such that* $\partial y \cap \partial A_i \neq \phi$, $i = 1,2$, *the notation may be chosen so that* $A_1 \subset \mathrm{Fr}\, X_1$, *where* X_1 *is a component of* $f^{-1}(\Sigma')$ *and for* $Y_1 = X_1 \cap \partial M$, *the pair* (X_1, Y_1) *is an I-pair and not an* S^1*-pair, and* $A_2 \subset \mathrm{Fr}X_2$, *where* X_2 *is a component of* $f^{-1}(\Sigma')$ *and for* $Y_2 = X_2 \cap \partial M$, *the pair* (X_2, Y_2) *is an* S^1*-pair*.

This follows immediately from Assertion 2 and the refined part of Assertion 3.

Now, consider the components of $f^{-1}(\mathfrak{Q}')$. If $\tilde{Q}$ is such a component, $\tilde{Q}$ is a tube ($\tilde{Q} \approx D^2 \times S^1$) and then *a priori* the possibilities for $\tilde{Q}$ are: $\tilde{Q}$ is contained in an I-pair of (Σ, Φ) that is *not* an S^1-pair, or $\tilde{Q}$ is contained in an S^1-pair of (Σ, Φ), or neither of the preceding (and hence, $\tilde{Q}$ contains a component Q of $\mathfrak{Q}$). It follows that the first two of these three possibilities contradicts Observation (6).

The conclusion is that for each component Q' of $\mathfrak{Q}'$, $\tilde{Q} = f^{-1}(Q')$ is connected and there exists a component Q of $\mathfrak{Q}$ such that $Q \subset \tilde{Q}$. Now, using covering space arguments, as before, it follows that there is a deformation such that $Q = f^{-1}(Q')$; and f splits at $\mathrm{Fr}\mathfrak{Q}'$. So, f splits at $\mathrm{Fr}\mathfrak{n}' \cup \mathrm{Fr}\mathfrak{p}' \cup \mathrm{Fr}\mathfrak{Q}' = \mathrm{Fr}\Sigma'$.

ASSERTION 8 <u>There is a deformation such that</u> (i) $f|\overline{(M-\Sigma)}: \overline{(M-\Sigma)} \to \overline{(M'-\Sigma')}$ <u>is a homeomorphism and</u> (ii) $f|\Sigma: \Sigma \to \Sigma'$ <u>is a homotopy equivalence</u>.

It is straight forward to show that there is a deformation such that $f|\mathfrak{p} \cup \mathfrak{Q}: \mathfrak{p} \cup \mathfrak{Q} \to \mathfrak{p}' \cup \mathfrak{Q}'$ is a homeomorphism, $f|\mathfrak{n}: \mathfrak{n} \to \mathfrak{n}'$ is a homotopy equivalence and $f|\Sigma: \Sigma \to \Sigma'$ is a homotopy equivalence. Hence, it is necessary to prove that there is a deformation moving points only in a small neighborhood of $\mathfrak{n}$ such that $f|\mathfrak{n}: \mathfrak{n} \to \mathfrak{n}'$ is a homeomorphism. This is the most interesting part of this proof. I need what could be called a generalization of Lemmas 1.2 and 1.4. At least, it is a generalization of the ideas, since a generalization of Lemma 1.4 (to this relative situation) using the language of ends of pairs of groups is impossible.

So, let N be a component of $\mathfrak{n}$. There exists a unique component N' of $\mathfrak{n}'$ such that $f|N: N \to N'$ is a homotopy equivalence with homotopy inverse $f'|N'$. Let $T = N \cap \partial M$ and let $T' = N' \cap \partial M'$. The pairs (N,T) and (N',T') are simple, both T and T' are incompressible, no component of T or T' is an annulus

(Observation 3.2 (4)), and $f|FrN\colon FrN \to FrN'$ is a homotopy equivalence with homotopy inverse $f'|FrN'$. So, after a deformation, moving points only in a small neighborhood of FrN, I can assume that $f|FrN\colon FrN \to FrN'$ is homeomorphism. Notice that $FrN = \overline{(\partial N-T)}$, $FrN' = \overline{(\partial N' - T')}$, and each component of $\overline{(\partial N - T)}$, $\overline{(\partial N' - T')}$, is either an incompressible annulus or torus.

If A is a annulus parametrized as $A = S^1 \times I$, then define $\tau_s\colon S^1 \times I \to S^1 \times I$ as $\tau_s(x,r) = (x,(1-r)s + r(1-s))$, $0 \le s \le 1$. The homotopy τ_s, $0 \le s \le 1$, is a "flip" homotopy.

3.3. LEMMA: *Suppose that (N,T) and (N',T') are Haken-manifold pairs such that no component of T or T' is an annulus and each component of $\overline{(\partial N' - T')}$ is either an incompressible annulus or torus. Suppose that $f\colon(N, \overline{(\partial N-T)}) \to (N', \overline{(\partial N'-T')})$ is a map of pairs such that $f|\overline{(\partial N-T)}\colon \overline{(\partial N-T)} \to \overline{(\partial N'-T')}$ is a homeomorphism. If f is a homotopy equivalence and (N,T) is a simple pair, then there exists a homotopy $f_s\colon (N, \overline{(\partial N-T)}) \to (N', \overline{(\partial N' - T')})$, $0 \le s \le 1$, such that*

(i) $f_0 = f$,

(ii) f_1 *is a homeomorphism and*

(iii) $f_s|\overline{(\partial N-T)}$ *is equal to f on each component that is not an annulus and is equal to $f \circ \xi_s$ on each component that is an annulus, where ξ_s is either the identity or τ_s, $0 \le s \le 1$.*

In order to prove Lemma 3.3, I first prove that if t is a component of T, then there is a deformation (rel ∂t) of $f|t$ to a map taking t into $\partial N'$. Of course, it will then follow that there is a deformation (rel$\overline{(\partial N - T)}$) of f to a boundary preserving map; however, in order to obtain part (iii) of the conclusion, I will need to prove more.

Let $\tilde{N}'$ be the covering space of N' corresponding to the conjugacy class of $f_*(\pi_1(t))$ in $\pi_1(N')$ and let $q: \tilde{N}' \to N$ denote the covering projection. There is a lifting $\widetilde{f|t}: (t, \partial t) \to (\tilde{N}', \partial\tilde{N}')$ of $f|t$ (and $\widetilde{f|t}: \partial t \to \partial\tilde{N}'$ is an embedding). Using standard techniques, I find a neighborhood $\tilde{U}$ of $\widetilde{f|t}(t)$ such that $\pi_1(\tilde{U}) \to \pi_1(\tilde{N}')$ is an epimorphism, $\tilde{U} \cap \partial\tilde{N}'$ is a regular neighborhood of $\widetilde{f|t}(\partial t)$ in $\partial\tilde{N}'$ and $Fr\tilde{U}$ is incompressible in $\tilde{N}'$. It follows that $\pi_1(\tilde{U}) \to \pi_1(\tilde{N}')$ is an isomorphism.

If $\tilde{L}$ is a component of $Fr\tilde{U}$, then $\pi_1(\tilde{L}) \to \pi_1(\tilde{N}')$ is an isomorphism. If $\tilde{L}$ is closed, then this is quite easy. If $\tilde{L}$ is not closed, then $\partial\tilde{L} \neq \emptyset$; and (up to isotopy in $\partial\tilde{N}'$) $\partial\tilde{L}$ consists of a subcollection of the curves in $\widetilde{f|t}(\partial t)$. Using a homology argument and the fact that $\partial\tilde{L}$ bounds in $\tilde{N}'$, it follows that $\pi_1(\tilde{L}) \to \pi_1(\tilde{N}')$ is an epimorphism; and therefore, an isomorphism.

Now, each component $\tilde{L}$ of $Fr\tilde{U}$ separates $\tilde{N}'$. The next thing to prove is that each component $\tilde{L}$ of $Fr\tilde{U}$ separates $\tilde{N}'$ into two components and one of them has compact closure.

Let $\tilde{N}_t$ be the covering space of N corresponding to the conjugacy class of $\pi_1(t)$ in $\pi_1(N)$ and let $p: \tilde{N}_t \to N$ denote the covering projection. There is a component $\tilde{t}$ of $p^{-1}(t)$ such that $p|\tilde{t}: \tilde{t} \to t$ is a homeomorphism and $\pi_1(\tilde{t}) \to \pi_1(\tilde{N}_t)$ is an isomorphism. The manifold $\tilde{N}_t$ admits a manifold compactification to the product $t \times I$ via a homeomorphism taking $\tilde{t}$ to $t \times \{0\}$. I assume that a parametrization of $\tilde{N}_t$ has been chosen so that $\tilde{N}_t$ corresponds to $t \times I$ with a closed subset of $t \times \{1\}$ missing and $\tilde{t}$ corresponds to $t \times \{0\}$. There is a lifting $\tilde{f}: \tilde{N}_t \to \tilde{N}'$ of $f \circ p$ such that $\tilde{f}|\tilde{t} = \tilde{f}|t \circ p|\tilde{t}$ and $\tilde{f}$ is a <u>proper</u> homotopy equivalence.

Let $\tilde{L}$ be a component of $\text{Fr}\tilde{U}$. Since $\tilde{f}$ is a proper homotopy equivalence, there is a deformation (rel $\partial\tilde{N}_t$) such that $(\tilde{f})^{-1}(\tilde{L})$ is a compact, incompressible, two-sided 2-manifold in $\tilde{N}_t$.

Suppose that $\tilde{F}$ is a component of $(\tilde{f})^{-1}(\tilde{L})$. Then $\partial\tilde{F} \cap \partial\tilde{N}_t \subset \partial t \times I \cup t \times \{1\}$ in the compactification of $\tilde{N}_t$ to $t \times I$. However, what is more important is the fact that $\partial\tilde{F} \cap p^{-1}\overline{(\partial N\text{-}T)}$ is contained in a neighborhood of $\partial\tilde{t}$ in $\partial t \times I$. This follows from the hypothesis that $f|\overline{(\partial N\text{-}T)}: \overline{(\partial N\text{-}T)} \to \overline{(\partial N'\text{-}T')}$ is a homeomorphism; and so, for the covering $\tilde{N}_t$ of N (with covering projection $p: \tilde{N}_t \to N$) and the covering $\tilde{N}'$ of N' (with covering projection $q: \tilde{N}' \to N'$) the map $\tilde{f}|p^{-1}\overline{(\partial N\text{-}T)}: p^{-1}\overline{(\partial N\text{-}T)} \to q^{-1}\overline{(\partial N'\text{-}T')}$ is an embedding into $\partial\tilde{N}'$. In particular, if a component of $\partial\tilde{F} \cap \partial\tilde{N}_t$ is not in a neighborhood of $\partial\tilde{t}$ in $\partial t \times I$, then it is contained in a component of $p^{-1}(T) - \tilde{t}$.

The components of $p^{-1}(T) - \tilde{t}$ can be described very easily.

Since the pair (N,T) is simple, a component of $p^{-1}(T) - \tilde{t}$ is either simply connected or has infinite cyclic fundamental group, and is in the same component of $\partial \tilde{N}_t$ as $\tilde{t}$ and can be deformed through an annulus in $p^{-1}\overline{(\partial N-T)}$ into $\partial \tilde{t}$. It follows that $\tilde{F}$ is either a disk and is parallel in $\tilde{N}_t$ into a component of $p^{-1}(T)-\tilde{t}$, or an annulus and is parallel in $\tilde{N}_t$ into either a component of $p^{-1}(T) - \tilde{t}$ or a component of $p^{-1}\overline{(\partial N-T)}$ meeting $\partial \tilde{t}$ or a homeomorphic copy of $\tilde{t}$ and is parallel in $\tilde{N}_t$ into $\tilde{t}$. The conclusion is that $(\tilde{f})^{-1}(\tilde{L})$ is compact and separates $\tilde{N}_t$ into a finite number of components having compact closures and one component with non-compact closure. Since $\tilde{f}$ is a proper homotopy equivalence, this implies that $\tilde{L}$ separates $\tilde{N}'$ into two components and one of them has compact closure, as was to be shown. Hence, by $\pi_1(\tilde{L}) \to \pi_1(\tilde{N}')$ an isomorphism, it follows that $\tilde{L}$ is parallel into $\partial \tilde{N}'$. Since $\tilde{L}$ was an arbitrary component of $Fr\tilde{U}$, the neighborhood $\tilde{U}$ is parallel into $\partial \tilde{N}'$; and so, $f|\tilde{t}$ deforms (rel ∂t) into $\partial \tilde{N}'$.

So, for any component t of T, there is a deformation (rel ∂t) of $f|t$ to a map taking t into $\partial N'$. I may assume that $f|\partial N$: $\partial N \to \partial N'$, while keeping $f|\overline{(\partial N-T)}$: $\overline{(\partial N-T)} \to \overline{(\partial N'-T')}$ a homeomorphism.

Now, since $f|t$ induces an injection on $\pi_1(t)$ and no component t of T is an annulus, the map $f|T$ deforms (rel ∂T) to a covering map onto a surface in $\partial N'$. At this point the map $f|\partial N$: $\partial N \to \partial N'$ maynot be a covering map. However, it is very close. For suppose that t_1 and t_2 are distinct components of T and $f(t_1) \cap f(t_2) \neq \phi$. Then $f(t_1) \cap f(t_2)$ contains a component of T' or a component of $\overline{(\partial N'-T')}$.

Suppose that t' is a component of T' and $t' \subset f(t_1) \cap f(t_2)$. Let t_i' be a component of $f^{-1}(t') \subset t_i$, $i = 1,2$. Set $H = f_*^{-1}(\pi_1(t'))$. Then both $\pi_1(t_1')$ and $\pi_1(t_2')$ have finite index in H. This implies either H is infinite cyclic or the pair (N,T) is not simple. Both cases present a contradiction. So $f(t_1) \cap f(t_2)$ does not contain a component of T'. Now, by using the "flip" homotopy τ_s, $0 \leq s \leq 1$, and an inductive argument on the number of components of T, allows the conclusion that there is a deformation f_s, $0 \leq s \leq 1$, such that $f_0 = f$ and $f_1|\partial N\colon \partial N \to \partial N'$ is a homeomorphism with $f_s|\overline{(\partial N-T)}$ equal to f on components that are not annuli and equal to $f \circ g_s$ on components that are annuli, where g_s is either the identity or the "flip" homotopy τ_s, $0 \leq s \leq 1$. This concludes the proof of Lemma 3.3 and establishes Assertion 8.

ASSERTION 9: <u>There is a deformation such that</u> (i) $f|\overline{(M-\Lambda)}\colon \overline{(M-\Lambda)} \to \overline{(M'-\Lambda')}$ <u>is a homeomorphism and</u>

(ii) $f|\Lambda\colon \Lambda \to \Lambda'$ <u>is a homotopy equivalence.</u>

If (σ,φ) is a component of (Σ,Φ) and (σ,φ) is not peripheral, then $\varphi = \emptyset$. However, by Assertion 8, the map $f|\sigma: (\sigma,\partial\sigma) \to (\sigma',\partial\sigma')$ where (σ',φ') is a component of (Σ',Φ'); furthermore, $f|\partial\sigma: \partial\sigma \to \partial\sigma'$ is an embedding. It follows that $f|\sigma$ may be deformed (rel $\partial\sigma$) to a homeomorphism.

This establishes the proof of Theorem 3.1.

BIBLIOGRAPHY

[J] W. Jaco, Lectures on Three-manifold Topology, CBMS 43, to appear.

[J-S] W. Jaco and P. Shalen, "Seifert fibered spaces in 3-manifolds", Memoirs, Amer. Math. Soc., V.21, No. 220, Providence, Rhode Island (1979)

[Jo_1] K. Johannson, "Homotopy equivalences in knot spaces", preprint.+

[Jo_2] K. Johannson, "Homotopy equivalences in 3-manifolds with boundary", preprint.+

[W_1] F. Waldhausen, "On irreducible 3-manifolds which are sufficiently large", Annals of Math. 87 (1968), 56-88.

[W_2] F. Waldhausen, "On some recent results in 3-dimensional topology", Proc. Sym. in Pure Math. Amer. Math. Soc. 32 (1977), 21-38.

+ cf. also K. Johannson, "Homotopy Equivalences of 3-manifolds with boundary", Lect.Notes in Math. 761, Springer, Heidelberg-Berlin -New York 1979

Department of Mathematics
Rice University
Houston, Texas 77001
U.S.A.

Weaving Patterns and Polynomials

by Louis H. Kauffman

1. Introduction

In [1] (see also [2] and [3]) John H. Conway introduced a polynomial invariant of oriented knots and links in three-space that is determined by three simple axioms:

A1. To each oriented link $K \subset S^3$ there is a polynomial $\nabla_K(a) \in \mathbb{Z}[a]$ such that $\nabla_K(a) = \nabla_{K'}(a)$ whenever K is ambient isotopic ($\approx$) to K'.

A2. If K is an unknotted circle, then $\nabla_K(a) = 1$.

A3. Let K, $\bar{K}$, L be links that differ only at a single crossing as indicated in the diagram below. Then $\nabla_K(a) - \nabla_{\bar{K}}(a) = a\nabla_L(a)$.

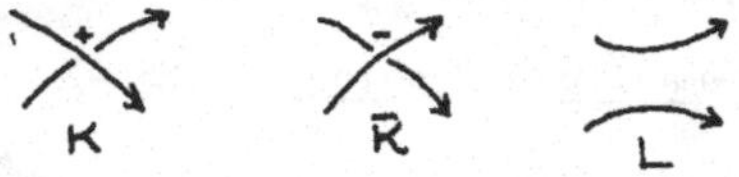

These axioms are sufficient to calculate $\nabla_K(a)$ for any knot or link in the three-sphere S^3. The consistency of the axioms can be demonstrated with the aid of a model.

The purpose of this paper is to illustrate calculations of the Conway polynomial and applications to the study of recursive weaving patterns. Consider the transformation illustrated in Figure 1. A knot Λ is given with unknown tying within the black triangle, and a particular configuration at the base of the triangle. The transform $R\Lambda$ is a knot of the same form. Thus we can contemplate the process: $\Lambda, R\Lambda, R^2\Lambda, R^3\Lambda, \ldots$
In this case it is possible to calculate that $\nabla_{R^{n+1}\Lambda} = \nabla_{\bar{\Lambda}} - a^2\nabla_{R^n\Lambda}$ (See section 2.). The knot $\bar{\Lambda}$ is obtained from Λ by changing one crossing as indicated in Figure 1. Thus one is tempted to let $K_\infty = R^\infty\Lambda$ so that $RK_\infty = K_\infty$ and $\nabla_{K_\infty} = \nabla_{\bar{\Lambda}} - a^2\nabla_{K_\infty}$ whence $(1 + a^2)\nabla K_\infty = \nabla_{\bar{\Lambda}}$ or $\nabla_{K_\infty} = \nabla_{\bar{\Lambda}}/(1+a^2)$, expressing an invariant of this "infinite" knot.

Rather than going into the usual category of wild knots for this infinite construction , I shall instead replace knots by infinite sequences of knots. In the example above, K_∞ will be formally represented by

$$K_\infty = (\Lambda, R\Lambda, R^2\Lambda, R^3\Lambda, \ldots).$$

Invariance is obtained by defining $R(A,B,C,D,\ldots) = (A,RA,RB,RC,RD,\ldots)$. Thus $RK_\infty = K_\infty$. This construction is convenient because Λ does not disappear "off to infinity". It formalizes the intuitive idea that $K_\infty = R^\infty\Lambda$ represents $RRR\ldots RR\Lambda$ where the number of R's has become "too large to count". Taking the limit has the effect of picking up a general pattern about the weaving process that is independent of specific values of n.

The paper is organized as follows: Section 2 sketches basic calculations and computes the example discussed above. Section 3 introduces the sequence category and its properties. Section 4 is a rapid discussion of one model for the Conway axioms.

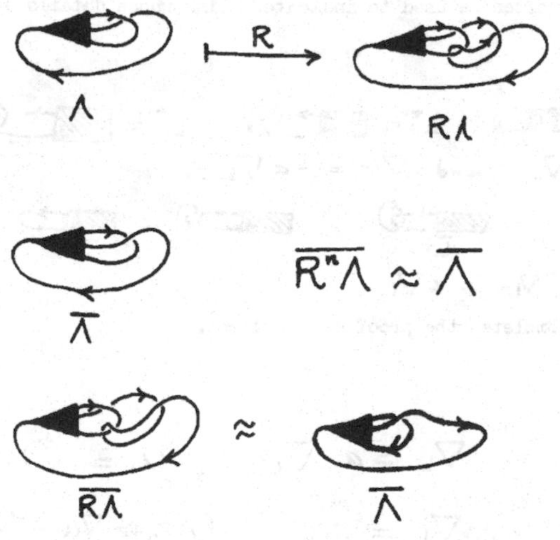

Figure 1

2. Calculation

In order to perform calculations using the axioms A1 , A2 , A3 (see section 1) it is useful to note that the **Conway polynomial vanishes for split links.**

Definition 2.1. A link $L \subset S^3$ is a split link if $L = L_1 \cup L_2$ where L_1 and L_2 are disjoint nonempty sub-links that can be separated by disjoint three-balls embedded in S^3.

Lemma 2.2. If $L \subset S^3$ is a split link, then $\nabla_L = 0$.

Proof. Let L_1 and L_2 be positioned as in Figure 2. Then there are the associated links (or knots) K and $\overline{K}$ as indicated in the same figure. Since K and $\overline{K}$ are ambient isotopic (use a 360° twist), we have

$0 = \nabla_K - \nabla_{\overline{K}} = a\nabla_L$ and hence $\nabla_L = 0$.

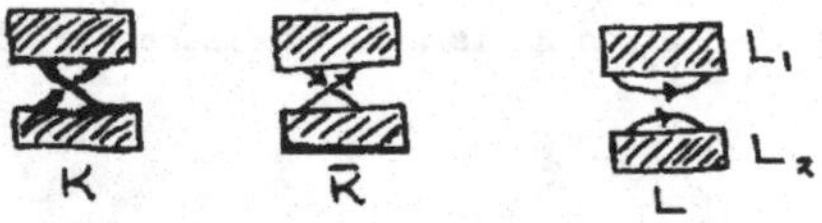

Figure 2

A shaded box will often be used to indicate a link whose details are unknown.

Lemma 2.3. If $L =$  , $L^+ =$  , $L^- =$ 

then $\nabla_{L^+} = a\nabla_L$ and $\nabla_{L^-} = -a\nabla_L$.

Proof. 

Since $\overline{L}^+$ is split, $\nabla_{L^+} = a\nabla_L$.

Similar work on L^- completes the proof of the lemma.

Examples:

1.  $= L \Rightarrow \nabla_L = a\nabla_u$, $u =$ 

$\therefore \nabla_L = a \qquad (A2 \Rightarrow \nabla_u = 1)$.

2. 

$K \qquad \bar{K} \approx U \qquad L$

$$\nabla_K = \nabla_U + a\nabla_L$$
$$\therefore \nabla_K = 1 + a^2.$$

3. 

$E \qquad U \qquad L'$

$$\nabla_{L'} = -a \ (\text{by Lemma 1.})$$
$$\therefore \nabla_E = 1 - a^2.$$

4. Let $\Lambda \longmapsto R\Lambda$ be the operation indicated in Figure 1 . Then as indicated in Figure 1 , $\overline{R^n\Lambda} \approx \bar{\Lambda}$.

Claim: $\nabla_{R^n\Lambda} = \nabla_{\bar{\Lambda}} - a^2\nabla_{R^{n-1}\Lambda}$.

Proof. Let $X_n = R^n\Lambda$. Thus $\bar{X}_n \approx \bar{\Lambda}$.

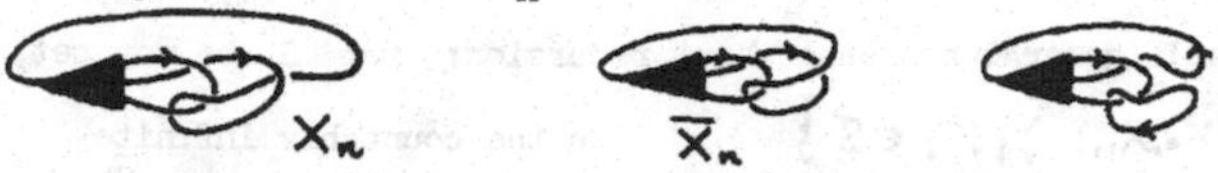

$X_n \qquad \bar{X}_n \qquad L_n$

By Lemma 2.2, $\nabla_{L_n} = -a\nabla_{X_{n-1}}$.

$$\therefore \nabla_{X_n} = \nabla_{\bar{X}_n} + a\nabla_{L_n} = \nabla_{\bar{\Lambda}} - a^2\nabla_{X_{n-1}}.$$

For example, let Λ be a trefoil knot. Then $\bar{\Lambda}$ is an unknot , and we have (letting $\nabla_n = \nabla_{X_n}$) $\nabla_0 = 1 + a^2$, $\nabla_n = 1 - a^2\nabla_{n-1}$.

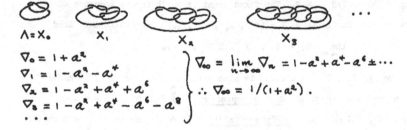

$\Lambda = X_0 \qquad X_1 \qquad X_2 \qquad X_3 \qquad \cdots$

$$\left.\begin{aligned} \nabla_0 &= 1 + a^2 \\ \nabla_1 &= 1 - a^2 - a^4 \\ \nabla_2 &= 1 - a^2 + a^4 + a^6 \\ \nabla_3 &= 1 - a^2 + a^4 - a^6 - a^8 \\ &\cdots \end{aligned}\right\} \begin{aligned} \nabla_\infty &= \lim_{n\to\infty} \nabla_n = 1 - a^2 + a^4 - a^6 \pm \cdots \\ \therefore \nabla_\infty &= 1/(1+a^2). \end{aligned}$$

It is tempting to try to make sense out of this by assigning ∇_∞ as an extended Conway invariant for an infinite knot K_∞ such that $RK_\infty \cong K_\infty$. Then the recursion would become $\nabla_{RK_\infty} = \nabla_{\bar{\Lambda}} - a^2\nabla_{K_\infty}$.

Hence
$$\nabla K_\infty = 1 - a^2 \nabla K_\infty$$
$$(1+a^2)\nabla K_\infty = 1$$
$$\nabla K_\infty = 1/(1+a^2) .$$

However, K_∞ is not the wild knot

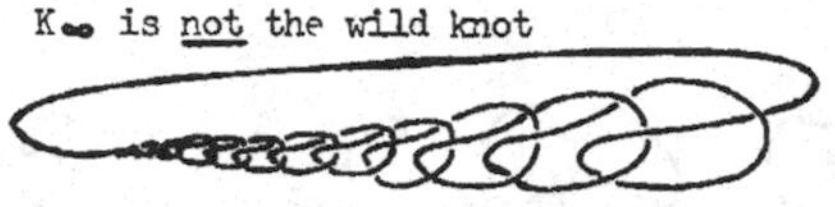

W!

It is the case that $RW \approx W$, but $\overline{W}$ remains a wild arc - it is not unknotted. In fact, there is no clear justification for extending Conway's axioms to wild knots. The next section will make precise the appropriate notion of infinite knot.

3. Infinite Knots

First, a little general nonsense about recursion: Let S be any set, and let $\mathcal{S} = \{(X_0, X_1, \ldots) \mid X_i \in S\}$ be the countably infinite cartesian product of copies of S. Given a mapping $T: S \longrightarrow S$, define its extension $T: \mathcal{S} \longrightarrow \mathcal{S}$ by the formula
$$T(X_0, X_1, \ldots) = (X_0, TX_0, TX_1, \ldots).$$
Note that X_0 is retained while all other components are transformed and shifted. As pointed out in the introduction $\mathcal{S}$ provides fixed points for any mapping $T: S \longrightarrow S$. The fixed point is built from a sequence of iterates under T. Letting $X = (X_0, X_1, X_2, \ldots)$ we obtain the fixed point $T^\infty(X) = \lim_{n \to \infty} T^n(X) = (X_0, TX_0, T^2X_0, \ldots)$.

Since $(T^nX_0, T^{n+1}X_0, \ldots)$ is also a fixed point for T, it makes sense to form a category of stable sequences, $\hat{\mathcal{S}} = \mathcal{S}/\sim$, where two elements $X, Y \in \mathcal{S}$ are stably equivalent $(X \sim Y)$ if they agree term for term after a finite shift and finite amputation of initial terms. Let $[X]$ denote the stable equivalence class of X. Since
$(X_0, TX_0, T^2X_0, \ldots) \sim (TX_0, T^2X_0, \ldots) \sim (T^2X_0, T^3X_0, \ldots) \sim \ldots$
we see that $[T^\infty(X)]$ is an appropriate formalization of $TTT\ldots TTX_0$ (an indefinite number of T's).

Any notion of pointwise addition on S extends to a coordinatewise addition on $\mathcal{S}$. S embeds in $\mathcal{S}$ by taking the constant sequence : $s \longmapsto (s, s, s, \ldots)$. This embedding extends to $\hat{\mathcal{S}}$ and allows well-defined addition between constant sequences and arbitrary elements of $\hat{\mathcal{S}}$.

Now transpose these ideas to knot theory. Let $\mathcal{K}$ denote the collection of knots and links in S^3 with equivalence relations $=$ (pointwise equality) and $\approx$ (ambient isotopy). Let $\mathcal{S}(\mathcal{K})$ and $\hat{\mathcal{S}}(\mathcal{K})$ denote the sequence and stable sequence categories respectively. Both $\mathcal{S}(\mathcal{K})$ and $\hat{\mathcal{S}}(\mathcal{K})$ inherit the relations $=$ and $\approx$ (via coordinatewise equivalence). If $\Lambda \longmapsto R\Lambda$ is a weaving operation as in Figure 1 then $K_\infty = [R^n\Lambda] = [\Lambda, R\Lambda, R^2\Lambda, R^3\Lambda, \ldots]$ represents an infinite knot that is a fixed point for the operation R. We may informally write $K_\infty \equiv R^\infty\Lambda \equiv R\cdots R\Lambda$, or in a diagram as below:

K_∞.

(Here Λ is a trefoil.)

Let $\mathcal{S}(P)$ and $\hat{\mathcal{S}}(P)$ be the sequence categories corresponding to the polynomial ring $P = Z[a]$. Addition and scalar multiplication are inherited from P. Define Conway maps $\nabla : \mathcal{S}(\mathcal{K}) \longrightarrow \mathcal{S}(P)$ and $\nabla : \hat{\mathcal{S}}(\mathcal{K}) \rightarrow \hat{\mathcal{S}}(P)$ by $\nabla(K_0, K_1, \ldots) = (\nabla K_0, \nabla K_1, \ldots)$.

It now becomes clear that such things as $1/(1+a^2)$ are appropriate <u>names</u> for certain elements of $\mathcal{S}(P)$ or $\hat{\mathcal{S}}(P)$. For example, when $\nabla_\Lambda = 1$ and $\nabla_{R^n\Lambda} = 1 - a^2\nabla_{R^{n-1}\Lambda}$, then

$$\nabla_{K_\infty} = [\nabla_\Lambda, \nabla_{R\Lambda}, \nabla_{R^2\Lambda}, \ldots] = [1, 1-a^2\nabla_\Lambda, 1-a^2\nabla_{R\Lambda}, \ldots]$$
$$= [1, 1, \ldots] + [0, -a^2\nabla_\Lambda, -a^2\nabla_{R\Lambda}, \ldots]$$
$$= 1 + [-a^2\nabla_\Lambda, -a^2\nabla_{R\Lambda}, \ldots] = 1 - a^2[\nabla_\Lambda, \nabla_{R\Lambda}, \ldots] = 1 - a^2\nabla_{K_\infty}.$$

Thus $\nabla_{K_\infty} = 1 - a^2\nabla_{K_\infty}$ in $\hat{\mathcal{S}}(P)$. This is ample justification for naming $\nabla_{K_\infty} \equiv 1/(1+a^2)$. (Note that different elements of $\hat{\mathcal{S}}(P)$ could, by the same justification, have *identical names*.)

Let $T($$) =$ , $T($$) =$ , $\Lambda =$ .
Then $T^{\infty}\Lambda = (\Lambda, T\Lambda, \ldots)$ and $\nabla_{T^n\Lambda} = a^n$ so that
$\nabla_{T^{\infty}\Lambda} = (a, a^2, \ldots)$. Thus $a\nabla_{T^{\infty}\Lambda} = (a^2, a^3, \ldots) \sim \nabla_{T^{\infty}\Lambda}$.
Hence $aB = B$. In this case it is appropriate to name $B \equiv a^{\infty}$.
$(B = \nabla[T^{\infty}\Lambda])$

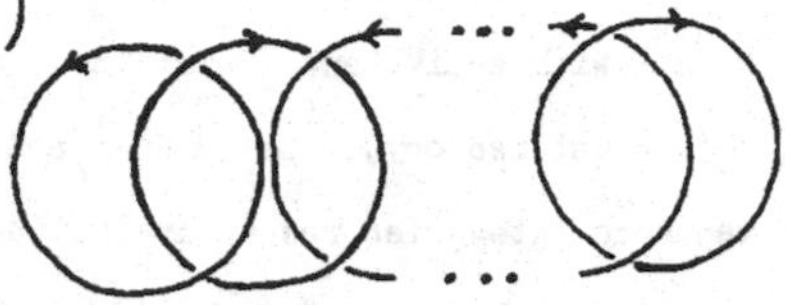

$\overset{\nabla}{\longmapsto} a^{\infty}$

4. The Model

Let $K \subset S^3$ be an oriented link and $F \subset S^3$ an oriented , connected spanning surface for K. Let $\Theta : H_1(F) \times H_1(F) \longrightarrow Z$ denote the Seifert pairing (see [4]) defined by the formula $\Theta(x,y) = Lk(x^+,y)$. Here Lk denotes linking number in S^3 and x^+ is the result of pushing x into S^3-F along the positive normals to F.

The Seifert pairing is not an invariant of K , but different choices of surface F and/or ambient isotopy of K result in s-equivalent Seifert matrices (A Seifert matrix is a matrix of the Seifert pairing with respect to a basis for $H_1(F)$.). Two matrices are s-equivalent if one can be obtained from the other by a sequence of congruences ($A \longmapsto PAP'$, P unimodular , P' = the transpose of P.) and elementary moves:

$$A \longleftrightarrow \left(\begin{array}{c|cc} A & \alpha & 0 \\ \hline 0 & 0 & 1 \\ 0 & 0 & 0 \end{array}\right) \quad \alpha \text{ a column vector.}$$

$$A \longleftrightarrow \left(\begin{array}{c|cc} A & 0 & 0 \\ \hline \beta & 0 & 0 \\ 0 & 1 & 0 \end{array}\right) \quad \beta \text{ a row vector.}$$

(See [5]).

Define the potential function $\Omega_K(x) = \mathrm{Det}(x\theta - x^{-1}\theta') \in \mathbb{Z}[x, x^{-1}]$ where θ is a Seifert matrix for K. Since

$$\mathrm{Det}\left(x\begin{bmatrix}0&1\\0&0\end{bmatrix} - x^{-1}\begin{bmatrix}0&0\\1&0\end{bmatrix}\right) = \mathrm{Det}\begin{bmatrix}0&x\\-x^{-1}&0\end{bmatrix} = 1,$$

it follows that $\Omega_K(x)$ is a invariant of ambient isotopy of K , and that $\Omega_K(x) = 1$ when K is the unknot.

Theorem 3.1. $\nabla_K(a) = \Omega_K(x) = \mathrm{Det}(x\theta - x^{-1}\theta')$ where $x - x^{-1} = a$.

Proof. See [3].

The proof involves showing that $\Omega_K(x)$ satisfies axiom A3 when $a = x - x^{-1}$. Thus Ω_K provides a model for the axioms.

Since $x = a + \frac{1}{x} = a + \frac{1}{a + 1/x} = \cdots$ we may take $x = a + \frac{1}{a} + \frac{1}{a} + \frac{1}{a} + \cdots$

or $x = \frac{1}{2}(a + \sqrt{a^2+4})$ and $\nabla_K(a) = \Omega_K(a + \frac{1}{a} + \frac{1}{a} + \frac{1}{a} + \cdots)$.

Example. Let K_n denote a torus link of type $(2, n+1)$:

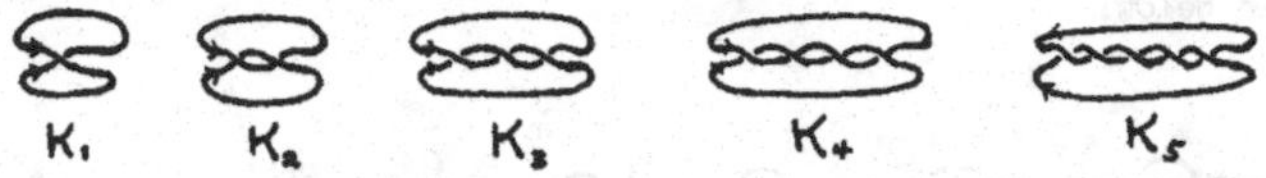

Then $\nabla K_n = a\nabla K_{n-1} + \nabla K_{n-2}$, $\nabla K_1 = 1$, $\nabla K_2 = a$.

Hence $\nabla K_3 = 1 + a^2$, $\nabla K_4 = a^3 + 2a$, $\nabla K_5 = a^4 + 3a^2 + 1, \ldots$

Note that $a + \frac{1}{a} = \frac{a^2+1}{a}$, $a + \frac{1}{a + \frac{1}{a}} = \frac{a^3+2a}{a^2+1}$, ...

Thus $\nabla K_n = Num\left(\underbrace{a + \frac{1}{a} + \frac{1}{a} + \frac{1}{a} + \cdots + \frac{1}{a}}_{n \text{ terms}}\right)$

where Num denotes the numerator of this continued fraction.

Since a Seifert matrix for K_n is given by the nxn matrix

$$\Theta_n = \begin{bmatrix} 1 & 1 & & & \\ & 1 & 1 & & 0 \\ & & 1 & 1 & \\ & 0 & & \ddots & 1 \\ & & & & 1 \end{bmatrix},$$

we may use Theorem 3.1 to conclude that

$$\underbrace{\mathrm{Num}\left(a+\frac{1}{a}+\frac{1}{a}+\cdots+\frac{1}{a}\right)}_{n\text{-terms}} = \mathrm{Det}\underbrace{\begin{bmatrix} a & x & & & \bigcirc \\ \bar{x} & a & x & & \\ & \bar{x} & a & \ddots & \\ & & \bar{x} & \ddots & x \\ \bigcirc & & & \bar{x} & a \end{bmatrix}}_{n\times n}.$$

$$x+\bar{x}=a\,,\quad x\bar{x}=-1$$

This identity can also be verified directly.

Other methods assign these same polynomials to the rational knots of type (see [1]):

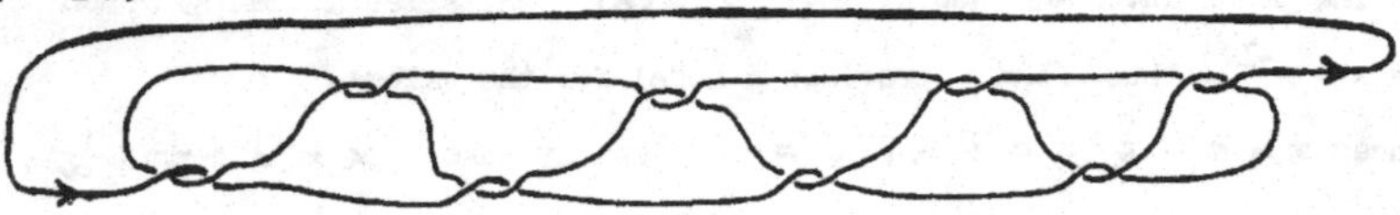

In fact these continued fraction knots unwind into (2 , n +1) torus knots as illustrated below:

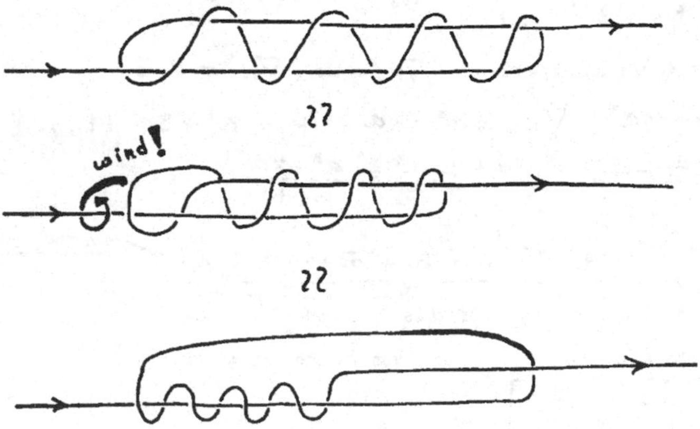

References

1. J. H. Conway (1970), An enumeration of knots and links and some of their algebraic properties. Computational Problems in Abstract Algebra, Pergamon Press, N.Y., 329-358.

2. C. Giller (1979), A family of links and the Conway Calculus, (to appear).

3. L. H. Kauffman (1979), The Conway polynomial, (to appear)

4. J. Levine (1969), Knot cobordism groups in codimension two, Comm. Math. Helv. 44, 229-244.

5. J. Levine (1970), An algebraic classification of some knots of codimension two, Comm. Math. Helv. 45, 185-198.

Department of Mathematics
University of Illinois at Chicago Circle
Chicago, Illinois 60680

VECTOR FIELDS ON (4q+2)-MANIFOLDS

by ULRICH KOSCHORKE

In this lecture I would like to make a little bit of propaganda for the singularity approach to existence questions of framefields I will outline an obstruction theory based on this approach, and a few consequences For more details and applications, see the forthcoming SLN-volume [3]

Let M be a closed connected smooth n-dimensional manifold, and let k be a natural number such that $n > 2k$. We define a k-field on M to be a system $u = (u_1, \ldots, u_k)$ of k tangent vectorfields on M The singularity of u is the set of all the points $x \in M$ where the tangent vectors $u_1(x), \ldots, u_k(x) \in T_x(M)$ fail to be linearly independant. We would like to decide when M allows a k-field without singularity, i e k linearly independant vector fields.

Note that any k-field can also be interpreted as a vector bundle homomorphism from the trivial bundle $\underset{\sim}{\mathbb{R}}^k = M \times \mathbb{R}^k$ to the tangent bundle TM (mapping (x, i th unit vector of $\mathbb{R}^k$) to $u_i(x)$), or, equivalently, as a section of the homomorphism bundle $\underset{\sim}{\mathrm{Hom}}(\underset{\sim}{\mathbb{R}}^k, TM)$. The fiber of this bundle at a point x in M is the space of all linear maps from $\mathbb{R}^k$ to T_xM, it can be decomposed into the smooth submanifolds ("strata") consisting of all linear maps of a given fixed rank (e g the "stratum" A_x^0 corresponding to the rank 0 consists only of the zero map, the "stratum" A_x^k corresponding to rank k is the set of all monomorphisms). Similarly, the total space of $\underset{\sim}{\mathrm{Hom}}(\underset{\sim}{\mathbb{R}}^k, TM)$ is decomposed into the "strata" $A^0, A^1, \ldots, A^k$, where e.g A^0 is just the zero section.

The idea is now to make our section avoid the strata $A^0, A^1, \ldots,$ and thereby increase the minimum rank of u successively Actually, our assumption on k, as well as the high codimension of the strata $A^0, \ldots, A^{k-2}$, guarantee that any k-field can be approximated by a smooth k-field u such that

(i) $\quad \operatorname{rank}(u_x : \mathbb{R}^k \longrightarrow T_xM) \geq k-1 \quad$ for all $x \in M$,

and

(ii) $\quad u$, considered as a section in $\underset{\sim}{\mathrm{Hom}}(\underset{\sim}{\mathbb{R}}^k, TM)$, is transverse to A^{k-1}.

It remains to analyse the singularity

(1) $S = \{x \in M \mid \text{rank } u_x = k-1\}$

in order to see whether it can be removed alltogether.

First of all, S is a $(k-1)$-dimensional submanifold of M. At each point $x \in S$, the kernel of u_x is a line in $\mathbb{R}^k$, or, equivalently, a point in projective space P^{k-1}. Thus, we get a "kernel map" which, together with the inclusion $S \subset M$, defines a continuous map

(2) $$g : S \longrightarrow P^{k-1} \times M$$

Moreover, the normal bundle $\nu(S, M)$ can be expressed canonically in terms of the kernel and cokernel bundles of $u|S$, after some stabilizations and simplifications this leads to an isomorphism

(3) $\bar{g}$ $$TS \oplus g^*((\lambda \otimes TM) \oplus \underset{\sim}{\mathbb{R}}) \cong g^*(k\lambda \oplus TM)$$

where λ denotes the canonical line bundle over P^{k-1} (pulled back to $P^{k-1} \times M$)

Denote the vector bundles $\lambda \otimes TM \oplus \underset{\sim}{\mathbb{R}}$ and $k\lambda \oplus TM$ over $P^{k-1} \times M$ by ϕ_M^+ and ϕ_M^- for short. We obtain the triple $(S, g, \bar{g})$ consisting of a smooth closed $(k-1)$-manifold S, a continuous map g into the target space $P^{k-1} \times M$, and an isomorphism $\bar{g}$ expressing the stable normal bundle of S as a pullback of the "virtual vector bundle" $\phi_M = \phi_M^+ - \phi_M^-$ over the target space. The bordism group of such triples is denoted by $\Omega_{k-1}(P^{k-1} \times M; \phi_M)$. It is not hard to see that our analysis of the behaviour of u around its singularity leads to a well-defined invariant

$$\omega_k(M) = [S, g, \bar{g}] \in \Omega_{k-1}(P^{k-1} \times M, \phi_M).$$

<u>Theorem A.</u> <u>There are k linearly independant vectorfields on M if and only if $\omega_k(M) = 0$</u>

In the proof (see [3], § 3) the assumption $n > 2k$ is crucial for embedding, destabilization and unlinking arguments The basic idea is as follows Given a zero bordism

$$(\mathcal{S}\ , G = (G_1, G_2) \quad \mathcal{S} \longrightarrow P^{k-1} \times M, \quad \bar{G})$$

of the tripel $(S, g, \bar{g})$ (so that $\partial \mathcal{S} = S$, $G|S = g$ etc), the information contained in G_2 allows us to represent $\mathcal{S}$ as a submanifold of $M \times I$ such that

$$\mathcal{S} \pitchfork M \times \{0\} = S \times \{0\} \quad \text{and} \quad \mathcal{S} \cap M \times \{1\} = \emptyset \ .$$

Furthermore, the data G_1 and $\bar{G}$ make it possible to extend u (defined over $M = M \times \{0\}$) over a tubular neighborhood of $\mathcal{S}$ After some unlinking, we end up with a vector bundle homomorphism

$$u^I \ : \ (M \times I) \times \mathbb{R}^k \longrightarrow \pi_1^* (TM)$$

over all of $M \times I$ whose singularity lies at $\mathcal{S}$. Thus it is monomorphic over $M = M \times \{1\}$ and leads to the desired k linearly independant vector fields.■

We may consider the obstruction $\omega_k(M)$ as a <u>characteristic class</u> lying in a ("normal bordism") group which depends on M If we replace the inclusion g_2 . $S \subset M$ by a classifying map g_2' $S \longrightarrow BO(n)$ of $TM|S$, we obtain in an obvious way the <u>"characteristic number"</u>

$$\omega_k'(M) = [S, (g_1, g_2'), \bar{g}] \in \Omega_{k-1}(P^{k-1} \times BO(n), \phi)$$

lying in a fixed group which is independant of M ; here ϕ denotes the virtual vector bundle $\lambda \otimes \gamma \oplus \mathbb{R} - k\lambda \otimes \gamma$, γ being the (pullback of the) universal n-plane bundle over $BO(n)$

<u>Theorem A'.</u> <u>The invariant $\omega_k'(M)$ depends only on the bordism class of M in B . Reinhart's refined sense. M is bordant in this sense to a manifold M' with k linearly independant vector fields if and only if $\omega_k'(M)$</u> = 0 .

Reinhart [5] defines M and M' to be bordant if there is a bordism B (in the classical sense) from M to M' with a nowhere vanishing tangent vector field on it which points inward along M and outward along M' This bordism relation is only slightly finer than the classical one, the only additional invariant (in the unoriented theory) being the Euler number.

It is clear why this refinement should be relevant here we can extend TM and TM' to an n-plane bundle over all of B (namely the complement of the vectorfield in TB), and hence we can extend our singularity data accordingly. The proof of theorem A' is simpler than that of theorem A, since we don't have to embed or unlink in M x I , but we can attach suitable handles instead.■

The construction of $\omega_k(M)$ can be generalized easily to arbitrary compact manifolds, equipped with a nonsingular k-field u^∂ at the boundary ; our singularity obstruction then measures wheather u^∂ can be extended without singularity over the whole manifold E.g. we can interpret every element in $\pi_{n-1}(V_{n,k})$ as a homotopy class of nonsingular k-fields

$$u^\partial \quad S^{n-1} \times \mathbb{R}^k \longrightarrow S^{n-1} \times \mathbb{R}^n = TD^n|S^{n-1}$$

and hence assign to it the obstruction to extending u^∂ over the whole n-disk D^n. This leads to the "singularity isomorphism" σ below which identifies the homotopy group of the Stiefel manifold $V_{n,k}$ with a normal bordism group.

Here is a survey of the obstructions and groups mentioned so far

$$\begin{array}{ccccc}
 & & \{\text{index}(u) \in\} \; \pi_{n-1}(V_{n,k}) & & \\
 & & \sigma \big\downarrow \cong & & \\
 & & \Omega_{k-1}(P^{k-1} \times D^n;\ \phi_D) & & \\
 & \swarrow^{\Theta_M} & & \searrow^{\Theta} & \\
\Omega_{k-1}(P^{k-1} \times M,\ \phi_M) & & \xrightarrow{(id \times \tau)_*} & & \Omega_{k-1}(P^{k-1} \times BO(n),\ \phi) \\
\Psi & & & & \Psi \\
\omega_k(M) & & & & \omega'_k(M)
\end{array}$$

The arrows are induced in an obvious way by some fixed embedding $D^n \subset M$ and by classifying maps of TD^n and TM .

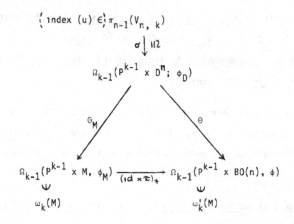

If M allows a k-field u with finite singularity, we may assume that all the singular points have already been isotoped into the interior of D^n (M is connected!). Then u, restricted $\partial D^n = S^{n-1}$, defines an element in $\pi_{n-1}(V_{n,k})$ called the <u>index</u> of u. Moreover, $\omega_k(M)$ lies clearly in the image of Θ_M.

Conversely, if $\omega_k(M)$ lies in this image, and hence $\omega_k(M) = \Theta_M \cdot \sigma [u^\partial]$ for some nonsingular k-field u^∂ over ∂D^n, then the obstruction to extending u^∂ nonsingularly over all of $M - \mathring{D}^n$ vanishes, and therefore M carries a k-field with just one singular point (the center of D^n).

The arrows in the commuting diagram (4) map our invariants into one another. It turns out that these relations between our invariants are extremely useful even if we are only interested in knowing one of them (usually $\omega_k(M)$) thoroughly.

As an illustration, we consider the case $n \equiv 2(4)$ in some detail. The following is just one of many possible applications of our obstruction theory.

<u>Theorem B.</u> <u>Let M be a closed connected smooth manifold of dimension</u> $n \geq 10$, $n \equiv 2(4)$, <u>such that the map</u>

$$H^2(M; \mathbb{Z}_2) \xrightarrow{(w_1(M)\cdot + Sq^1,\ Sq^2)} H^3(M, \mathbb{Z}_2) \oplus \mathrm{Hom}(H_4(M; \mathbb{Z}), \mathbb{Z}_2)$$

<u>is injective.</u>

<u>Then: M has a 4-field with finite singularity if and only if the Stiefel-Whitney classes</u> $w_{n-2}(M) \in H^{n-2}(M, \mathbb{Z}_2)$ <u>and</u> $W_{n-3}(M) \in H^{n-3}(M, \tilde{\mathbb{Z}})$ <u>vanish.</u>

<u>M has four linearly independant vectorfields if and only if</u> $w_{n-2}(M)$, $W_{n-3}(M)$ <u>and the Euler number</u> $\chi(M)$ <u>vanish</u>.

<u>Example</u>. For $q \geq 2$ complex projective space $\mathbb{C}P(2q+1)$ has a 4-field with finite singularities if and only if q is odd (i.e. $\dim \mathbb{C}P(2q+1) \equiv 6(8)$). On the other hand, $\mathbb{C}P(2q+1)$ has not even one nonzero vectorfield, let alone four independant ones.

<u>Corollary</u>. <u>Let M be a closed connected smooth n-manifold,</u> $n \geq 10$, $n \equiv 2(4)$, <u>such that</u> $H^2(M, \mathbb{Z}_2)$ <u>is generated by</u> $w_1(M)^2$, <u>and</u> $w_1(M)^3 \neq 0$.

<u>Then M has four independant vectorfields if and only if</u> $W_{n-3}(M) = 0$

<u>and</u> $\chi(M) = 0$.

Indeed, the assumptions here imply the injectivity required in theorem B; if in addition $\chi(M) = 0$, then also $w_{n-2}(M) = 0$ (since

$$\begin{aligned} w_1(M)^2\, w_{n-2}(M) &= w_2(M)\, w_{n-2}(M) + Sq^2(w_{n-2}(M)) \\ &= w_n(M) \end{aligned} \tag{5}$$

by Wu's formulas).

<u>Proof of theorem B.</u> Let Y stand for D^n, M or $BO(n)$, and let η denote the n-plane bundle TD^n, TM or γ respectively.

First we have to study the obstruction group $\Omega_3(P^3 \times Y; \phi_Y)$. The Thom-Pontryagin construction allows to interpret it as a homotopy group, and presumably one can investigate it by homotopy-theoretic methods. However, it is also possible to employ the singularity approach once more.

Indeed, consider the homomorphism

$$f \ : \ \Omega_3(P^3 \times Y, \phi_Y) \longrightarrow \Omega_3(P^3 \times Y)$$

which forgets about the vector bundle isomorphism $\bar{g}$ (cf. (3)) and retains only the underlying orientation information (i.e $f\,[S, g, \bar{g}] = [S, g]$, where S is oriented by $\Lambda^{n+k}\,\bar{g}$). We can study the kernel and the cokernel of such a forgetful map by trying to construct the necessary vector bundle morphisms and by analyzing the singularities which pop up. In this way we can fit f into an exact sequence where the third term is again a normal bordism group. Iterating this procedure, we get the following natural cross of exact sequences (see [3], theorem 9.3, for more details)

(6)

$$\begin{array}{ccccccc}
 & & 0 & & & & \\
 & & \downarrow & & & & \\
 & & \mathbb{Z}_2 & & & & \\
 & & \downarrow & & & & \\
\Omega_4(P^3 \times Y) & \xrightarrow{\sigma \circ j} & \Omega_2(P^3 \times Y \times BO(2);\ \phi_Y + \Gamma) & \xrightarrow{\delta} & \Omega_3(P^3 \times Y,\ \phi_Y) & \xrightarrow{f} & \Omega_3(P^3 \times Y) \\
 & & \downarrow f' & & & & \cong \\
 & & & & & & \mathbb{Z} \oplus H_2(Y,\mathbb{Z}_2) \oplus H_3(Y;\mathbb{Z}) \\
 & & \left(\ker : H_2(P^3 \times Y, \mathbb{Z}_2) \xrightarrow{w_1(\lambda)^2 + w_1(\lambda) w_1(\eta)} \mathbb{Z}_2\right) \oplus \mathfrak{N}_2 \oplus \mathbb{Z} & & & & \\
 & & \downarrow & & & & \\
 & & 0 & & & &
\end{array}$$

Thus the obstruction group $\Omega_3(P^3 \times Y,\ \phi_Y)$ is built up from two parts: from the image of f on one hand, and from the much more delicate part $\ker f = \operatorname{im} \delta \cong \operatorname{coker} \sigma \circ j$ on the other hand. Fortunately, the homomorphism $f' \circ \sigma \circ j$, which involves singularities, can be computed easily, and so we can get some control over $\operatorname{coker} \sigma \circ j$.

If $Y = M$, then it can be shown that f maps the singularity invariant $\omega_4(M)$ to the tripel of classical invariants

$$(\chi(M),\ PD(w_{n-2}(M)),\ PD(W_{n-3}(M)))$$

where PD denotes suitable Poincaré duality isomorphisms Moreover, the injectivity condition in theorem B guarantees that the first component of $f' \circ \sigma \circ j$ is onto, and hence all elements in the image of δ come already from elements in $\Omega_3(P^3 \times D^n,\ \phi_D)$ via Θ_M(see (4))

Now assume that $w_{n-2}(M)$ and $W_{n-3}(M)$ vanish. Then the mod 2 Euler number $w_n(M)\ [M]$ is also trivial (see the Wu relations (5)), and $f(\omega_4(M))$ is of the form $(2\ s,\ 0,\ 0)$. But this element comes already from $\Omega_3(P^3 \times D^n;\ \phi_D)$ (in fact, it equals $f \circ \Theta_M\ (s\ z)$ where z corresponds to the index of a 4-field with finite singularities on S^n). It follows from the remark above that $\omega_4(M)$ lies in the image of Θ_M, and hence M allows a 4-field with finite singularity.

Assume in addition that $\chi(M) = 0$ It follows from the Wu relations

$$w_1(M)^2\, w_{n-2}(M) = Sq^1\,(w_1(M)\, w_{n-2}(M))$$
$$= w_1(M)\, w_{n-1}(M)$$

that all Stiefel-Whitney numbers of M, containing $w_{n-3}(M)$, $w_{n-2}(M)$, $w_{n-1}(M)$ or $w_n(M)$ as a factor, vanish. Hence, according to Stong ([6],7.2), M is bordant in $\mathfrak{N}_n$ to a manifold M' which fibers over the torus $(S^1)^4$, and therefore has four linearly independant vector fields. Since $\chi(M) = \chi(M') = 0$, M and M' are even bordant in B. Reinhart's refined sense, and we conclude from theorem A' that $\omega_4'(M) = 0$. Thus $\omega_4(M) = \Theta_M(v)$ for some $v \in \ker\Theta$ (see also the diagram (4)). We will see presently that $\ker\Theta = 0$, therefore $\omega_4(M) = 0$, and M allows four independant vector fields, by theorem A.

It is not hard to see that the homomorphism Θ fits into a commuting diagram of the form

$$\begin{array}{ccccc}
\pi_{n-1}(V_{n,5}) & \xrightarrow{proj_*} & \pi_{n-1}(V_{n,4}) & \xrightarrow{\partial} & \pi_{n-2}(S^{n-5}) \\
\sigma\downarrow & & \sigma\downarrow\cong & & \downarrow \\
\Omega_4(P^4 \times D^n, \phi_{D,5}) & \xrightarrow{\Delta} & \Omega_3(P^3 \times D^n, \phi_{D,4}) & \xrightarrow{d} & \Omega_3(D^n, \text{trivial}) \\
 & & \Theta\downarrow & & \downarrow \\
 & & \Omega_3(P^3 \times BO(n), \phi) & \longrightarrow & \Omega_3(BO(n), \text{trivial}) \\
 & & & & \downarrow \\
 & & & & \Omega_3(\text{point}, \text{trivial}) .
\end{array}$$

The top line is part of the exact homotopy sequence of the fibration $proj : V_{n,5} \to V_{n,4}$, and the two lower vertical arrows to the right compose to give an isomorphism Thus

$$\ker\Theta \subset \ker d = \sigma\,(proj_*(\pi_{n-1}(V_{n,5}))) .$$

It follows from the tables of Paechter [4] that the group to the right is infinite cyclic Since it contains the element z (correponding to the index of a 4-field on S^n) and since $f(\Theta(z)) = (2, o, o) \in \mathbb{Z} \oplus H_2(BO(n), \mathbb{Z}_2) \oplus H_3(BO(n), \mathbb{Z})$ (see the discussion of (6)), we conclude that, indeed, $\ker \Theta$ vanishes. ■

In the last proof, we first showed that $\omega_4'(M) = 0$ (solving the problem only up to bordism) and deduced then that $\omega_4(M) = 0$ While this approach is often the only way to calculate certain parts of $\omega_k(M)$, the following result present an alternate method in the particular case of theorem B

Proposition C. Assume $n > 2k+1$ If an element v of $\pi_{n-1}(V_{n,k})$ occurs as the index of a k-field with finite singularities on some closed connected smooth n-manifold M, then v lies in the image of the obvious homomorphism $\mathrm{proj}_*\ \pi_{n-1}(V_{n,k+1}) \longrightarrow \pi_{n-1}(V_{n,k})$

This was also noted by M Crabb

Proof Consider the exact homotopy sequence

$$\cdots \longrightarrow \pi_{n-1}(V_{n,k+1}) \xrightarrow{\mathrm{proj}_*} \pi_{n-1}(V_{n,k}) \xrightarrow{\partial} \pi_{n-2}(S^{n-k-1}) \longrightarrow \cdots .$$

If we identify $\pi_{n-2}(S^{n-k-1})$ with the framed bordism group Ω_{k-1}(point;trivial) via the Thom-Pontrjagin construction, we can describe $\partial(v)$ as follows. Given u' . $S^{n-1} \times \mathbb{R}^k \hookrightarrow TD^n|S^{n-1}$ representing v, let ξ be a complement of the image of u' and let $Z \subset S^{n-1}$ be the zero set of a generic section of ξ. Then Z, together with the stable framing given by

$$TZ \oplus \xi|Z \oplus \underline{\mathbb{R}} = TZ \oplus \nu(Z,S^{n-1}) \oplus \underline{\mathbb{R}} = TD^n|Z \cong \underline{\mathbb{R}}^k \oplus \xi|Z,$$

represents $\partial(v)$.

If v occurs as the index of a k-field u with finite singularity (which we may assume to lie in $D^n \subset M$), then $u|M-\mathring{D}^n$ provides a zero bordism for the whole situation above, and clearly $v \in \ker \partial = \mathrm{proj}_*(\pi_{n-1}(V_{n,k+1}))$. ■

Theorem D. Assume that $1 \le k \le 6$, $n > 2k$ and $n \equiv 2(4)$. Let M be a closed connected smooth n-manifold allowing a k-field with finite singularity.

Then M has k linearly independant vectorfields if and only if the Euler number $\chi(M)$ vanishes.

Proof. For k = 1 this is a special case of the Poincaré-Hopf theorem. If k=2, our claim follows from work of E. Thomas (see table 1 in [7]) when M is orientable, and it can be very easily deduced by singularity techniques of M is not orientable.

So assume that $3 \leq k \leq 6$. Then it follows from Paechter's tables [4] that $\mathrm{proj}_*(\pi_{n-1}(V_{n,k+1}))$ is infinite cyclic and generated by the index z of a suitable k-field on S^n (e.g. for k = 3 we can deduce this from the following exact sequence

$$\begin{array}{ccccccc} \pi_{n-1}(V_{n,4}) & \xrightarrow{\mathrm{proj}_*} & \pi_{n-1}(V_{n,3}) & \to & \pi_{n-2}(S^{n-4}) & \to & \pi_{n-2}(V_{n,4}) \\ \wr\wr & & \wr\wr & & \wr\wr & & \| \\ \mathbb{Z}_{12} \oplus \mathbb{Z} z & & \mathbb{Z} z \oplus \mathbb{Z}_2 & & \mathbb{Z}_2 & & 0 \end{array}$$
).

This, and proposition C imply that the index y of a k-field with finite singularity on M was the form $y = rz$ for some $r \in \mathbb{Z}$. If we apply the obvious homorphism $pr_*:\pi_{n-1}(V_{n,k}) \to \pi_{n-1}(S^{n-1}) = \mathbb{Z}$, we get indices of 1-fields. Therefore

$$\chi(M) = pr_*(y) = r\,\chi(S^n) = 2r$$

vanishes if and only the index y does, i.e., if and only if M carries a k-field without any singularity.■

Since $\pi_{n-2}(V_{n,3}) = 0$ for $n \equiv 2(4)$ (see [4]), the classical secondary obstruction vanishes, and we conclude that M allows a 3-field with finite singularity if and only if $w_{n-2}(M) = 0$.

Corollary E. Let M be a closed connected smooth n-manifold, $n \equiv 2(4)$, $n > 6$.

Then M has three linearly independant vectorfields if and only if $w_{n-2}(M)$ *and* $\chi(M)$ *vanish.*

For orientable M this was previously proved by Atiyah and Dupont [1].

An alternate proof, based entirely on the singularity method, can be found in [3], § 14

References

[1] M. Atiyah and J. Dupont, Vectorfields with finite singularities, Acta Math 128 (1972), 1-40.

[2] U. Koschorke, Framefields and nondegenerate singularities, Bull. AMS 81 (1975), 157-160.

[3] ______________, A singularity approach to vector fields and other vector bundle morphisms, to appear as a volume in the Springer Lecture Notes series.

[4] G. F. Paechter, The groups $\pi_r(V_{n,m})$, Quart. J Math. Oxford (2) 7 (1956), 249-268.

[5] B. Reinhart, Cobordism and the Euler number, Topology 2 (1963), 173-177.

[6] R. Stong, On fibering of cobordism classes, Trans AMS 178 (1973), 431-447.

[7] E. Thomas, Vectorfields on manifolds, Bull. AMS 75 (1969), 643-683.

Simplices of maximal volume in hyperbolic space, Gromov's norm, and Gromov's proof of Mostow's rigidity theorem (following Thurston).

Hans J. Munkholm

Odense University

§0 Introduction

In my lecture at the conference I gave a relatively detailed proof of the following theorem, which represents joint work with U. Haagerup, and which had been conjectured by Milnor, [2].

Theorem 1 In hyperbolic n-space H^n a geodesic n-simplex σ is of maximal volume if and only if σ is ideal and regular.

Here ideal means that all vertices are on "the sphere at infinity" S_∞^{n-1} . And regular means that all faces of σ are congruent modulo the isometries of H^n .

I also outlined, very briefly, how this result can be used in a proof of Mostow's rigidity theorem for hyperbolic manifolds.

Theorem 2 (Mostow) Any homotopy equivalence $f:M\to N$ between closed, orientable, hyperbolic n-manifolds with $n\geq 3$ is homotopic to an isometry.

The proof that I refer to was given by Thurston (who attributed it to Gromov) in his 1977/78 Princeton University lecture notes, [4]. Thurston considered only the case $n=3$ because the validity of theorem 1 was unknown for $n>3$.

Since the lecture notes are not easily accessible, and since, at the conference, there was considerable interest in some of the details of Gromov's proof (especially what is below called step 3) I have decided to write down a rather detailed exposition of Gromov's argument. The proof of theorem 1 will then appear elsewhere.

It follows that I claim absolutely no originality concerning the material in this note. It is nothing but my interpretation and expansion of one of Thurston's lectures.

§1 Outline of Gromov's proof

In this section we outline Gromov's proof of Mostow's theorem. Details are given in later sections. Thus let a homotopy equivalence $f:M\to N$ be given. It fits into the commutative diagram

$$\begin{array}{ccc} H^n & \xrightarrow{\ \tilde{f}\ } & H^n \\ \downarrow{\scriptstyle p} & & \downarrow{\scriptstyle p} \\ M=\Gamma\backslash H^n & \xrightarrow[\ f\ ]{} & \Theta\backslash H^n=N \end{array}$$

where p denotes universal covering maps. Also $\tilde{f}$ is φ-equivariant where $\varphi:\Gamma\to\Theta$ is the isomorphism of fundamental groups induced by f .

Step 1 $\tilde{f}:H^n \to H^n$ is a pseudo isometry, i.e. there are constants a,b such that

$$a^{-1}d(x,y)-b \le d(\tilde{f}(x),\tilde{f}(y)) \le ad(x,y)$$

for all $x,y \in H^n$.

Step 2 Any pseudo isometry g of H^n gives rise to a continuous map $g_+:S_\infty^{n-1} \to S_\infty^{n-1}$ on the sphere at infinity. This association is such that $\tilde{f}_+$ is still φ-equivariant.

Step 3 If $v_o,v_1,\ldots,v_n \in S_\infty^{n-1}$ span a geodesic n-simplex of maximal volume then so do $\tilde{f}_+(v_o),\tilde{f}_+(v_1),\ldots,\tilde{f}_+(v_n)$.

Step 4 $\tilde{f}_+=h_+$ for some isometry $h:H^n \to H^n$.

Let us see how this finishes the proof. It is well known that an isometry h of H^n is completely determined by $h_+:S_\infty^{n-1} \to S_\infty^{n-1}$. Therefore, the above $h:H^n \to H^n$ is φ-equivariant. And the map $\bar{h}:\Gamma\backslash H^n \to \Theta\backslash H^n$ that it covers is the desired isometry. $\bar{h}$ is homotopic to f because it induces φ on fundamental group level, at least up to conjugacy.

§2 Proof of step 1

We may assume that f is simplicial w.r.t. triangulations of M and N. Then f satisfies a Lipschitz condition. Hence so does $\tilde{f}$, i.e.

(2.1) $$d(\tilde{f}(x),\tilde{f}(y)) \le ad(x,y) .$$

We may also choose a homotopy inverse f_1 covered by an $\tilde{f}_1$ which satisfies (increase a, if need be)

(2.2) $d(\tilde{f}_1(x),\tilde{f}_1(y)) \leq ad(x,y)$

(2.3) $\tilde{f}_1\tilde{f}$ is Γ-equivariantly homotopic to 1_{H^n} .

On a compact set the homotopy involved in (2.3) moves any x only a bounded distance. By equivariance, and compactness of M , the same holds on all of H^n , i.e. for some b_1

(2.4) $d(\tilde{f}_1\tilde{f}(x),x) \leq b_1$.

Now one has

$$d(x,y) \leq 2b_1 + d(\tilde{f}_1\tilde{f}(x),\tilde{f}_1\tilde{f}(y)) \leq 2b_1 + ad(\tilde{f}(x),\tilde{f}(y))$$

which implies, with $b=2b_1/a$

(2.5) $d(\tilde{f}(x),\tilde{f}(y)) \geq a^{-1}d(x,y)-b$

Q.E.D.

§3 Proof of step 2

The main ingredient is the following proposition which states that pseudo isometries "almost preserve" geodesics and "almost preserve" normal geodesic hyperplanes. If γ is a geodesic in H^n we let P_γ denote the orthogonal projection onto γ .

Proposition 3.1 If $g:H^n \to H^n$ is a pseudo isometry then there exists a constant r so that

(i) Any geodesic γ has $g(\gamma)$ contained in a tubular neighbourhood $N_r(\bar{\gamma})$ of radius r around a unique geodesic $\bar{\gamma}$.

(ii) For any geodesic γ and any geodesic hyperplane Q orthogonal to γ the image $P_{\bar{\gamma}}(g(Q))$ is a segment of length $\leq r$.

Before we outline a proof let us apply the proposition. If we call two geodesics equivalent when $d(\gamma_1(t),\gamma_2(t))$ is bounded for $t\to\infty$ (t a natural parameter) then S_∞^{n-1} is the set of equivalence classes of geodesics (as a set). It is easily seen that $\gamma\to\bar{\gamma}$ respects the equivalence relation. Hence $\gamma\to\bar{\gamma}$ induces a function $g_+:S_\infty^{n-1}\to S_\infty^{n-1}$. Note that when g is an isometry we may take $r=0$ and we recover the usual extension of isometries over S_∞^{n-1} . Also if g is equivariant w.r.t. to some $\varphi:\Gamma\to\Theta$ (where Γ and Θ are isometry groups) then so is g_+ .

To check continuity of g_+ at $z\in S_\infty^{n-1}$ we argue in the upper half space model. Then $S_\infty^{n-1}=R^{n-1}\cup\infty$ and we arrange coordinates so that both z and $g_+(z)$ are $\neq\infty$. Let z be determined by γ which passes through ∞ . We may arrange coordinates so that $\infty\in\bar{\gamma}$. Now any neighbourhood of $g_+(z)$ contains a disc D which is "boundary" for a geodesic hyperplane Q orthogonal to $\bar{\gamma}$.

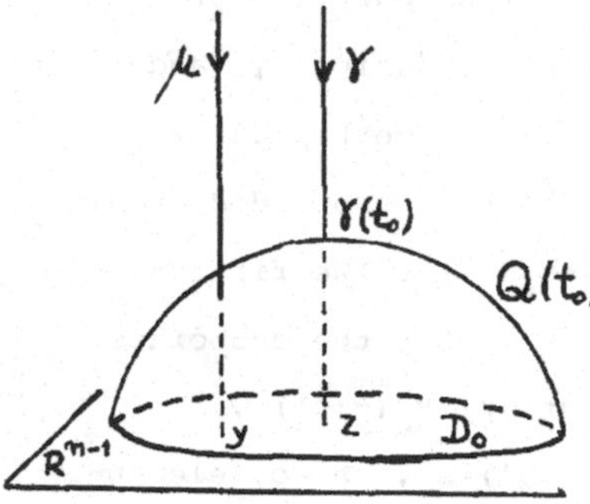

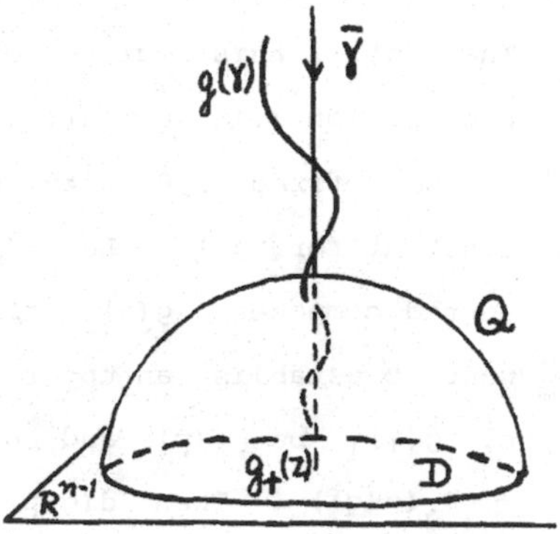

Let $H^+(Q)$ and $H^-(Q)$ be the half spaces determined by Q. One easily checks that

$$d(P_{\bar{\gamma}}(g\gamma(t)),H^-(Q))\to\infty$$

as $t\to\infty$. Hence for suitable t_o

(3.1) $$d(P_{\bar{\gamma}}(g\gamma(t)),H^-(Q))>2r\ , \quad \text{for} \quad t\geq t_o\ .$$

If $Q(t)$ is the geodesic hyperplane orthogonal to γ through $\gamma(t)$ then (ii) and (3.1) imply that

(3.2) $$d(P_{\bar{\gamma}}(g(Q(t))),H^-(Q))>r\ , \quad \text{for} \quad t\geq t_o\ .$$

Let D_o be the disc in R^{n-1} "bounding" $Q(t_o)$. We finish the proof of continuity by showing that $g_+(D_o)\subseteq D$. In fact let $y\in D_o$ be determined by μ. If $g_+(y)\notin D$ then $\bar{\mu}(t)$, and hence $P_{\bar{\gamma}}(\bar{\mu}(t))$, must be in $H^-(Q)$ for all $t\geq$ some t_1. Since $P_{\bar{\gamma}}$ decreases distances it follows that

$$d(P_{\bar{\gamma}}(g(\mu(t))),H^-(Q))\leq r$$

for arbitrarily large values of t. But this contradicts (3.2).

The rest of this section contains a proof of part (i) of proposition 3.1. We start by considering geodesics γ and ρ and a fixed $s>0$ with $\cosh(s)>a^2$ (a = the Lipschitz constant for g). Let ℓ be the length of a bounded, connected component $g(\gamma)_1$ of $g(\gamma)\cap(H^n-N_s(\rho))$. We first want to establish an upper bound for ℓ. Let the endpoints of $g(\gamma)_1$ be $g(p)$ and $g(q)$ and put $p'=P_\rho(g(p))$, $q'=P_\rho(g(q))$. Then $d(g(p),p')=d(g(q),q')=s$. Also, elementary hyperbolic gemoetry shows that $P_\gamma|H^n-N_s(\rho)$ decreases lengths by a factor $\leq \cosh(s)^{-1}$. Therefore

$$a^{-1}d(p,q)-b\leq d(g(p),g(q))$$
$$\leq 2s+\ell\cosh(s)^{-1}$$
$$\leq 2s+ a\cosh(s)^{-1}d(p,q) .$$

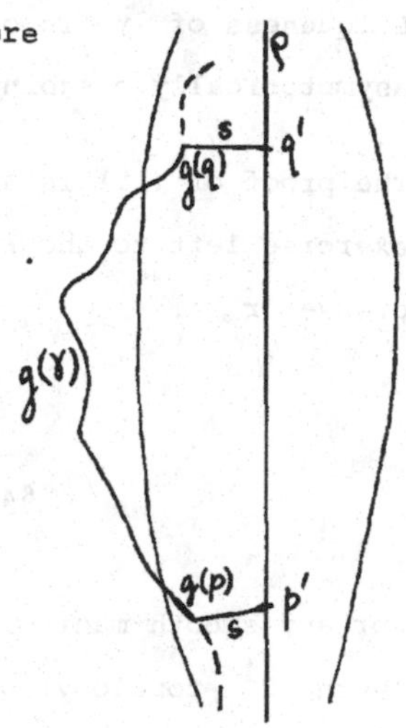

It follows that

$$d(p,q)\leq k=\frac{(2s+b)\ a\cosh(s)}{\cosh(s)-a^2}$$

and, by the Lipschitz condition,

$$\ell\leq ak .$$

Now take $r=s+ak$. We then have

(3.3) If $g\gamma(p)$ and $g\gamma(q)$ lie on ρ then $g\gamma[p,q]\subseteq N_r(\rho)$.

In fact, if $g(\gamma(t))$ leaves $N_s(\rho)$ for some $t\in[p,q]$ then it must return to $N_s(\rho)$ before arc length has grown by ak , so it cannot leave $N_r(\rho)$.

For fixed γ we now let ρ_n be the geodesic through $g(\gamma(0))$ and $g(\gamma(n))$. Since $d(g(\gamma(n)),g(\gamma(0)))\to\infty$ as $n\to\infty$ (3.3) implies that the angle $v_{n,m}$ between ρ_n and ρ_{n+m} at $g(\gamma(0))$ goes to zero as $n\to\infty$ (any $m>0$), see the figure .

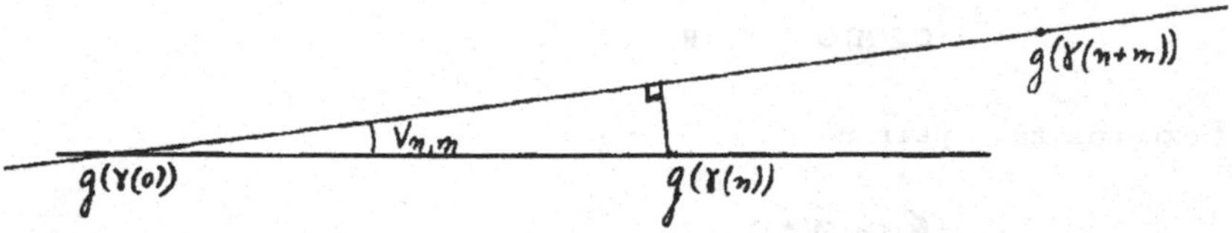

Hence ρ_n has a limit geodesic $\bar{\gamma}$ as $n\to\infty$. And one may show that $g(\gamma)\subseteq N_r(\bar{\gamma})$.

Uniqueness of $\overline{\gamma}$ is clear since $N_r(\overline{\gamma})$ and $N_r(\overline{\overline{\gamma}})$ are asymptotically disjoint in at least one end, if $\overline{\gamma} \neq \overline{\overline{\gamma}}$.

The proof of (ii) is another relatively simple geometric exercise left to the reader (one may of course have to increase r) .

§4 Gromov's norm

For any smooth manifold M let $C^1(\Delta(k),M)$ be the space (with C^1 topology) of C^1 maps $\sigma:\Delta(k)\to M$ of the standard k-simplex $\Delta(k)$ into M . Let $\mathcal{C}_k(M)$ be the real vector space of compactly supported Borel measures μ of bounded total variation $\|\mu\|$, on the space $C^1(\Delta(k),M)$. The various face inclusions $\eta_i:\Delta(k-1)\to\Delta(k)$ induce maps $\eta_i{}^*:C^1(\Delta(k),M)\to C^1(\Delta(k-1),M)$ and homomorphisms $\partial_i=(\eta_i{}^*)_*:\mathcal{C}_k(M)\to\mathcal{C}_{k-1}(M)$. $\partial=\Sigma(-1)^i\partial_i:\mathcal{C}_k(M)\to\mathcal{C}_{k-1}(M)$ makes $\mathcal{C}_*(M)$ into a chain complex.

If $C_*(M)$ is the real, singular chain complex, based on $C^1(\Delta(k),M)$, then there is an obvious natural transformation $\iota:C_*(M)\to\mathcal{C}_*(M)$. <u>On homology ι induces an isomorphism</u>. Moreover, if $\Lambda^*(M)$ is the deRham cochain complex then the usual pairing

$$<,>:C_*(M)\otimes\Lambda^*(M)\to R$$

extends to a pairing

$$<,>:\mathcal{C}_*(M)\otimes\Lambda^*(M)\to R$$

defined by

$$\langle\mu,\omega\rangle=\int_{\sigma\in C^1(\Delta(k),M)}\left(\int_{\Delta(k)}\sigma^*(\omega)\right)d\mu$$

Now let M be a closed, oriented, hyperbolic n-manifold with volume form Ω_M. If $\mu\in\mathcal{C}_n(M)$ is a cycle then μ represents $\langle\mu,\Omega_M\rangle V(M)^{-1}[M]$, where $V(M)$ is the volume of M and $[M]$ the orientation class.

Definition 4.1 For a closed, oriented n-manifold M one defines Gromov's norm to be

$$\|M\| = \inf\{\|\mu\| \mid \mu \text{ a cycle representing } [M]\} .$$

Theorem 4.2 (Gromov) For any closed, oriented, hyperbolic n-manifold M one has

$$\|M\| = V(M)/V_n$$

(V_n = maximal volume of a geodesic n-simplex in H^n) .

Proof We include a proof because it is very nice and because it is used in the next section.

If $\sigma\in C^1(\Delta(k),H^n)$ we have another simplex $s(\sigma)\in C^1(\Delta(k),H^n)$ which is affine and has the same vertices as σ. Obviously this defines a continuous map

$$s:C^1(\Delta(k),H^n)\to C^1(\Delta(k),H^n)$$

which is homotopic to the identity. Represent M as $\Gamma\backslash H^n$ with universal covering projection $p:H^n\to M$. One easily checks that there is a unique map $\bar{s}:C^1(\Delta(k),M)\to C^1(\Delta(k),M)$, homotopic to the identity, which has $p_*s=\bar{s}p_*$. The induced chain map

$$S_M=\bar{s}_*:\mathcal{C}_*(M)\to\mathcal{C}_*(M)$$

is chain homotopic to the identity, and of course

$$p_*S_{H^n}=S_M p_*:\mathcal{C}_*(H^n)\to\mathcal{C}_*(M) \quad .$$

Now let μ be a cycle representing $[M]$. Then so does $S_M(\mu)$ so, if $\tilde{\sigma}\in C^1(\Delta(n),H^n)$ lifts σ ,

$$\begin{aligned} V(M) &= \langle S_M(\mu),\Omega_M\rangle \\ &= \int_{\tau\in C^1(\Delta(n),M)}\left(\int_{\Delta(n)}\tau^*(\Omega_M)\right)d(\bar{s}_*\mu) \\ &= \int_{\sigma\in C^1(\Delta(n),M)}\left(\int_{\Delta(n)}\bar{s}(\sigma)^*(\Omega_M)\right)d\mu \\ &= \int_{\sigma\in C^1(\Delta(n),M)}\left(\int_{\Delta(n)}s(\tilde{\sigma})^*p^*(\Omega_M)\right)d\mu \\ &= \int_{\sigma\in C^1(\Delta(n),M)}\left(\int_{\Delta(n)}s(\tilde{\sigma})^*(\Omega_{H^n})\right)d\mu \quad . \end{aligned}$$

Since $s(\tilde{\sigma})$ is affine one has

$$\begin{aligned} \left|\int_{\Delta(n)}s(\tilde{\sigma})^*(\Omega_{H^n})\right| &= \int_{s(\tilde{\sigma})(\Delta(n))}\Omega_{H^n} \\ &= V(s(\tilde{\sigma})(\Delta(n))) \\ &\le V_n \quad . \end{aligned}$$

Hence

$$V(M)\le\int V_n\, d|\mu| = V_n\|\mu\|$$

and we have proved that

$$\|M\| \ge V(M)/V_n \quad .$$

To prove the opposite inequality we need an explicit construction of a cycle representing $[M]$ and of total variation close to $V(M)/V_n$. It proceeds as follows. We have a map of principal K bundles, where K is a maximal compact subgroup of the orientation preserving isometry group $I_+(H^n)$

$$\begin{array}{ccc} K & = & K \\ \downarrow & & \downarrow \\ I_+(H^n) & \longrightarrow & \Gamma\backslash I_+(H^n)=D(M) \\ \downarrow & & \downarrow \\ H^n & \underset{p}{\longrightarrow} & \Gamma\backslash H^n=M \end{array}$$

and the horizontal maps are principal Γ bundles. As a topological space $I_+(H^n)=K\times H^n$ and the Haar measure h_o on $I_+(H^n)$ is the product of the one on K and the volume form Ω_{H^n}. Since h_o is left invariant and $I_+(H^n)\to D(M)$ is a locally trivial Γ-bundle, there is a unique measure h_M on $D(M)$ such that $I_+(H^n)\to D(M)$ is locally measure preserving. Since, locally, h_M is the product of the Haar measure on K and the volume form Ω_M one has

(4.1) $\qquad h_M(D(M))=V(M)$.

One now defines a function

$$\alpha: C^1(\Delta(k),H^n)\to\mathcal{C}_k(M)$$

as follows. Given $\sigma:\Delta(k)\to H^n$ there is a continuous map

$$\varphi_\sigma: D(M)\to C^1(\Delta(k),M)$$

given by

$$\varphi_\sigma(\Gamma g)=pg\sigma \ , \quad g\in I_+(H^n) \ .$$

We let

$$\alpha(\sigma)=\varphi_{\sigma*}(h_M)\in\mathcal{C}_k(M) .$$

It is then easy to check the following properties.

<u>Lemma 4.2</u>

(i) $\alpha(\sigma)=\alpha(g\sigma)$, all $g\in I_+(H^n)$

(ii) $\alpha\left(\sigma^{(i)}\right)=\partial_i\alpha(\sigma)$, $\sigma^{(i)}=i^{th}$ face of σ

(iii) $\|\alpha(\sigma)\|=V(M)$ if $\sigma\in C^1(\Delta(n),H^n)$

(iv) If $\sigma\in C^1(\Delta(n),H^n)$ then $\langle\alpha(\sigma),\Omega_M\rangle=V(\sigma)V(M)$ where $V(\sigma)=\int_{\Delta(n)}\sigma^*(\Omega_{H^n})$.

In fact (ii) is purely formal, (iii) a restatement of (4.1), (i) a consequence of the right invariance of h_M under $I_+(H^n)$, and (iv) is seen by the following computation

$$\begin{aligned}
&\langle\alpha(\sigma),\Omega_M\rangle = \\
&=\int_{\tau\in C^1(\Delta(n),M)}\left(\int_{\Delta(n)}\tau^*(\Omega_M)\right)d(\varphi_{\sigma*}(h_M)) \\
&=\int_{\Gamma g\in D(M)}\left(\int_{\Delta(n)}\varphi_\sigma(\Gamma g)^*(\Omega_M)\right)dh_M \\
&=\int_{\Gamma g\in D(M)}\left(\int_{\Delta(n)}\sigma^*g^*p^*(\Omega_M)\right)dh_M \\
&=\int_{\Gamma g\in D(M)}\left(\int_{\Delta(n)}\sigma^*(\Omega_{H^n})\right)dh_M \\
&=\left(\int_{\Delta(n)}\sigma^*(\Omega_{H^n})\right)V(M) .
\end{aligned}$$

Note that when $\sigma:\Delta(n)\to H^n$ is affine with image set $\overline{\sigma}\subseteq H^n$ then $V(\sigma)=\pm V(\overline{\sigma})$ where the sign depends on the orientation character of σ .

For any affine $\sigma \in C^1(\Delta(n),M)$ let $\zeta(\sigma)=\alpha(\sigma)-\alpha(\sigma_-)\in \mathscr{C}_n(M)$ where $\sigma_- = \sigma$ followed by a reflection in one of $\bar{\sigma}$'s faces. Properties (i) and (ii) above immediately imply that $\zeta(\sigma)$ is a cycle ($\sigma^{(i)}$ and $\sigma_-^{(i)}$ are congruent modulo $I_+(H^n)$ even though σ and σ_- are not). Also $\|\zeta(\sigma)\| = \|\alpha(\sigma)\| + \|\alpha(\sigma_-)\| = 2V(M)$ by (iii) (and because $\alpha(\sigma)$, $\alpha(\sigma_-)$ are disjointly supported). And, because of (iv), $\zeta(\sigma)$ represents $2V(\sigma)[M]$. It follows that $\|M\| \le V(M)/V(\bar{\sigma})$, and since $V(\bar{\sigma})$ can be chosen arbitrarily close to V_n this implies that $\|M\| \le V(M)/V_n$.

Q.E.D.

§5 Proof of step 3

Assume that $v_0,v_1,\dots,v_n \in S^{n-1}_\infty$ span a geodesic simplex of maximal volume, but $\tilde{f}_+(v_0),\dots,\tilde{f}_+(v_n)$ do not. We may find neighbourhoods U_i of v_i , <u>in H^n</u> , and an $\varepsilon>0$ so that

(5.1) If $v_i \in U_i$ and σ is the geodesic simplex spanned by $v_0,\dots,v_n$ then $V(s\tilde{f}_+(\sigma)) \le V_n-\varepsilon$.

Here s is the "straightening" map introduced in section 4. Note that (5.1) deals only with geodesic simplices in H^n, no ideal vertices are involved any more.

For smaller neighbourhoods $V_i (\subseteq U_i \subseteq H^n)$ of v_i consider the condition

(5.2) $v_i \in V_i$, $i=0,1,\dots,n$
$\Downarrow$
$g(v_i) \in U_i$.

It is easily seen that V_i may be chosen so that

$$D_1(M)=\{\Gamma g\in D(M) \mid g \text{ satisfies } (5.2)\}$$

has measure

(5.3) $\quad h_M(D_1(M))=h_1>0$.

Now choose a positively oriented affine simplex $\sigma_o \in C^1(\Delta(n),M)$ with vertices in the neighbourhoods V_i and with

(5.4) $\quad V(\bar{\sigma}_o)>V_n-\delta$.

By (5.1) and the definition of $D_1(M)$ one has

(5.5) If $\Gamma g\in D_1(M)$ then
$V(s\tilde{f}_+(g\sigma_o))\leq V_n-\varepsilon\leq V(\sigma_o)+\delta-\varepsilon$.

Also

(5.6) If $\Gamma g\notin D_1(M)$ then
$V(s\tilde{f}_+(g\sigma_o))\leq V_n\leq V(\sigma_o)+\delta$.

We go on to compute which multiple of $[N]$ is represented by $S_N f_*\zeta(\sigma_o)$. We have

$$\langle S_N f_*\alpha(\sigma_o),\Omega_N\rangle=$$
$$=\int_{\tau\in C^1(\Delta(n),N)}\left(\int_{\Delta(n)}\tau^*(\Omega_N)\right)d(\bar{s}_*f_*\varphi_{\sigma_o*}(h_M))$$
$$=\int_{\rho\in C^1(\Delta(n),M)}\left(\int_{\Delta(n)}(\bar{s}(f\rho))^*(\Omega_N)\right)d(\varphi_{\sigma_o*}(h_M))$$
$$=\int_{\Gamma g\in D(M)}\left(\int_{\Delta(n)}(\bar{s}(fpg\sigma_o))^*(\Omega_N)\right)dh_M$$
$$=\int_{\Gamma g\in D(M)}\left(\int_{\Delta(n)}(\bar{s}(p\tilde{f}g\sigma_o))^*(\Omega_N)\right)dh_M$$
$$=\int_{\Gamma g\in D(M)}\left(\int_{\Delta(n)}(s(\tilde{f}g\sigma_o))^*p^*(\Omega_N)\right)dh_M$$
$$=\int_{\Gamma g\in D(M)}\left(\int_{\Delta(n)}(s(\tilde{f}g\sigma_o))^*(\Omega_{H^n})\right)dh_M$$

$$=\int_{\Gamma g\in D(M)} V(s(\tilde{f}g\sigma_o))\,dh_M$$

$$\leq h_1(V(\sigma_o)+\delta-\varepsilon)+(V(M)-h_1)(V(\sigma_o)+\delta)$$

$$=(V(\sigma_o)+\delta)V(M)-\varepsilon h_1 .$$

Now choose $\delta<\varepsilon h_1/V(M)$. Then one gets

$$<S_N f_* \alpha(\sigma_o),\Omega_N><V(\sigma_o)V(M) .$$

Similarly

$$-<S_N f_* \alpha(\sigma_{o-}),\Omega_N><-V(\sigma_{o-})V(M)=V(\sigma_o)V(M) .$$

Also $V(M)=V(N)$, because $f_*([M])=[N]$ and $\|f_*[M]\| \leq \|M\|$. Hence it follows that $S_N f_*(\zeta(\sigma_o))$ represents a multiple $A[N]$ with $A<2V(\sigma_o)$ and this contradicts the fact that $\zeta(\sigma_o)$ represents $2V(\sigma_o)[M]$.

§6 Proof of step 4

It is in this part that theorem 1 enters the picture. It permits one to translate the result of step 3 into

(6.1) If $v_o,v_1,\ldots,v_n\in S^{n-1}_\infty$ span an ideal, regular, geodesic n-simplex in H^n then so do $\tilde{f}_+(v_o),\ldots,\tilde{f}_+(v_n)$.

The rest of the argument is conveniently illustrated in the Poincaré (unit disc) model.

We may compose $\tilde{f}_+$ with an isometry h to obtain that $h_+\tilde{f}_+$ fixes all vertices of some regular, ideal n-simplex, say ABC.

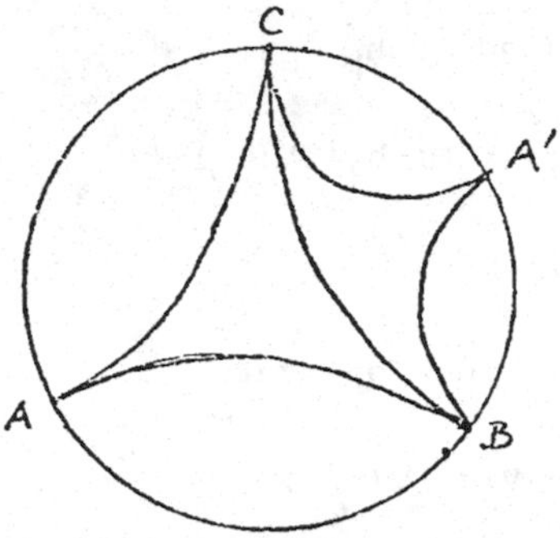

But then it must also fix the reflection of each vertex in the opposite face, such as A' (because, for n>2, there are only two ideal regular n-simplices containing the given face, and $\tilde{f}_+$ is injective). Repeating this procedure ad infinitum we see that $h_+\tilde{f}_+$ fixes a dense subset of S_∞^{n-1}. By continuity $h_+\tilde{f}_+=id$, i.e. the original $\tilde{f}_+=h_+^{-1}$.

References.

1. U. Haagerup and H.J. Munkholm, Simplices of maximal volume in hyperbolic n-space, to appear (available as preprint from Odense University, Denmark).

2. J.W. Milnor, Computation of volume, chapter 7 in [4].

3. G.D. Mostow, Strong rigidity of locally symmetric spaces, Ann.Math.Studies, vol. 78, Princeton University Press, Princeton N.J. 1973.

4. W.P. Thurston, The geometry and topology of 3-manifolds, lecture notes, Princeton University 1977/78.

AN INVARIANT OF PLUMBED HOMOLOGY SPHERES

Walter D. Neumann

If A is a ring, a 3-dimensional A-sphere will mean a closed oriented 3-manifold M^3 with the same A-homology as S^3. An A-homology bordism between two A-spheres M_1^3 and M_2^3 is a compact oriented 4-manifold W^4 with $\partial W = M_1 + (-M_2)$, such that the inclusions $M_i \hookrightarrow W$ induce isomorphisms in A-homology. Here, and in the following, $+$ means disjoint union, $-$ means reversed orientation, and all manifolds will be assumed smooth, compact, and oriented. Diffeomorphisms should preserve orientation.

The set of A-homology bordism classes of 3-dimensional A-spheres forms a group under connected sum, which we denote Θ_3^A. We are most interested in $A = \mathbb{Z}/2$ or $\mathbb{Z}$. Every $\mathbb{Z}$-sphere is a $\mathbb{Z}/2$-sphere and every $\mathbb{Z}/2$-sphere is a rational sphere.

If M^3 is a $\mathbb{Z}/2$-sphere, then its μ-invariant (see for instance [3], where the opposite sign convention is used) can be defined as

$$\mu(M^3) = \operatorname{sign}(W^4) \pmod{16},$$

where W^4 is any simply connected parallelizable compact manifold with $\partial W^4 = M^3$. This invariant is a $\mathbb{Z}/2$-homology bordism invariant and its range of values is summed up in the commutative diagram

$$\begin{array}{ccc} \Theta_3^{\mathbb{Z}} & \xrightarrow{\ \mu\ } & 8\mathbb{Z}/16\mathbb{Z} \\ \downarrow & & \downarrow \\ \Theta_3^{\mathbb{Z}/2} & \xrightarrow{\ \mu\ } & 2\mathbb{Z}/16\mathbb{Z} \end{array} .$$

The groups Θ_3^A are of interest in their own right, but also because of applications. The most important application is the result of Galewski and Stern [2] and Matumoto [5] that for any

---

Research partially supported by the NSF. The hospitality of the Sonderforschungsbereich 40 in Bonn during the preparation of this paper is also gratefully acknowledged.

$n \geq 5$, all closed TOP n-manifolds are simplicially triangulable if and only if $\Theta_3^{\mathbb{Z}}$ contains a subgroup $\mathbb{Z}/2$ on which μ is an isomorphism.

Another application is to the classical knot concordance group C_3. Given a knot (S^3,K) one can do a p/q-Dehn surgery on it to get a homology lens space $M^3_{p/q}(S^3,K) = M^3_{p/q}$ with $H_1(M^3_{p/q};\mathbb{Z}) = \mathbb{Z}/p$. Thus if p is invertible in A then $M^3_{p/q}$ is an A-sphere. The map $C_3 \longrightarrow \Theta_3^A$ given by $(S^3,K) \longmapsto [M^3_{p/q}(S^3,K)] - [L(p,p-q)]$ is well defined, though possibly not a homomorphism. Thus information about Θ_3^A could yield new knowledge of C_3. Other similar constructions can be used to the same purpose.

Our main results are the following time-dependent theorem and the non time-dependent consequences, theorems 6.1 to 6.3, of section 6 below.

THEOREM. There exists an invariant $\bar{\mu}(M^3) \in \mathbb{Z}$ with the following properties:

(i). $\bar{\mu}(M^3)$ is defined for $\mathbb{Z}/2$-spheres M^3 which are plumbed manifolds (= graph manifolds in Waldhausen's sense [10]) and is an oriented diffeomorphism type invariant.

(ii). $\bar{\mu}(M^3)$ (mod 16) $= \mu(M^3)$.

(iii). For all known (May 1979) $\mathbb{Z}/2$-homology bordisms of such $\mathbb{Z}/2$-spheres, $\bar{\mu}$ is invariant.

If $\bar{\mu}$ is in fact a $\mathbb{Z}/2$-homology bordism invariant and if it can be defined for any $\mathbb{Z}/2$-sphere M^3, then of course we have a negative answer to the triangulation problem. If on the other hand $\bar{\mu}$ turns out not to be a $\mathbb{Z}/2$-homology bordism invariant, it will hopefully be due to some new idea in constructing such bordisms which then might lead to an example which solves the triangulation problem positively. Moreover, $\bar{\mu}$ might be invariant under some stronger bordism relation of geometric significance; we discuss this briefly in the last section.

Some consequences of our results are: plumbed $\mathbb{Z}/2$-spheres with non-zero μ-invariant admit no orientation reversing homeomorphisms; the Kummer surface cannot be pasted together from two simply connected plumbed 4-manifolds. The latter (see 6.2) depends on a generalization of $\bar{\mu}$ described in section 4.

2. The invariant.

For the purpose of this paper we define a connected plumbing graph Γ to be a connected graph with no cycles, each of whose vertices, which we label $i = 1,2, \ldots ,s$, carries an integer weight e_i. The oriented 4-manifold $P(\Gamma)$ obtained by plumbing according to Γ is then defined in the usual way, namely, for each vertex i we let E_i be the D^2-bundle over S^2 of euler number e_i and then plumb these together according to the edges of Γ (see for instance [3]). We call $M(\Gamma) = \partial P(\Gamma)$ the oriented 3-manifold obtained by plumbing according to Γ.

More generally, we allow disconnected plumbing graphs Γ whose components are as above by making the convention that if $\Gamma = \Gamma_1 + \Gamma_2$ is the disjoint union of Γ_1 and Γ_2, then $P(\Gamma) = P(\Gamma_1) \natural P(\Gamma_2)$ (boundary connected sum), so $M(\Gamma) = M(\Gamma_1) \# M(\Gamma_2)$ (connected sum). We allow also the empty graph by defining $P(\emptyset) = D^4$ and $M(\emptyset) = S^3$.

It is more customary, when plumbing, to allow graphs with cycles and bundles over surfaces of higher genus, but, since such plumbing could never yield homology spheres, our more restricted plumbing graphs include all we want. They are precisely the plumbing graphs for which $P(\Gamma)$ is simply connected.

If Γ is a plumbing graph in our sense, recall that the matrix

$$A(\Gamma) = (a_{ij})_{i,j=1, \ldots ,s}$$

$$a_{ij} = \begin{cases} e_i & \text{if } i = j \\ 1 & \text{if } i \text{ is connected to } j \text{ by an edge} \\ 0 & \text{otherwise} \end{cases},$$

is the intersection matrix for the natural basis of $H_2(P(\Gamma);\mathbb{Z})$, namely the basis represented by the zero-sections of the plumbed bundles.

THEOREM 2.1. *If Γ is a plumbing graph as above then $M(\Gamma)$ is a $\mathbb{Z}/2$-sphere if and only if $\det A(\Gamma)$ is odd. In this case there exists a unique subset $S \subset \mathrm{Vert}(\Gamma) = \{1, \ldots ,s\}$ such that the following condition holds for any $j = 1, \ldots ,s$:*

(*) $$\sum_{i\in S} a_{ij} \equiv a_{jj} \pmod 2 .$$

<u>The integer</u>

$$\bar{\mu}(M(\Gamma)) = \operatorname{sign} A(\Gamma) - \sum_{j\in S} a_{jj} \quad (\underline{\text{recall}}\ a_{jj} = e_j)$$

<u>then only depends on</u> $M(\Gamma)$ <u>and not on</u> Γ. <u>Its modulo</u> 16 <u>reduction is the usual</u> μ<u>-invariant</u> $\mu(M(\Gamma))$.

<u>Proof</u>. Everything except the oriented diffeomorphism type invariance of $\bar{\mu}(M(\Gamma))$ was proved in [9]. We describe the main ingredient however, since we need it later.

Suppose M^3 is a $\mathbb{Z}/2$-sphere and $M^3 = \partial W^4$ where W^4 is oriented. Recall that an <u>integral Wu class</u> for W^4 is a class $d \in H_2(W;\mathbb{Z})$ such that (dot represents intersection number):

$$d.x \equiv x.x \pmod 2 \quad \text{for all} \quad x \in H_2(W;\mathbb{Z}) .$$

We assume that such a class d exists and moreover that it can be chosen to be <u>spherical</u>, that is, it is representable by a smoothly embedded sphere in W. Then the μ-invariant can be computed as

$$\mu(M^3) = \operatorname{sign}(W) - d.d \pmod{16} .$$

It is easily seen that d does exist if $H_1(W;\mathbb{Z})$ has no even torsion and is unique up to even multiples of elements of $H_2(W;\mathbb{Z})$ if $H_1(W;\mathbb{Z}) = 0$. In particular, if Γ is as in the theorem, then there is a unique subset $S \subset \operatorname{Vert}(\Gamma)$ such that $d = \sum_{j\in S} \alpha_j$ is a Wu class for $P(\Gamma)$, where $\alpha_1, \ldots, \alpha_s$ is the natural basis of $H_2(P(\Gamma);\mathbb{Z})$. The defining property of d translates to condition (*) of the theorem, so this set is the set S of the theorem. Condition (*) implies that no two adjacent vertices of Γ can both be in S. It follows that d is spherical and $d.d = \sum_{j\in S} a_{jj}$, so $\bar{\mu}(M(\Gamma)) = \operatorname{sign} P(\Gamma) - d.d \equiv \mu(M(\Gamma))$ modulo 16.

To see $\bar{\mu}(M(\Gamma))$ only depends on $M(\Gamma)$ we need the following proposition.

PROPOSITION 2.2. *If Γ_1 and Γ_2 are two plumbing graphs as defined above, then $M(\Gamma_1) \cong M(\Gamma_2)$ if and only if Γ_1 can be obtained from Γ_2 by a sequence of the following moves and their inverses.*

1a. *Delete a component of Γ_1 consisting of an isolated vertex with weight ± 1.*

1b. 

1c. 

2. 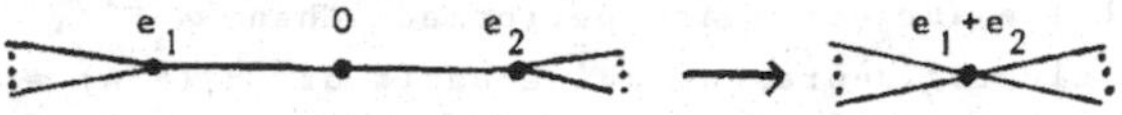

3. 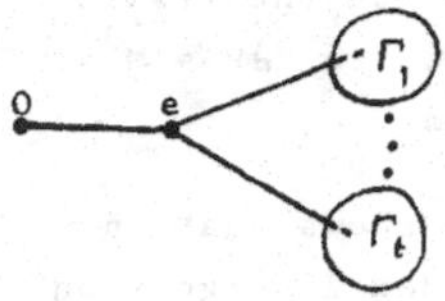

$\longrightarrow \Gamma_1 + \ldots + \Gamma_t$ (*disjoint union*).

Move 1 is called *blowing down*, and its *inverse blowing* up.

This proposition is proved in [8, theorem 3.2] and I understand a forthcoming paper of Bonahon and Siebenmann will also contain a proof.

To complete the proof of theorem 2.1 we need only verify that $\bar{\mu}(M(\Gamma))$ is invariant under the above moves. This is a trivial computation, so we leave it to the reader.

COROLLARY 2.3. *The formulae of [9, theorem 6.2] for the μ-invariants of Seifert manifolds which are $\mathbb{Z}/2$-spheres are actually formulae for $\bar{\mu}$, if one does not reduce modulo 16.*

3. Examples.

A quite general method of generating homology spheres which are homology null-bordant is given by the following simple lemma.

LEMMA3.1. *Let* A *be a principal ideal domain and let* V^4 *be a connected oriented 4-manifold with connected boundary* $\partial V^4 = N^3$, *such that*

$$H_1(V;A) = A^s \quad , \quad H_2(V;A) = H_3(V;A) = 0 \quad .$$

Let M^3 *be an A-sphere obtained by performing* s *index 2 surgeries on* N. *Then* M *represents zero in* Θ_3^A.

Proof. Let $\alpha_i : S^1 \times D^2 \longrightarrow N$, $i = 1, \ldots, s$, be the embeddings on which the surgeries are performed. Then $\alpha_i S^1$, $i = 1, \ldots, s$, clearly represent an A-basis of $H_1(N;A) = H_1(V;A) = A^s$, the first equality here being by the exact homology sequence for the pair (V,N). Let W^4 be the result of adding s 2-handles to V along the α_i, so $\partial W = M^3$. Then W is clearly A-acyclic, proving the lemma.

Applied to $V^4 = S^1 \times D^3$, this lemma shows that any A-sphere M^3 which results by a single index 2 surgery on $S^1 \times S^2$ represents zero in Θ_3^A. Such manifolds are called *Mazur manifolds*.

PROPOSITION 3.2. *Let* Γ_1 *and* Γ_2 *be the following two plumbing graphs:*

$\Gamma_1 =$ 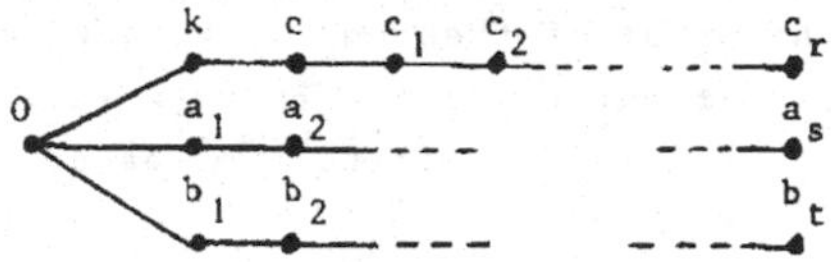

$\Gamma_2 =$ 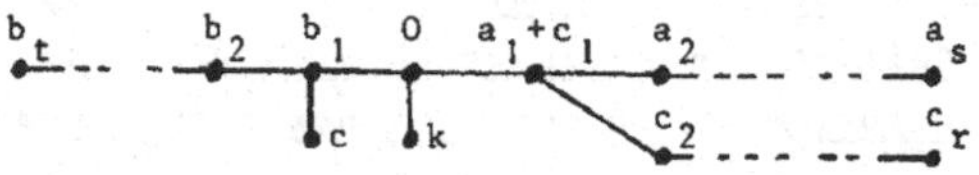

Then $M(\Gamma_1)$ *results by doing a single index 2 surgery on* $M(\Gamma_2)$ *and vice versa. Thus if one of them is* $S^1 \times S^2$, *then the other, if an A-sphere, is a Mazur manifold and represents zero in* Θ_3^A.

PROPOSITION 3.3. If M is a plumbed $\mathbb{Z}/2$-sphere which is shown to be a Mazur manifold by proposition 3.2 then $\bar{\mu}(M) = 0$.

Before giving proofs we make some remarks. Firstly it is not hard to write down many propositions of the form of 3.2. The above was the most productive one I found. Secondly it is a very easy matter to check if a plumbing graph Γ represents $S^1 \times S^2$: it follows from [8] that this is so if and only if Γ can be reduced by the moves of proposition 2.2 (without using the inverse moves) to an isolated vertex with weight 0 . For Seifert manifold plumbing graphs it is even easier.

Recall from [9] that the plumbing graph

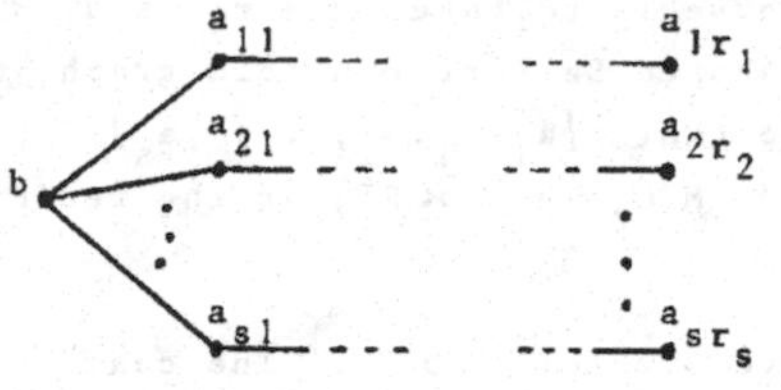

yields the Seifert manifold

$$M(0;(1,-b),(\alpha_1,\beta_1), \ldots ,(\alpha_s,\beta_s))$$

(unnormalized Seifert invariants, as in [6],[9]) with

$$\alpha_i/\beta_i = [a_{i1}, \ldots ,a_{ir_i}] \quad , \quad i = 1, \ldots ,s .$$

We are using the continued fraction notation

$$[x_1, \ldots ,x_r] = x_1 - \cfrac{1}{x_2 - \cfrac{1}{\ddots - \cfrac{1}{x_r}}} ,$$

and we assume $\alpha_i \neq 0$ for all i .

This Seifert manifold is $S^1 \times S^2$ if and only if: $\alpha_i = \pm 1$ for at least $s-2$ of the indices i , and $b - \sum \beta_i/\alpha_i = 0$.

It is thus a simple but tedious matter to list all Seifert manifolds which are shown to be Mazur manifolds by proposition

3.2. One obtains all the examples of Casson and Harer [1] plus the following three parameter family of Seifert manifolds:

$$M(0;(1\,,\,1),(pq-q+1,q),(qr-r+1,r),(rp-p+1,p))\ .$$

Here p,q,r are arbitrary integers whose product is not -1 and $H_1(M;\mathbb{Z})$ has order $(pqr+1)^2$ (by convention a Seifert pair (α,β) with α negative is taken to mean $(-\alpha,-\beta)$).

It is remarkable that, although there are about 100 cases to check, each of which yields a 3-parameter family of examples, they all seem to give one of Casson and Harer's four families or the above family. The calculations are exceedingly tedious however, and I have not checked every case. One way of obtaining the family mentioned above is to take $c = -1$ and $r = 1$, so Γ_2 reduces to a star shaped Seifert manifold graph, and then choose the weights so that $[a_1+c_1,a_2,\ \dots\ ,a_s]$ is the reciprocal of an integer and $M(\Gamma_2) = S^1\times S^2$; as the reader can check, with patience.

We shall just prove 3.2 for $k>0$. The case $k<0$ then follows by reversing orientations and the case $k = 0$ is given by suitably interpreting the proof, though it can be seen much more easily directly.

First note that by blowing up (-1)-vertices directly to the right of the k-weighted vertex one obtains the equivalent graph to Γ_1 (after k such blow-ups)

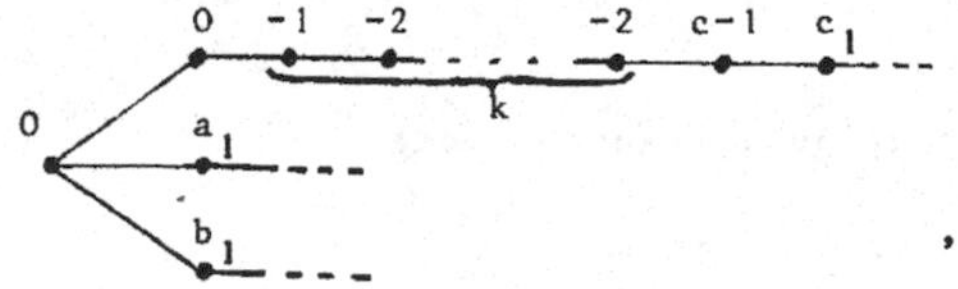

,

which is equivalent, by move 2 of 2.2, to the graph

$\Gamma_1' =$ 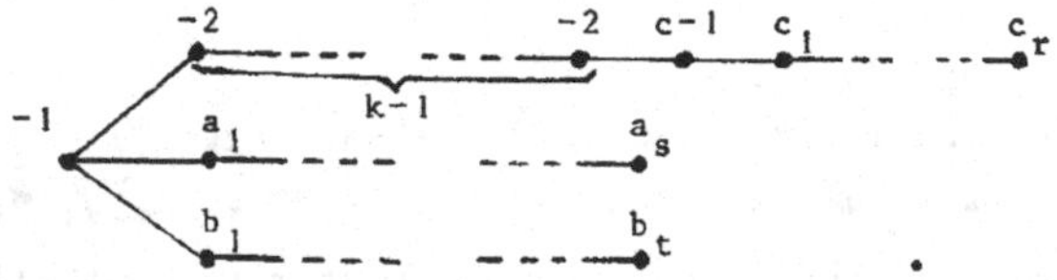

.

This can be represented in Kirby's link calculus [4] by the framed link

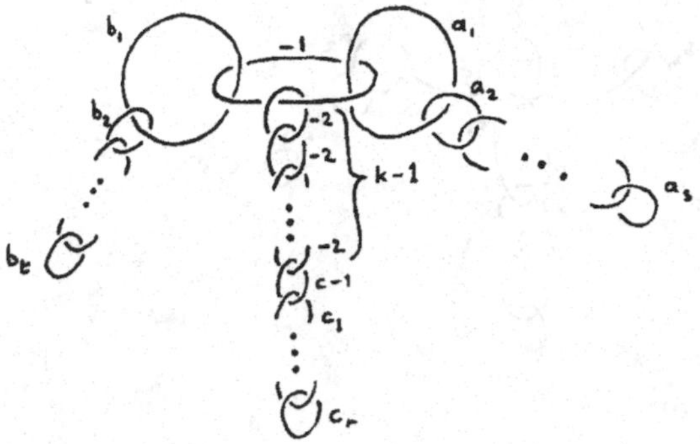

.

By sliding the handles which are linked with the (-1)-framed handle over this handle we obtain

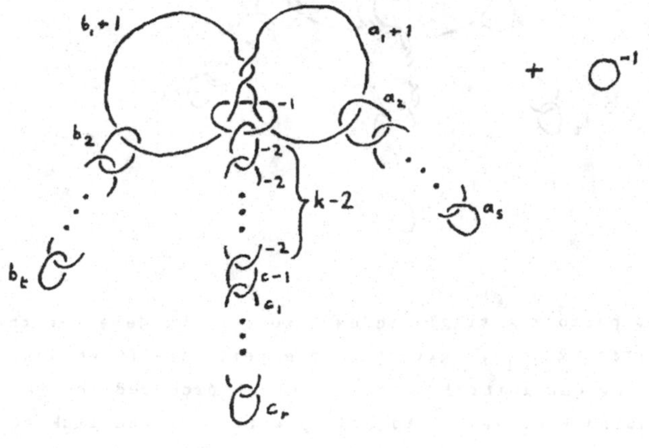

and by iterating this procedure, we eventually obtain the first link on the next page. We then slide the (a_1+k)-framed handle over the c_1-framed handle to obtain the second link on the next page.

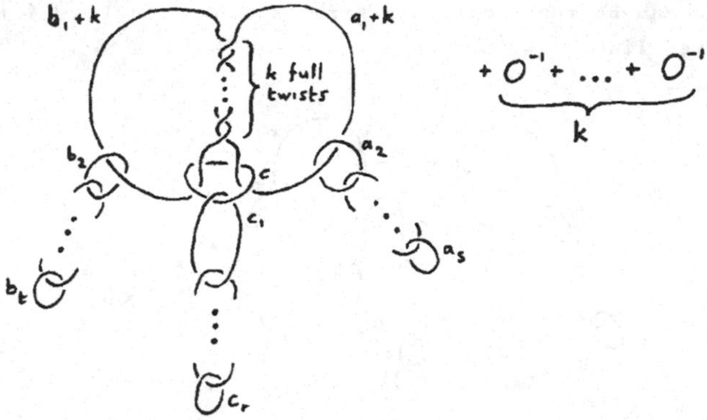

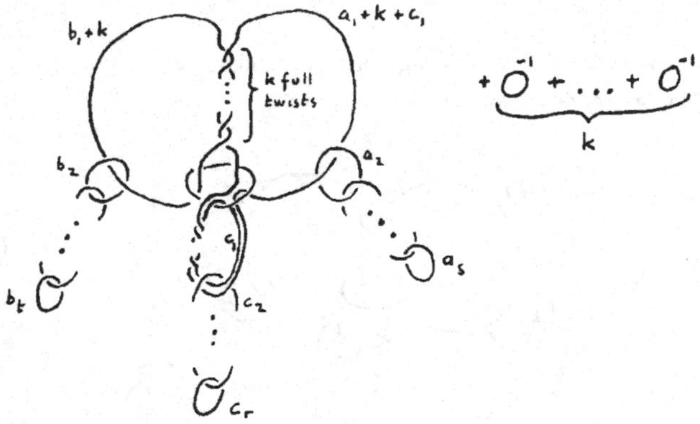

We now perform a single index 2 surgery by deleting the handle with framing c_1, giving us the next link (next page). Then by reversing our initial procedure which produced the k twists we can unwind them again, to end up with a framed link represented by the graph on the next page, called Γ_2'. Finally, the same argument which showed that Γ_1 is equivalent to Γ_1' shows that Γ_2' is equivalent to Γ_2, completing the proof.

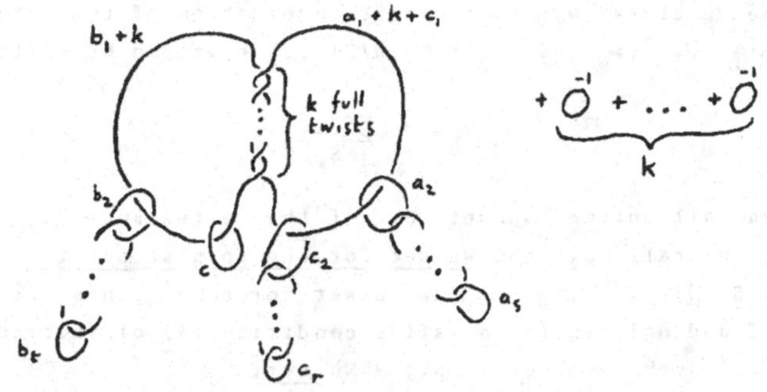

$\Gamma_2' =$

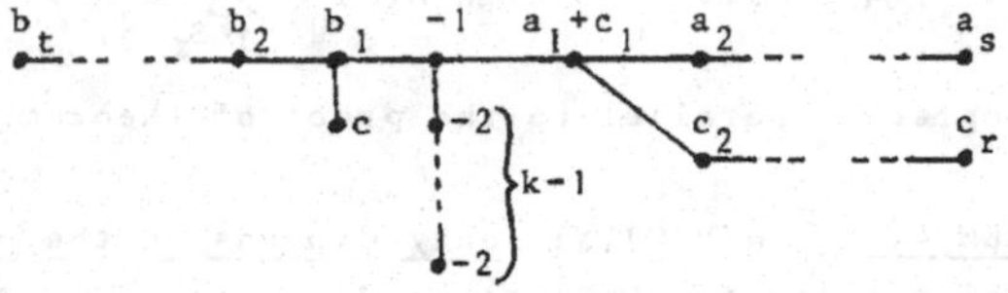

For the proof of 3.3 we need an extension of our invariant.

4. $\bar{\mu}$ for spin structures.

The μ-invariant is actually defined for any closed oriented 3-dimensional spin manifold. It is well defined for a $\mathbb{Z}/2$--sphere M^3 without spin structure because such an M admits a unique spin structure (in general the various spin structures on a spin manifold X are classified by $H^1(X;\mathbb{Z}/2)$).

Let Γ be a plumbing graph as in section 2 but suppose $M(\Gamma)$ is not necessarily a $\mathbb{Z}/2$-sphere. Then to any spin structure γ on $M(\Gamma)$ one has an obstruction in $H^2(P(\Gamma),M(\Gamma);\mathbb{Z}/2)$ to extending γ over $P(\Gamma)$. The Poincaré dual $w_\gamma \in H_2(P(\Gamma);\mathbb{Z}/2)$ of this obstruction will be called the <u>homology Wu class</u>. A class $w \in H_2(P(\Gamma),\mathbb{Z}/2)$ is the homology Wu class of some spin structure on $M(\Gamma)$ if and only if w satisfies

$$w.x = x.x \quad \text{for all } x \in H_2(P(\Gamma),\mathbb{Z}/2) .$$

The Wu class w_γ is a linear combination of the natural basis $\alpha_1, \dots, \alpha_s$ of $H_2(P(\Gamma);\mathbb{Z}/2)$, so it can be written as

$$w_\gamma = \sum_{j\in S_\gamma} \alpha_j$$

for some well defined subset S_γ of the vertex set $\{1, \dots, s\}$ of Γ. We call S_γ the _Wu set for the spin structure_ γ. A subset $S \subset \{1, \dots, s\}$ is the Wu set for some spin structure on $M(\Gamma)$ if and only if it satisfies condition (*) of theorem 2.1, so we call such a subset simply a _Wu set_.

In analogy to the definition in section 2 we define

$$\bar{\mu}(M(\Gamma),\gamma) = \text{sign } A(\Gamma) - \sum_{j\in S_\gamma} e_j .$$

A proof completely parallel to the proof of theorem 2.1 shows:

THEOREM 4.1. $\bar{\mu}(M(\Gamma),\gamma)$ _only depends on the oriented spin diffeomorphism type of_ $(M(\Gamma),\gamma)$ _and not on_ Γ. _Its modulo_ 16 _reduction is the usual_ μ_-invariant of the spin manifold_ $(M(\Gamma),\gamma)$.

This involves strengthening proposition 2.2 to include the spin structure, but this is not hard to do by the same method as was used to prove it without spin structure in [8].

We shall need two simple lemmas on bilinear spaces, that is finite dimensional vector spaces equipped with bilinear forms.

LEMMA 4.2. _If_ V _is a non-degenerate symmetric bilinear space over_ $\mathbb{Z}$ _and_ $V_1 \subset V$ _has codimension_ 1, _then_ V_1 _is degenerate (with induced form) if and only if_ sign V = sign V_1 .

LEMMA 4.3. _If_ V _is a non-degenerate symmetric bilinear space over_ $\mathbb{Z}/2$ _and_ $V_1 \subset V$ _has codimension_ 1, _then_ V_1 _is degenerate if and only if the Wu class_ w _of_ V _is in_ V_1. _The Wu class is the unique element with_ $w.x = x.x$ _for all_ $x \in V$.

Proofs. We prove 4.3 and leave 4.2 to the reader. V_1 degenerate means $V_1^\perp \subset V_1$ where $\perp$ is orthogonal complement. Now $V_1^\perp$ is 1-dimensional, so $V_1^\perp = \{0,x\}$ for some x. Since $w.x = x.x = 0$ we have $w \in (V_1^\perp)^\perp = V_1$. Conversely if V_1 is non-degenerate then $V = V_1 \oplus V_1^\perp$ and $w = w(V) = w(V_1) \oplus w(V_1^\perp)$ is not in

V_1 since $w(V_1^{\perp}) \neq 0$.

We can now prove 3.3. We use the notation of proposition 3.2 and its proof and we assume $M(\Gamma_1)$ is a $\mathbb{Z}/2$-sphere and $M(\Gamma_2) = S^1 \times S^2$. $P(\Gamma_1')$ is diffeomorphic to the 4-manifold V_1 obtained by adding 2-handles to D^4 along the framed link in $S^3 = \partial D^4$ first pictured in the proof of 3.2. $H_2(V_1;\mathbb{Z})$ has non-degenerate intersection form (in fact this is true modulo 2) since $\partial V_1 = M(\Gamma_1)$ is a $\mathbb{Z}/2$-sphere. As we alter the link in the proof of 3.2 we do not alter V_1 until the point where we delete a component of the link. This corresponds to removing a handle, and gives us a manifold V_2 diffeomorphic to $P(\Gamma_2')$. Thus $H_2(V_2;\mathbb{Z})$ is a codimension 1 subspace of $H_2(V_1,\mathbb{Z})$. But $H_2(V_2,\mathbb{Z})$ has degenerate intersection form, since $\partial V_2 = M(\Gamma_2) = S^1 \times S^2$. Thus $\operatorname{sign} V_1 = \operatorname{sign} V_2$ by lemma 4.2, that is, $\operatorname{sign} A(\Gamma_1') = \operatorname{sign} A(\Gamma_2')$, whence also follows easily: $\operatorname{sign} A(\Gamma_1) = \operatorname{sign} A(\Gamma_2)$.

Similarly one can use lemma 4.3 to follow what happens to the Wu set $S \subset \operatorname{Vert}(\Gamma_1)$ as one goes through the proof of 3.2. In particular one sees that the vertices weighted a_1 and c_1 are either both in S or both not and that there is a Wu set $S_2 \subset \operatorname{Vert}(\Gamma_2)$ which contains precisely the vertices of Γ_2 which correspond to the vertices in $S \subset \operatorname{Vert}(\Gamma_1)$, with the possible exception of the vertices weighted 0 and k . But by considering condition (*) of theorem 2.1 at the 0-weighted vertex one then sees that the k-weighted vertices are also either in both Wu sets S and S_2 or in neither. Thus $\sum_{j\in S} e_j = \sum_{j\in S_2} e_j$, so

$$\bar{\mu}(M(\Gamma_1)) = \operatorname{sign} A(\Gamma_2) - \textstyle\sum_{j\in S_2} e_j .$$

Thus $\bar{\mu}(M(\Gamma_1))$ equals the $\bar{\mu}$-invariant of some spin structure on $M(\Gamma_2) = S^1 \times S^2$. But $S^1 \times S^2$ can be given by the graph consisting of a single point with weight 0 , so both spin structures on $S^1 \times S^2$ have $\bar{\mu} = 0$. Thus the proof of 3.3 is complete for this case. The case that $M(\Gamma_2)$ is a $\mathbb{Z}/2$-sphere and $M(\Gamma_1)$ is $S^1 \times S^2$ is completely analogous.

5. Seifert spheres.

We recall some facts from [9]. Let $\alpha_1, \ldots, \alpha_n$ be pairwise coprime integers, each $\alpha_i \geq 2$. Then there exists a unique Seifert manifold whose unnormalized Seifert invariants are $(0,(\alpha_1,\beta_1), \ldots, (\alpha_n,\beta_n))$ and satisfy

$$\alpha_1 \ldots \alpha_n (\sum \beta_i/\alpha_i) = 1 .$$

This manifold will be denoted $\Sigma(\alpha_1, \ldots, \alpha_n)$. It is a $\mathbb{Z}$-sphere and every Seifert manifold which is a $\mathbb{Z}$-sphere is diffeomorphic to such a manifold after reversing orientation if necessary. Various other descriptions of these manifolds are given in [7] and [9], for instance as links of certain complex surface singularities.

Note that

$$\Sigma(\alpha_1, \ldots, \alpha_n) = \Sigma(\alpha_1', \ldots, \alpha_n')$$

if and only if $(\alpha_1', \ldots, \alpha_n')$ is a permutation of $(\alpha_1, \ldots, \alpha_n)$. We define

$$\Sigma(\alpha_1, \ldots, \alpha_n, 1) = \Sigma(\alpha_1, \ldots, \alpha_n)$$

in order to permit $\alpha_i = 1$ in the definition.

No closed formula for $\bar{\mu}(\Sigma(\alpha_1, \ldots, \alpha_n))$ is known and in practice the formula of [9, theorem 6.2] seems the fastest way of computing it (see corollary 2.3 above). The only general formulae for it that I know are the following.

<u>THEOREM 5.1.</u> (i). <u>If</u> $\alpha_n \equiv \alpha_n' \mod 2\alpha_1 \ldots \alpha_{n-1}$ <u>then</u>

$$\bar{\mu}(\Sigma(\alpha_1, \ldots, \alpha_n)) = \bar{\mu}(\Sigma(\alpha_1, \ldots, \alpha_{n-1}, \alpha_n')) .$$

(ii). <u>If</u> $\alpha_n \equiv -\alpha_n' \mod 2\alpha_1 \ldots \alpha_{n-1}$ <u>then</u>

$$\bar{\mu}(\Sigma(\alpha_1, \ldots, \alpha_n)) = -\bar{\mu}(\Sigma(\alpha_1, \ldots, \alpha_{n-1}, \alpha_n')) .$$

<u>Proof</u>. (i). It suffices to prove (i) for $\alpha_n' = \alpha_n + 2\alpha_1 \ldots \alpha_{n-1}$. Let

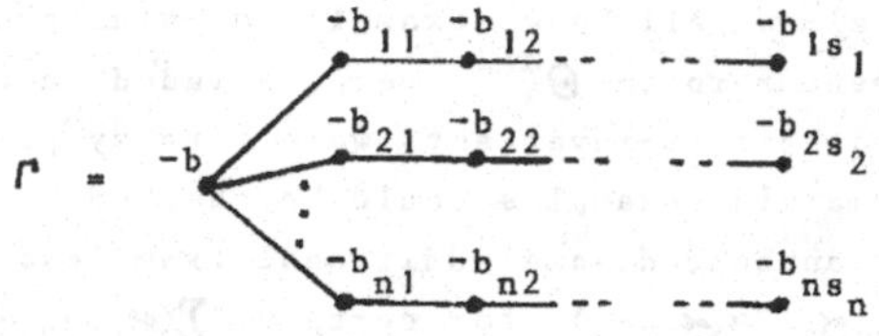

be the normal form plumbing graph in the sense of [8] (called the canonical graph in [9]) for $\Sigma = \Sigma(\alpha_1, \ldots, \alpha_n)$. It is characterized, among plumbing graphs for Σ, by the condition that all b_{ij} be ≥ 2. As shown in [9], its intersection form $A(\Gamma)$ is negative definite.

A reasonably efficient algorithm for computing Γ is given in [9], and a simple computation with that algorithm shows that the normal form plumbing graph Γ' for $\Sigma' = \Sigma(\alpha_1, \ldots, \alpha_{n-1}, \alpha_n')$ is

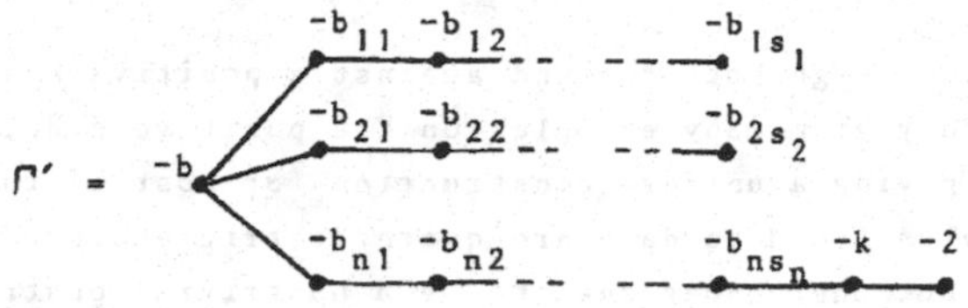

with $k = [\alpha_1 \ldots \alpha_{n-1}/\alpha_n] + 2$.

Now let S be the Wu set for Γ. Considering S as a subset of $\mathrm{Vert}(\Gamma')$ one sees that either S itself or S plus the (-2)-weighted vertex of Γ' is a Wu set S' for Γ', and hence, by uniqueness, <u>the</u> Wu set. Since sign $A(\Gamma)$ and sign $A(\Gamma')$ differ by 2 and $\sum_{j \in S} e_j$ and $\sum_{j \in S'} e_j$ differ by at most 2, we see that $\bar{\mu}(\Sigma)$ and $\bar{\mu}(\Sigma')$ differ by at most 4. But they are both divisible by 8, so they are equal.

It is enough to prove (ii) for $\alpha_n < 2\alpha_1 \ldots \alpha_{n-1}$ and $\alpha_n' = 2\alpha_1 \ldots \alpha_{n-1} - \alpha_n$, because then one can apply (i). The argument proceeds like case (i) after verifying that, if Γ is as above, then the above Γ' with minus signs deleted and k reduced by 1 is a plumbing graph for $\Sigma(\alpha_1, \ldots, \alpha_{n-1}, \alpha_n')$.

The above proof is valid, with suitable interpretation, also for $\alpha_n = 1$. For example, $\Sigma(\alpha_1, \alpha_2, 1)$ is always the 3-sphere, so the theorem implies $\bar{\mu}(\Sigma(\alpha_1, \alpha_2, 2m\alpha_1\alpha_2 \pm 1)) = 0$ for all $m > 0$.

Until recently, almost all known examples of plumbed $\mathbb{Z}/2$--spheres which represent zero in $\Theta_3^{\mathbb{Z}/2}$ were included in Casson and Harer's lists, so their $\bar{\mu}$-invariants were zero by proposition 3.3, while the remaining examples could be checked by hand. Recently R. Stern has announced some additional examples. They are of the form $\Sigma(\alpha_1,\alpha_2,2\alpha_1\alpha_2\pm\alpha_3)$ for certain $\Sigma(\alpha_1,\alpha_2,\alpha_3)$ in the Casson Harer lists and hence have $\bar{\mu}$ equal to zero by the above theorem.

6. Final comments.

PROBLEM 1. Is $\bar{\mu}$ a $\mathbb{Z}/2$-homology bordism invariant? Equivalently, does every plumbed $\mathbb{Z}/2$-sphere M^3 which bounds a $\mathbb{Z}/2$-acyclic 4-manifold have $\bar{\mu}(M) = 0$?

Our results weigh both for and against a positive answer, for although they give many examples on the positive side, they also show, by giving a uniform construction for most of them, that the examples found to date are quite restricted in their construction. But there does seem to be a nontrivial obstruction to finding counterexamples: by analyzing the proof of 3.3 I found constructions which at first seemed certain to yield some, but they only gave rational homology spheres and no $\mathbb{Z}/2$-spheres.

A property of the examples found so far is that the $\mathbb{Z}/2$-acyclic manifold W with $\partial W = M$ can always be chosen such that $\pi_1(M) \longrightarrow \pi_1(W)$ is onto. If M is a $\mathbb{Z}$-sphere, W can even be chosen contractible. This again points out the probable restricted nature of these examples. However these conditions on W lead to stronger concepts of homology bordism which are still of interest and the answer to problem 1, if negative, might still be positive for such a relation. Non-triviality with respect to this stronger homology bordism relation can sometimes be detected by the Casson Gordon invariant, even when the μ-invariant and the only other known homology bordism invariant, the Witt class of the torsion linking form of M^3, vanish.

If a counterexample to problem 1 exists, the new idea in its construction might be applicable to construct other sought for objects. One was mentioned in the introduction. Another might be a closed simply connected 4-manifold with nontrivial even definite intersection form.

In this context we mention the following problem, which seems unlikely to have a positive answer, though it would be nice if it did.

PROBLEM 2. If M is a plumbed $\mathbb{Z}/2$-sphere and $M = \partial V^4$ where V is simply connected and has even definite form, is then $\bar{\mu}(M) =$ sign V ? What if one weakens simply connected to $H_1(V;\mathbb{Z}/2) = 0$?

A positive answer to the second part would answer problem 1 for such M and suggest a faint hope for a general definition of $\bar{\mu}$.

PROBLEM 3. Can one extend the definition of $\bar{\mu}$ to arbitrary $\mathbb{Z}/2$-spheres, keeping its properties (whatever they are)?

If one just wants an oriented diffeomorphism invariant, additive for connected sum, satisfying $\bar{\mu}(-M) = -\bar{\mu}(M)$, whose mod 16 reduction is μ , then the answer is probably yes (it is equivalent to theorem 6.1 below holding for any $\mathbb{Z}/2$-sphere), although the answer to the same question for spin manifolds is no. Indeed the Lie-invariant spin structure on T^3 satisfies $\mu(T^3,\gamma) = 8 \pmod{16}$ and $(T^3,\gamma) \cong -(T^3,\gamma)$, so 6.1 does not extend to arbitrary spin manifolds.

One might hope that for arbitrary framed links representing Z/2-spheres there is a formula like the one of theorem 2.1, but this hope must fail. Indeed, the link

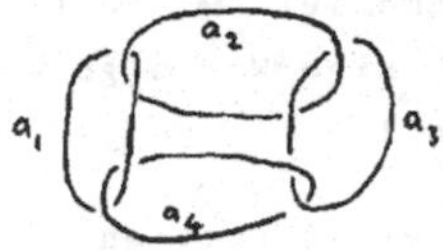

, all a_i odd,

if some a_i is -1 , represents the Seifert manifold $M = M(0;(1,-1),(a_{i+1}+2,1),(a_{i-1}+2,1),(a_{i+2}+1,1))$ (indices mod 4) , which has $\bar{\mu}(M) = -(a_1+a_2+a_3+a_4) - 4 + \text{sign } A$, where A is the intersection form determined by the link. But if $a_1 = a_3 = 1$ and $a_2 = a_4 = 3$ it gives $\sum(2,3,7)$, whose $\bar{\mu}$-invariant is $+8$ rather than -8 , which the above formula would give.

PROBLEM 4. Is the Browder Livesay invariant of a free involution on a plumbed $\mathbb{Z}/2$-sphere M always equal to $\bar{\mu}(M)$?

The answer is yes for Seifert manifolds, at least if the orbit space is either sufficiently large or a Seifert manifold.

Despite the many questions about $\bar{\mu}$, the invariant does have applications.

THEOREM 6.1. *If M^3 is the boundary of a simply connected plumbed 4-manifold, then for any spin structure γ on M with $\mu(M,\gamma)$ nontrivial, (M,γ) admits no orientation reversing spin homeomorphism.*

Proof. Clearly $\bar{\mu}(M,\gamma) \neq 0$, so $\bar{\mu}(-(M,\gamma)) = -\bar{\mu}(M,\gamma) \neq \bar{\mu}(M,\gamma)$.

As already remarked, this theorem fails for more general 3-dimensional spin manifolds. However, at this conference, L. Siebenmann announced that it is true for all sufficiently large 3-dimensional $\mathbb{Z}$-spheres. In this case, indeed for $\mathbb{Z}/2$-spheres, the spin structure is unique and can thus be disregarded.

THEOREM 6.2. *If the closed oriented spin manifold W^4 can be written as the union $W_1 \cup (-W_2)$ of two simply connected plumbed 4-manifolds, pasted along their boundaries, then sign $W^4 = 0$.*

Proof. Let $\partial W_1 = \partial W_2 = M$ and let γ be the restriction of the spin structure to M. Then sign W = sign W_1 - sign W_2 = $\bar{\mu}(M,\gamma) - \bar{\mu}(M,\gamma) = 0$ by theorem 4.1.

Let $\Sigma = \Sigma(\alpha_1, \ldots, \alpha_n)$ be as in section 5. We can write Σ as the boundary of a 4-manifold V with negative definite intersection form by considering Σ as the link of a complex surface singularity and taking the minimal resolution of this singularity. In [9] a list was given for n = 3 and low values of α_1 and α_2 for which this V has even form. It turned out that α_3 was then also restricted to be small. This is explained by the following theorem.

THEOREM 6.3. The above V never has even form if $\alpha_n > \alpha_1\alpha_2 \ldots \alpha_{n-1}$.

We just sketch the proof. Suppose $\alpha_n > \alpha_1 \ldots \alpha_{n-1}$ and V has even form. Let $\bar{V}$ be constructed like V, but using the minimal good resolution. Then $\bar{V} \cong P(\Gamma)$ as smooth manifolds with Γ as in the proof of 5.1. Let V_m and $\bar{V}_m$ be the corresponding manifolds for $\Sigma_m = \Sigma(\alpha_1, \ldots, \alpha_{n-1}, \alpha_n + 2m\alpha_1 \ldots \alpha_{n-1})$ and let Γ_m be the corresponding graph. Γ_m is obtained from Γ by adding a chain of $2m$ (-2)-weighted vertices to Γ, as in the proof of 5.1 (note our assumption on α_n implies $k = 2$ in the proof of 5.1).

V is obtained from $\bar{V}$ by blowing down (in the sense of complex manifolds) some exceptional curves of $\bar{V}$, and it follows that the Wu set S for Γ consists only of vertices corresponding to exceptional curves which are blown down in this process. In particular, the vertex weighted $-b_{ns_n}$ is not in S, for if it were, then in passing from $\bar{V}_m$ to V_m, the whole chain of new (-2)-weighted exceptional curves would blow down, and this would contradict negative definiteness of the intersection form of V_m for m sufficiently large. The proof of 5.1 now shows that $\bar{\mu}(\Sigma_m) = \bar{\mu}(\Sigma) - 2m$, which is a contradiction to 5.1.

Added September 1979: Jonathon Wahl has obtained examples of rational homology spheres which bound rational homology balls as a by-product of work on smoothing singularities. Namely let p,q,r be positive integers and $\zeta = \exp(2\pi i/(pqr+1))$. Then $G = \mathbb{Z}/(pqr+1)$ acts on $\mathbb{C}^3$ by $\zeta(x,y,z) = (\zeta x, \zeta^{-p}y, \zeta^{pq}z)$ and this action is free on $\mathbb{C}^3 - \{0\}$ and leaves invariant the polynomial $f: \mathbb{C}^3 \to \mathbb{C}$ given by $f(x,y,z) = x^p y + y^q z + z^r x$. Thus f induces $g: \mathbb{C}^3/G \to \mathbb{C}$ and the fibers $g^{-1}(t)$, $t \neq 0$, are smoothings of the isolated surface singularity $(g^{-1}(0), 0)$. Wahl showed that the "Milnor fiber" $W^4 = g^{-1}(t) \cap (D/G)$ (D the unit disc in $\mathbb{C}^3$ and $0 < |t|$ small) satisfies $H_1(W) = G$, $H_2(W) = H_3(W) = 0$, whence also $\partial W^4 = M^3$ satisfies $|H_1(M)| = (pqr+1)^2$, $H_2(M) = 0$. (This is because the Milnor fiber V^4 of f is homotopy equivalent to a wedge of (pqr) 2-spheres, and $W = V/G$.)

In fact, using the methods of [9] one can verify that M^3 is the Seifert manifold $M(0;(1,1),(pq-q+1,q),(qr-r+1,r),(rp-p+1,p))$ mentioned in section 3 . It is surprizing, given the difference of Wahl's construction from earlier ones, that his examples are not "new", but the same approach could well give many other examples.

References

1. A. Casson, J. Harer, Some homology lens spaces which bound homology balls, Preprint.
2. D.E. Galewski, R.J. Stern, Classification of simplicial triangulations of topological manifolds, Bull. Amer. Math. Soc. 82(1976), 916-918.
3. F. Hirzebruch, W.D. Neumann, S.S. Koh, Differentiable Manifolds and Quadratic Forms, Lecture Notes in Pure and Appl. Math. No. 4 (Marcel Dekker, New York, 1971).
4. R. Kirby, A calculus for framed links in S^3, Invent. Math. 45 (1978), 35-56.
5. T. Matumoto, Variétés simpliciales d'homologie et variétés topologiques métrisables, Thesis (Univ. Paris 1976).
6. W.D. Neumann, S^1-actions and the α-invariant of their involutions, Bonner Math. Schriften 44 (Bonn 1970).
7. W.D. Neumann, Brieskorn complete intersections and automorphic forms, Invent. Math. 42(1977), 285-293.
8. W.D. Neumann, A calculus for plumbing applied to the topology of complex surface singularities and degenerating complex curves, To appear.
9. W.D. Neumann, F. Raymond, Seifert manifolds, plumbing, μ-invariant and orientation reversing maps, Alg. and Geom. Topology, Proc. Santa Barbara 1977, Springer Lect. Notes 664 (1978), 163-196.
10. F. Waldhausen, Eine Klasse von 3-dimensionalen Mannigfaltigkeiten, Invent. Math. 3(1967), 308-333, and 4(1967), 87-117.

<u>October</u>: L.C. Siebenmann, in "On vanishing of the Rohlin invariant and nonfinitely amphicheiral homology 3-spheres", these Proceedings, [illegible]ows $M \cong -M \Rightarrow \mu(M) = 0$ for a very large class of $(\mathbb{Z}/2)$-spheres. He also gives a different definition of $\bar{\mu}$ and interesting examples.

Some topology of Zariski surfaces

by Richard Randell

This paper will be concerned with topological and birational invariants of branched covers of algebraic surfaces. The basic construction is as follows Let W be a smooth, compact complex algebraic surface, and let C be an algebraic curve on W. Corresponding to conjugacy classes of subgroups of $\pi_1(W - C)$ we have topological coverings $p : F \to W - C$. If the cover happens to be finite, the Riemann-Enriques-Grauert-Remmert-Grothendieck existence theorem [4] says that one may "complete" the space F to obtain a compact algebraic surface B, which is a branched cover of W with branching locus contained in C. The surface B may have singularities, so we resolve these and obtain a non-singular model $\tilde{B}$ of B. Thus we have a diagram

$$\begin{array}{ccc} & & \tilde{B} \\ & \text{if } p \text{ finite} & \big\downarrow \pi\text{, a resolution} \\ F & \subset & B \\ \big\downarrow p\text{, cover} & & \big\downarrow p\text{, branched cover} \\ W - C & \subset & W \supset C . \end{array} \tag{0.1}$$

Here to be specific we form $\tilde{B}$ by taking the minimal good resolution [9] of each singular point of B.

This construction is of course classical in algebraic geometry, especially when $W = P^2$, the (complex) projective plane. The surfaces B or $\tilde{B}$ are then called multiple planes. About fifty years ago these were studied in a series of papers by O. Zariski [19]-[21] (see also the short sketch by M. Artin and B. Mazur [23]) If $C \subset P^2$ is irreducible and of degree

n , then $H_1(P^2 - C, \mathbb{Z}) \cong \mathbb{Z}/n\mathbb{Z}$, and corresponding to the kernel of φ $\pi_1(P^2 - C) \to H_1(P^2 - C)$ one has the cyclic multiple plane. Several of Zariski's striking early results consider these surfaces, which can reasonably be called Zariski surfaces.

Now Zariski was interested in the function theory and birational geometry of these surfaces. These aspects of the surfaces, and their additional topological properties, depend solely on C and its embedding. As we will see, even for relatively simple curves in P^2 this is not well understood, and the topology of the situation can be quite rich. The analogy with classical knot theory is close, of course, though the algebraicity enters strongly In this article we primarily wish to illustrate the difficulty of the problems involved, and give some idea of some useful techniques. Thus we will not try to be as general as we could be. This means usually letting $W = P^2$ and taking $C \subset W$ to be irreducible with only the simplest sort of singularities. A general treatment which uses the framework to be given here is in preparation.

I. Some history and a particularly interesting case

A natural question to ask about (0 1) is "What surfaces can be thus realized?" The answer, well known to algebraic geometers, is "All." Specifically, embed the smooth surface in P^N for some large N , and project to a suitable plane $H \cong P^2$. A general projection will be a finite map. (For this, see [17, p. 52].) A different answer says to embed the surface

V in P^5 (as can always be done) and project to a general P^3. The image V' has only "ordinary" singularities, i.e. a double curve containing finitely many triple points and "pinch points" (see [22, p. 131]). Then project V' to a P^2 in P^3, this can be done so that the branch curve has only <u>nodes</u> and <u>cusps</u> as singularities.

<u>Definition</u>. Let $p \in C \subset W$. The point p is a <u>node</u> (resp, <u>cusp</u>) of C if there is a local analytic coordinate system (x,y) on W, centered at p, so that locally $C = \{xy = 0\}$ (resp., $= \{x^2 = y^3\}$).

We want to consider an opposite type of question. Namely, given C we should identify B or $\tilde{B}$. This leads to the following questions.

<u>Q1</u>. Are all curves of the same "type" as C actually isotopic to C (in P^2)?

<u>Q2</u>. What is $\pi_1(P^2 - C)$?

<u>Q3</u>. What is $\pi(P^2 - C)$?

By "type" in Q1 we mean having the same degree and the same singularities. In Q3, π is the algebraic fundamental group, built from <u>finite</u> covers. It is the profinite completion of π_1.

The particularly interesting case referred to in the section heading is when C is allowed to have no singularities other than nodes (a "nodal curve"). Then we can refine the above questions.

To do this we introduce the variety $V_{n,d}$ of curves of degree n with d nodes. Any curve C of degree n in P^2 is given by a complex polynomial of degree n in 3 variables. The space of all such curves is thus P^N, where $N = n(n + 3)/2$, and the coefficients of the monomials of the defining polynomial give coordinates on P^N. The variety $V_{n,d}$ is defined as the smallest closed algebraic subset of P^N which contains all curves with d nodes. A curve with exactly d nodes is a smooth point of $V_{n,d}$ of dimension $n(n+3)/2 - d$. See [10, pp. 90-97] as a general reference.

Then for nodal curves there are the following special versions of Q1-Q3.

SQ1. $V_{n,d}$ is irreducible.

SQ2. $\pi_1(P^2 - C)$ is abelian.

SQ3. All finite covers of $P^2 - C$ are abelian.

Here the status is SQ1 ⇒ SQ2 ⇒ SQ3 , and SQ2 is known to hold. The first implication is due to Zariski [19] while the second is of course trivial. The statement SQ2 is the "Zariski conjecture on nodal curves" This was given as a theorem in [19], as a consequence of SQ1 ⇒ SQ2 . This, however, relied on Severi's argument [16, Anhang F] for SQ1, and this argument is now known to be false [22, D. Mumford's appendix to Chap. 8]. Thus SQ1 is open. For SQ2 and SQ3, the situation has recently been resolved affirmatively More specifically, W. Fulton [3] has proved

Theorem. Let C be a nodal curve in P^2 Then any covering of P^2 branched along C (tamely ramified if the characteristic is not zero) is abelian.

This in fact is what Zariski wanted in [19]. The techniques used in [3] complete a program initiated by Abhyankar in [1]. Even more recently, P. Deligne has "topologized" the key part of Fulton's argument to obtain a proof of the Zariski conjecture!

Theorem. Let C be a nodal curve in P^2 Then $\pi_1(P^2 - C)$ is abelian.

To complete this discussion I want to briefly mention an interesting numerical result concerning SQ1. As background, note that for an irreducible nodal curve we always have $4d \leq 2(n-1)(n-2)$ (Plucker formula) Then if $4d \geq 2n^2 - 9n + 4$, then SQ1 holds [2].

In the next section we will investigate curves with arbitrary singularities

II. Cyclic covers branched over general irreducible curves

In this section we will combine and extend ideas of Oka [8] and Prill [11] to make an analysis of fundamental groups. For simplicity, in this part we will assume $W = P^2$ and C is irreducible. All the results we state have analogues in more general settings.

Suppose C has degree n. Then as noted above, $H_1(P^2 - C, \mathbb{Z}) \cong \mathbb{Z}/n\mathbb{Z}$ (by Alexander duality and intersection theory). Let F be the covering space corresponding to the kernel of the Hurewicz map $\varphi : \pi_1(P^2 - C) \to H_1(P^2 - C, \mathbb{Z})$. Thus $\pi_1(F)$ is isomorphic to the commutator subgroup of $\pi_1(P^2 - C)$. Suppose $f(x,y,z)$ is the irreducible polynomial defining C. Then $F \cong f^{-1}(1) \subset \mathbb{C}^3$ and $B \cong \{f(x,y,z) = w^n\} \subset P^3$, as in [8]. The branch locus C of $p : B \to P^2$ sits inside B as $B \cap \{w = 0\}$, and $p(x:y:z:w) = (x:y:z)$. There is a short exact sequence

$$(2.1) \qquad 1 \to \pi_1(F) \to \pi_1(P^2 - C) \to \mathbb{Z}/n\mathbb{Z} \to 1 .$$

We will prove the following

<u>Theorem 1.</u> There is a map $\alpha \quad B - C \to \widetilde{B}$ which induces epimorphisms $\alpha_* \quad \pi_1(B - C) \to \pi_1(\widetilde{B})$ and $\alpha_* \quad H_1(B - C, \mathbb{Z}) \to H_1(\widetilde{B}, \mathbb{Z})$ and an isomorphism $\alpha_* \quad H_1(B - C, Q) \to H_1(\widetilde{B}, Q)$.

<u>Proof.</u> Let Σ be the singular set of C . We first prove that the inclusion $B - C \xrightarrow{i} B - \Sigma$ induces an isomorphism on integral first homology. Note that B is singular only on Σ , so that duality gives a natural diagram

$$\text{(2 2)} \qquad \begin{array}{ccc} H_1(B - C) & \xrightarrow{i_*} & H_1(B - \Sigma) \\ \uparrow \cong & & \uparrow \cong \\ H^3(B,C) & \xrightarrow[i^*]{} & H^3(B,\Sigma) \end{array}$$

The exact cohomology sequence of (B,C,Σ) is

$$\begin{array}{ccccccccccc} \to H^2(B,\Sigma) & \xrightarrow{i^*} & H^2(C,\Sigma) & \to & H^3(B,C) & \to & H^3(B,\Sigma) & \to & H^3(C,\Sigma) & \to . \\ \cong \downarrow i^* & & \cong \downarrow i^* & & & & \cong \downarrow i^* & & \cong \downarrow i^* & \\ H^2(B) & \longrightarrow & H^2(C) \cong \mathbb{Z} & & & & H^3(B) & & H^3(C) \cong 0 & \end{array}$$

The vertical isomorphisms are trivial, because Σ is just a finite set of points. Since C is an irreducible curve , $H^2(C) \cong \mathbb{Z}$ and $H^3(C) \cong 0$. The diagram

$$\begin{array}{ccc} H^2(B) & \xrightarrow{j^*} & H^2(C) \\ p^* \uparrow & & \uparrow \text{Id} \\ H^2(P^2) & \xrightarrow{j^*} & H^2(C) \\ 1 & \longrightarrow & n \end{array}$$

and the fact that B is the branched cyclic n-fold cover can be used to show that the upper j^* is onto. Thus $i^*: H^3(B,C) \to H^3(B,\Sigma)$ is an isomorphism, and thus $i_*: H_1(B - C, \mathbb{Z}) \to H_1(B - \Sigma, \mathbb{Z})$ is as well.

Next we note that $\pi^{-1}: B - \Sigma \xrightarrow{\cong} \widetilde{B} - \pi^{-1}(\Sigma)$, so that we must consider the inclusion $j: \widetilde{B} - \pi^{-1}(\Sigma) \to \widetilde{B}$. Let N be a regular neighborhood of $\pi^{-1}(\Sigma)$ in $\widetilde{B}$. There is a Mayer-Vietoris sequence

(2.3) $$\to H_1(\partial N) \to H_1(N) \oplus H_1(\widetilde{B} - N) \to H_1(\widetilde{B}) \to 0 .$$

Also, in the exact sequence

$$\to H_2(N) \xrightarrow{j_*} H_2(N,\partial N) \to H_1(\partial N) \xrightarrow{i_*} H_1(N) \to H_1(N,\partial N)$$

we observe that j_* has finite cokernel, since its matrix is just the negative definite intersection form of the resolution [7]. Since $H_1(N,\partial N) \cong H^3(N) \cong H^3(\pi^{-1}(\Sigma)) \cong 0$, the map i_* is an isomorphism of the free part of $H_1(\partial N)$ with $H_1(N)$. Thus we may define the map $\alpha = j \circ \pi^{-1} \circ i$. Thus the results on H_1 follow.

For the fundamental group, note that α embeds $B - C$ as the complement of $C \cup \pi^{-1}(\Sigma)$ in the manifold $\widetilde{B}$. Since $C \cup \pi^{-1}(\Sigma)$ has real dimension two and $\widetilde{B}$ has real dimension four, the desired epimorphism follows by general position. □

If C is non-singular, then $H_1(B - C, \mathbb{Z}) \cong 0$. In general the effect of singularities of C on $H_1(B - C)$ has local and global aspects. The following definitions and theorem deal with local properties.

Suppose p_i is a singular point of C, with local equation $g_i(x,y)$ in some coordinate system centered at $p_i \in P^2$. The surface B also has a singular point at $p_i \in C$, and here the local equation is $g_i(x,y) = w^n$.

A small 5-sphere centered at $p_i \in C \subset B \subset P^3$ intersects B in a 3-manifold $K_i = \{g_i(x,y) - w^n = 0\} \cap S^5_\varepsilon$. But K_i is an n-fold branched cover of $S_i = \{g_i(x,y) - w = 0\} \cap S^5_\varepsilon$, where the covering sends (x,y,w) to (x,y,w^n) and normalizes. Finally, S_i is diffeomorphic to the 3-sphere $S^3 = \{(x,y) \in C^2 \mid \|(x,y)\| = 1\}$, and the branching is over the link $L_i = S^3 \cap \{g_i(x,y) = 0\}$ contained therein.

Suppose $g_i(x,y)$ has r_i analytically irreducible factors. Then L_i is a link of r_i components, and $H_1(S^3 - L_i) \cong \mathbb{Z}^{r_i}$. It easily follows that $H_1(K_i - L_i)$ contains a subgroup isomorphic to $\mathbb{Z}^{r_i}$. Let $h_i : H_1(F_i;\mathbb{Z}) \to H_1(F_i,\mathbb{Z})$ be the local monodromy for $g_i(x,y)$. That is, by [6], $S^3 - L_i$ fibers over S^1 with fiber F_i, and h_i is induced by the characteristic map of this fibration. Therefore, $K_i - L_i$ also fibers over S^1 with fiber F_i, and the characteristic map induces h_i^n.

There are then Wang sequences [6, Lemma 8.4]

$$0 \to H_2(K_i - L_i) \to H_1(F_i) \xrightarrow{I - h_i^n} H_1(F_i) \to H_1(K_i - L_i) \to \mathbb{Z} \to 0$$

$$0 \to H_2(S^3 - L_i) \to H_1(F_i) \xrightarrow{I - h_i} H_1(F_i) \to H_1(S^3 - L_i) \to \mathbb{Z} \to 0 .$$

Thus, $H_1(S^3 - L_i) \cong \mathbb{Z}^{r_i}$ implies $\operatorname{coker}(I - h_i) \cong \mathbb{Z}^{r_i - 1}$. Similarly, $\operatorname{coker}(I - h_i^n)$ has a $\mathbb{Z}^{r_i - 1}$ subgroup.

This explains somewhat the hypothesis of the following theorem.

<u>Theorem 2</u>. Suppose C is an irreducible algebraic curve of degree n in P^2, and suppose that for each singular point p_i of C one has

$$(*) \qquad \operatorname{coker}[(I - h_i^n) \quad H_1(F_i, \mathbb{Z}) \to H_1(F_i; \mathbb{Z})] \cong \mathbb{Z}^{r_i - 1}$$

Then $H_1(B - C, \mathbb{Z}) \cong 0$.

<u>Corollary 1</u>. When (*) holds, $\pi_1(P^2 - C)$ is an extension of a perfect group by $\mathbb{Z}/n\mathbb{Z}$. Further, $H_1(\tilde{B}, \mathbb{Z}) \cong 0$.

For further corollaries, we wish to know when (*) holds. We give the following sample list, here h_i is given with respect to some basis, and except for $x^3 + xy^3$ can be obtained from [6, Chap. 9].

$g_i(x,y)$	h_1	n	$\operatorname{coker}(I - h_1^n)$	$H_1(K_1, \mathbb{Z})$	(*)
$x^2 + y^2$	(1)	all n	$\mathbb{Z}$	$\mathbb{Z}/n\mathbb{Z}$	Yes
$x^2 + y^3$ A_2	$\begin{pmatrix} 0 & 1 \\ -1 & 1 \end{pmatrix}$	$\equiv \pm 1(6)$	0	0	Yes
		$\equiv \pm 2(6)$	$\mathbb{Z}/3\mathbb{Z}$	$\mathbb{Z}/3\mathbb{Z}$	No
		$\equiv 3(6)$	$(\mathbb{Z}/2\mathbb{Z})^2$	$(\mathbb{Z}/2\mathbb{Z})^2$	No
		$\equiv 0(6)$	$\mathbb{Z}^2$	$\mathbb{Z}^2$	No
$x^2 + y^{2k}$ A_{2k-1}	$\begin{pmatrix} 0 & 0 & \cdot & \cdot & \cdot & 0 & 1 \\ -1 & 0 & \cdot & \cdot & \cdot & 0 & 1 \\ 0 & -1 & 0 & \cdot & \cdot & 0 & \cdot \\ \cdot & 0 & -1 & \cdot & \cdot & \cdot & \cdot \\ \cdot & \cdot & \cdot & \cdot & \cdot & \cdot & \cdot \\ \cdot & \cdot & \cdot & \cdot & \cdot & \cdot & \cdot \\ 0 & 0 & \cdot & \cdot & \cdot & -1 & 1 \end{pmatrix}$ $(2k-1) \times (2k-1)$	$(n,2k) = 1$	$\mathbb{Z}$	$\mathbb{Z}/n\mathbb{Z}$	Yes
$x^3 + y^3$ D_4	$\begin{pmatrix} 0 & 0 & 0 & 1 \\ 0 & 0 & -1 & 1 \\ 0 & -1 & 0 & 1 \\ 1 & -1 & 1 & 1 \end{pmatrix}$	$\equiv \pm 1(3)$	$\mathbb{Z}^2$	$(\mathbb{Z}/n\mathbb{Z})^2$	Yes
		$\equiv 0(3)$	$\mathbb{Z}^4$	$\mathbb{Z}^2 \oplus \mathbb{Z}/n\mathbb{Z} \oplus \mathbb{Z}/\frac{n}{3}\mathbb{Z}$	No
$x^3 + xy^3$ E_7	$\begin{pmatrix} 0&0&0&0&0&1&1 \\ 0&0&0&0&0&-1&0 \\ 1&0&0&0&0&0&-1 \\ 0&1&0&0&0&1&1 \\ 0&0&1&0&0&-1&0 \\ 0&0&0&1&0&0&-1 \\ 0&0&0&0&1&1&1 \end{pmatrix}$	$n \equiv \pm 1(3)$	$\mathbb{Z}$	$\mathbb{Z}_n$	Yes
		$n \equiv \pm 3(9)$	$\mathbb{Z} \oplus (\mathbb{Z}/3\mathbb{Z})^2$	$\mathbb{Z}/n\mathbb{Z} \oplus (\mathbb{Z}/3\mathbb{Z})^2$	No
		$n \equiv 0(9)$	$\mathbb{Z}^7$	$\mathbb{Z}^6$	No

Of particular interest in this table are rows for which (*) holds but $H_1(K_1, \mathbb{Z}) \not\cong 0$. From the table one has, for example,

Corollary 2. Suppose C is irreducible of degree n, where $n = 6m \pm 1$.

If every singularity of C is equivalent to one of $x^2 + y^{2k}$, $(n,k) = 1$, $x^2 + y^3$, $x^3 + y^3$, or $x^3 + xy^3$, then $H_1(B - C, \mathbb{Z}) \cong H_1(\tilde{B},\mathbb{Z}) \cong 0$.

Proof of Theorem 2. Let N_i be an open regular neighborhood of L_i in K_i , and let $M_i = K_i - N_i$. We first show that (*) implies

(**). $\quad i_* \quad H_1(\partial M_i) \to H_1(M_i) \cong H_1(K_i - L_i)$ is onto.

We use integral coefficients throughout this proof.

To prove (**) , we write the exact sequence of $(M_i,\partial M_i)$. $\to H_3(M_i)$ $\to H_3(M_i,\partial M_i) \to H_2(\partial M_i) \to H_2(M_i) \to H_2(M_i,\partial M_i) \to H_1(\partial M_i) \to H_1(M_i) \to H_1(M_i,\partial M_i)$ $\to H_0(\partial M_i) \to H_0(M_i) \to 0$.

Now $H_3(M_i) \cong 0$, $H_3(M_i,\partial M_i) \cong \mathbb{Z}$, $H^2(\partial M_i) \cong \mathbb{Z}^{r_i}$ since ∂M_i is just r_i tori, $H_1(\partial M_i) \cong \mathbb{Z}^{2r_i}$, $H_0(\partial M_i) \cong \mathbb{Z}^{r_i}$. But (*) implies $H_1(M_i) \cong \mathbb{Z}^{r_i}$ Thus $H^1(M_i) \cong H_2(M_i,\partial M_i) \cong \mathbb{Z}^{r_i}$ and $H^2(M_i) \cong H_1(M_i,\partial M_i)$ is free, as is $H_2(M_i)$. Since K_i is a three-manifold, $\chi(K_i) = \chi(M_i) +$ $+ \chi(N_i) - \chi(\partial M_i) = 0$. Since $\chi(N_i) = \chi(\partial M_i) = 0$, $\chi(M_i) = 0$, and $H_2(M_i) \cong \mathbb{Z}^{r_i-1}$.

Thus the exact sequence becomes

$$\to 0 \to \mathbb{Z} \to \mathbb{Z}^{r_i} \to \mathbb{Z}^{r_i-1} \to \mathbb{Z}^{r_i} \to$$
$$\to \mathbb{Z}^{2r_i} \to \mathbb{Z}^{r_i} \to \mathbb{Z}^{r_i-1} \to \mathbb{Z}^{r_i} \to \mathbb{Z} \to 0 .$$

Then by algebra, this splits into three short exact sequences, and $H_1(\partial M_i) \to H_1(M_i)$ is onto, for all i .

To use (**) , we take a regular neighborhood N of C in B , and consider the Mayer-Vietoris sequence

$$\to H_1(N - C) \xrightarrow{(i_1,i_2)} H_1(N) \oplus H_1(B - C) \to H_1(B) \to ,$$

where i_1 and i_2 are induced by inclusion. We want to show that i_1 and i_2 are onto. Consider the branched cover $p: B \to P^2$, take $N(C)$ a regular neighborhood of C in P^2. Then the maps $(i_1)_*: \pi_1(N(C) - C) \to \pi_1(N(C))$ and $(i_2)_*: \pi_1(N(C) - C) \to \pi_1(P^2 - C)$ are onto, the former by general position in P^2, and the latter by [11, Proposition 2]. (The surjectivity of $(i_2)_*$ is vital, it is due to the fact that $C = f^{-1}(0)$ for some holomorphic function f.) If we choose $N = p^{-1}(N(C))$, the fact that $\pi_1(N - C)$ (resp. $\pi_1(B - C)$) has index n in $\pi_1(N(C) - C)$ (resp. $\pi_1(P^2 - C)$) shows that $(i_2)_*: \pi_1(N - C) \to \pi_1(B - C)$ is onto, and hence so is $i_2: H_1(N - C) \to H_1(B - C)$. By general position, i_1 is onto.

Next we note that $\pi_1(B) \cong 1$. To see this, we set $K = \{f(x,y,z) - w^n = 0\} \cap S^7 \subset C^4$. By [6, Theorem 5.2], $\pi_1(K) \cong 1$. As usual, since $f(x,y,z) - w^n$ is homogeneous, the Hopf action on $C^4 - \{0\}$ restricts to an S^1 action on K, and $B \cong K/S^1$. Thus $\pi_1(B) \cong 1$, and $H_1(B) \cong 0$.

Thus, it follows algebraically from the Mayer-Vietoris sequence above that $H_1(B - C) \cong i_2(\ker i_1)$.

We next use (**) to analyze $\ker i_1$. We have N, a regular neighborhood of C in B. For each singular point p_i of C, choose a regular neighborhood $N(i)$ of p_i in B so that $N_i \cap C = N(i,C)$ is a regular neighborhood of p_i in C, $\partial(N(i)) = K_i$, $\partial(N(i)) \cap C = L_i$, and $N - \cup_i N(i)$ is a D^2 bundle over $C - \cup_i N(i,C)$. See figure 1.

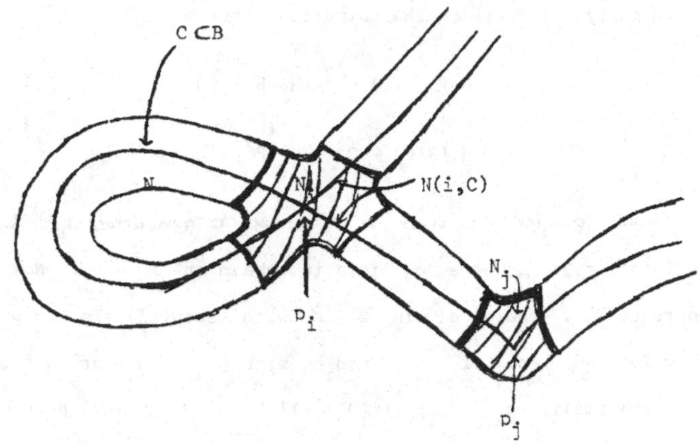

Figure 1.

Since we may assume C has at least one singular point, $H^2(C - \cup N(i,C)) \cong 0$, and the D^2 bundle is trivial. We pick a trivialization and think of the total space as $C' \times D^2$, where $C' = C - \cup_i N(i,C)$.

We now write a diagram of Mayer-Vietoris sequences, the second for $N = \overline{N - \cup_i N(i)} \cup \overline{\cup_i N(i)}$, and the first the same, except C is deleted. This gives a commuting diagram

$$\begin{array}{ccccccc} \to \oplus_i H_1(\partial M_i) & \longrightarrow & (\oplus_i H_1(M_i)) \oplus H_1(C' \times (D^2 - 0)) & \to & H_1(N - C) & \longrightarrow & 0 \\ \downarrow & & \downarrow & & \downarrow i_1 & & \\ \to \oplus_i H_1(L_i \times D^2) & \to & (\oplus_i H_1(N_i)) \oplus H_1(C' \times D^2) & \longrightarrow & H_1(N) & \to & 0 \end{array}$$

Let $a \in H_1(C' \times (D^2 - 0))$ be the class of a loop generating $H_1(D^2 - 0)$. Then $a' = i_*(a) \in H_1(N - C)$ is clearly in the kernel of i_1. Since $H_1(\partial M_i) \to H_1(M_i)$ is onto for all i, by (**), a diagram chase shows that in fact a' generates $\ker i_1$.

Finally, we consider the commuting diagram

$$\begin{array}{ccc} H_1(N - C) & \xrightarrow{i_2} & H_1(B - C) \\ \downarrow p & & \downarrow p \\ H_1(N(C) - C) & \rightarrow & H_1(P^2 - C) \cong \mathbb{Z}/n\mathbb{Z} . \end{array}$$

We have $p \circ i_2(a') = n \cdot b = 0$ (where b generates $H_1(P^2 - C)$). Thus $p \circ i_2(a') = \partial c_2$, where c_2 is a two-chain in $P^2 - C$. Now one can construct a 2-chain c_2' in $B - C$ with $\mathrm{supp}(c_2') = p^{-1}(\mathrm{supp}(c_2))$ and $\partial c_2' = i_2(a')$. For, if $c_2 = \Sigma n_i \sigma_i$, with σ_i singular 2-simplices which are small, $c_2' = \Sigma n_i p^{-1}(\sigma_i)$ will be a 2-chain. Here small means $p^{-1}(\sigma_i)$ is a disjoint union of singular 2-simplices. Thus $i_2(a') = 0 \in H_1(B - C)$, so that $i_2(\ker i_1) \cong 0$, and the proof is complete

Remarks.

i) In general, α_* is not injective on π_1 or H_1. Here is an example, due to Zariski [19]. Let C be the triscuspidal quartic (with equation $f(x,y,z) = x^2y^2 + y^2z^2 + z^2x^2 - 2xyz(x + y + z))$. Then $\pi_1(B - C) \cong \mathbb{Z}/3\mathbb{Z} \cong H_1(B - C, \mathbb{Z})$ But $\pi_1(\tilde{B}) \cong 0$, a fact which can be proved as follows. To form $\tilde{B}$ for this C we resolve three singularities with local equation $x^2 + y^3 = w^4$. These are rational singularities, and as such the neighborhood of the proper transform in the resolution is diffeomorphic to the Milnor fiber [18]. Thus $\tilde{B}$ is formed from B by removing the three singular points and sewing back in the Milnor fiber. It can be shown, therefore, that $\tilde{B}$ is isomorphic to the surface formed by branching along a non-singular curve of degree four in P^2, so that $\pi_1(\tilde{B}) \cong 1$

ii) By combining the above ideas with techniques of [11], one can get the same conclusion, so long as not too many singularities which fail to satisfy (*) are allowed. Also, the degree to which (*) fails provides an "upper bound" on $H_1(B - C)$. We will treat these more general situations in a forthcoming paper.

III. Other invariants of $\tilde{B}$.

One ultimate goal is to understand $\tilde{B}$ as an algebraic surface in terms of C, i.e. in terms of its degree, singularities, and their position. Here we make some comments about some birational invariants of $\tilde{B}$. The basic invariants are the irregularity, geometric genus, and plurigenera, defined as follows.

On an algebraic surface one has the structure sheaf $\mathcal{O}$ of germs of holomorphic functions and the sheaves Ω^p of germs of p forms. As usual, let $h^{p,q} = \dim H^q(X,\Omega^p)$. Then the irregularity q is defined by $q = h^{0,1}$. By Hodge theory and the fact that X is Kahler, $h^{0,1} = h^{1,0}$, and $b_1 = \dim H^1(X,\mathbb{C}) = h^{0,1} + h^{1,0}$. Thus $q = b_1/2$.

The geometric genus p_g of X is $h^{0,2} = h^{2,0}$, and the arithmetic genus is $\chi = 1 - q + p_g$.

The plurigenera $p_n = \dim H^0(X,(\Omega^2)^{\otimes n})$. Thus $p_1 = p_g$.

Finally, we note that for nonsingular algebraic surfaces, π_1 is a birational invariant. This follows from [17, p. 212], where it is shown that any two birationally equivalent smooth surfaces blow up to a common surface. Since blowing up is topologically connected sum with $-P^2$, it does not change π_1 .

Our primary interest is the irregularity. We can thus rephrase Theorem 1 as a statement that the irregularity may be computed from $B - C$; $q = 1/2 \dim H_1(B - C, \mathbb{Q})$.

These results are basically local in nature A result more global in nature, depending on the degree n of C and not on its singularities, is the following.

<u>Theorem.</u> (Zariski [20]). If $n = p^{\alpha}$, p prime, then $q = 0$. This may be generalized using essentially the same proof, to

<u>Theorem.</u> $2q \leq \#\{n$ -th roots of unity not of prime power order$\}$. This is probably not very sharp in general.

The consideration of q is important, since the arithmetic genus $\chi = 1 - q + p_g$ is calculable, and so q gives the geometric genus p_g .

Here is how to find χ . First, note that any curve C is the limit of a family of non-singular curves. Thus B is the limit of surfaces branched over non-singular curves, or in other words, B is the limit of a family $\{B_t\}$ of non-singular hypersurfaces of degree n in P^3 . By [15], χ does not vary in such a family. Further, $\chi(B_1)$ is easily determined; $\chi(B_1) = 1 + (n-1)(n-2)(n-3)/6 = \chi(B)$. By [5, p. 46, (4)], $\chi(\tilde{B}) = \chi(B) - \sum_{p_i} p_g^{\ell}(p_i)$. Here the sum is over the singular points p_i of B_1 and p_g^{ℓ} is the "local geometric genus", defined as follows. Let N_i be a Stein neighborhood of p_i , and let $\pi_i : M_i \to N_i \cap B$ be a resolution of the singularity at p_i . Then $p_g^{\ell}(p_i) = \dim H^1(M_i, \mathcal{O})$. In [5, Theorem 1], Laufer obtains a formula for p_g^{ℓ} in terms of topological invariants, so that

$$(3.1) \qquad \chi(\tilde{B}) = 1 + (n-1)(n-2)(n-3)/6 - \sum_{p_i} p_g^{\ell}(p_i)$$

is calculable. If C has only nodes, then the singularities of B are given locally by $f(x,y,w) = xy + w^n$, and $p_g^{\ell} = 0$.

Example ([19]). Let C be the sextic curve with six cusps on a conic. Then B has six singular points, each with local equation $x^2 + y^3 + w^6$, and $p_g^{\ell} = 1$ for each. Further, $\pi_1(B - C)$ is the free group on two generators, so $q = 1$. Thus $\chi(\tilde{B}) = 1 + 5 \cdot 4 \cdot 3/6 - 6 = 5$, and $p_g = \chi - 1 + q = 5$.

The complex 2-manifold $\tilde{B}$ also has topological invariants, the most simple of which are the euler characteristic e and signature σ . For the euler characteristic there is the easily derived formula [13, Theorem 5.11):

$$(3.2) \qquad e(B) = n(n^2 - 4n + 6) - (n-1)\sum_{p_\nu} \mu_\nu$$

where μ_i is the Milnor number of the singular point p_i on C. Knowledge of the resolution then gives $e(\tilde{B})$.

Letting c_i denote the Chern classes of the tangent bundle of $\tilde{B}$ one has the formulas

(3.3.1) $$e(\tilde{B}) = c_2[\tilde{B}]$$

(3.3.2) $$12X(\tilde{B}) = (c_1^2 + c_2)[\tilde{B}] \quad \text{(Hirzebruch Riemann-Roch)}$$

(3.3.3) $$3\sigma(\tilde{B}) = (c_1^2 - 2c_2)[\tilde{B}] \quad \text{(Hirzebruch signature formula)}$$

Thus $e(\tilde{B})$ gives $c_2[\tilde{B}]$, and $X(\tilde{B})$ then gives $c_1^2[\tilde{B}]$. Thus one can compute $\sigma(\tilde{B})$ as well.

Example (cont.). Let C again be the sextic with six cusps on a conic. One has $e(B) = e(\tilde{B}) = 36$, by (3.2) and the fact that the singularity $x^2 + y^3 + w^6$ may be resolved with a single elliptic curve as exceptional set. Thus $c_2[\tilde{B}] = 36$, $c_1^2(\tilde{B}) = 24$, and $\sigma(\tilde{B}) = -16$.

References

1. S. Abhyankar, Tame coverings and fundamental groups of algebaic varieties, II, Amer. J. Math. 82 (1960), 120-178.

2. D. Alibert and G. Maltsiniotis, Groupe fondamental du complementaire d'une courbe à points doubles ordinaires, Bull. Soc. Math. France 102 (1974), 335-351.

3. W. Fulton, On the fundamental group of the complement of a node curve, preprint, 1979.

4. H. Grauert and R. Remmert, Komplexe Räume, Math. Ann. (136), 1958, 245-318.

5. H. Laufer, On μ for surface singularities, Proc. Sym. Pure Math., Vol. XXX part 1, Amer. Math. Soc., Providence, 1977, pp. 45-50.

6. J. Milnor, Singular points of complex hypersurfaces, Ann. Math. Studies 61, Princeton, 1968.

7. D. Mumford, The topology of normal singularities of an algebraic surface and a criterion for simplicity, Publ. Math. I.H.E.S., no. 9 (1961), 5-22.

8. M. Oka, The monodromy of a curve with ordinary double points, Inventiones math. 27 (1974), 157-164.

9. P. Orlik and P. Wagreich, Isolated singularities of algebraic surfaces with C^* -action, Ann. Math. 93 (1971), 205-228.

10. H. Popp, Fundamentalgruppen algebraischer Mannigfaltigkeiten, Lecture notes in math., no. 176, Springer, Berlin, Heidelberg, New York, 1970.

11. D. Prill, The fundamental group of the complement of an algebraic curve, manuscripta math. 14 (1974), 163-172.

12. D. Prill, Branched coverings of curves and surfaces, unpublished manuscript, 1975.

13. R. Randell, On the topology of nonisolated singularities, in Geometric Topology, J. Cantrell, ed., Academic Press, New York, 1979, pp. 445-474.

14. R. Randell, On the fundamental group of the complement of a singular plane curve, Quarterly J. Math., to appear

15. O. Riemenschneider, Über die Anwendung algebraischer Methoden in der Deformationstheorie komplexer Räume, Math. Ann. 187 (1970), 40-55.

16. F. Severi, Vorlesungen über algebraische Geometrie, Leipzig, Teubner, 1921

17. I. R. Shafarevich, Basic Algebraic Geometry, Die Grundlehren der mathematischen Wissenschaften, Band 213, Springer, New York, 1974 .

18. G. N. Tjurina, Resolution of singularities for flat deformations of rational double points, Funkcional Anal. i Priložen 4 (1970), 77-83 = Functional Anal. Appl. 4 (1970), 68-73.

19. O. Zariski, On the problem of existence of algebraic functions of two variables possessing a given branch curve, Amer. J. Math., 51 (1929), 305-328.

20. O. Zariski, On the linear connection index of the algebraic surfaces $z^n = f(x,y)$, Proc. Nat. Acad. Sci. (15), 1929, 494-501.

21. O. Zariski, On the irregularity of cyclic multiple planes, Ann. Math. 32 (1931), 485-511.

22. O. Zariski, Algebraic Surfaces, 2nd Supplemented Edition, Springer, Berlin-Heidelberg-New York, 1971.

23. O. Zariski, Collected works, Vol. III, MIT Press, Cambridge, 1978.

A FIBRATION FOR $\mathrm{Diff}\Sigma^n$

Nigel Ray & Erik K. Pedersen

0. Introduction

Let M^n be an oriented smooth manifold, and let $\mathrm{Diff}M^n$ be its group of orientation preserving diffeomorphisms. Let $\mathrm{Diff}(M^n,M^n_o)$ and Diff_*M^n be the two subgroups (with the C^∞ topology) which consist of diffeomorphisms fixing $M^n_o = M^n - D^n$ (for some embedded disc D^n) and the base point $* \in D^n \subset M^n$ respectively.

It is a common procedure when studying $\pi_*(\mathrm{Diff}M^n)$ to restrict attention to either one of the above subgroups; it is therefore of some interest to study the homotopy braid of the triple $(\mathrm{Diff}M^n, \mathrm{Diff}_*M^n, \mathrm{Diff}(M^n,M^n_o))$.

In fact this is equivalent to considering the following 'comparison diagram' of principal fibrations, constructed by the usual techniques.

(0.1) Diagram.

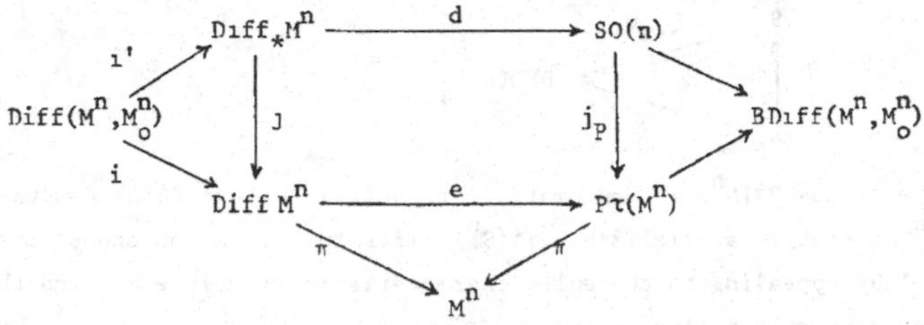

Here d is the derivative map at $*$, $P\tau(M^n)$ is the total space of the oriented smooth principal tangent bundle to M^n, e maps $f \in \mathrm{Diff}M^n$ to the derivative at $*$ of the restriction $f|D^n$, and π is evaluation at $*$.

The only fibration on display above which is not well documented is the lower horizontal one, i.e.

$$\mathrm{Diff}(M^n,M^n_o) \xrightarrow{\;i\;} \mathrm{Diff}M^n \xrightarrow{\;e\;} P\tau(M^n) \qquad (F)$$

Its existence follows from the fibration

$$\mathrm{Diff}(M^n,M^n_o) \xrightarrow{\;i\;} \mathrm{Diff}M^n \xrightarrow{\;e'\;} \mathcal{E}(D^n,M^n) \;,$$

where $\mathcal{E}(\;,\;)$ denotes the space of orientation preserving smooth embeddings, and e' is evaluation on D^n. But $\mathcal{E}(D^n,M^n)$ may be identified with $P\tau(M^n)$, since

$$\mathcal{E}_*(D^n,M^n) \longrightarrow \mathcal{E}(D^n,M^n) \longrightarrow M^n$$

is a model for the principal tangent bundle of M^n via the derivative map.

The homotopy braid of (0.1) gives rise to two interlinked problems. Firstly, to what extent does the 'linearization' j_P determine the map j; and secondly, to what extent does d determine e?

We shall discuss the second of these questions in the special case of M^n an exotic sphere Σ^n. In this case, (F) generalizes an exact sequence described by

*This talk was given by the first author

R. Schultz [S]. In particular, we study the difference between the boundary maps associated to d and to e, and reduce the detection of a certain class of 'stable' homotopy elements so arising to an interesting, but apparently unsolved, problem in the homotopy groups of spheres.

Throughout, we shall write S^n_β for the exotic sphere given by an element $\beta \in \pi_n(\mathrm{Top}/O)$, $n \geq 7$. Such a β arises from an isotopy class of diffeomorphisms $\beta \in \mathrm{Diff}S^{n-1}$, so any S^n_β can be presented as $D^n_o \cup_\beta D^n_c$, where $0 \in D^n_o$ is the base point, and D^n_c is the complementary disc.

We are grateful to Dick Lashof for a helpful letter.

1. The fibration

Our fibration (F) of §0 can be further simplified when $M^n = S^n_\beta$. For it has long been known that, whatever β, $P\tau(S^n) \doteq SO(n+1)$. It is most convenient to describe this fact by means of

(1.1) Lemma. There is a homeomorphism of degree 1, say $h: S^n_\beta \longrightarrow S^n$, such that the diagram

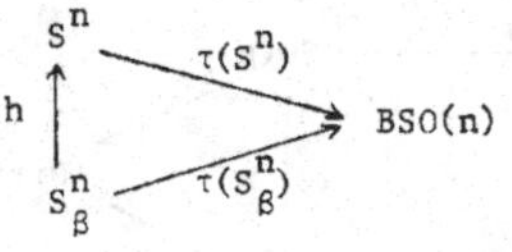

homotopy commutes. Thus $P\tau(S^n_\beta)$ is homeomorphic to $SO(n+1)$ via an $SO(n)$ equivariant map.

Proof. Since S^n_β is stably parallelizable, $\tau(S^n_\beta)$ lifts to S^n. We can choose the lift h to have degree 1 by appealing to the euler characteristic if n is even, and the Kervaire semi-characteristic if n is odd. □

(1.2) Note. The resulting homeomorphism $\bar{h}\cdot P\tau(S^n_\beta) \longrightarrow SO(n+1)$ is defined only up to alteration by any map $S^n_\beta \longrightarrow SO(n)$. □

We can now construct our special version of (0.1) as follows:

(1.3) Diagram.

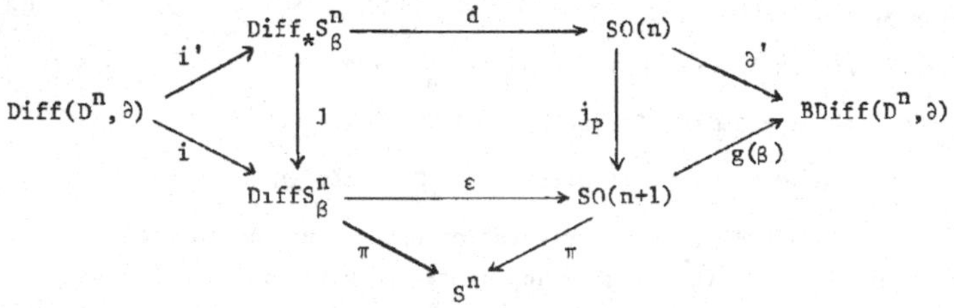

Note that we have labelled $\bar{h}\bullet e$ as ε, and the classifying map of ε as $g(\beta)$. Of course, j_P is precisely the standard inclusion, and the homotopy commutativity of the central square is assured by construction.

(1.4) Definitions.

(i) Let $W_*(S^n) \subset \pi_*(\mathrm{Diff}S^n)$ be the graded subgroup $\mathrm{Im}\, i'_* \cap \mathrm{Ker}\, j_*$.

(ii) Let $X_*(S^n) \subset \pi_*(SO(n+1)$ be the graded set of elements x with the property that $\pi_*(x) \neq 0 \neq g(\beta)_*(x)$. □

Thus $W_*(S^n_\beta)$ is a measure of the extent to which j_p fails to determine j, and $X_*(S^n_\beta)$ is a measure of the extent to which ∂' fails to determine $g(\beta)$. Also, $0 \neq w \in W_*(S^n_\beta)$ yields $(i')^{-1}(w) \in X_*(S^n_\beta)$.

We investigate $X_*(S^n_\beta)$ below. Note that, if $k < n$, then $W_k(S^n_\beta) = 0$ and $X_k(S^n_\beta) = \emptyset$. Also, $W_*(\ )$ and $X_*(\ )$ are defined for arbitrary M^n.

If S^n_β is the standard sphere S^n, then (1.3) 'collapses'. For the symmetry of S^n allows a splitting $\mathrm{Diff}S^n \longleftarrow SO(n+1)$ of (F), which restricts to a splitting of the upper fibration. Thus $W_*(S^n) = 0$ and $X_*(S^n) = \emptyset$.

Hence the cardinalities of $W_*(\ )$ and $X_*(\ )$ in some sense reflect the asymmetry of S^n_β. We develop below a detection procedure for 'stable' elements in $X_*(\ )$.

2. Detecting elements in $X_*(\ )$

We first summarize some information from [S] concerning the map ∂' of (1.3). Before so doing, however, it is convenient to recall some familiar notation which will also be useful for the remainder of this section.

We shall write $\mathrm{Top}S^n$ for the group of orientation preserving homeomorphisms of S^n, so that we have the following commutative diagram:

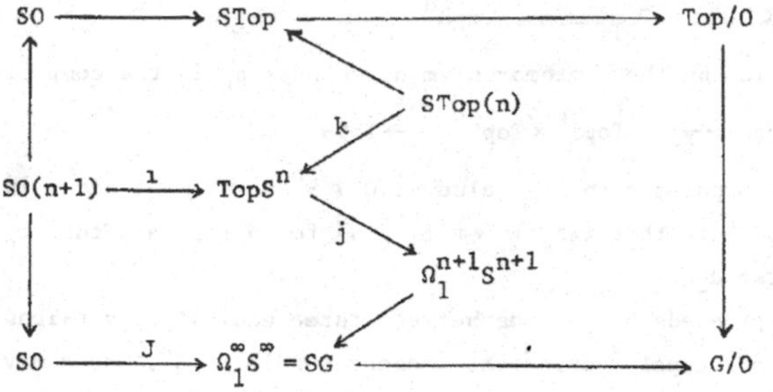

Here i is the standard inclusion, j is suspension, and k is one-point compactification.

Now according to [BL], the composite $\pi_*(\mathrm{STop}(n)) \longrightarrow \pi_*(\mathrm{Top}/O)$ is epic, so we may suppose that S^n_β is represented by a map $S^n \longrightarrow \mathrm{STop}(n)$ (tantamount to choosing a framing for S^n_β); or, via k_*, a map $\bar{\beta}: S^n \longrightarrow \mathrm{Top}S^n$. Continuing around the diagram, we also obtain $\tilde{\beta} \in \pi_n(G/O)$ as the image of both $\bar{\beta}$ and β. Moreover, $\tilde{\beta}$ lies in the summand $\pi_n(G)/\mathrm{Im}J$.

We can now state

(2.1) Theorem (R. Schultz). Given $\alpha \in \pi_k(SO(n))$, then $\partial'_*\alpha \neq 0$ in $\pi_k(B\mathrm{Diff}(D^n, \partial))$ whenever $\tilde{\beta}J(\alpha) \neq 0$ in $\pi^S_{n+k}/\mathrm{Im}J$. □

Our result concerning $X_*(\ )$ is in the same spirit, and can be stated thus:

<u>(2.2) Theorem.</u> Given $x \in \pi_k(SO(n+1))$, then $x \in X_k(S^n_\beta)$ whenever $\tilde{\beta}^2.H(x) \neq 0$ in $\pi^S_{n+k}/\mathrm{Im}J$. □

Here $H(x) \in \pi_k(S^n)$ is represented by the composite $S^k \xrightarrow{x} SO(n+1) \xrightarrow{\pi} S^n$ (π being the usual projection), which is proven in [Ke] to be the Hopf invariant (up to sign and suspension) of the element $J(x) \in \pi_{k+n+1}(S^{n+1})$.

Elements so detectible constitute a stable subset $SX_*(S^n_\beta) \subset X_*(S^n_\beta)$. Unfortunately, we know of no non-zero classes of the form $y^2.H(x)$ in any stem mod ImJ. The experts seem to regard this as a potentially accessible, but unsolved, problem of homotopy theory. Certainly, all $x \in \pi_{n+t}(SO(n+1))$ give $y^2.H(x) = 0$ for all y when t is small.

There are two main steps involved in establishing (2.2): these follow below as (2.3) and (2.4).

The first includes generalizing a diagram of [S p.240]. Since in (1.3) we have set up a map $\varepsilon: \mathrm{Diff}S^n \longrightarrow SO(n+1)$ (which depends on the choice of $\bar{h}$ in (1.2)), it is important to relate ε with the standard inclusion of both $\mathrm{Diff}S^n$ and $SO(n+1)$ in $\mathrm{Top}S^n$. This is done by

<u>(2.3) Lemma.</u> The following diagram is homotopy commutative:

$$\begin{array}{ccc} \mathrm{Diff}S^n_\beta & \subset\!\!\longrightarrow & \mathrm{Top}S^n_\beta \\ e\downarrow & & \downarrow\chi_h \\ P\tau(S^n) & \quad\varepsilon\ (\mathrm{Diff}S^n_\beta \to SO(n+1)) & \mathrm{Top}S^n \\ \bar{h}\downarrow & & \downarrow c_\beta \\ SO(n+1) & \overset{i}{\subset\!\!\longrightarrow} & \mathrm{Top}S^n \end{array}$$ □

Here χ_h is conjugation by the homeomorphism h, whereas c_β is the composition

$$\mathrm{Top}S^n \xrightarrow[\pi\times 1]{} S^n \times \mathrm{Top}S^n \xrightarrow[\tilde{\beta}\times 1]{} \mathrm{Top}S^n \times \mathrm{Top}S^n \xrightarrow[\mu]{} \mathrm{Top}S^n$$

where π projects a homeomorphism onto its value at $0 \in S^n = \mathbb{R}^n \cup \{\infty\}$, and μ is composition of functions. Note that $\pi \circ c_\beta(f) = \pi(f) \in S^n$ for all f, and that c_β is a homeomorphism with inverse $c_{-\beta}$.

The proof of (2.3) proceeds by passing between three equivalent versions of $P^{Top}\tau(S^n)$, the oriented principal topological tangent bundle of S^n. These may be displayed by the commutative diagram of (principal) fibrations

$$\begin{array}{ccccc} STop(n) & \xrightarrow{\sigma_1} & Top_* S^n & \xrightarrow{\sigma_2} & \mathcal{E}^{Top}_*(D^n, S^n) \\ \downarrow & & \downarrow & & \downarrow \\ P^{Top}\tau(S^n) & \xrightarrow{\phi_1} & \mathrm{Top}S^n & \xrightarrow{\phi_2} & \mathcal{E}^{Top}(D^n, S^n) \\ \pi\downarrow & & \pi\downarrow & & \downarrow\pi' \\ S^n & \xrightarrow{1} & S^n & \xrightarrow{1} & S^n \end{array}$$

The maps σ_1 and ϕ_1 are induced by compactification, and σ_2 and ϕ_2 by restricting a homeomorphism of $\mathbb{R}^n \cup \{\infty\}$ to D^n.

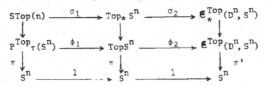

To introduce the second ingredient in the proof of (2.2), let us return to our fibration of (1.3). Suppose that $\alpha: Y \longrightarrow SO(n+1)$ is a map of some reasonable space into the base. If α does not factor through $\varepsilon:\mathrm{Diff}S^n \longrightarrow SO(n+1)$ and does not lift to $SO(n)$, then it represents a class with the properties we are seeking for $X_*(S^n_\beta)$ (in case Y is a sphere). We thus wish to discuss obstructions to lifting α to $\mathrm{Diff}S^n_\beta$.

We may assume without loss of generality that the suspension $S^1{\wedge}Y$ is given as an open subset of some euclidean space $\mathbb{R}^n$; in other words as an open smooth manifold. Now let $\alpha^*\gamma_\beta$ be the topological S^n bundle over $S^1{\wedge}Y$ which arises by adjointing the composite

$$\begin{array}{ccccc} Y & \xrightarrow{\ \alpha\ } & SO(n+1) & \xrightarrow{\ \iota\ } & \mathrm{Top}S^n \\ & & & \searrow^{\gamma_\beta} & \downarrow c_{-\beta} \\ & & & & \mathrm{Top}S^n \end{array}$$

Then if α lifts to $\mathrm{Diff}S^n_\beta$, (2.3) tells us that the total space $E(\alpha^*\gamma_\beta)$ admits a smoothing which restricts to β on each fibre.

In fact it is most useful to work universally, and to consider the case of $Y = SO(n+1)$ and α the identity map. Then $E(\gamma_\beta)$ can be constructed by first choosing the 'core' plus a single fibre, i.e. $S^n \cup CSO(n+1)$, where the attaching map is $\pi\circ\gamma_\beta = \pi$. To this we must further attach a cone on the join $S^{n-1}*SO(n+1)$ by a suitable map η. We therefore have a cofibre sequence

$$S^n{\wedge}SO(n+1) \xrightarrow{\ \ \eta\ \ } S^n\cup_\pi CSO(n+1) \xrightarrow{\ \ \theta\ \ } E(\gamma_\beta) \qquad \text{(C)}$$

But $E(\gamma_\beta)$ is a topological manifold, fibred by S^n's and over a smooth base. As such, it admits a $STop(n)$ bundle of tangents along the fibres, say

$${}^{\mathrm{Top}}\tau_F : E(\gamma_\beta) \longrightarrow BSTop(n)\ .$$

Our aim is to determine the extent to which $E(\gamma_\beta)$ admits a smoothing fibred by S^n_β's, or equivalently to which it carries an n-plane bundle τ_F, agreeing with ${}^{\mathrm{Top}}\tau_F$ topologically and restricting to $\tau(S^n_\beta)$ on each fibre.

Now from §1, $\tau_\beta : S^n_\beta \longrightarrow BSO(n)$ extends to some bundle $\bar{\tau}_\beta$ over $S^n_\beta\cup_\pi CSO(n+1)$. Thus we may construct τ_F at least over $S^n\cup_\pi CSO(n+1)$, by composing $\bar{\tau}_\beta$ with $\bar{h}^{-1}$.

Returning to our cofibration (C), we can consider $\eta^*\tau_F$ over $S^n{\wedge}SO(n+1)$. This is topologically trivialized by the existence of ${}^{\mathrm{Top}}\tau_F$, so we have a map $\sigma(\beta): S^n{\wedge}SO(n+1) \longrightarrow Top(n)/O(n)$ which fits into the following homotopy commutative diagram

$$\begin{array}{ccccc} S^n{\wedge}SO(n+1) & \xrightarrow{\ \ \eta\ \ } & S^n\cup_\pi CSO(n+1) & \xrightarrow{\ \ \theta\ \ } & E(\gamma_\beta) \\ \downarrow{\scriptstyle \sigma(\beta)} & & \downarrow{\scriptstyle \tau_F} & & \downarrow{\scriptstyle {}^{\mathrm{Top}}\tau_F} \\ Top(n)/O(n) & \longrightarrow & BSO(n) & \longrightarrow & BSTop(n) \end{array}$$

So $\sigma(\beta)$, which we shall confuse with its adjoint $SO(n+1) \longrightarrow \Omega^n(Top(n)/O(n))$, is the obstruction to extending τ_F over the whole of $E(\gamma_\beta)$.

Thus in terms of our original $\alpha^*\gamma_\beta$ we deduce that if $\alpha: Y \longrightarrow SO(n+1)$ lifts to $\mathrm{Diff}S^n_\beta$ then the composite

$$Y \xrightarrow{\alpha} SO(n+1) \xrightarrow{\sigma(\beta)} \Omega^n(Top(n)/O(n))$$

is nul-homotopic.

We have shown only that this map is a necessary obstruction to lifting α. In the light of the celebrated Morlet equivalence $BDiff(D^n, \partial) \simeq \Omega^n(Top(n)/O(n))$ (e.g. see [KS]) it seems highly likely that $\sigma(\beta)$ and $g(\beta)$ of (1.3) are the same map. Note that on homotopy groups $\sigma(\beta)$ induces a non-bilinear extension to $\pi_k(SO(n+1))$ of the Milnor pairing $(\ ;\beta):\pi_k(SO(n)) \longrightarrow \pi_{k+n}(Top/O)$.

For calculational purposes, and given the current state of the art, unstable results such as we have obtained are not especially helpful. We must therefore show

(2.4) Lemma. The stabilization

$$s\sigma(\beta):SO(n+1) \xrightarrow{\sigma(\beta)} \Omega^n(Top(n)/O(n)) \longrightarrow \Omega^n(Top/O)$$

may be described as

$$SO(n+1) \xrightarrow[\gamma_\beta]{} TopS^n \xrightarrow[J]{} \Omega^{n+1}S^{n+1} \xrightarrow[\Omega^{n+1}\beta]{} \Omega^{n+1}B(Top/O) \quad \square$$

This formula follows simply from stabilizing the bundles in our discussion above.

To complete the proof of (2.2), we must choose $Y = S^k$ and α to represent a class $x \in \pi_k(SO(n+1)$ such that $\pi_*(x) \neq 0$ in $\pi_k(S^n)$. Then by (2.4), $x \in X_k(S^n)$ if

$$S^k \xrightarrow[x]{} SO(n+1) \xrightarrow[\sigma(\beta)]{} \Omega^n(Top/O)$$

is not nul-homotopic. The usual detection procedure for such a map is then to pass to $\Omega^n(G/O)$, and to compute its value in the summand $\pi^S_{n+k}/ImJ \subset \pi_{k+n}(G/O)$.

In our case the maps involved can be unravelled to give $s\sigma(\beta)_*x$ modulo ImJ as

$$S^k \xrightarrow{x} SO(n+1) \xrightarrow{\pi\times J} S^n \times G \xrightarrow{-\tilde{\beta}\times 1} G \times G \xrightarrow{\circ} G \xrightarrow{\circ\tilde{\beta}} \Omega^n G\ .$$

This represents $\tilde{\beta}\circ(J(x) - \tilde{\beta}(\pi\circ x))$ in π^S_{n+k}/ImJ. But $\pi\circ x = H(x)$, whilst Novikov [N] and Kosinski [Ko] have shown that $\tilde{\beta}\circ J(x) \in ImJ$ whenever $k > \frac{1}{2}n + 1$ (which is certainly the case here).

We can now deduce our detection formula (2.2), in the form

$$s\sigma(\beta)_*x = \pm\tilde{\beta}^2.H(x) \quad \text{in } \pi^S_{n+k}/ImJ\ .$$

Note that if S^n_β bounds a parallelizable manifold, then $\tilde{\beta} = 0$ by definition. So $SX_*(S^n_\beta) = \emptyset$. We conclude with a result which is a more subtle version of this same fact.

(2.5) Proposition. Let $S'X_*(S^n_\beta)$ be the intermediate set $SX_*(S^n_\beta) \subset S'X_*(S^n_\beta) \subset X_*(S^n_\beta)$ of elements detected by $\sigma(\beta)_*x \in \pi_{k+n}(Top/O)$. Then $S'X_*(S^n) = \emptyset$ if S^n_β bounds a parallelizable manifold.

Proof. By choice, $\beta \in \pi_{n+1}(B(Top/O))$ lifts to $\pi_n(G/Top)$. But localized at 2, G/Top is a product of Eilenberg-MacLane spaces, and at odd primes is equivalent to BO. In either case $\beta\circ f = 0$ for any $f \in \pi_{k+n+1}(S^{n+1})$ $\square$

This may be one more way of saying that such S^n_β's are the most symmetric of exotic spheres.

REFERENCES

[BL] D. Burghelea & R. Lashof, 'The homotopy type of the space of diffeomorphisms II' Trans. Amer. Math. Soc. 196 (1974), 37-50.

[Ke] M. Kervaire, An interpretation of G. Whitehead's generalization of H. Hopf's invariant', Ann. of Math. 69 (1959), 345-365.

[KS] R.C. Kirby & L.C. Siebenmann, 'Foundational essays on topological manifolds, smoothings and triangulations', Ann. of Math. Studies no. 88, Princeton Univ. Press (1977).

[Ko] A. Kosinski, 'On the inertia group of π - manifolds', Amer. Jour. of Math. 89 (1967), 227-248.

[N] S.P. Novikov, 'Differentiable sphere bundles', Izv. Akad. Nauk. SSSR Mat. 29 (1965), 1-96.

[S] R. Schultz, 'Improved estimates for the degree of symmetry of certain homotopy spheres', Topology 10 (1971), 227-235.

Matematisk Institut
Odense Universitet
5000 Odense
Denmark

Mathematics Department
The University
Manchester M13 9PL
England.

ON VANISHING OF THE ROHLIN INVARIANT

AND NONFINITELY AMPHICHEIRAL HOMOLOGY 3-SPHERES

by

L. Siebenmann

at 91405-Orsay, France

§ 1. Introduction.

I shall discuss the following conjecture and make some contributions to its proof.

CONJECTURE (A) *Let M^3 be an oriented Z_2-homology 3-sphere, i.e. a closed oriented C^∞ smooth 3-manifold with $H_*(M^3,Z_2) \cong H_*(S^3;Z_2)$. If M is amphicheiral, i.e. M is degree +1 diffeomorphic to (-M), then the Rohlin invariant $\rho(M) \in Z_{16} = Z/16Z$ is zero.*

Recall that $\rho(M)$ is by definition the signature $\sigma(W^4)$ modulo 16 of any smooth compact parallelisable 4-manifold W^4 with boundary $\partial W^4 = M^3$. That $\rho(M)$ is well-defined depends on Rohlin's theorem stating that $\sigma(W^4) \equiv 0$ modulo 16 if $\partial W^4 = S^3$, see [Ro] [La], it also depends on $H_*(M^3,Z_2) \cong H_*(S^3;Z_2)$ to assure that M^3 has only one parallelization on the complement of a point, equivalently only one spin structure ; the Rohlin invariant is defined for oriented closed spin 3-manifolds *. (There is a generalisation of ρ to all dimensions 4k-1 [EK] [HNK], which was inspired by J. Milnor's exotic spheres ; it is appropriately called the μ-invariant.)

Note that $M^3 \cong -M^3$ certainly implies that $\rho(M^3) = \rho(-M^3) = -\rho(M^3)$, whence $2\rho(M^3) = 0$, i.e. $\rho(M^3)$ is 0 or 8 modulo 16.

Conjecture (A) for Z-homologyspheres was proposed by Casson in [Ki, § 3.43] as a prelude to a deeper conjecture (B) to which (A) would bar an easy approach.

CONJECTURE (B). There exists a Z-homology 3-sphere M^3 with $\rho(M^3) \neq 0$ such that M^3 is Z-homology cobordant to $(-M^3)$.

A decade of work by several authors has led to a proof that (B) is equivalent to the (simplicial) triangulability of high dimensional manifolds (see [Matu] [GaS_1] for example), say in the form

CONJECTURE (T). Every metrizable topological manifold of dimension ≥ 5 (without boundary) is a simplicial complex.

Conjecture (A) was perhaps first suggested by the observation that the Klein-Poincaré Z-homology 3-sphere $M^3 = SO(3)/A_5$ with $\rho(M) \neq 0$ and indeed all Seifert fibered Z-homology 3-spheres (except S^3) admit no orientation reversing diffeomorphism. Bur real progress began in 1978 with :

THEOREM 1. ([Bi] [GaS_1] [HsP]). Conjecture (A) is true if M^3 admits a smooth, orientation reversing involution.

The proof in [GaS_2] can be described as exploiting the mapping cylinder of the involution to build a $W^4 \cong -W^4$ to define $\rho(M)$.

At Ron Stern's request, J. Van Buskirk and I constructed examples [SiV] , cf. [Bi], to which this theorem applies ; perceiving other reasons why their Rohlin invariants vanished I proved the somewhat special.

THEOREM 2. Conjecture (A) is true if M^3 is of plumbing type.

A closed 3-manifold is of <u>plumbing type</u> if it is the boundary of a regular neighborhood of smooth, normally immersed surface in an (oriented) 4-manifold. These manifolds are called (closed) <u>graph manifolds</u> by Waldhausen, who classified them in [Wa_1] (see [Or] or [Si_2, II] for complementary information).

My proof of theorem 2 runs as follows (some further details are supplied in § 2). Montesinos observed in [Mo_1], cf. [Mo_2][BoS_2], that on any Z_2-homology 3-sphere M of plumbing type, there exists a prefered class I_M of orientation preserving involutions such that, for $\tau \in I_M$, the orbit space M/τ is S^3 and the fixed point set is a knot $K_\tau \subset S^3$ that is said to be of <u>plumbing type</u> [Si_1] [Bo S] or equivalently <u>algebraic</u> [Co_1]. One can observe further , using Waldhausen's classification, that if τ' is also in I_M, the knot $K_{\tau'}$ is related to K_τ by a sequence of mutations of Conway [Co_1]. A <u>mutation</u> is the 180° rotation of a tangle within the knot as in figure 1-a.

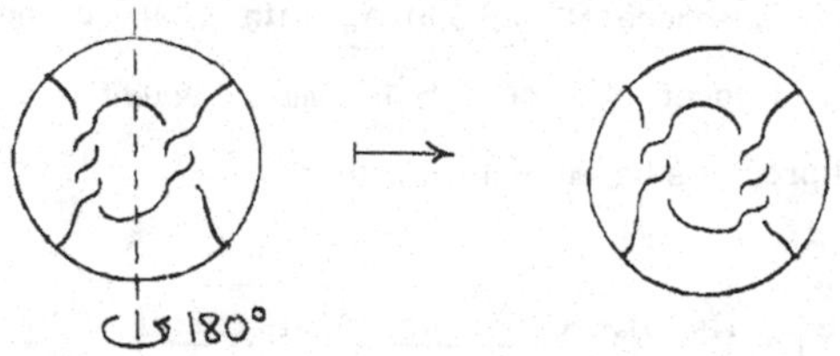

Figure 1-a

(Of course, other tangle strings are allowed in place of the ones illustrated, provided they exit at four points as illustrated).

Let $F^2_\tau \subset S^3$ be any orientable, Seifert surface for the knot K_τ $(= \partial F^2_\tau)$. Push the interior of F^2_τ radially into the 4-ball B^4, $(\partial B^4 = S^3)$, and pass to the 2-fold branched cyclic covering W^4_τ of B^4, branched along F_τ. This W^4_τ is a manifold with boundary M^3, and it is parallelizable (for example because $W^4_\tau - F_\tau$ certainly is parallelizable and W^4_τ embeds in $W^4 - F_\tau$). Thus $\rho(M^3) = \sigma(W^4_\tau) \bmod 16$. But the integer $\sigma(W^4_\tau)$ is by definition the well-known signature invariant of the knot $(S^3, K_\tau) \cong (M/\tau, K_\tau)$, see [Go].

As Conway first observed (read between the lines of [Co_1, §7], or request [Co_2]) knot signature is a mutation invariant (a geometrical proof is indicated in § 2). Thus the integers $\sigma(W^4_\tau)$, $\tau \in I_M$, all coincide. Call this common value $\tilde{\rho}(M)$; certainly $\tilde{\rho}(M) \equiv \rho(M)$ modulo 16.

One has $\tilde{\rho}(-M) = -\tilde{\rho}(M)$, since $\sigma(-W^4_\tau) = -\sigma(W^4_\tau)$. Thus $M \cong -M$ implies $\tilde{\rho}(M) = -\tilde{\rho}(M) \in Z$, and $\tilde{\rho}(M) = 0 = \rho(M)$, proving Theorem 2.

<u>Remark</u>. W. Neumann [Ne_1] announced that there is a lifting of $\rho(M)$ to an integer invariant $\bar{\mu}(M) \in \mathbf{Z}$. Clearly, the plausible equation $\bar{\mu}(-M) = -\bar{\mu}(M)$ would then imply Theorem 2. Thus it became clear that Neumann had an earlier proof of Theorem 2. I have nevertheless presented $\tilde{\rho}$ because I understand that his methods are different (see these proceedings).

<u>Remark</u>. Because signature is a cobordism invariant of knots, it is clear that $\tilde{\rho}$ is invariant under Z-homology cobordism arising from knot cobordism between algebraic knots by passage to 2-fold branched covering of the knot cobordism. However the natural domain of definition of $\tilde{\rho}$ and its invariance under Z-homology cobordism are matters for conjecture.

<u>Remark</u>. Note that <u>any</u> mutation invariant of an algebraic knot K_τ associated to the homology sphere M^3 of plumbing type gives a diffeomorphism invariant of M^3. These include (a fortiori) all skein invariants of Conway [Co_2], in particular the Alexander polynomial Δ. According to D. Cooper (see end of § 3) they include all signatures σ_ω where $\Delta(\omega) \neq 0$ and ω is complex of norm 1 [Coo].

Recently L. Contreras-Caballero (at Cambridge) and A. Kawauchi (at IAS) have proved .

<u>THEOREM</u> 1^* [Cont] [Ka]. <u>Conjecture</u> (A) <u>is true if</u> M^3 <u>admits an orientation reversing diffeomorphism</u> f <u>of finite order</u>.

If f has order $n2^k$ with n odd , then $g = f^n$ is an orientation reversing diffeomorphism of order 2^k . Thus, given Theorem 1, it remained to consider the case where f has order 2^k , $k \geq 2$.

The proof by Contreras [Cont] (for **Z**-homology 3-spheres only) is quite simple ; the trick amounts to considering the knot that is the fixed point set of $f^{(a/2)}$, $a = 2^k$; this is an amphicheiral knot living in $M/f^{a/2}$ and its 2-fold branched cover is M . Kawauchi's proof [Ka] , which is considerably more difficult, has the advantage of applying equally well to Z_2-homology 3-spheres.

This theorem is quite powerful. It applies for example to any amphicheiral Z_2-homology sphere M that is hyperbolic , indeed, an orientation reversing diffeomorphism of M can be replaced by a homotopic isometry via G. Mostow's hyperbolic rigidity theorem (see [Mos] [Th , chap. 5]) , and the isometry group of any compact (or finite volume complete) hyperbolic manifold has finite order.

Thus encouraged, one might conjecture that Theorem 1^* applies directly to prove Theorem 2, or even conjecture (A) in general. However, by construction of examples, I shall show that this is not so.

Recall that an prime 3-manifold is one in which every (smoothly embedded) separating 2-sphere bounds a 3-ball.

<u>FACT</u> 3. There exist prime amphicheiral Z-homology 3-spheres M^3 of plumbing type such that .

(a) Every orientation reversing diffeomorphism of M has infinite order, even up to homotopy, and

(b) no knot with 2-fold branched covering M is amphicheiral.

FACT 4. There exist prime amphicheiral Z-homology 3-spheres M^3 containing a family T of just two 2-tori, so that M^3 - T is complete hyperbolic of finite volume such that every homeomorphism M → M (except the identity) has infinite order (cf. [RaT]), and every map M → M (except one homotopic to the identity) has infinite order up to homotopy.

These constructions, which are perhaps as instructive as the theorems, are presented in § 4 and § 5 . In the intervening § 3, I pause to describe a reasonably convenient way to list all the homology 3-spheres of plumbing type and those that are amphicheiral.

In spite of facts 3 and 4 , one can go a long way towards a proof of conjecture (A) by exploiting Theorem 1^* in conjunction with a caracteristic splitting along 2-tori into Seifert pieces and hyperbolic pieces. I shall prove

THEOREM 5. Conjecture (A) is true in case M is prime and sufficiently large or Seifert or hyperbolic, or a connected sum (# sum) of such manifolds.

Perhaps all closed 3-manifolds verify these hypotheses (but see [Th, chap. 4]). No homotopy 3-sphere except S^3 does !

The proof of Theorem 5 is given in § 5 and § 6 . but can be read immediately. It incidentally provides still another proof of Theorem 2 (which can avoid mention of plumbings and knots). For M^3 not of plumbing type, the proof makes use of W. Thurston's hyperbolization theorem . To be independent of this deep theorem of [Th] (whose proof is not yet available) one can reformulate Theorem 5 as follows.

THEOREM 5^* . Conjecture (A) is true in case M^3 is a connected sum $M_1 \# \dots \# M_k$ where each M_i^3 contains a family T_i^2 of disjoint 2-tori such that each component of $M_i^3 - T_i^2$ admits a Seifert fibration or a complete hyperbolic metric of finite volume.

I am endebted to Lucia Contreras-Caballero and Andrew Casson for raising the question whether Theorem 1^* implies Conjecture (A) and more particularly whether every amphicheiral homology 3-sphere has a finite order degree -1 diffeomorphism.

Also I want to thank Francis Bonahon for discussions concerning counter-examples of plumbing type (fact 3) ; some more refined counterexamples he constructed are mentioned in § 4 I have made §§2-5 depend to some extent on [BoS] .

§ 2. - KNOTS OF PLUMBING TYPE AND MUTATION

This section indicates alternative proofs for two of the known results used in § 1 to prove Theorem 2 , namely :

THEOREM 2.1 Every oriented Z_2-homology 3-sphere of plumbing type M^3 is the 2-fold branched cyclic covering of S^3 branched along a knot K of plumbing type (= algebraic), and M^3 determines K up to isotopy and Conway mutations.

THEOREM 2.2 (Conway). If knots K , K' in S^3 are related by a Conway mutation, they have the same signature.

Theorem 2.1 was first proved in terms of Conway's tangles as a corollary of [$Mo_{1,2}$] and [Wa_1]. Here we will sketch the proof in terms of the formalism of band plumbing proposed in [Si_1] .

Consider an integrally weighted planar tree such as (†)

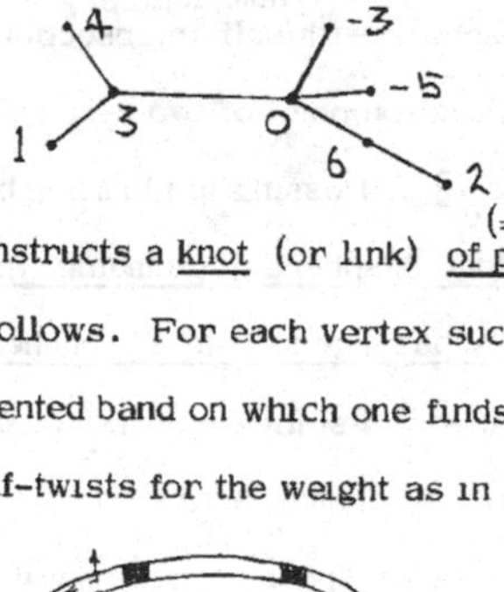

Figure 2-a

From it, one constructs a knot (or link) of plumbing type (=algebraic) in S^3 well-defined up to isotopy -as follows. For each vertex such as 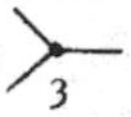 in figure 2-a, take an unknotted oriented band on which one finds in cyclic order gluing patches for the valences and half-twists for the weight as in figure 2-b .

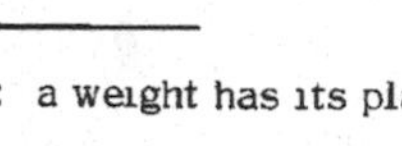

Figure 2-b

(†) Curious fact : a weight has its place in the cyclic order at a vertex, and this may count in the following knot construction.

Then plumb all these bands together in S^3 (see figure 2-c) as the tree demands.

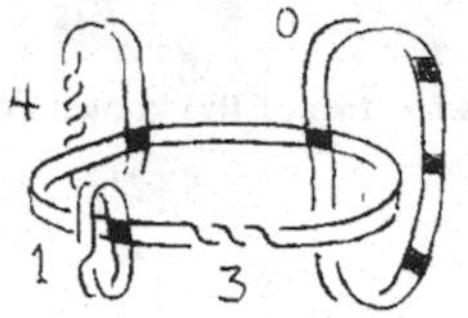

Figure 2-c

One takes care to match band core orientation to normal orientation and visa versa, there results a surface P^2 in S^3 whose boundary $\partial P = K$ is the knot (or link) associated to the graph. This K is by definition of plumbing type. (No string orientation on K is specified ; P^2 is not orientable if odd weights occur.)

Next push the interior of the above plumbed surface P^2 radially into B^4 ($\partial B^4 = S^3$) so that $P \cap S^3 = \partial F$ and then form the 2-fold branched cyclic covering P^4 of B^4 branched along (pushed) P^2.

OBSERVATION 2.3 (of Gonzales-Acuña and the author). P^4 is the plumbing of 2-disc bundles over S^2 corresponding to the given weighted tree.

In other words P^4 is an (abstract) regular neighborhood of a collection of 2-spheres embedded in a 4-manifold, that intersect as the plumbing graph describes : a vertex with weight x corresponds to a 2-sphere with homological self-intersection number x and an edge represents a transversal intersection point of two 2-spheres.

For a proof, see [Si_1] or [Mo_3] . Passing to plumbing boundaries, we conclude that every oriented Z_2-homology 3-sphere of plumbing type arises from a (connected) knot of plumbing type in S^3 by passage to 2-fold branched cyclic covering. At this point, one should recall the Smith theorem stating that the fixed point set of an automorphism of order 2 (or p^k with p prime) acting on a homology sphere for coeficients Z_2 (or Z_p) is itself a homology sphere for the same coeficients.

It remains to show that the homology sphere determines the knot up to mutation. For this, we use

PROPOSITION 2.4. Let K, $K' \subset S^3$ be two knots or links of plumbing type that have degree +1 diffeomorphic 2-fold branched cyclic coverings. Then, there exist plumbed surfaces P, $P' \subset S^3$ with boundaries K, K' that arise from weighted planar trees Γ, Γ' that are isomorphic as abstract weighted trees

This is proved in [BoS_2] by modifying P (and similarly P') in elementary fashion, via certain "flyping" moves known to the 19th century knot tabulators (Tait Kirkman Little) but equally related to blowing up and down algebraic curves, until the weighted tree, viewed as an abstract weighted tree, is almost canonical (see § 3.4); then, by Waldhausen [Wa_1], it then turns out to be an invariant of the oriented 2-fold branched covering.

The proof of 2.1 can now be finished off with

OBSERVATION 2.5. Suppose two knots or links K, $K' \subset S^3$ arise from weighted planar trees Γ, Γ' that are isomorphic as abstract weighted trees. Then K and K' are related by a sequence of mutations (and isotopies).

The proof is an exercice. For example, the two knots of figure 2-d

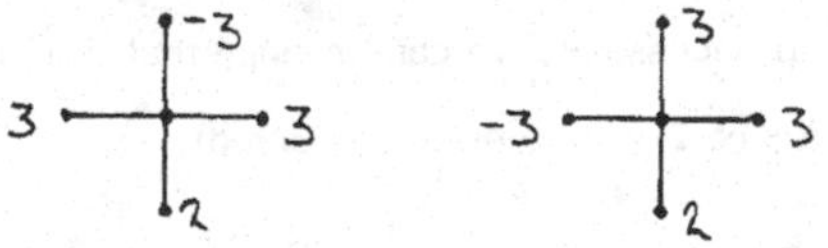

Figure 2-d

are related by one mutation.

Mutation leaves signature unchanged.

We now give simple minded geometric proof of Conway's Theorem 2.2.

Consider a tangle in a knot K, for convenience we can think of it as cut out by the hemisphere B^3_+ of $S^3 = \partial B^4$ where K meets the equator $S^2 = \partial B^3_+$ transversally in four equidistant points on a great circle. A mutation alters K to K' by 180° rotation μ of B^3_+ about the vertical dotted axis in figure 2-e. Of

Figures :

2-e 2-f 2-g 2-h

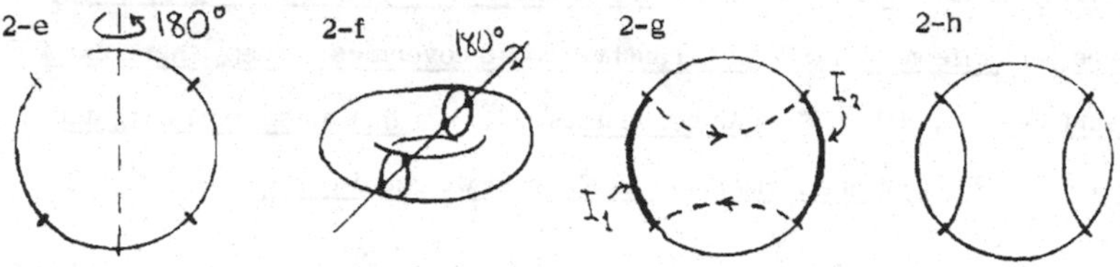

course we could use rotation μ' about the horizontal axis. The crucial point is that the motion $\mu \mid S^2$ used should respect the four exit points and satisfy a triviality condition (*) on the covering automorphism $\tilde{\mu}$ of the 2-torus T^2 that is the 2-fold branched covering of S^2 (branched at the four points), namely :

(*) Modulo the covering translation τ of T^2 (see figure 2-f), the automorphism $\tilde{\mu}$ is isotopic to the identity , equivalently, $\tilde{\mu}$ acts as $\pm$ identity on $H_1(T^2)$.

It is known that up to isotopy fixing the four points this criterion leaves just a group $V_4 \cong Z_2 \oplus Z_2$ giving 3 choices for μ ; V_4 is generated by the two rotations pointed out, cf. [BoS$_2$] .

This clarifies the meaning of mutation.

Reparametrizing S^2 if necessary, we can arrange that $K \cap B^3_+$ joins the four points $K \cap S^2$ as in figure 2-g (orientation counted).

<u>Assertion</u>. In this situation, there exists an (oriented) Seifert surface F^2 for K that meets S^2 transversally so that $F^2 \cap B^3_+$ consists of the two arcs I_1, I_2 in figure 2-g.

To prove this one can, for example, adopt the projection to paper given by figure 2-g positioning K so that its projection on the paper has only the inevitable four points of intersection with the circle delimiting the image of B^3_+ . Then Seifert's own rule [Sei$_2$] produces F . □

At this point, choosing μ to be a standard $180°$ rotation in V_2, we note that $F \cap B^3_-$ and $\mu(F \cap B^3_+)$ fit together to form a Seifert surface F' for the mutated knot K'.

Now think of B^3_+ as a hemisphere again ; and let B^3_0 be the equatorial 3-disc of B^4 that is the cone on $\partial B^3_+ = S^2$. This B^3_0 cuts B^4 into two half-balls. B^4_+, B^4_- with $B^4_\pm \cap S^3 = B^3_\pm$.

After pushing the interior of F^2 radially into B^4 so that $F^2 \cap \partial B^4 = K$, note that $F \cap B^3_0$ is the trivial tangle of figure 2-h.
Thus the 2-fold branched covering W^4 of B^4 branched over F splits as $W^4 = \tilde{B}^4_+ \cup \tilde{B}^4_-$ with $\tilde{B}^4_+ \cap \tilde{B}^4_- = \tilde{B}^3_0$ the 2-fold covering of the trivial tangle of figure 2-g ; this $\tilde{B}^3_0$ is a solid torus illustrated in 2-f.

The rotation μ extends canonically to B^4 and we can assume that the push of F keeps K μ-equivariant near B^3_0. Then pushed F' is $(F \cap B^4_-) \cup \mu(F \cap B^4_+)$ and the 2-fold branched cyclic cover of B^4 branched along F' is :

$$W' = \tilde{B}^4_- \cup_{\tilde{\mu}} \tilde{B}^4_+$$

where $\tilde{\mu} : \tilde{B}^3_0 \to \tilde{B}^3_0$ is either automorphism covering the $180°$ rotation $\mu | B^3_0$. It follows that this $\tilde{\mu}$ is isotopic to the identity on the solid torus $\tilde{B}^3_0$ (or to the covering translation) ; hence, $W^4 = \tilde{B}^4_- \cup \tilde{B}^4_+ \cong \tilde{B}^4_- \cup_{\tilde{\mu}} \tilde{B}^4_+ = W'^4$. The signatures of these two 4-manifolds are a fortiori the same and so Theorem 2.2 is proved. □

<ins>Remark.</ins> I have heard that Daryl Cooper (at Warwick) has a simple proof that mutation even preserves the S-equivalence class of the Seifert matrix of a knot.

<ins>Question.</ins> Does mutation preserve the cobordism class of a knot ?

§ 3. CENSUS OF HOMOLOGY 3-SPHERES OF PLUMBING TYPE.

The aim is to help make practical the theoretical census of [Wa$_1$]

The census of Z_2-homology 3-spheres given below arose in the classification of algebraic knots,† see [Co$_1$] [Si$_1$] [BoS$_2$] [CauS] [Cau] . But it should also be of interest to algebraic geometers since the language relates directly to configurations of algebraic curves , see 3.4 below.

For Z-homology 3-spheres the method of census to be discussed revealed Fact 3 (see § 4) and is perhaps of most interest in testing Conjecture (B) of § 1.

In a 'practical' census one ultimately wants to avoid sorting through huge numbers of manifolds that are not of interest. Next best is to be able to do such sorting very efficiently.

A closed oriented 3-manifold M^3 <u>of plumbing type</u> (= a graph manifold) is one that is the boundary of the regular neighborhood P^4 of a normally immersed closed unoriented surface F^2 (perhaps not connected) in an oriented 4-manifold. Normally immersed means that F^2 is smoothly embedded away from finitely many isolated normal crossings with model

$$(R^2 \times 0) \cup (0 \times R^2) \subset R^4.$$

---

† Algebraic knots were so named by Conway [Co$_1$] . They were renamed knots of plumbing type in § 2 because algebraic geometers have reserved the term algebraic knots for the iterated torus knots met as links of algebraic curve singularities. However Conway's terminology can be defended with more algebraic geometry [BoS$_2$] .

In order for M^3 to be a Z_2-homology 3-sphere, i.e. for $H_1(M^3;Z_2)$ to be zero where Z_2 is $Z/(2Z)$, it is easily seen to be necessary that each component of F be an <u>embedded</u> 2-<u>sphere</u>. (Indeed if x, y are curves in F with a Z_2-intersection $x.y \neq 0$, then putting x and y in general position, taking the circle bundle X over x in ∂P^4, and lifting y to y' in ∂P^4, we get a nonzero Z_2-homology intersection $X.y'$ in ∂P^4.) Further the naturally defined <u>graph</u> Γ with one <u>vertex</u> for each 2-sphere component of F and one <u>edge</u> for each normal crossing of F cannot contain a cycle, i.e., it must be a <u>tree</u>; in fact Γ can be embedded as a retract of $M^3 = \partial P^4$.

We therefore make henceforth the

<u>Assumption</u> : All components of F^2 are 2-spheres, and the self-intersection graph Γ is a tree.

<u>Remark</u>. The same assumption is justified if we are considering Q-homology spheres not Z_2-homology spheres. But there is a pitfall to avoid. The above argument shows that Γ is a tree and that each component F_v of F satisfies $H_1(F_v;Q) = 0$. Beware: this leaves the (genuine!) possibility that $F_v \cong RP(2)$; fortunately Waldhausen [Wa_1, § 3] has shown that in this case, without changing M^3 up to diffeomorphism, F_v can be replaced by several (three) 2-sphere components; a pretty explanation of this in terms close to the language of plumbings is given in [Mo_2].

It is now clear for homological reasons (cf. [HNK]) that

(3.1) $H_1(M^3; Z_2) = 0$ <u>if and only if the intersection form on</u> $H_2(P^4; Z)$ <u>has odd determinant</u>.

Since large determinants are notoriously awkward to evaluate, I point out two practical criteria in terms of the above tree, in which each vertex is weighted by its homological self-intersection number (its normal euler number).

(3.2) $H_1(M^3; Z_2) = 0$ if and only if (one or every) 2-dimensional plumbed surface P^2 in R^3 corresponding, by the rules in § 2, to the same weighted tree (made planar) has connected boundary ∂P^2.

This is because the 2-fold branched covering of S^3 branched over ∂P^2 is M^3 (see § 2, and use Smith theory).

(3.3) $H_1(M^3; Z_2) = 0$ if and only if the weighted tree above can be reduced to the single weighted vertex $\bullet \pm 1$ by the moves catalogued in [NeW] or in [BoS_2] together with replacement of even weights by 0 and of odd weights by ± 1.

Given [NeW] or [BoS_2] the proof of 3.3 is an easy exercise. (Moves splitting the graph are unnecessary, cf. 3.9 below which proves an integral version in detail).

Since we can now test for $H_1(M^3; Z_2) = 0$, it will suffice in principle to give a census of oriented diffeomorphism types of manifolds arising from weighted trees. This (and more) is done in [Wa_1]; a practical reformulation of Waldhausen's classification can be described as follows, (refer again to [BoS_2] for details).

An integrally weighted tree Γ is almost canonical if, when all polyvalent vertex stars are removed from Γ, producing Γ_L say, each component of Γ_L is a linear chain •—•—• . . → with ≥ 1 vertex, such that the weights are nonzero of alternating sign, with end weight(s) of norm ≥ 2.

THEOREM 3.4. (see proof in [BoS]). Prime closed 3-manifolds of plumbing type distinct from S^3 and $S^1 \times S^2$, and arising from integrally weighted trees as above are classified up to orientation preserving diffeomorphism by all almost canonical trees as defined above modulo the following two 'theft' operations and their inverses :

1) - - - $\overset{x}{\bullet}$—$\overset{2}{\bullet}$ $\mapsto$ - - - $\overset{x-1}{\bullet}$—$\overset{-2}{\bullet}$

2) - - $\overset{x}{\bullet}$—$\overset{2}{\bullet}$—$\overset{y}{\bullet}$ - - - $\mapsto$ - - $\overset{x-1}{\bullet}$—$\overset{-2}{\bullet}$—$\overset{y-1}{\bullet}$ - - - .

Here dashes at a vertex indicate a continuation of the tree making the vertex

<u>polyvalent</u>, e.g. $---\overset{x}{\bullet}\!\!-\!\!\!-\!\!\!\overset{2}{\bullet}$ <u>may mean</u> $3, 3, -2$ attached to x, with $x \;—\; 2$ <u>but not</u> $\overset{3}{\bullet}\!-\!\overset{x}{\bullet}\!-\!\overset{2}{\bullet}$ <u>nor</u> $\overset{x}{\bullet}\!-\!\overset{2}{\bullet}$.

Frequently these theft operations create little ambiguity. When there is just one weighting in the theft equivalence class giving least total absolute weight, we call the tree with that weighting <u>perfectly canonical</u>.

<u>Warning</u> There do exist Z_2-homology 3-spheres not corresponding to a perfectly canonical tree. For example:

And it is just an exercise (given 3.6 or § 4 below) to give examples that are Z-homology 3-spheres.

If the reader desires a canonical tree in all cases, he can neuter and emprison the 2's where thefts occur above, as follows :

(1) $---\overset{x}{\bullet}\!-\!\!-\!\overset{\pm 2}{\bullet} \;\mapsto\; ---\overset{x \mp 1/2}{\bullet}\!-\!\!-\!\text{②}$

(2) $---\overset{x}{\bullet}\!-\!\!-\!\overset{\pm 2}{\bullet}\!-\!\!-\!\overset{y}{\bullet}-- \;\mapsto\; ---\overset{x \mp 1/2}{\bullet}\!-\!\!-\!\text{②}\!-\!\!-\!\overset{y \mp 1/2}{\bullet}--$

The result, strange, is absolutely unique !

A naive census for Z_2-homology 3-spheres of plumbing type could now run as follows. List all almost canonical trees in order of total absolute weight discarding ones giving $H_1(M ; Z_2) \neq 0$, and organize into theft equivalence classes.

Amphicheirality then means that changing the sign of all weights gives the same weighted tree up to isomorphism and thefts ; for perfectly canonical trees, this means that changing the signs of all weights yields a strictly isomorphic weighted tree.

This census is still not practical since one will find that, in the long run, most almost canonical trees give $H_1(M ; Z_2) \neq 0$, i.e. most of our effort is wasted.

To avoid this wastage, it is perhaps best to first list trees with weights in $Z/2 = Z_2$ (not Z) that give $H_1(M ; Z_2) = 0$ -say with the help of 3.3 or the $Z_{(2)}$ version of 3.9 below. (By 3.1, 3.2, or 3.3, one sees that the weights mod 2 decide whether $H_1(M ; Z_2)$ is zero). Then one can go on to lift the $Z/2$ weightings to special integral weightings.

RECOGNIZING AND CONSTRUCTING Λ-HOMOLOGY 3-SPHERES ($Z \subset \Lambda \subset Q$)
(§ 3 continued)

When homology coeficients are not specified, understand integer coeficients. Λ will denote a unitary subring of the rational numbers Q.

I am most interested here in Z-homology 3-spheres, but I keep the extra generality (given by using Λ in place of Z) where it costs nothing.

Note that $Z_2 = (Z/2)$ homology 3-spheres are precisely Λ-homology 3-spheres where Λ is the ring $Z_{(2)}$ of rational numbers with odd denominators.

Given a unitary ring $\Lambda \subset Q$, note that there is a well-defined set $p = \{p_1, p_2, \ldots\}$ of distinct prime numbers > 1 in Z such that Λ is precisely the ring $Z_{(p)}$ of rational numbers with denominators prime to $p_1, p_2, \ldots$. For example $Q = Z_{(p)}$ when $p = \emptyset$ and $Z = Z_{(p)}$ when $p = \{\text{all primes}\}$.

For any compact 3-manifold N^3 with boundary $\partial N \cong T^2$, the kernel of the inclusion induced map $i_* : H_1(\partial N) \to H_1(N^3)$ is infinite cyclic. (Hint : Consider non-zero x in $H_1(\partial N)$; there exists y in $H_1(\partial N)$ with $x.y \neq 0$. If $x = \partial X$ for X in $H_2(N, \partial N)$, one has homology intersections $X.i_* y = x.y \neq 0$ proving $i_* y \neq 0$.)

Consider a compact manifold N^3 obtained by modifying $B^2 \times S^1$ (according to Dehn) by removing the interiors of disjoint sub-tubes $\overset{o}{B}{}^2_i \times S^1$, $i = 1, \ldots, r$, then replacing via gluing maps $f_i : \partial(B_i^2 \times S^1) \to (B^2 - \cup \overset{o}{B}_i) \times S^1$ so that f_i maps $\partial B_i^2 \times S^1$ onto itself and

$$(\#) \qquad f_i(\partial B_i^2) = \alpha_i \ \partial B_i^2 + \beta_i S^1$$

in $H_1(\partial B_i^2 \times S^1)$, where α_i, β_i are coprime integers.

<u>LEMMA 3.5. The infinite cyclic kernel of</u> $i_* : H_1(\partial N) \to H_1(N^3)$ <u>has generator of the form</u> $\alpha \partial B^2 + \beta S^1$ <u>in</u> $H_1(\partial N)$, $\partial N = \partial B^2 \times S^1$, <u>where</u>

$$\frac{\beta}{\alpha} = \frac{\beta_1}{\alpha_1} + \ldots + \frac{\beta_r}{\alpha_r},$$

<u>provided no</u> $\alpha_i = 0$, <u>while</u> $\beta/\alpha = (\pm 1)/0$ <u>if some</u> $\alpha_i = 0$.

<u>Proof of 3.5.</u> We can assume no α_i is zero since the result is clear if some α_i is zero. In $H_1(N)$, let $a, a_1, a_2, \ldots, a_r$ be represented by the boundary components of the punctured disc $B^2 - \bigcup_i \overset{o}{B}{}^2_i$. And let b in $H_1 N$ be represented by the given circle factor of ∂N. Then $a + a_1 + \ldots + a_r = 0$, and $\alpha_i a_i - \beta_i b = 0$, $i = 1, \ldots, r$. Thus, combining we get :

$$\ell cm(\alpha_1,\ldots,\alpha_r)a + \sum_{i=1}^{r} \beta_i \ell cm(\alpha_1,\ldots,\alpha_r)\,\alpha_i^{-1} b = 0$$

where $\ell cm(\alpha_1, \ldots, \alpha_r)$ is the least common multiple of $\alpha_1, \ldots, \alpha_r$. This proves 3.5. □

Remark. Looking more closely at the proof of 3.5, one can see that the kernel generator is (up to sign) precisely :

$$\ell cm(\alpha_1,\ldots,\alpha_r)\, a + \sum_{i=1}^{r} \beta_i \ell cm(\alpha_1, \ldots ,\alpha_r)\,\alpha_i^{-1} b ,$$

not a submultiple. Thus the kernel generator is in general divisible ; take for example $r = 3$ and $(\alpha_1,\beta_1) = (3,1)$, $(\alpha_2,\beta_2) = (2,1)$, $(\alpha_3,\beta_3) = (2,-1)$; the kernel generator is then : $-6a + 2b = 2(-3a + b)$.

LEMMA 3.6. With the above data, $H_1(N^3,\Lambda) \cong H_1(S^1,\Lambda) = \Lambda$ if and only if the integers $\alpha_1,\ldots,\alpha_r$ are pairwise coprime in Λ in the sense that, whenever $i \neq j$, the sum of principal ideals $\alpha_i\Lambda + \alpha_j\Lambda$ is Λ, equivalently the greatest common divisor $\alpha_i \cap \alpha_j$ is divisible by no prime in p $(\Lambda = \mathbf{Z}_{(p)})$.

Proof of 3.6. One calculates (for example by a Mayer-Vietoris sequence, cf. [Wa$_3$]) that $H_1(N^3;\Lambda)$ is the Λ-module with generators x , $x_1,\ldots,x_r$ and relations

$$x = \alpha_1 x_1 = \ldots = \alpha_r x_r .$$

Here x is represented by the factor S^1 in $\partial N = \partial B^2 \times S^1$, and x_i is represented by the core of the i-th tube (reglued).

Suppose that $H_1(N,\Lambda) \cong \Lambda$ and consider the (cyclic !) quotient of $H_1(N^3;\Lambda)$ got by killing x , namely

$$\Lambda/(\alpha_1\Lambda) \oplus \ldots \oplus \Lambda/(\alpha_r\Lambda) .$$

Since this sum of cyclic modules over the euclidean domain Λ is cyclic, $\alpha_1,\ldots,\alpha_r$ must be pairwise coprime in Λ .

Conversely, supposing $\alpha_1,\ldots,\alpha_r$ pairwise coprime in Λ , there must exist $\delta_1,\ldots,\delta_r$ in Λ so that $\Sigma(\delta_i\alpha_1\ldots\alpha_r)\alpha_i^{-1} = 1$. We claim that $z = \delta_1 x_1 + \ldots + \delta_r x_r$ is a generator of $H_1(N^3,\Lambda)$. Certainly :

$$\alpha_1 \ldots \alpha_r z = \sum_{i=1}^{r} (\delta_i \alpha_1 \ldots \alpha_r)\, \alpha_i^{-1} x = x .$$

Consider the two Λ-cyclic submodules $(x) \subset (z) \subset H_1(N;\Lambda)$. Since we have abstract isomorphism $(z)/(x) = \Lambda/(\alpha_1\ldots\alpha_r\Lambda) \cong H_1(N,\Lambda)/(x)$, it follows that the inclusion of torsion modules $(z)/(x) \subset H_1(N;1)/(x)$ is an isomorphism and $(z) = H_1(N,\Lambda)$. $\square$

Remark. Integers $\alpha_1,\ldots,\alpha_r$ are pairwise coprime in Q if and only if at most one of $\alpha_1,\ldots,\alpha_r$ is zero.

COMPLEMENT 3.7. For any compact 3-manifold N^3 verifying $H_1(N^3,\Lambda) \cong \Lambda$, and having boundary ∂N a 2-torus T^2, the pair $(N^3,\partial N^3)$ has the same exact Λ-homology sequence as $(B^2,\partial B^2) \times S^1$.

Proof of 3.7. It suffices to show (in view of exactness) that the inclusion induced map $i_*\ H_1(\partial N;\Lambda) \to H_1(N;\Lambda) \cong \Lambda$ of the sequence is surjective. Aiming to show this, consider a generator x of the (cyclic !) kernel of i_* . Since $H_1(N;\Lambda) \cong \Lambda$, the element x must be indivisible , hence for some y in $H(\partial N;\Lambda)$ we have $x.y = 1$. Since $i_*(x) = 0$, there is X in $H_2(N,\partial N,\Lambda)$ with boundary $\partial X = x$. Then $X.i_*y = \partial X.y = 1$, which reveals that i_*y is not divisible in $H_1(N,\Lambda) \cong \Lambda$. Thus i_* is onto $\square$

Consider now the plumbing P^4 associated to an integrally weighted finite tree Γ . The above two lemmas will provide a very efficient way to decide whether $\partial P^4 = M^3$ is a Λ-homology 3-sphere. Also it will give a way to alter Γ, P, M without changing $H_*(M^3;\Lambda)$, this is how the examples of § 4 below were built.

For a 2-torus T^2 in a Λ-homology 3-sphere M^3 , we shall often use the elementary fact that for each of the two closed complementary components C_1, C_2 of T^2 in M^3 $(\partial C_1 = T^2)$, the pair $(C_1,\partial C_1)$ has the integral homology sequence of the solid torus $(B^2 \times S^1, \partial B^2 \times S^1)$, cf. 3.7.

In the weighted tree Γ, fix attention on a vertex v and the edges emanating from it.

For such an edge e , there is a 2-torus $T(e)$ in $M^3 = \partial P^4$; it is the corner set (before smoothing) arising from the plumbing for e (figure 3-a) . This $T(e)$ gets a standard homology basis $a, b \in H_1(T(e))$, where a corresponds to the base of the 2-disc bundle $E(v)$ over S^2 and b to the circle fiber of $\partial E(v)$. More precisely, if B^2 is the oriented disc in the base 2-sphere such that $E(v)|B^2$ is the patch plumbed for edge e and if $E(v)|B^2$ is $p_1 : B^2 \times D^2 \to B^2$, then

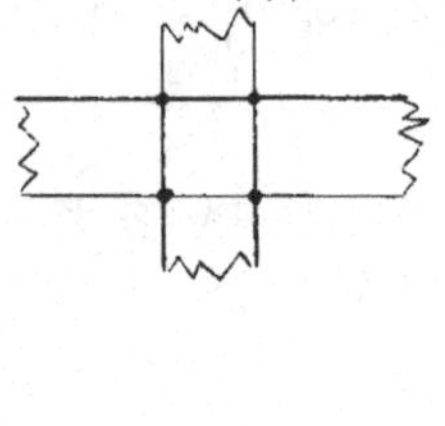

Figure 3-a

$T(e)$ is $\partial B^2 \times \partial D^2$, and, in $H_1(T(e))$, the generator a is $\partial B^2 \times$ (point) while b is (point) $\times \partial D^2$.

Write $M(t)$ for the part of $M^3 = \partial P^4$ arising from the branch t of Γ attached by e to v : 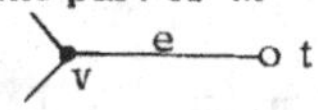

; it has boundary $T(e)$. Also let

$$\pm(\alpha a + \beta b) \in H_1(T(e)), \text{ with } \alpha, \beta \text{ coprime in } Z ,$$

be the generator of the kernel $i_* : H_1(T(e)) \to H_1(M(t))$. Call β/α the <u>characteristic fraction</u> (for vertex v and emanating edge e), it lies in Q .

Now 3.5, 3.6, 3.7 yield by a simple pursuit of definitions the :

<u>FOLDING LEMMA 3.8</u>. <u>With the above conventions, consider a collection of edges</u> $e_1, \ldots, e_r$ <u>emanating from vertex</u> v <u>leading to branches</u> $t_1, \ldots, t_r$ <u>and having characteristic fractions</u> β_i/α_i , $i + 1, \ldots, r$.

<u>Suppose that the denominators</u> $\alpha_1, \ldots, \alpha_r$ <u>are pairwise coprime in</u> Λ .

<u>Then the following change in the graph</u> Γ <u>produces no change in the</u> Λ<u>-homology of the corresponding plumbing boundary</u> $\partial P^4 = M^3$: <u>replace the branches</u> $t_1, \ldots, t_r$ <u>by a single branch</u> t <u>attached to</u> v <u>by an edge</u> e <u>where the characteristic franction</u> β/α <u>for the new branch is</u> :

$$\frac{\beta}{\alpha} = \frac{\beta_1}{\alpha_1} + \ldots + \frac{\beta_r}{\alpha_r}$$

(or $\pm 1/0$ if some $\alpha_i = 0$) . □

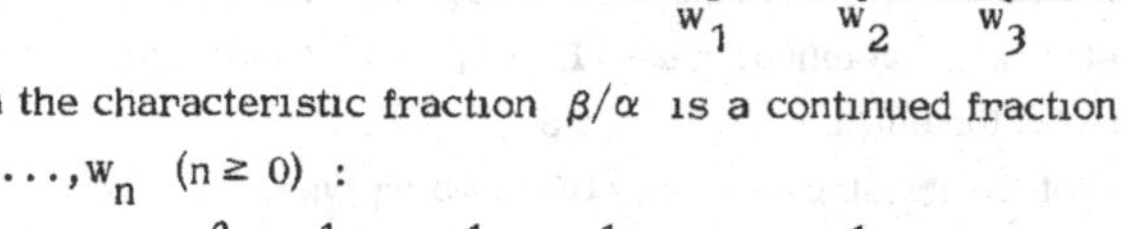

To apply this, we recall the well-known fact (see [HNK]) that if t is the weighted linear branch 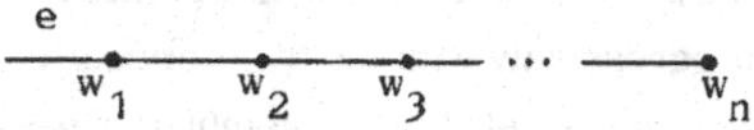

,

then the characteristic fraction β/α is a continued fraction in terms of the weights $w_1, \ldots, w_n$ $(n \geq 0)$:

$$(*) \qquad \frac{\beta}{\alpha} = \frac{1}{w_1} + \frac{1}{(-w_2)} + \frac{1}{w_3} + \ldots + \frac{1}{(\pm w_n)} .$$

Conversely, given β/α , there is a unique expression $(*)$ where the integers w_i verify the normalizing conditions

(i) $w_i \neq 0$ for $i \geq 2$.

(ii) the signs of $w_1, w_2, \ldots, w_n$ alternate.

(iii) $|w_n| \geq 2$ if $n \geq 2$.

This follows from a euclidean division algorithm (cf. [Si]) ; for example $\beta/\alpha = 54/11$ yields

11	54	11	10	1
	0	4	1	10

, so that $w_1, w_2, \ldots$ is $0, -4, 1, -10$.

At any polyvalent vertex v of Γ there is a well-defined <u>neat folding</u> process as follows. Let $t_1, \ldots, t_r$ be the linear branches emanating from v ; for M^3 to be a Λ-homology 3-sphere it is necessary that the denominators $\alpha_1, \ldots, \alpha_r$ of the corresponding characteristic fractions be pairwise coprime in Λ. In this case we can, without affecting $H_1(M^3;\Lambda)$, replace $t_1, \ldots, t_r$ by a single linear branch t ; further t is unique if we subject it to the normalizing conditions above.

If M^3 is a Λ-homology 3-sphere we can perform well-defined neat foldings on Γ simultaneously at all polyvalent vertices. If Γ is not linear, there is ≥ 1 polyvalent vertex having ≤ 1 <u>non</u> -linear branch , hence this neat folding process reduces the number of polyvalent vertices, and repeating this must result finally in a linear graph. This whole process is uniquely defined. A weighted linear graph

gives the lens space $L^3(\alpha,\beta)$ with β/α given by equation (*) above. Since $H_1(L^3;\Lambda) = \Lambda/(\alpha\Lambda)$, this lens space is a Λ-homology 3-sphere precisely if α is a unit in Λ. To resume, we have :

<u>PROPOSITION 3.9.</u> <u>The weighted tree Γ gives a Λ-homology sphere if and only if the above canonical iterated neat folding process on Γ yields a lens space that is a Λ-homology sphere.</u>

Working backwards from linear weighted trees that are Λ-homology spheres and unfolding progressively with the Λ-coprimality condition of 3.6 on denominators, we clearly can progressively list all weighted trees Γ that give Λ-homology spheres (and no others). This is useful for building examples as in § 4.

As a classification of Λ-homology spheres of plumbing type, this seems rather clumsy. So we turn now to another way of classifying these Λ-homology spheres suggested by the constructions of § 6. It will give a very satisfying classification for $\Lambda = Z$.

CLASSIFICATION OF Z-HOMOLOGY 3-SPHERES OF PLUMBING TYPE

Given an oriented closed 3-manifold M^3 and an oriented 2-torus $T^2 \subset M$ separating M, there are, as explained in § 6 unoriented essential embedded circles x, y in T^2 unique up to isotopy that die in rational homology respectively on the inside and the outside of T^2 in M. (At this point jump ahead to read § 6 up to 6.1.)

Adjusting x and y by isotopy to meet transversally in the least possible number α of points, we obtain a configuration $x,y \subset T^2$ that is well-defined up to isotopy of T^2.

We assume henceforth that M^3 is a Λ-homology 3-sphere (to assure that $\alpha \neq 0$).

Together x and y determine and are determined up to deg + 1 automorphism of T^2 by a rational number modulo 1 written $\beta/\alpha \in Q/Z$, and defined as follows : α is the number of intersection points $x \cap y$; the points of $x \cap y$ successive on y are distance β apart on x. To get β well-defined (modulo α) we agree that x and y be (temporarily) oriented so that $x.y$ is negative. This convention assures that M is the lens space $L^3(\alpha,\beta)$ if the two components of $M^3 - T^2$ are solid tori.

Then up to isotopy the only degree + 1 diffeomorphisms of T^2 mapping the (ordered but unoriented) pair x, y to itself are the identity and the 'central' involution τ of § 2 that acts as multiplication by -1 on homology. This fact permits us to define next a <u>splicing process</u> for Λ-homology spheres inverse to the splitting process of § 6.

Consider two situations as above distinguished by indices 1 and 2 (for the same $\Lambda \subset Q$). Suppose the <u>outside</u> of $T_1 \subset M_1$ is a solid torus while the <u>inside</u> of $T_2 \subset M_2$ is a solid torus. Also suppose $\beta_1/\alpha_1 = \beta_2/\alpha_2$ in Q/Z, which means there is a degree + 1 diffeomorphism $\theta : T_1 \to T_2$, unique up to isotopy and composition with τ, sending the pair x_1, y_1 to x_2, y_2. Then if M_1^- is the inside of T_1 and M_2^+ is the outside of T_2, we can form a spliced (oriented) Λ-homology 3-sphere

$$(M_1,T_1) * (M_2,T_2) = M_1^- \cup M_2^+/\theta$$

where θ identifies $\partial M_1^- = T_1$ to $\partial M_2^+ = -T_2$. This spliced manifold is well-defined up to degree +1 diffeomorphism <u>provided</u> that the involution τ extends over M_1^- or M_2^+. Fortunately it does extend (to both) <u>provided</u> that M_1 is of <u>plumbing type</u> and T_1 is one of the 2-tori arising (as in figure 3-a) from the plumbing process corresponding to a tree Γ. (Indeed the quotient of M_1, say, by such an extension is any algebraic knot (S^3,K) arising as in § 2 by band plumbing from the same tree made planar and T_1 gives a 2-sphere regularly surrounding a cut across one of the bands, far from the plumbings.)

Observe that the result of splicing depends only on (M_i,N_i) where N_i is the solid torus with boundary $\pm T_i$, $i = 1,2$. And the operation is commutative: expressed in symbols

$$(M_1,N_1) * (M_2,N_2) \cong (M_2,N_2) * (M_1,N_1) \quad .$$

The reader will find it amusing and not difficult to give arithmetic rules for splicing in terms of almost canonical trees, exploiting the moves of [NeW] (or BoS_2]). Here is an example to try out ($\Lambda = Z$):

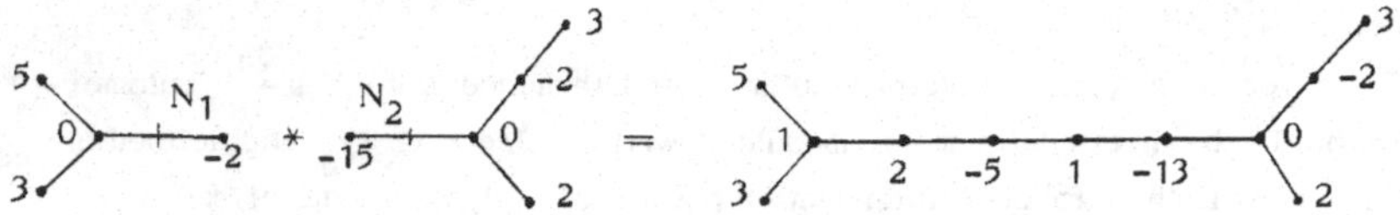

Splicing lets one classify the Λ-homology 3-spheres M^3 of plumbing type in terms of those with stellar graph (the Seifert fiber spaces).

In geometric terms, it runs as follows. Suppose first that M^3 is prime and corresponds to the almost canonical graph Γ. For each polyvalent vertex v of Γ, form a Λ-homology sphere $M^3(v)_+$ as in § 6, distinguishing in it the solid tori $N(v,\ell_i)$, $i = 1,\ldots,r$, that arise from linear segments ℓ_i of Γ that each join two polyvalent vertices of Γ. From these $M^3(v)_+$ with the distinguished solid tori we retrieve M^3 by (commuting disjoint) splicing operations, one for each linear segment ℓ in Γ joining two polyvalent vertices v, v'. The linear segment ℓ serves to associate for splicing a solid torus in $M(v)_+$ to one in $M(v')_+$. The union of splicing tori (one for each ℓ) is characteristic in M^3 by [Wa], i.e. invariant up to isotopy under self-diffeomorphisms of M^3; hence the collection of manifolds $M(v)_+$ with distinguished solid tori in them paired off for splicing is a complete invariant for M^3.

Conversely, starting with a tree Γ^* with vertices v and edges ℓ, suppose that for each vertex v we have a specified stellar Λ-homology 3-sphere M_v with distinguished disjoint solid tori $N(v;\ell_i)$, $i = 1,\ldots,r(v)$, where the ℓ_i are the edges emanating from v in Γ^*. We ask under what conditions this is a canonical splice factorization as above for a prime Λ-homology 3-sphere of plumbing type. Here are the conditions in geometric terms.

1) There exists a stellar plumbing graph yielding M_v so that the given distinguished solid tori in M_v correspond to some or all of the arms. All arms with singular characteristic fraction $\pm 1/0$ do correspond to distinguished solid tori.

2) For each edge ℓ of Γ^* a splicing is possible (this is a condition of coincidence of fractions in Q/Z which always holds if $\Lambda = Z$).

3) Splicing for edge ℓ does not lead to a manifold described by a stellar graph with distinguished arms as in 1) (but rather to a graph with two polyvalent vertices).

This geometrical classification can be translated into a purely arithmetical classification of prime Λ-homology 3-spheres of plumbing type. For $\Lambda = \mathbf{Z}$, the arithmetical version is particularly satisfactory and runs as follows.

Recall first that (excepting S^3) those with stellar almost canonical graph (the Seifert fiber spaces) are classified by giving a determinant sign ± 1 and a collection $\alpha_1, \ldots, \alpha_r$ of pairwise coprime integers > 1, which are the denominators for the characteristic fractions for the arms of the almost canonical graph, see [Sei$_1$] or [Si$_2$].

THEOREM 3.10. Oriented prime Z-homology 3-sphere ($\neq S^3$) of plumbing type are classified (via splicing and folding) up to degree +1 diffeomorphism by the following suitably equipped abstract trees Γ (forgetting all equipment, Γ is naturally homeomorphic to the almost canonical tree of 3.4).

(a) No vertex of Γ is bivalent.

(b) Each valence at each polyvalent vertex carries an integer weight ≥ 0 (it is the denominator of the characteristic fraction for the corresponding arm of the canonical tree of 3.4). At any one polyvalent vertex these valence weights are pairwise coprime. A valence of weight 0 or 1 never leads to a linear arm of Γ.

(c) Each polyvalent vertex with no valence of weight 0 has an attached sign + or − (the determinant sign for the corresponding splice factor).

(d) About no edge does one have a configuration

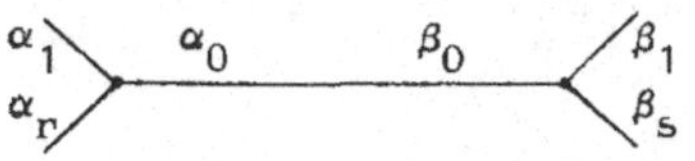

where $r \geq 2$, $s \geq 2$ and $\alpha_0 = \beta_1 \ldots \beta_s$, $\beta_0 = \alpha_1 \ldots \alpha_r$ while the two vertices have the same sign or both no sign. □

For example, the graph classifying the Z-homology sphere illustrating splicing above is :

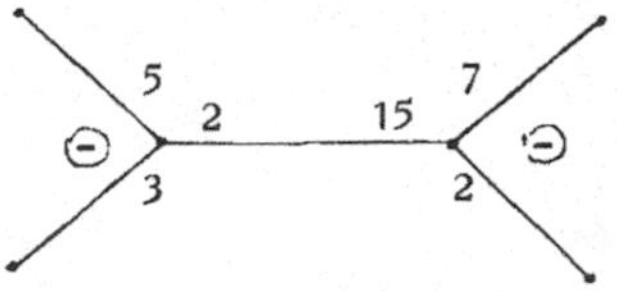

4. NON-FINITE AMPHICHEIRALITY FOR HOMOLOGY SPHERES OF PLUMBING TYPE.

As promised in the introduction, we give examples to establish

FACT 3. There exist amphicheiral prime $\mathbb{Z}$-homology 3-spheres M^3 of plumbing type such that :

a) Every orientation reversing diffeomorphism of M has infinite order, even up to homotopy, and

b) no knot with 2-fold branched covering M is amphicheiral.

We now prepare for a somewhat technical lemma based on surface topology that will be used to assure that all degree -1 diffeomorphisms of M^3 are of infinite order.

Start with two trivial circle bundles $N^3_+ = G^2_+ \times S^1$ and $N^3_- = G^2_- \times S^1$ where G^2_+ and G^2_- are compact planar surfaces each with ≥ 5 boundary components. Then form oriented N^3 from $N_+ \amalg N_-$ by identifying one torus boundary component T^2_+ of N^3_+ to a boundary component T^2_- of N^3_- via a diffeomorphism $\varphi : T_+ \to T_-$ of degree -1 such that the φ-image of a circle fiber of N^3_+ is not homologous to $\pm$ a fiber of N^3_-.

Such an N^3 will be a part of the examples M^3 establishing Fact 3.

A permutation of the boundary components of $N^3_\pm \cong G^2_\pm \times S^1$ will be called rotary if it can be induced by $\varphi \times (\mathrm{id}\,|\,S^1)$ where $\varphi : G^2_\pm \to G^2_\pm$ is a rotation, i.e. is the restriction of a finite order homeomorphism conjugate to a rotation, on the 2-sphere obtained by coning off each boundary component of $G^2_\pm$.

TECHNICAL LEMMA 4.1.

In the context presented above, consider a degree -1 diffeomorphism $f : N^3 \to N^3$ giving by restriction diffeomorphisms $f_+ : N_+ \to N_+$ and $f_- : N_- \to N_-$

<u>such that the induced permutations of boundary components of</u> N_+ <u>and of</u> N_- <u>are both non-rotary.</u>

<u>Then at least one of</u> f_+ <u>and</u> f_- <u>is of infinite order up to homotopy.</u>

COMPLEMENT 4.2.

<u>Suppose</u> $f : N^3 \to N^3$ <u>given in 4.1 is the restriction of a diffeomorphism</u> $F : M^3 \to M^3$ <u>where</u> $N^3 \subset \text{int } M^3$. <u>Suppose that inclusion gives an injection</u> $\pi(N^3) \rightarrowtail \pi_1(M^3)$.

<u>Then</u> $F : M^3 \to M^3$ <u>is of infinite order up to homotopy.</u>

<u>Proof of 4.2 granting 4.1.</u>

At least one of f_+ and f_-, say f_+, is of infinite order up to homotopy (granting 4.1).

By the Van Kampen theorem, one sees that $\pi_1(M^3)$ is a free product with amalgamation, of $\pi_1(N_+)$ and groups $\pi_1(X_i)$, where X_i, $i = 1,\ldots, r$ are the components of $M\text{-int } N_+$, the amalgamations being given by the injections $\pi_1(N_+^3) \longleftarrow \pi_1(N_+^3 \cap X_i) \longrightarrow \pi_1(X_i)$. Since the injections into $\pi_1(N_+^3)$ are onto proper subgroups, it is an elementary fact about such products that the normalizer of $\pi_1 N_+$ in $\pi_1(M)$ is $\pi_1(N_+)$ itself. (The geometric methods in [Ma] can serve).

Choosing base point $*$ in N_+ and making f_+ and F fix $*$, we get a commutative diagram

$$\begin{array}{ccc} F_* : \pi_1(M) & \longrightarrow & \pi_1(M) \\ \uparrow & & \uparrow \\ (f_+)_* : \pi_1(N_+) & \longrightarrow & \pi_1(N_+). \end{array}$$

As N_+ is a $K(\pi, 1)$ space (aspherical), and no power f_+^s, $s \neq 0$, is homotopic to the identity, we conclude that no power $(f_+)_*^s$, $s \neq 0$, is an inner automorphism of $\pi_1(N_+)$.

Since $\pi_1(N_+)$ is its own normalizer in $\pi_1(M)$, it follows that $(f_+)_*^s$, $s \neq 0$, is not the restriction of an inner automorphism of $\pi_1(M)$; in particular F_*^s, $s \neq 0$, is not an inner automorphism of $\pi_1(M)$; hence F^s, $s \neq 0$, is not homotopic to the identity. □

Proof of 4.1.

In search of a contradiction, let us suppose that f_+ and f_- are both of finite order up to homotopy, i. e., for each, some power is homotopic to the identity. Raising f to a suitable odd power we can arrange without loss of generality that f_+ and f_- each have order up to isotopy that is some power of two.

According to a verticalisation lemma of Waldhausen [Wa$_1$, p. 319], $f_\pm$ can be isotoped to map circle fibers to circle fibers Then $f_\pm$ induces a diffeomorphism $\varphi_\pm \cdot G_\pm^2 \circlearrowleft$ of the circle bundle base.

Then both φ_+ and φ_- are of finite order up to homotopy, for $\varphi_\pm \times (\mathrm{id}|R)$ is homotopic respecting boundary to the covering $\bar{f}_\pm \cdot G_\pm^2 \times R \longrightarrow G_\pm^2 \times R$, of $f_\pm : G_\pm^2 \times S^1 \longrightarrow G_\pm^2 \times S^1$.

Since $\varphi_\pm$ is of finite order up to homotopy it is also of finite order up to isotopy. In fact any homeomorphism $h : G_\pm^2 \longrightarrow G_\pm^2$ that is homotopic to the identity is isotopic to the identity. Indication of proof : Since h is homotopic to the identity and $G_\pm^2$ is not an annulus one can show that h maps each boundary component to itself preserving orientation. Any homeomorphism of a planar surface with this last property can be isotoped to the identity by a synthetic or hyperbolic procedure that makes h respect a system of cuts reducing $G_\pm^2$ to a disc. See [ZCV] or [Th].

Since $\varphi_\pm$ is of finite order up up to isotopy, a theorem of J. Nielsen shows that $\varphi_\pm$ is isotopic to a diffeomorphism $\psi_\pm$ of finite order. This is classically proved using moduli and Smith theory as follows (here G_+^2 can be any compact surface).

The set of smooth curvature -1 (hyperbolic) metrics on $G^2_\pm$ making the boundary totally geodesic is naturally a space "of moduli" $\mathcal{M} = \mathcal{M}(G^2_\pm)$ homeomorphic to an open convex set of an euclidean space R^N, provided we identify two such metrics whenever they are isometric by a diffeomorphism isotopic to the identity. Indeed it is an elementary matter to set up the homeomorphism into R^N with the help of cuts reducing S^2_o to a disc. See [ZCV] or, better, [Th]. In the case at hand, if $G^2_\pm = B^2$-(k discs) we can use $k-1$ disjoint cuts, and N will be $2k-1$, with k parameters giving length of boundary components and $(k-1)$ more the length of the cuts when isotoped to be geodesic and perpendicular to boundary.

The group of isotopy classes of diffeomorphisms of $G^2_\pm$ (= G say) acts continuously on $\mathcal{M}$. In particular $\varphi_\pm$ gives a homeomorphism $\Phi : \mathcal{M}\cup\infty \longrightarrow \mathcal{M}\cup\infty \approx S^N$ whose order is a power of 2. By Smith theory (see [Br]) the fixed point set of Φ has the Čech homology (with Z_2 coefficients) of a sphere, and in particular fixes ≥ 1 point in $\mathcal{M}$. This means that there exists a metric μ on G^2 as described such that $\varphi_\pm : G^2 \longrightarrow G^2$ is isotopic to an isometry ψ with respect to the metric μ on G^2. This isometry ψ necessarily has finite order (see § 1 below Theorem 1*).

At this point, revising notation, we can assume that φ_+ and φ_- have finite order. Now $\varphi_\pm$ extends canonically, by coning, to a finite order homeomorphism $\hat{\varphi}_\pm : S^2 \longrightarrow S^2$. This $\hat{\varphi}_\pm$ must therefore be topologically conjugate to a rotation or a reflection (consider the orbit space $S^2/\hat{\varphi}_\pm$ to show this).

Since the permutation of boundary components of $N_\pm$ induced by $f_\pm$ is not rotary, the permutation of boundary components of $G^2_\pm$ induced by $\varphi_\pm$ is not rotary, i.e. $\hat{\varphi}_\pm$ cannot be (conjugate to) a rotation. Hence both $\hat{\varphi}_+$ and $\hat{\varphi}_-$ are both have degree -1.

Since $\varphi_\pm : G^2_\pm \hookleftarrow$ and $f_\pm \cdot N^3_\pm \hookleftarrow$ both have degree -1, it follows that $f_\pm$ has degree $+1$ on the circle fibers of $N_\pm$. Now restrict attention to $T = N_+ \cap N_-$ on which the restrictions of f, f_+, f_- are all isotopic as maps

$T \longrightarrow T$. Since the fibers of N_+ and N_- are supposed non-isotopic in T and hence independent in $H_1(T)$ we conclude that the identity map (up to isotopy) is induced on T. At the same time it must have degree -1. This is the contradiction saught and completes the proof of Lemma 4.1. □

We now turn to the construction of prime Z-homology 3-spheres M^3 of plumbing type establishing Fact 3.

M^3 is the boundary of the 4-manifold obtained by plumbing together 2-disc bundles over S^2 according to an integrally weighted abstract tree Γ. We describe the tree Γ gradually. Firstly because it is big, and secondly because there is much freedom in it's construction. Γ has the form in Figure 4-a

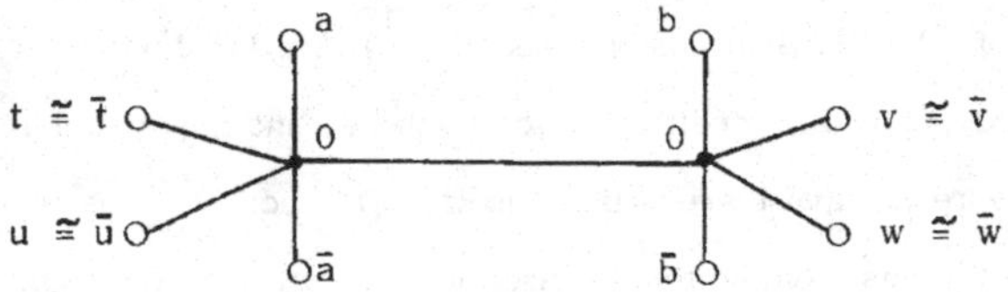

Figure 4-a

In this figure each baloon O— is a weighted tree with one open valence ; it is called a <u>branch</u> ; there are 8 of these $a, \bar{a}, b, \bar{b}, t, u, v, w$. If x is a branch, then $\bar{x}$ is derived from it by multiplying all weights by -1. The branches a and b can be

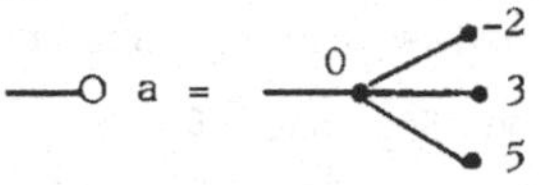

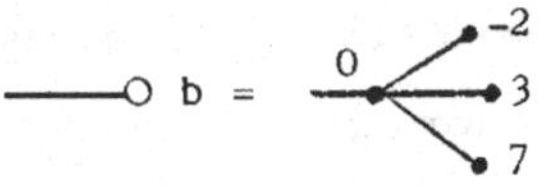

The branches t and v can both be

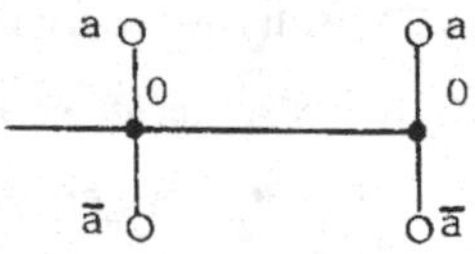

The branches u and w can both be

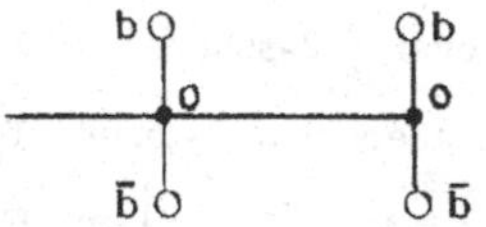

('neat folding')

Under the Z-homology equivalence move of § 3.9 the branches a and b become respectively  and  while the branches t, u, v, w all become . Then under the same equivalence moves the tree Γ of 4-a becomes  showing that the manifold M^3 derived from the tree Γ is a Z-homology sphere.

M^3 is prime because Γ is perfectly canonical.

M^3 is amphicheiral, i. e. $M^3 \cong -M^3$, because $\Gamma \cong -\Gamma$ $(= \bar{\Gamma})$.

However there is only one isomorphism $\theta : \Gamma \longrightarrow -\Gamma$. It fixes the two black nodes (vertices) seen in Figure 4-a ; it exchanges a and $\bar{a}$ also b and $\bar{b}$; it respects each of t, u, v, w.

Since θ cannot preserve (up to sign) any cyclic order on the branches at either black node seen in Figure 4-a there exists no amphicheiral algebraic knot with 2-fold branched covering M^3. See $[BoS_2]$ for explanation.

Every knot with 2-fold branched cyclic covering M^3 is algebraic. This is because all involutions on a sufficiently large manifold of plumbing type (graph manifold) are of standard type, i.e. fibered on the Seifert fibrations mentioned in the footnote below. See $[BoS_2]$, incidentally F. Bonahon was first to prove this.

It remains only to show that any degree -1 diffeomorphism $F : M^3 \to M^3$ is of infinite order up to homotopy. To prove this we appeal to Waldhausen's results in $[Wa_1]$† to isotop F until it realizes $\theta : \Gamma \longrightarrow -\Gamma$ in the following sense.

† This amounts to applying the characteristic variety theorem of Jaco-Shaler and Johanson (stated in § 7) to make F respect a certain characteristic 2-torus family and then make F fiber preserving with respect to Seifert fibrations of each component of the complement of this 2-torus family See $[BoS_2]$.

As constructed by plumbing, M^3 contains, for each vertex v of Γ a submanifold M^3_v that is a trivial circle bundle over a 2-sphere with as many holes as v has valences in Γ ; also for each edge e there is a 2-torus T_e in M^3 such that if e joins v and v' one has $T_e = M_v \cap M_{v'}$. For F to realize θ means that $F(M_v) = M_{\theta(v)}$ for each vertex v and $F(T_e) = T_{\theta(e)}$ for each edge e. Now let v and v' be the two black vertices (nodes) seen in Figure 4-a and fixed by θ ; let e be the edge joining them. We can then apply 4.1 and 4.2 with the substitutions:

$$N_+ \longmapsto M_v \quad ; \qquad N_- \longmapsto M_{v'} \quad ;$$

$$T \longmapsto T_e \quad ; \qquad N \longmapsto M_v \cup M_{v'} \, .$$

Knowing that the permutations of boundary components of M_v and $M_{v'}$ induced by F are non rotary, we conclude that $F : M^3 \longrightarrow M^3$ has infinite order up to homotopy as required. This establishes Fact 3.

<u>Concluding remark</u>. F. Bonahon pointed out that there are non-finitely amphicheiral prime Z-homology 3-spheres M^3 of plumbing type which nevertheless have an orientation-reversing diffeomorphism $F : M^3 \to M^3$ finite up to isotopy on $M^3 - T_e$ for some edge e (and hence on M_v for every vertex v). An example is given by the following tree, which is perfectly canonical. I leave the proof to the reader (cf. § 5 end).

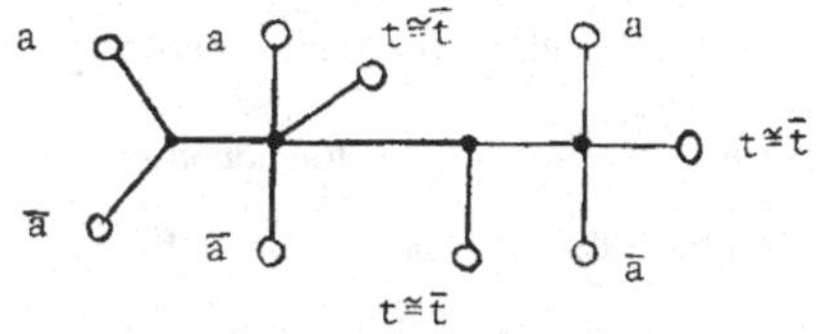

§5. NON-FINITE AMPHICHEIRALITY FOR HOMOLOGY SPHERES OF SPLIT HYPERBOLIC TYPE.

We say that a compact 3-manifold M^3 is of split hyperbolic type if there is a finite collection T of disjoint 2-tori in int M^3 such that (int M^3)-T is complete hyperbolic of finite volume. If T has k components we say M^3 has k splits.

This section is devoted to proving

FACT 4. There exist prime amphicheiral Z-homology 3-spheres M^3 of split hyperbolic type having no periodic homeomorphism (except the identity).

Further M^3 can be chosen to have just two splits and so that every map $M \to M$ not homotopic to the identity has infinite order up to homotopy.

The hyperbolic manifolds we use are complements of algebraic knots and links, because a convenient calculation of their symmetry group up to isotopy has been made [BoS]. The somewhat limited choice of symmetry groups here seems to force us to have ≥ 2 splits in M^3. So the reader may wish to try to find a simpler example with a single split.

Here are the materials for the construction of M^3.

(I) A knot complement N such that N is compact with 2-torus boundary, int N is complete hyperbolic of finite volume, and every diffeomorphism of N is isotopic to the identity.

Assertion (I) N can be the complement of the algebraic knot of figure 5-a

Figure 5-a

Remark on (I). According to [Th , Chap. 5] , almost every Dehn surgery on such a knot will yield a closed hyperbolic 3-manifold with trivial symmetry group (trivial because the surgered knot becomes the shortest closed geodesic).

Also we need

(II) The closed complement M_0 of a link of two homologically unlinked components, such that the compact manifold M_0^3 with ∂M_0 two 2-tori admits a complete hyperbolic structure of finite volume on its interior, and such that the diffeomorphism group up to isotopy is restricted as follows:

The subgroup of all diffeomorphisms that are either degree +1 and respect both boundary components , or are degree -1 and exchange boundary components is (up to isotopy) a Z_4 generated by an orientation reversing diffeomorphism of period 4 (pointwise !) whose square is isotopic on the boundary to the identity.

Assertion (II) M_0 can for example be the closed complement $S^3 - \overset{\circ}{N}(L)$ of the algebraic link L of figure 5-b.

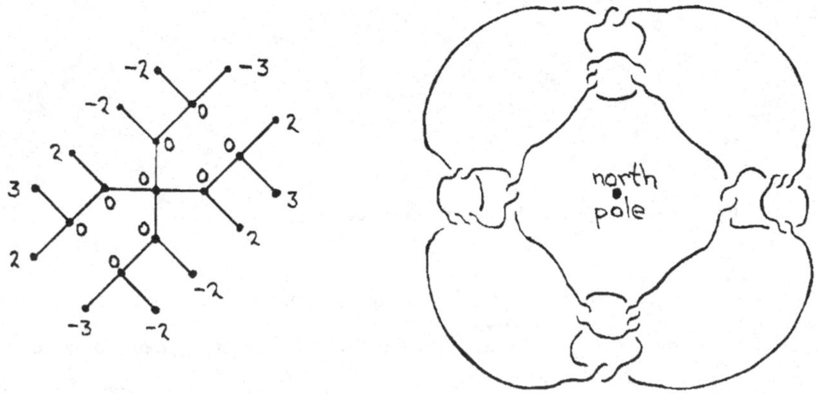

Figure 5-b

Think of the pictured link as lying (nearly) on $S^2 = \partial B^3$ (the earth's surface) of which we see a polar stereographic projection. Then the announced symmetry group up to isotopy of M_0^3 is generated by composition of 90° rotation about the earth's axis composed with reflection in the equator ; name this order 4 rotary reflection ρ .

We pause to check the assertions (I) and (II).

N is prime because S^3 is prime and the knot is connected. M_0 is prime because the 2-fold branched cyclic covering of S^3 branched along L is prime (its graph 5-b is almost canonical as described in §3).

To assure that these examples N and M_0 have complete hyperbolic interiors of finite volume, we have, according to Thurston's hyperbolisation theorem [Th], combined perhaps with the torus theorem announced by Waldhausen in 1969 (now proved in [Feu], [JaS], [Jo_1], [Sc])*, only to show that, for each of these manifolds, neither does it contain an embedded incompressible torus non-parallel to boundary, nor is it a Seifert fiber space. It is shown in [BoS_2] that an almost canonical graph (as described in §3) yields an algebraic knot or link with no incompressible T^2 in its complement non-parallel to boundary, provided the graph contains no configuration as in this figure :

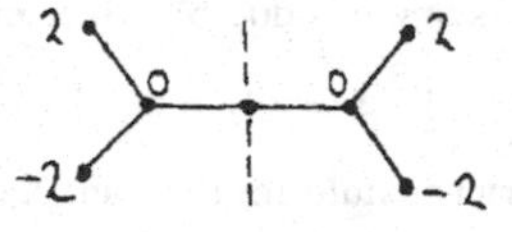

Figure 5-c

The dashes in 5-c indicate that the rest of the tree connects to the central vertex. Also one can note, see [Mo_2], that, if the complement of such a knot or link is a Seifert fiber, space then we are dealing with a torus knot or link ; this happens for only a few

---

* I suspect this difficult torus theorem will prove unnecessary when Thurston's complete argument for hyperbolisation is understood.

star shaped special graphs. Thus our specific examples of N and M_0 do have complete hyperbolic interiors of finite volume.

We turn next to verifying, for the examples N and M_0, the statements in (I) and (II) concerning diffeomorphisms The following two lemmas are crucial.

PROPOSITION 5 1. Any degree +1 diffeomorphism of a complete hyperbolic knot complement of finite volume extends up to isotopy to a diffeomorphism of S^3 respecting the knot.

Proof of 5 1. Any diffeomorphism of a knot complement preserves longitude up to sign for homological reasons. Also it is of finite order up to isotopy by Mostow's rigidity theorem [Th, Chap. 5] or Johannson's finiteness theorem [Jo_2] . Hence, as it preserves orientation, it preserves meridian up to sign ; indeed it is isotopic on the torus boundary to the identity or to the change of sign. Thus it extends to a S^3 diffeomorphism of S^3 respecting the knot. □

Remark on 5.1 (and 5.2). By using Johannson's finiteness theorem [Jo_2] in the above proof, we get a proof of 5.1 without assuming any hyperbolic structure. This finiteness theorem states that if $f : X^3 \to X^3$ is a diffeomorphism of a compact orientable irreducible and boundary irreducible 3-manifold such as a knot complement, then some power f^s, $s \geq 1$ a function of X, is up to isotopy a composition of Dehn twists along incompressible tori in int X and along incompressible annuli meeting ∂X in their boundary When $\partial X = T^2$ we conclude that $f^s | . \partial X \to \partial X$ is isotopic to the identity, because for any oriented incompressible annulus $(A , \partial A) \subset (X , \partial X)$ the two oriented circles of $\partial A \subset \partial X$ are parallel and oppositely oriented in ∂X (since X is orientable)

The same argument would let us get by without hyperbolic structure on the link complement in 5.2 below, provided no two boundary components of the closed link complement are joined by an incompressible annulus .

This remark came up in discussing Johannson's work with F. Waldhausen.

PROPOSITION 5.2. Consider a link L in S^3 such that each component is null-homologous in the complement of the rest (equivalently such that all homological linking numbers of the components are zero) Suppose the closed link complement $X^3 = S^3 - \overset{o}{N}(L)$ is complete hyperbolic of finite volume. On it consider any diffeomorphism $f : X \to X$ such that there is known to be a diffeomorphism $g : X \to X$, extending to a diffeomorphism $(S^3,L) \to (S^3,L)$, where g has the same degree as f and g induces the same permutation of boundary components as f.

In this situation f also extends to a diffeomorphism $(S^3,L) \to (S^3,L)$.

Proof of 5.2. The homological hypothesis guarantees that for each torus boundary component T_1 of the (compact) link complement X, the inclusion induces a map $H_1(T_1) \to H_1(X)$ with kernel of rank 1 generated, say, by $(\pm \ell_1)$. Then the set of "longitudes" $\pm \ell_1$ must be respected by diffeomorphisms of X.

Note that f extends if and only if the composition $h = g^{-1}f$ extends But h is orientation preserving and respects each boundary component of X, and in each boundary component T_1 it must respect ℓ_1 (up to sign). Now, just as in the proof of 5.1 we show that h respects each meridian up to sign ; therefore h extends. □

To show for our example N of Assertion (I) that every diffeomorphism is isotopic to the identity, first note that there are no orientation reversing diffeomorphisms. Indeed the knot signature σ is non-zero ; but knot signature switches sign with orientation, and depends only on the knot complement (by [Mi] for example). That $\sigma \neq 0$ is a tabulated fact : $\delta = |\delta|_1^{\sigma}$ $(= \delta_0 1^{\sigma})$ [Co, p 340], and $\delta = -9-14-6-2 = -31$, our knot being (3,2)(3,2-) of [Co, p. 352]. But a direct calculation is not difficult and yields $\sigma = +6$.

Secondly, apply Lemma 5.1 to show that every diffeomorphism of N^3 extends to a pair diffeomorphism of (S^3, K). But every pair diffeomorphism of (S^3, K) is (pair) isotopic to the identity by [BoS$_2$] (more explanation presently)

Assertion (I) follows from this since the C^∞ space of tubular neighborhoods of any submanifold is contractible (as one can show by relative construction of tubular neighborhoods in a parameter-fibered situation). It would be convenient here to use instead the fact that, up to isotopy, the diffeomorphism group of a compact 3-manifold and that of its interior are isomorphic (by restriction) [Wa$_3$] cf. [Lau] ; but this requires $\Gamma_4 = 0$ (Cerf), even for the 3-ball.

To complete the proof of Assertion (II) it now suffices, in view of Lemma 5.2 and the fact about tubular neighborhoods used above, to know that the group of pair diffeomorphisms (up to isotopy) of (S^3, L) is $Z/4$ generated by the mentioned rotary reflection ρ.

For this fact we refer again to [BoS$_2$], where the pair diffeomorphism group up to isotopy is calculated for all "sufficiently large" algebraic knots and links. Suffice it here to quote a corollary saying that if as in 5-a and 5-b, the link (or knot) is determined by a perfectly canonical graph (see §3) and also has a pair of ajacent polyvalent vertices, then this diffeomorphism group is a subgroup of the group G of all degree ± 1 automorphisms of the graph as an abstract (non-planar) weighted graph. But for graph 5-a, this group G is 0, and for 5-b this group G is Z_4, realized by ρ.

We are finally in full possession of manifolds N and M_0 with properties (I) and (II), and so procede to

CONSTRUCTION OF M^3.

The closed manifold M^3 establishing Fact 4 is made from two copies N_1, N_2 of N and a single copy of M_0 whose two (torus) boundary components we call T_1 and T_2.

For convenience we add to M_0 outwards collars $C_1 = T_1 \times [0,1]$ and $C_2 = T_2 \times [0,1]$, identifying T_1 to $T_1 \times 0$ and T_2 to $T_2 \times 0$. Set $M_0' = M_0 \cup C_1 \cup C_2$.

Form $M^3 = (M_0' \amalg N_1 \amalg N_2)/\sim$ where the equivalence $\sim$ glues $\partial(N_1 \cup N_2)$ to $\partial M_0'$ by diffeomorphisms $\partial N_i \xrightarrow{f_i} T_i \xrightarrow{\times 1} T_i \times 1$, $i = 1, 2$, as follows Choose the diffeomorphism $f_1 : \partial N_1 \longrightarrow T_1$ such that, if ℓ_1 is the knot longitude in ∂N_1, then

$$(1) \qquad f_{1*}(\ell_1) = \pm m_1' + k\ell_1' \quad \text{in} \quad H_1(T_1)$$

where k is any integer and m_1', ℓ_1' are link meridian and longitude. (The link longitude ℓ_1' can here be well-defined up to sign, as remarked in the proof of 5.2). Then choose $f_2 : \partial N_2 \longrightarrow T_2$ to be $x \longmapsto \rho f_1 \theta^{-1}(x)$ where $\theta : N_1 \longrightarrow N_2$ is the canonical identification. Since ρ extends to an automorphism of S^3, one has

$$(2) \qquad f_{2*}(\ell_2) = \rho\, f_{1*}(\ell_1) = \pm m_2' + k\ell_2' \quad \text{in} \quad H_1(T_2)$$

where m_2' and ℓ_2' are meridian and longitude of the second link component.

Give to M the orientation compatible with M_0 (and N_1 and $-N_2$).

<u>Remark</u>. For distinct integers k, we obtain diffeomorphically distinct manifolds $M^3 = M^3(k)$. We leave this as an exercise for the reader with the arguments to follow, cf. [BoS$_2$]

M^3 is a Z-homology 3-sphere by equations (1) and (2) together with a Mayer-Vietoris sequence.

M^3 is prime because M_0', N_1 and N_2 are prime and not solid tori ; an innermost 2-disc argument reveals this.

M^3 is amphicheiral, since there is an extension $\bar{\rho} : M \longrightarrow M$ of the degree -1 diffeomorphism $\rho : M_0 \longrightarrow M_0$. Indeed, we can define $\bar{\rho} : N_1 \xrightarrow{\theta} N_2$ and $\bar{\rho} : N_2 \xrightarrow{\theta^{-1}} N_1$ to be the canonical identifications, then it remains to extend over C_1 and C_2. On C_1 the map $(\rho | T_1) \times [0,1]$ from C_1 to C_2 will do. But on C_2

the proposed degree -1 diffeomorphism of collars

$$\bar{\rho}\,| : C_2 \longrightarrow C_1$$

must give $\rho\,| : T_2 \times 0 \longrightarrow T_1 \times 0$ at one end, and on the other end be the map ψ making commutative the square

$$\begin{array}{ccc} T_2 \times 1 & \xrightarrow{\ \psi\ } & T_1 \times 1 \\ \uparrow & & \uparrow f_1 \\ \partial N_2 & \xrightarrow{\ \theta^{-1}\ } & \partial N_1 \end{array} \qquad f_2 = \rho f_1 \theta^{-1}$$

This ψ is none other than $\rho^{-1}| : T_2 \times 1 \to T_2 \times 1$. Fortunately, although ρ itself has order 4 on all of M_o, the restricted square $\rho^2 | \partial M_o$ is, by (II), isotopic to the identity, hence ρ is isotopic to ρ^{-1} on ∂M_o. Thus $\bar{\rho}$ exists as desired to establish amphicheirality.

Consider any map $g : M^3 \to M^3$ that is not homotopic to the identity. We propose to show that g has infinite order up to homotopy. By [Wa$_2$] we can assume g is a diffeomorphism.

It follows from the remarkable characteristic variety theorem of [JaS] [Jo$_1$] (which is stated carefully in § 7) that by an isotopy we can arrange that g respects $C_1 \cup C_2$ and hence also M_o. Then, by (I) and (II), after a further isotopy g is a power of ρ on M_o and on $N_1 \cup N_2$ it is either the permutation of the two copies of N or the identity. (Note that the non-amphicheirality of N_1 forces $g\,| : M_o \longrightarrow M_o$ to exchange boundary components if degree of g is -1, cf. (II).)

We shall leave to the reader the rather straightforward case where $g\,|M_o =$ identity.

If $g\,|M_o \neq$ identity, we can assume that $g\,|M_o = \rho^2$ (for if $g\,|M_o = \rho$ we can replace g by g^2 without loss). Parametrizing T_1^2 suitably as R^2/Z^2 with $R^2 = \mathbb{C} \supset R^1$, note that $\rho^2 | T_1^2$ is $x + Z^2 \longmapsto x + \frac{1}{2} + Z^2$, a rotation of 180° on one of the circle factors. Now by an isotopy* of $g\,| : C_1 \longrightarrow C_1$ fixing $g\,|\partial C_1$, found by classical methods we arrange that $g\,| : C_1 \longrightarrow C_1$ is a Dehn twist. This means that, identifying $C_1 = (R^2/Z^2) \times [0,1]$, the restriction $g\,|C_1$ is a map

* Homotopy suffices.

$g| = g_1 : (R^2/Z^2) \times [0,1] \longrightarrow (R^2/Z^2) \times [0,1]$ sending $(x+Z^2,t) \longmapsto (x + tn_1 + Z^2,t)$ where $n_1 = (\frac{n_{11}}{2}, n_{12}) \in (\frac{1}{2}Z) \times Z$ is some vector with $n_{11} \in Z$ odd .

We shall show that, for all $s \neq 0$, the power g^{2s} induces an automorphism g_*^{2s} of $\pi_1(M^3)$ that is not inner ; it will follow that g^s is not homotopic to the identity.

By Van Kampen's theorem [Mas], the group $\pi_1(M^3)$ is the direct limit of injections $\pi_1 N_1 \longleftarrow \pi_1 T_1^2 \longrightarrow \pi_1 M_o \longleftarrow \pi_1 T_2^2 \longrightarrow \pi_1 N_2$, a free product with amalgamation, on which g_*^{2s} acts as follows. On $\pi_1(N_1) \subset \pi_1(M)$ it is inner automorphism by $2sn_1 \in Z^2 = \pi_1(T_1^2) \subset \pi_1(N_1)$, and on $\pi_1(M_o)$ it is the identity. Note that $2sn_1 = (sn_{11}, 2sn_{12})$ is non zero since n_{11} is odd. Note also that each of $\pi_1(N_1)$, $\pi_1(N_2)$ and $\pi_1(M_o)$ is its own normaliser in $\pi_1(M)$ (an elementary fact about such free products with amalgamation, cf. §4 end.)

Aiming for a contradiction suppose g_*^{2s} is inner. Then it is inner automorphism by an element z in the intersection $\pi_1 N_1 \cap \pi_1 N_2 \cap \pi_1 M_o$. But z would be central in $\pi_1 M_o$ and hence the identity (recalling that M_o is hyperbolic or using [Wa_3]). Yet g_*^{2s} is not the identity because $2sn_1 \neq (0,0)$ and $\pi_1 N_1$ is centerless (recalling that N_1 is hyperbolic or using [Wa_3]). This is the contradiction showing that g^s, $s \neq 0$, is not homotopic to the identity.

Finally, consider a periodic homeomorphism $h : M^3 \to M^3$ that is homotopic to the identity. It remains to prove that h is necessarily the identity map. Consider the covering $\overline{M}$ of M with fundamental group $\pi_1\overline{M} = \pi_1 N_1 \subset \pi_1 M$. By a boundary theorem explained in [JaS] , $\overline{M}$ is N_1 with a boundary collar added and thus is complete hyperbolic of finite volume. Also h lifts to a unique periodic homeomorphism $\bar{h}$ of $\overline{M}$, homotopic to the identity. Via the homotopy, $\bar{h}$ lifts to a well-defined homeomorphism $\tilde{h}$ of hyperbolic 3-space (Poincaré model) $H^3 = \text{int } B^3$, that is $\pi_1\overline{M}$-equivariantly homotopic to the identity, hence fixing the sphere S^2 at infinity [Th] . A power $\tilde{h}^s$ covers the identity, hence is a covering translation (an isometry !) , which, as it fixes S^2 , must be the identity. Thus $\tilde{h}$ is a periodic homeomorphism of B^3 fixing S^2 ; so $\tilde{h}$ is the identity by Newman's theorem or Smith theory (see [Br]). This shows h = identity. This argument is moderately general.

Fact 4 is now established. □

§ 6. - AN ADDITIVITY FORMULA

In an oriented connected closed 3-manifold M^3 we consider a compact 2-submanifold T (unoriented), each component of which is a 2-torus that separates M^3 into two pieces ; T is called a torus family. Let $N(T) \cong T \times [-1,1]$ be a regular (bicollar) neighborhood of T in M^3 .

The pair $(M,N(T))$ determines a finite graph Γ : a vertex v of Γ corresponds to a (connected) component of $M - \mathring{N}(T)$, called $M(v)$; an edge e of Γ corresponds to a component of $N(T)$ called $M(e)$ which is just a thickening of a torus $T(e)$ of T ; the edge e joins those two vertices v, v' such that $M(e)$ meets $M(v)$ and $M(v')$. This graph Γ can easily be embedded as a retract of M so that, for each edge e , the torus $T(e)$ is the transversal preimage of the midpoint of e . Since each $T(e)$ by hypothesis separates M , the graph Γ is in fact a tree.

Consider any oriented 2-torus $T^2 \subset M^3$ splitting M^3 into two pieces M' , M'' with $\partial M' = T^2 = -\partial M''$ (so that M' is on the inside of T^2 and M'' is on the outside).

It is an elementary fact that the kernel K of $i_* : H_1(T^2;Q) \to H_1(M'',Q)$ has Q-dimension 1 (cf. above 3.5). Let $\pm m \in H_1(T^2;Z)$ be the generator of

$$K \cap H_1(T^2;Z) \subset H_1(T^2;Q) \quad .$$

It is unique up to sign, and, being indivisible, is represented by an unoriented circle $\pm m$ unique up to isotopy. Call $\pm m = \pm m(T^2)$ the generator dying rationally on the outside.

In case M^3 is a Λ-homology 3-sphere, where Λ is a unitary subring of the rational numbers Q , one calculates readily that :

$$H_*(M';\Lambda) \cong H_*(S^1;\Lambda) \cong H_*(M'';\Lambda)$$

and further that $(M',\partial M')$, $(M'',\partial M'')$ have the same Λ-homology sequence as $(B^2,\partial B^2) \times S^1$.

Returning to the situation with given torus family $T \subset M^3$, we can now naturally augment $M(v)$, for $v \in \Gamma^{(0)}$ a vertex, to a <u>closed manifold</u> $M(v)_+$ by plugging each (oriented) boundary component T^2 of $M(v)$ with a solid torus in which the boundary of a meridian disc is identified to the generator $\pm m(T^2)$ dying rationally on the outside. Similarly define $M(e)_+$, for $e \in \Gamma^{(1)}$ an edge ; it is clearly a lens space.

$M(v)_+$ is well-defined up to diffeomorphism fixing $M(v)$, more generally a diffeomorphism $f: (M, M(v)) \to (M, M(v'))$ induces a diffeomorphism $f_+ : M(v)_+ \to M(v')_+$ extending $f(M(v))$. Similarly for $M(e)_+$.

In case M^3 is a Λ-homology 3-sphere, one now readily sees that $M(v)_+$ and $M(e)_+$ are Λ-homology 3-spheres ; we have merely replaced some Λ-homology solid tori in M by real solid tori.

Recall here that $\mathbf{Z}_2$-homology n-spheres, where $\mathbf{Z}_2 = \mathbf{Z}/2\mathbf{Z}$, are precisely $\mathbf{Z}_{(2)}$ homology n-spheres, where $\mathbf{Z}_{(2)}$ is the ring of rational numbers with odd denominators. For such oriented homology 3-spheres the Rohlin invariant ρ is well-defined in $\mathbf{Z}/16\mathbf{Z}$.

<u>ADDITIVITY THEOREM</u> 6.2. <u>Let</u> M^3 <u>be a</u> Z_2<u>-homology</u> 3-<u>sphere and</u> $T \subset M$ <u>be a</u> 2-<u>torus family determining a graph</u> Γ. <u>With the notations of this section</u> :

$$\rho(M) = \sum_{v \in \Gamma^{(0)}} \rho(M(v)_+) - \sum_{e \in \Gamma^{(1)}} \rho(M(e)_+) .$$

Proof of 6.2.

By induction on the number of components in T , we can reduce to the case where T is a single 2-torus (this reduction is left to the reader). Thus we can assume henceforth that Γ has 2 vertices v , v' and a single edge e . We must show that :

$$\rho(M) = \rho(M(v)_+) + \rho(M(v')_+) - \rho(M(e)_+) .$$

For this it suffices to produce a parallelizable manifold W^4 with signature zero whose boundary is ∂W^4

$$\partial W^4 = M(v)_+ \amalg M(v')_+ \amalg (-M) \amalg (-M(e)_+) .$$

We let W^4 be :

$$W^4 = ([-1,0] \times M^3) \cup ([0,1] \times M(v)_+) \cup ([0,1] \times M(v')_+) , \text{ (see figure 6-a) } ;$$

the creases at $0 \times \partial M(v)$ and $0 \times \partial M(v')$ have to be smoothed out.

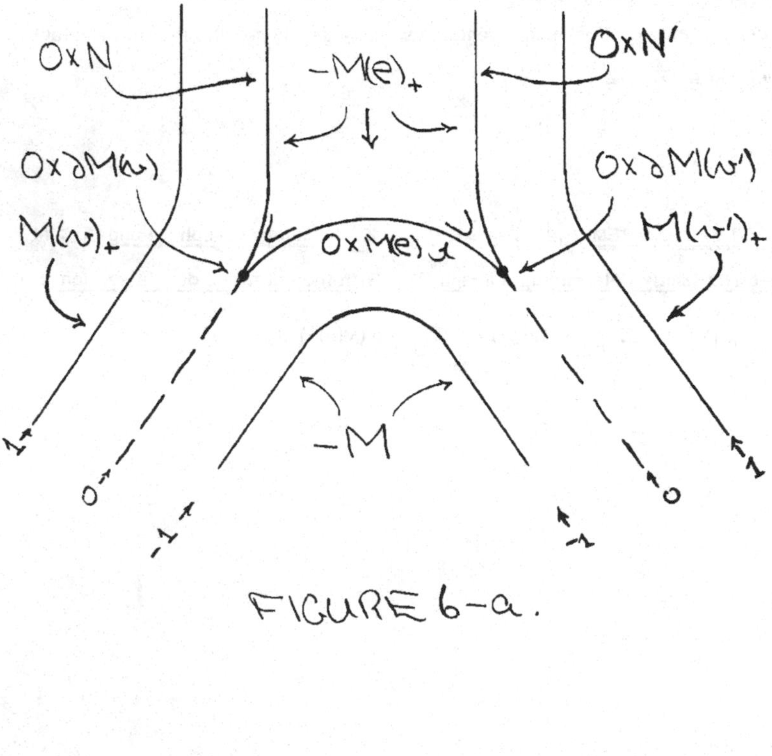

FIGURE 6-a.

The boundary component marked $-M(e)_+$ is merely isomorphic to $-M(e)_+$. The other three boundary components are canonically identified to $M(v)_-$, $M(v)_+$ and $-M$ as labelled.

One can verify that W has the Z_e-homology intersection ring of $S^2 \times S^2$ with four points removed ; hence W is parallelizable with signature 0 .

Indications. By construction, the solid torus $N' = M(v')_+ - \mathring{M}(v')$ has meridian 2-disc D' whose boundary $\partial D'$ is a generator of the 2-torus $\partial M(v')$, which dies on the outside in the sense that it is also boundary of a Z_e-cycle δ in $M(v) \cup M(e)$. Now consider :

$$V^3 = 0 \times (M(v) \cup M(e) \cup N') \;;$$

it has the Z_e-homology of $S^2 \times S^1$ with 2-dimensional generator x represented by $\delta + (-D')$. Similarly and symmetrically construct $V' = 0 \times (M(v') \cup M(e) \cup N)$ with 2-dimensional generator x' represented by $\delta' + (-D)$. Certainly W has $V \cup V'$ as deformation retract, and $V \cap V' = 0 \times M(e) \simeq T^2$; then the Mayer-Vietoris sequence of $(W ; V, V')$ shows that W has the Z_e-homology of $(S^2 \times S^1) \cup (S^1 \times S^2) \subset S^2 \times S^2$ which is $S^2 \times S^2$ minus four open discs. Next $x \cdot x = 0 = x' \cdot x'$ since V , V' are collared on one side in W . Finally $x \cdot x' \in Z_e$ coincides with $\pm r_*(\partial D) \cdot r_*(\partial D') \in Z$ where $r . M(e) \to T \cong T^2$ is a retraction ; but this is an odd integer since the lens-space $M(e)_+$ is a Z_2-homology 3-sphere. □

This completes the proof of the additivity theorem 6.2. □

§ 7. - AMPHICHEIRALITY KILLS ρ

Our aim is to prove Theorem 5.

First we reduce to the case of irreducible manifolds, by using the Kneser-Milnor uniqueness theorem [He] for expression of oriented closed M^3 as a connected sum $M_1 \# M_2 \# \ldots \# M_k$ of prime closed orientable 3-manifolds. This theorem asserts that M^3 determines the disjoint sum $M_1 \amalg M_2 \amalg \ldots \amalg M_k$ up to oriented diffeomorphism ($\cong$). Thus $M^3 \cong -M^3$ implies that $M_1 \amalg M_2 \amalg \ldots \cong -(M_1 \amalg M_2 \amalg \ldots)$ say by a degree -1 diffeomorphism f mapping M_i onto $M_{\varphi(i)} \cong -M_i$. By the usual trick, we may assume $\varphi : \{1,\ldots,k\} \hookleftarrow$ has order a power of 2.

Suppose that M satisfies the hypotheses of Theorem 5 and that Theorem 5 is true in the prime case. Then, for $i = \varphi(i)$, we know that $\rho(M_i) = 0$. For $i \neq \varphi(i)$, on the other hand, the orbit of M_i under f has even order (indeed a power of 2), write it $f(M_i), f^2(M_i), \ldots, f^{2n}(M_i) = M_i$. Since $f^s M_i = M_{\varphi^s(i)} \cong$ $\cong (-1)^s M_i$, we find that $\sum_{s=1}^{2n} f^s(M_i) = 0$. Adding over all φ-orbits we conclude that $\sum_i \rho(M_i) = 0$.

Since $\rho(M) = \sum_i \rho(M_i)$ (exercice), the reduction is complete.

The additivity result 6.2 makes the remainder of the proof of Theorem 5 somewhat similar. Henceforth we can (and do) assume M^3 prime.

In case M^3 is an amphicheiral Seifert manifold it admits an orientation reversing involution inducing a reflection on its base space S^2 (see the classification of Seifert manifolds in [Sei$_1$] or [Si$_2$]), hence Theorem 1 applies. Alternatively, since M^3 is of (stellar) plumbing type, Theorem 2 applies.

In case M^3 is hyperbolic, Theorem 1^* applies (as in § 1).

Hence we can now assume M^3 is irreducible and sufficiently large. A major structure theorem for closed irreducible sufficiently large 3-manifolds M^3 now applies. It asserts that there exists a family of 2-tori $T \subset M^3$ such that

(a) if N(T) is a regular (bicollar) neighborhood of T, each

$M^3 - \mathring{N}(T)$ is either a Seifert manifold, or a hyperbolic manifold with horospherical torus boundary components.

(b) T is characteristic in the sense that for any diffeomorphism $f : M^3 \hookleftarrow$ the image f(T) is (ambient) isotopic to T .

The discovery of the characteristic torus family T is due to Johannson and Jaco-Shalen (the techniques are based on Waldhausen's thesis [Wa_1], cf. [BoS_1])*. The discovery and proof that the non-Seifert components of $M - \mathring{N}(T)$ are hyperbolic is due to Thurston [Th].

This basic theorem does not require M^3 to be a Z_2-homology 3-sphere but we now reimpose this condition and consider the apparatus of § 6, surrounding the graph Γ associated to this characteristic $T \subset M^3$.

Let $f : M^3 \hookleftarrow$ be an orientation reversing diffeomorphism. By (b), we can assume (after isotopy of f) that $fN(T) = N(T)$. Then f clearly induces an automorphism $\varphi . \Gamma \hookleftarrow$ of the finite graph Γ. Raising f to an odd power we can (and do) assure that the order of φ is a power of 2 .

Let u be a vertex or edge of Γ such that $\varphi(u) = u$, then f gives an orientation reversing diffeomorphism $f_u : M(u) \hookleftarrow$. This extends to an orientation reversing diffeomorphism $f_{u+} . M(u)_+ \hookleftarrow$, because the construction of $M(u)_+$ was characteristic for the pair (M,M(u)) .

In case M(u) is Seifert (inevitable if u is an edge), it follows from construction that the closed manifold $M(u)_+ \cong -M(u)_+$ is either Seifert with fibration extended from M(u) , or at least a sum of Seifert manifolds (cf. [Si_2, §5] , or [Mo_2, §I]). Thus, we know that $\rho(M(u)_+) = 0$, using (say) Theorem 1.

---

* The exposition in [BoS_1] (or [Ha]) is intended to separate the matter from the difficult torus theorem announced by Waldhausen in 1969 ; compare the first footnote in §5.

Consider the alternative that $M(u)$ be hyperbolic of finite volume (here u must be a vertex). Then $M(u)$ is understood to have horocyclic torus boundary components, in the sense that $M(u)$ with an outward collar $\partial M(u) \times [0,\infty)$ added is complete hyperbolic of finite volume such that

(i) In the universal covering of $M(u) \cup (\partial M(u) \times [0,\infty))$, is <u>hyperbolic 3-space</u> H^3 say with conformal model $\overset{\circ}{B}{}^3 = H^3$, in such way that the boundary $\partial M(u)$ is covered by disjoint horospheres, i.e. punctured 2-spheres in B^3, punctured at a single point of tangency with ∂B^3.

(ii) All components of $\partial M(u)$ have the same area, say $\epsilon > 0$.

In this hyperbolic case, Mostow's rigidity theorem assures that after homotopy respecting boundary (or even isotopy, according to Waldhausen [Wa_2]), $f_u : M(u) \circlearrowleft$ is an isometry and (hence) of finite order. (The normalization (ii) is necessary here if f permutes the boundary components of $\partial M(u)$.)

<u>ASSERTION</u>. The isometry $f : M(u) \circlearrowleft$ extends to a diffeomorphism $f : M(u)_+ \circlearrowleft$ of finite order. Hence $\rho(M(u)_+) = 0$ by Theorem 1^*.

<u>Remark</u>. $M(u)_+$ may not be hyperbolic.

<u>Proof of assertion</u>. Consider any boundary component T_1 of $\partial M(u)$; it is Euclidean since the covering horocycles are isometric to $\mathbb{R}^2$. In T_1 we have the embedded generator $\pm m_1$ dying rationally on the outside (see § 6), tighten $\pm m_1$ to become geodesic and form the foliation of T_1 by geodesics parallel to $\pm m_1$ (covered by parallel lines in $\mathbb{R}^2$). Doing this for each boundary component, we get a well-defined foliation $\mathfrak{F}$ of $\partial M(u)$ by circles

Since the finite order isometry $f_u : M(u) \circlearrowleft$ extends to a diffeomorphism of M, the foliation $\mathfrak{F}$ of $\partial M(u)$ is respected by f_u. Thus f_u induces a finite order (orientation reversing) diffeomorphism of the manifold $M(u)/\mathfrak{F}$ obtained from M(u) by identifying each circle in $\mathfrak{F}$ to a point and smoothing suitably. But this manifold is clearly diffeomorphic to $M(u)_+$, even fixing M(u) minus a small collar of boundary. This proves the assertion. $\square$

Next consider a vertex or edge u of Γ such that $\varphi(u) \neq u$. Then the orbit of u has even order (indeed a power of 2) ; write it $u_i = \varphi^i(u)$, $i = 1,\dots,2n$. Since $f : (M,M(u_i)) \xrightarrow{\cong} -(M,M(u_{i+1}))$ with i in Z mod 2n, the map f induces $M(u_i)_+ \cong -M(u_{i+1})_+$. Thus $\sum_{i=1}^{2n} \rho(M(u_i)_+) = 0$.

We have now shown that if $\Phi = \{u_1, \varphi(u_1), \dots\}$ is the whole φ-orbit of any vertex or edge u_1 of Γ, then :

$$\Sigma \{\rho(M(u)_+) \mid u \in \Phi\} = 0$$

and this has been proved whether Φ has a single element or several. Summing over all orbits, with positive sign for vertex orbits and negative sign for edge orbits we get, in view of Theorem 6.1,

$$\rho(M) = \sum_{v \in \Gamma^{(0)}} \rho(M(v)_+) - \sum_{e \in \Gamma^{(1)}} \rho(M(e)_+) = 0 ,$$

which completes the proof of Theorem 5. $\square$

Bibliography

[Bi] J.S. BIRMAN, Orientation-reversing involutions on 3-manifolds, preprint, Columbia Univ., 1979.

[BoS_1] F. BONAHON and L. SIEBENMANN, Equivariant characteristic varieties in dimension 3, (to appear).

[BoS_2] F. BONAHON and L. SIEBENMANN, Les noeuds algébriques (à paraitre), see also proceedings of July 1979 topology conference, Bangor N. Wales, LMS lecture notes, to appear.

[Br] G. BREDON, Introduction to compact transformation groups, Academic Press, 1972.

[Cau] A. CAUDRON, L'arborescence dans la classification des noeuds et des enlacements, polycopié ENSET Tunis 1979.

[CauS] A. CAUDRON et L. SIEBENMANN, La description des noeuds algébriques par les arbres pondérés,

[Co_1] J. CONWAY, An enumeration of knots and links, and some of their related properties, pages 329-358 in Computational Problems in Abstract Algebra, ed. J. Leech Pergamon Press, 1970.

[Co_2] J. CONWAY, Skein theory (to appear).

[Cont] L. CONTRERAS-CABALLERO, Periodic transformations in homology 3-spheres and the Rohlin invariant, preprint, Cambridge Univ., 1979, to appear in Proc. London Math. Soc., cf. Bangor 1979 conference. LMS lecture notes, to appear.

[Coo] D. COOPER, manuscript, Warwick University.

[EK] J. EELLS and N. KUIPER, An invariant for smooth manifolds, Ann. of Math. Pura Appl., 60 (1962), 93-110.

[Feu] C.D. FEUSTEL, On the torus theorem, Trans. AMS 217 (1976), 1-57.

[GaS_1] D. GALEWSKI and R. STERN, Classification of simplicial triangulations of topological manifolds (to appear).

[GaS_2] D. GALEWSKI and R. STERN, Orientation-reversing involutions on homology 3-spheres, Math. Proc. Camb. Phil. Soc. 1979.

[Go] C. GORDON, Some aspects of classical knot theory, pages 1-60, in Knot Theory, Springer Lecture Notes in Math. N° 685 (1978).

[He] J. HEMPEL, 3-manifolds, Ann. of Math. Study 86 (1976), Princeton Univ. Press.

[HNK] F. HIRZEBRUCH, W. NEUMANN and S. KOH, Differentiable manifolds and quadratic forms, Dekker, math. lecture notes, vol. 4, 1972.

[HsP] W.C. HSIANG and P.S. PAO, Orientation reversing involutions on homology 3-spheres, see Notices Amer. Math. Soc. 26 (Feb 1979), page A-251.

[JaS] W. JACO and P. SHALEN, Seifert fibered spaces in 3-manifolds, I ; the mapping theorem, to appear , Memoirs AMS, 1979.

[Jo_1] K. JOHANNSON, Homotopy equivalences of knot spaces, preprint (U. of Bielefeld) 1975. Cf. Springer lecture Notes , 1979 .

[Jo_2] K. JOHANNSON, On the mapping class group of sufficiently large 3-manifolds, preprint (U. of Bielefeld) 1978.

[Ka] A. KAWAUCHI, Vanishing of the Rohlin invariants of some Z_2-homology 3-spheres, preprint, IAS, Princeton and Osaka city U., 1979. Cf. Bangor 1979 conference, LMS lecture notes, to appear.

[Lan] J. LANNES, Un faux espace projectif réel de dimension 4 (d'après Cappell et Shaneson). Sém. Bourbaki, 1978/79, exposé 527, Nov. 1978.

[Lau] F. LAUDENBACH, Dimension 3 homotopie et isotopie, Astérisque 12, 1974.

[Mas] W.S. MASSEY, Algebraic topology, Harcourt Brace and World, 1967.

[Matu] T. MATUMOTO, Triangulation of manifolds, Proc. Sympos. Pure Math. 32 (1978), Part II, pages 3-7.

[Mi] J. MILNOR, Infinite cyclic coverings, pages 115-133 in Topology of Manifolds, ed. J. Hocking, Prindle, Weber & Schmidt, 1968.

[Mo_1] J. MONTESINOS, Variedades de Seifert que son recubridores ciclicos ramificados de dos hojas, Bol. Soc. Mat. Mexicana 18 (1973), 1-32.

[Mo_2] J. MONTESINOS, Revetements doubles ramifiés de noeuds, variétés de Seifert et diagrammes de Heegard, (conférences 1976, polycopiés à Orsay, 1979) à paraître dans un volume d'Astérisque.

[Mo_3] J. MONTESINOS, 4-manifolds, 3-fold covering spaces and ribbons, preprint, 1977.

[Mos] G.D. MOSTOW, Strong Rigidity of Locally Symmetric Spaces, Ann. of Math. Study 78 (1976), Princeton U. Press.

[Ne_1] W. NEUMANN, An invariant of plumbed manifolds, Notices Amer. Math. Soc. 25 (Nov. 1978), page A-717 , and these proceedings.

[NeW] W. NEUMANN and S. WEINTRAUB, Four-manifolds constructed via plumbing, Math. Ann. 238 (1978), 71-78.

[OVZ] P. ORLIK, E. VOGT and H. ZIESCHANG, Zur topologie gefaserter dreidimensionalen Manigfaltigkeiten, Topologie 6 (1967), 49-64.

[Or] P. ORLIK, Seifert fiber spaces, Springer Lecture Notes Math. 291 (1972).

[RaT] F. RAYMOND and J. TOLLEFSON, Closed 3-manifolds which admit no periodic maps, Trans. Amer. Math. Soc. 221 (1976)

[Roh] V. ROHLIN, A new result in the theory of 4-dimensional manifolds, Doklady 8 (1952), 221-224.

[Rol] D. ROLFSEN, Knots and Links, Publishor Perish Inc, N° 7 (1976), Box 7108, Berkeley CA 94707.

[Sc] P. SCOTT, A new proof of the annulus and torus theorems, preprint, Univ. of Liverpool, 1978.

[Sei_1] H. SEIFERT, Topologie der dreidimensionalen gefaserten Raume, Acta Math. 60 (1933), 147-328.

[Sei_2] H. SEIFERT, Über das Geschlecht von Knoten, Math. Ann. 110 (1934), 571-592.

[Si_1] L. SIEBENMANN, Exercices sur les noeuds rationnels, polycopié Orsay, 1975.

[Si_2] L. SIEBENMANN, Introduction aux espaces fibrés de Seifert, Orsay 1977-78.

[SiV] L. SIEBENMANN and J. VAN-BUSKIRK, Construction of homology 3-spheres with orientation reversing involution, polycopied, Eugene, Oregon, 1979.

[Th] W. THURSTON, The Geometry and Topology of 3-Manifolds, preprints 1978 - , Princeton U.

[Wa_1] F. WALDHAUSEN, Eine Klasse von 3-dimensionalen Manigfaltigkeiten, Invent. Math. 3 (1967), 308-333, and 4 (1967), 87-117.

[Wa_2] F. WALDHAUSEN, On irreducible 3-manifolds that are sufficiently large, Ann. of Math. 87 (1968), 56-88.

[Wa_3] F. WALDHAUSEN, Gruppen mit Zentrum, und 3-dimensionale Manigfaltigkeiten, Topology 6 (1967), 505-517.

[ZCV] H. ZIESCHANG, E. VOGT, H.-D. COLDEWEY, Flachen und ebene discontinuierlichen Gruppen, Springer Lecture Notes Math. 122 (1970).

L. SIEBENMANN

Math. Orsay, 91405 France

András Szücs:

Cobordism of maps with simplest singularities

§1. Introduction

Denote Emb(n,k) — and Imm(n,k) — the cobordism group of embeddings — and immersions — of oriented n-dimensional manifolds into the sphere S^{n+k}, respectively.

The well-known Pontrjagin-Thom construction gives a homotopic description of the group Emb(n,k), namely

$$\mathrm{Emb}(n,k) \approx \pi_{n+k}(MSO(k)).$$

By the Hirsch theory there exists a homotopic description for the group Imm(n,k), too:

$$\mathrm{Imm}(n,k) \approx \pi_{n+k}(\Omega^{\infty} S^{\infty} MSO(k))$$

See [Wells]. Here $\Omega^{\infty}S^{\infty}MSO(k)$ denotes the infinite loop space of the infinite suspension of the Thom space MSO(k).

The aim of the present paper is to give a homotopic description for the cobordism group of S-maps which roughly speaking means maps with simplest singularities .

Definition of S-map (Haefliger)

A map f of an n-manifold into an (n+k)-manifold is an S-map if

1/ rank $df(x) \geq n-1$ at every point x,

2/ f has no triple points,

3/ the double points are not singular

4/ the self-intersections are transversal at every double point.

We will not use the following property 5 of S-maps:

5/ the first and second derivatives of f span the entire space R^{n+k} at every double point.

The cobordism group of S-maps from closed oriented n-manifolds into S^{n+k} can be defined in the usual way and will be denoted by S(n,k). The definition of S(n,k) can be obtained from the definition of Emb(n,k) replacing the world "embedding" by "S-map". Our MAIN RESULT is the construction of a space X(k) for which the isomorphism

$$S(n,k) \approx \pi_{n+k}(X(k)) \qquad \text{holds.}$$

Remark 1.

S-maps are of particular interest, because they form an open and dense subset of C^∞-maps with respect to the C^∞-topology, provided that $2k > n+1$ [Haefliger]. This latter fact allows us to obtain two interesting consequences of our construction.

Corollary I Denote Ω_i the Thom cobordism group of oriented i-dimensional manifolds. For $2k > n+1$ the groups Imm(n,k) can be described modulo the finite 2-primary component as follows:

If k is odd then $\mathrm{Imm}(n,k) \approx_{C_2} \Omega_n$

If k is even then $\mathrm{Imm}(n,k) \approx_{C_2} \Omega_n \oplus \Omega_{n-k}$

Here C_2 denotes the class of finite 2-primary groups and the sign $\approx_{C_2}$ means C_2-isomorhism.

Remark 2.

Let ε: $\mathrm{Imm}(n,k) \longrightarrow \Omega_n$ and δ: $\mathrm{Imm}(n,k) \longrightarrow \Omega_{n-k}$ be homomorphisms defined as follows: Denote [f] the cobordism class of an arbitrary immersion $f: M^n \longrightarrow S^{n+k}$ with transversal self-intersections and double point-set Δ^{n-k}. Then ε maps [f] into the cobordism class of M^n and δ maps [f] into the cobordism class of Δ^{n-k}. Then the C_2-isomorphisms in Corollary I are ε and $\varepsilon \oplus \delta$

Corollary II Denote D an arbitrary class of Abelian groups containing the class C_2. Let M^n be a manifold such that $H^i(M^n, Z) \in D$ for $i \equiv 1$ modulo 4. Denote $\alpha_j(M^n)$ the subgroup of the bordism group $\Omega_j(M^n)$ consisting of the elements which can be represented by immersions. Then for $j < \frac{2}{3}n$ the factor groups $\Omega_j(M^n)/\alpha_j(M^n)$ belong to the class D.

§2. Construction of the space X(k).

The space X(k) will be constructed from two blocks (spaces) gluing together by a map. We shall denote the first block by $\Gamma_2(k)$ and describe it in 2.a. The space $\Gamma_2(k)$ plays the same role for the cobordism of immersions without triple points as the Thom space MSO(k) does for cobordism of embeddings.

2.a. The first block $\Gamma_2(k)$ of the space X(k).

Definition. The cobordism group of immersions without triple points of oriented n-manifolds into the sphere S^{n+k} can be defined in the usual way and will be denoted by $Imm_2(n,k)$ (The bordism immersions have no triple points .)

Theorem. For any k there exists a space $\Gamma_2(k)$ such that $Imm_2(n,k) \approx \pi_{n+k}(\Gamma_2(k))$.

Proof: In [Szucs]. For the sake of completeness we outline the proof in § 4.

Here we recall the definition of the space $\Gamma_2(k)$. Denote MSO(k) the Thom space of the universal vector bundle with structure group SO(k) and denote S^∞ the infinite dimensional sphere. In the product

$$MSO(k) \times MSO(k) \times S^\infty$$

identify

1/ the points (m_1,m_2,s) with the points $(m_2,m_1,-s)$ for m_1, $m_2 \in MSO(k)$, $s \in S^\infty$ and
2/ the points $(m, *, s_1)$ with the points $(m, *, s_2)$ for $m \in MSO(k)$ $s_1,s_2 \in S^\infty$.

($*$ denotes the singular point of MSO(k).)

Remarks:

1/ $\Gamma_2(k)$ can be constructed using the Γ^+ functor of Barratt and Eccles [Barratt-Eccles]. For any space X the space $\Gamma^+(X)$ has a natural filtration $X = \Gamma_1(X) \subset \Gamma_2(X) \subset \ldots$ defined by means of "the length of the words". If X = MSO(k) then $\Gamma_2(X) = \Gamma_2(k)$.
2/ If $M^n \looparrowright N^{n+k}$ is an arbitrary immersion without triple points then it defines a map $N^{n+k} \longrightarrow \Gamma_2(k)$ which is unique up to homotopy. Conversely: an arbitrary map $N^{n+k} \longrightarrow \Gamma_2(k)$ defines

an immersion of an n-manifold into N^{n+k}, which has no triple points. This immersion is unique up to cobordism.

2.b. <u>The second block $D\zeta_k$ of the space $X(k)$.</u>

<u>Notations:</u> Denote $Z(k)$ the subgroup of $O(2k)$ consisting of matrices $\begin{pmatrix} A0 \\ 0A \end{pmatrix}$ and $\begin{pmatrix} 0A \\ A0 \end{pmatrix}$ $A \in SO(k)$.

Let $s: O(2k) \subset O(2k+1)$ be the inclusion defined by

$$s(B) = \begin{pmatrix} 1 & 0 & \dots & 0 \\ 0 & & & \\ \vdots & & B & \\ 0 & & & \end{pmatrix} \in O(2k+1).$$ Denote $SZ(k)$ the image of the group $Z(k)$ under s. Let θ_k and ζ_k be the universal vector bundles with structure group $Z(k)$ and $SZ(k)$, respectively. So $\dim \theta_k = 2k$, $\dim \zeta_k = 2k+1$ and $\zeta_k = 1 \oplus \theta_k$.

<u>Remark.</u> $Z(k) \approx SO(k) \times Z_2$.

For an arbitrary Euclidean vector bundle ξ denote by $D\xi$, $\partial D\xi$ and $M\xi$ the associated unit ball bundle, the unit sphere bundle and the Thom space, respectively.

The second block of the space $X(k)$ is $D\zeta_k$.

2.c. <u>The glueing map $\varrho: \partial D\zeta_k \to \Gamma_2(k)$.</u>

<u>Definition 1.</u> In the Euclidean space R^{2k+1}, with coordinates $w_1, \dots, w_{2k+1}$, we define a subset Q by

$$Q = \left\{ (w_1, \dots, w_{2k+1}) \;\middle|\; \begin{array}{l} w_1 \geq 0 \\ \text{and} \end{array} \left(\begin{array}{ll} w_i(1-\sqrt{w_1}) = w_{k+i}(1+\sqrt{w_1}) & \text{for } i=2,\dots,k+1 \\ \text{or } w_i(1+\sqrt{w_1}) = w_{k+i}(1+\sqrt{w_1}) & \text{for } i=2,\dots,k+1 \end{array} \right) \right\}$$

<u>Definition 2.</u> It is easy to see that Q is invariant under the action of the group $Z(k)$. Hence, for every $(2k+1)$-dimensional vector bundle $\xi: E(\xi) \to B(\xi)$ with structure group $Z(k)$ one can define the associated bundle $Q(\xi)$ with fibre Q over $B(\xi)$. $Q(\xi)$ is a subspace of $E(\xi)$.

<u>Definition</u> of the map ϱ.

The bundle ζ_k can be approximated by bundles $\zeta_{k|1} \subset \zeta_{k|2} \subset \dots$ the base space of which are finite dimensional manifolds and $\bigcup_N \zeta_{k|N} = \zeta_k$. Fix an arbitrary N and consider the space $\partial D(\zeta_{k|N})$. It contains the subset $\partial D(\zeta_{k|N}) \cap Q(\zeta_{k|N})$ which is the image of an immersion into the manifold $\partial D(\zeta_{k|N})$ and this immersion has no triple points. By Remark 2 in 2.a. we obtain a map $\varrho_N: \partial D(\zeta_{k|N}) \to \Gamma_2(k)$.

The maps ϱ_N can be chosen so that $\varrho_{N+1}\Big|_{\partial D(\zeta_{k|N})} = \varrho_N$.

So, we can define a map $\varrho: \partial D\zeta_k \to \Gamma_2(k)$ by the formula: $\varrho|_{\partial D(\zeta_{k|N})} = \varrho_N$ for every N.

2.d. <u>The space X(k).</u>

Identify the points e and $\varrho(e)$ in the disjoint union $\Gamma_2(k) \cup D\zeta_k$ for every $e \in \partial D\zeta_k$. The obtained space is $X(k)$.

The next consequences of the construction will be important in the following paragraphs

1/ $\Gamma_2(k) \subset X(k)$

2/ $X(k) / \Gamma_2(k) = M\zeta_k$

3/ The pair $(X(k), \Gamma_2(k))$ is 2k-connected. The space $\Gamma_2(k)$ is (k-1) -connected. Hence $\pi_i(X(k), \Gamma_2(k)) \approx \pi_i(M\zeta_k)$ for $i \leq 3k-1$.

§3. <u>Proof of Corollaries</u>

3.a. <u>Corollary I</u>

Consider the exact homotopy sequence of the pair $(X(k), \Gamma_2(k))$

$$\pi_{n+k+1}(X(k), \Gamma_2(k)) \to \pi_{n+k}(\Gamma_2(k)) \to \pi_{n+k}(X(k)) \to \pi_{n+k}(X(k), \Gamma_2(k))$$

By the Theorem in 2.a: $\pi_{n+k}(\Gamma_2(k)) \approx \mathrm{Imm}_2(n,k)$.

If $n < 2k-1$ then $\mathrm{Imm}_2(n,k) \approx \mathrm{Imm}(n,k)$.

By the Remark 1 in §1 $\pi_{n+k}(X(k)) \approx \Omega_n$.

By consequence 3, in 2.d. we have: $\pi_i(X(k), \Gamma_2(k)) \approx \pi_i(M\zeta_k)$ for $i \leq 3k-1$.

The equality $\zeta_k = 1 \oplus \theta_k$ implies $M\zeta_k = SM\theta_k$, hence $\pi_i(M\zeta_k) \approx \pi_{i-1}(M\theta_k)$ if $i < 4k-2$.

Hence, for $n < 2k-1$ we obtain the exact sequence

$$\to \pi_{n+k}(M\theta_k) \to \mathrm{Imm}(n,k) \to \Omega_n \xrightarrow{\varkappa} \pi_{n+k-1}(M\theta_k) \to$$

<u>Lemma.</u>

$$\pi_i(M\theta_k) \approx_{\mathcal{C}_2} \begin{cases} 0 & \text{, if } k \text{ is odd} \\ \Omega_{i-2k}, & \text{if } k \text{ is even} \end{cases}$$

The statement of Corollary I follows immediately from the Lemma if k is odd. To obtain the proof in the case when k is even we have to notice that the map $\varkappa$ occuring in the exact sequence can be considered, modulo C_2, as a map from Ω_n into Ω_{n-k-1}. Hence it is zero mod C_2 because $\Omega_n \in C_2$ or $\Omega_{n-k-1} \in C_2$.

By the mod p Hurewicz theorem the Lemma follows from the following

<u>Proposition.</u>

Let p be an arbitrary odd prime number and $j \leq 4k-1$.
Then

a/ $H^j(M\theta_k;Z_p) = 0$, if k is odd and

b/ the natural map $i: M\theta_k \subset MSO(2k)$ induces isomorphisms

$H^j(M\theta_k;Z_p) \approx H^j(MSO(2k);Z_p)$, if k is even.

<u>Proof.</u> If γ_k denotes the universal k-dimensional vector bundle with structure group SO(k) then the total space $E\,\theta_k$ of the bundle θ_k can be described as follows:

$$E\,\theta_k = E(\gamma_k \oplus \gamma_k) \underset{Z_2}{\times} S^\infty$$

where the group Z_2 acts

1/ on the total space $E(\gamma_k \oplus \gamma_k)$ of the vector bundle $\gamma_k \oplus \gamma_k$ changing the summands. Denote T the corresponding involution on $E(\gamma_k \oplus \gamma_k)$.

2/ Z_2 acts on the infinite dimensional sphere S^∞ as the multiplication by (-1) .

Hence there exists a fibration $(D\theta_k, \partial D\theta_k) \longrightarrow RP^\infty$ with fibre $(D(\gamma_k \oplus \gamma_k), \partial D(\gamma_k \oplus \gamma_k))$. Consider the spectral sequence of this fibration.

$$E_2^{**} = H^*(RP^\infty; \mathcal{H}^*(M(\gamma_k \oplus \gamma_k);Z_p)) \Longrightarrow H^*(M\theta_k;Z_p)$$

It is well known that if $\mathcal{A}$ is any local coefficient group without elements of order two or infinity, then

$$H^i(RP^\infty;\mathcal{A}) = 0 \quad \text{for } i>0$$

and $H^0(RP^\infty;\mathcal{A}) = \mathcal{A}^{Z_2}$ - the subgroup of $\mathcal{A}$ consisting of elements invariant under the action of the group $Z_2 = \pi_1(RP^\infty)$. Hence $H^i(M\theta_k;Z_p)$ is isomorphic to the subgroup of $H^i(M(\gamma_k \oplus \gamma_k);Z_p)$ consisting of the invariant elements of the involution T^*. By the Thom isomorphism $\bar{H}^*(M(\gamma_k \oplus \gamma_k);Z_p) \approx H^{*-2k}(BSO(k);Z_p) \smile U(\gamma_k \oplus \gamma_k)$ where $U(\gamma_k \oplus \gamma_k)$ is the Thom class of $\gamma_k \oplus \gamma_k$.

The action T^* is described as follows:

$$T^* \Big|_{H^*(BSO(k);Z_p)} = \text{identity}$$

$$T^*\left(U(\gamma_k \oplus \gamma_k)\right) = (-1)^k \cdot U(\gamma_k \oplus \gamma_k)$$

Thus if k is odd then $H^*(M\theta_k;Z_p) = 0$, which proves a/. If k is even then $H^*(M\theta_k;Z_p) \approx H^*(M(\gamma_k \oplus \gamma_k);Z_p)$ and this isomorphism is iduced by the natural map

$$M(\gamma_k \oplus \gamma_k) \to M\theta_k.$$

To prove b/ it will be enough to show that the natural map $\ell : M(\gamma_k \oplus \gamma_k) \to M\gamma_{2k}$ induces isomorphism of cohomology groups in dimension $j < 4k$. By the Thom isomorphism

$$\bar{H}^*(M(\gamma_k \oplus \gamma_k);Z_p) = H^{*-2k}(BSO(k);Z_p) \smile U(\gamma_k \oplus \gamma_k)$$

and $\bar{H}^*(M\gamma_{2k};Z_p) = H^{*-2k}(BSO(2k);Z_p) \smile U(\gamma_{2k})$.

The Thom class is natural, and so $\ell^* U(\gamma_{2k}) = U(\gamma_k \oplus \gamma_k)$.

The inclusion $BSO(k) \subset BSO(2k)$ induces isomorphism of cohomology groups in dimension $< 2k$. Part b/ is also proved.

3.b. <u>Proof of Corollary II.</u>

Consider the sequence of space $\Gamma_2(k) \subset X(k) \xrightarrow{pr} X(k)/\Gamma_2(k) = M\zeta_k$.

(pr is the natural projection onto the factorspace.)

For any CW complex K it gives an exact sequence of homotopy sets

$$[K,\Gamma_2(k)] \xrightarrow{\varkappa} [K,X(k)] \to [K,M\zeta_k]$$

If $\dim K \leq n$, then these homotopy sets are stable (and so they are groups) because all the spaces $\Gamma_2(k)$, $X(k)$ and $M\zeta_k$ are $(k-1)$-connected and $n \leq 2(k-1)$. If K is a manifold M^n then the group $[M^n, X(k)]$ can be identified with $\Omega_{n-k}(M^n)$ and the image of $\mathcal{K}$ can be identified with $\alpha_{n-k}(M^n)$. Hence the factorgroup $\Omega_{n-k}(M^n)/\alpha_{n-k}(M^n)$ is isomorphic with a subgroup of $[M^n, M\zeta_k]$.

It is well-known that there exists a spectral sequence with second member

$$E_2^{-p,q} = H^{q-2p}(M; \pi_{q-p} M\zeta_k)$$

which converges to the gratuated set of stable homotopy classes $\bigoplus_p \{M^n, M\zeta_k\}_p$. (See Mosher-Tangora, Ch. 14.), Especially in the group $[M^n, M\zeta_k]$ there exists a filtration

$0 = F_q \subset F_{q-1} \ldots \subset F_0 = [M^n, M\zeta_k]$ such that $F_{j-1} / F_j = E^{0,j}$

The groups $E_2^{0,j} = H^j(M; \pi_j(M\zeta_k))$ belong to the class D. (Indeed $\pi_j(M\zeta_k) \approx \pi_{j-1}(M\theta_k)$. If $j \not\equiv 1 \mod 4$ then $\pi_{j-1}(M\theta_k) \in C_2 \subset D$ by the Lemma in 3.a. If $j \equiv 1 \mod 4$ then $H^j(M;Z) \in D$ by the condition. Hence the groups $E^{0,j}$ also belong to D.

Corollary II is proved.

§4. <u>Proof of the isomorphism $\mathrm{Imm}_2(n,k) \approx \pi_{n+k}(\Gamma_2(k))$.</u>

First we give an alternative description of the space $\Gamma_2(k)$, due to M. Gromov.

It is easy to see that for an immersion $f: M^n \longrightarrow S^{n+k}$ without triple points (M^n is oriented) the normal bundle ν of the image of double points $f(\Delta(f))$ admits the group $Z(k)$ as a structure group. Let U be a small tubular neigbourhood of $f(\Delta(f))$ in S^{n+k}. Then the fibration $U \longrightarrow f(\Delta(f))$ can be induced by pulling back the universal fibration $D\theta_k \longrightarrow B\theta_k$ with structure group $Z(k)$ and fibre D^k via a map h_1:

$$\begin{array}{ccc} U & \xrightarrow{h_1} & D\Theta_k \\ \downarrow & & \downarrow \\ f(\Delta(f)) & \longrightarrow & B\Theta_k \end{array}$$

The subset $f(M) \cap (S^{n+k} \setminus U)$ is an embedded submanifold in S^{n+k}. By the usual Pontrjagin-Thom construction there exists a map $h_2: (S^{n+k} \setminus U) \longrightarrow M\gamma_k$ such that

$$h_2^{-1}(B\gamma_k) = f(M) \cap (S^{n+k} \setminus U)$$

Glue together the spaces $D\Theta_k$ and $M\zeta_k$ (yielding $\Gamma_2(k)$) in order to get a continuous map h of the whole sphere S^{n+k} into $\Gamma_2(k)$ such that

$$h\Big|_U = h_1 \quad \text{and} \quad h\Big|_{S^{n+k} \setminus U} = h_2$$

To define the glueing map, we first remark, that every fibre D_x^{2k} (where $x \in B\Theta_k$) in the bundle $D\Theta_k \longrightarrow B\Theta_k$ is decomposed into a product $D_{1x}^k \times D_{2x}^k$. But the order of the factors can be chosen only locally, but not globally. So in every fibre D_x^{2k} we have a bouquet $D_{1x}^k \vee {}_{2x}^k$. Denote $Y(k)$ the union of these bouquets for every fibre.

<u>Remark:</u>

The map h_1 can be chosen such that

$$h_1^{-1}(Y(k)) = f(M) \cap U$$

We shall consider the space $D\Theta_k$ as a finite dimensional manifold. /To be precise we ought to replace the fibration Θ_k by its finite dimensional approximation and argue as in 2.c./ Then $Y(k) \cap \partial D\Theta_k$ is an embedded submanifold in $\partial D\Theta_k$. Applying the Pontrjagin-Thom construction again, we obtain a map $r: \partial D\Theta_k \longrightarrow M\gamma_k$.

The space $\Gamma_2(k)$ is $D\Theta_k \cup_r M\gamma_k$ which means that in the disjoint union $D\Theta_k \cup M\gamma_k$ we identify the points e and $r(e)$ for any $e \in \partial D\Theta_k$.

We described not only the space $\Gamma_2(k)$ but also the map

$Imm_2(n,k) \longrightarrow \pi_{n+k}(\Gamma_2(k))$. The cobordism class of the immersion f corresponds to the homotopy class of the map h.

<u>Correctness of the map $Imm_2(n,k) \to \pi_{n+k}(\Gamma_2(k))$.</u>

We constructed a map $h : S^{n+k} \to \Gamma_2(k)$, by an immersion without triple points. In the same way we can construct a map $S^{n+k} \times I \longrightarrow \Gamma_2(k)$ by a bordism of immersions without triple points. Hence the maps corresponding to cobordant immersions are homotopic.

<u>The map $\pi_{n+k}(\Gamma_2(k)) \longrightarrow Imm_2(n,k)$</u>

Denote $D_\varepsilon \Theta_k$ the ε-ball bundle associated with the bundle Θ_k.

Hence $D_\varepsilon \Theta_k \subset D\Theta_k = D_1 \Theta_k$ if $\varepsilon < 1$.

Fix numbers α and β such that $0<\alpha<\beta<1$.
Now consider an arbitrary continuous map $h : S^{n+k} \longrightarrow \Gamma_2(k)$.
Denote U_ε, the set $h^{-1}(D_\varepsilon \Theta_k)$ for $\varepsilon = \alpha$ or β. The map h can be slightly deformed in such a way that

1/ the map $h|_{U_\beta}$ becomes transversal to $B\Theta_k$ in $D\Theta_k$; and

2/ the maps $h|_{\partial U_\alpha}$ and $h|_{U_\beta}$ become transversal to $Y(k)$ in $D\Theta_k$.

Then the set $h^{-1}(Y(k)) \cap U_\beta$ is the image of an immersion without triple points and with double points set $h^{-1}(B\Theta_k) \cap U_\beta$.

The set $\Gamma_2(k) \setminus D_\alpha \Theta_k$ can be retracted onto the set $MSO(k) = M\gamma_k \subset \Gamma_2(k)$. (The retraction $\tau:(\Gamma_2(k) \setminus D_\alpha \Theta_k) \to M\gamma_k$ is the composition of the projection $(D\Theta_k \setminus D_\alpha \Theta_k) \to \partial D\Theta_k$ along the radii, and the map $r: \partial D\Theta_k \longrightarrow M\gamma_k$. Compare with the definition of σ in §5.) Performing a slight deformation we may suppose that the map

$$g = \tau \circ h|_{(S^{n+k} \setminus U_\alpha)} : (S^{n+k} \setminus U_\alpha) \longrightarrow M\gamma_k$$

is transversal to $BSO(k)$ in $M\gamma_k$. Moreover, we can assume that this deformation keeps the points of the boundary $\partial(S^{n+k} \setminus U_\alpha) = \partial U_\alpha$ fixed because the restriction of h to ∂U_α is transversal to $Y(k)$, and so $\tau \circ h|_{\partial U_\alpha}$ is transversal to

$B\gamma_k = BSO(k)$. Then $g^{-1}(BSO(k))$ is an embedded submanifold in $S^{n+k} \setminus U_\alpha$. The union

$$\{h^{-1}(B\theta_k) \cap U_\beta\} \cup g^{-1}(BSO(k))$$

is the image of an immersion without triple points. The cobordism class of this immersion corresponds to the homotopy class of the map $h: S^{n+k} \longrightarrow \Gamma_2(k)$.

The correctness can be proved in a similar way as before. The constructed homomorphisms

$$\pi_{n+k}(\Gamma_2(k)) \rightleftarrows \mathrm{Imm}_2(n,k)$$

are two-side inverses of each other, which accomplishes the proof of the isomorphism $\pi_{n+k}(\Gamma_2(k)) \approx \mathrm{Imm}_2(n,k)$.

§5. Proof of the isomorphism $S(n,k) \approx \pi_{n+k}(X(k))$.

We construct two maps $\varphi: \pi_{n+k}(X(k)) \longrightarrow S(n,k)$ and $\psi: S(n,k) \longrightarrow \pi_{n+k}(X(k))$ which are inverses of each other.

5.a. Construction of the map $\varphi: \pi_{n+k}(X(k)) \longrightarrow S(n,k)$.

Let $g: S^{n+k} \longrightarrow X(k)$ be an arbitrary continuous map. Performing a slight deformation, if it is needed, we may suppose that g is transversal to the "submanifold" $B\zeta_k$ of the space $X(k)$. (Recall that $X(k) = D\zeta_k \cup_\varphi \Gamma_2(k)$ and $B\zeta_k \subset D\zeta_k$.) Then the preimage $g^{-1}(B\zeta_k)$ is a submanifold of dimensional $n-k-1$ in the sphere S^{n+k}. The normal bundle ν of $g^{-1}(B\zeta_k)$ in S^{n+k} has structure group $Z(k)$.

The total space $E(\nu)$ of ν can be identified with a small tubular neighbourhood V of $g^{-1}(B\zeta_k)$ in S^{n+k} and the bundle ν can be identified with the fibration V over $g^{-1}(B\zeta_k)$. Denote this fibration by $\mu: V \longrightarrow g^{-1}(B\zeta_k)$. Then μ will be a vector bundle with structure group $Z(k)$ and the map $g|_V$ is a map of $Z(k)$ bundles.

By Lemma A below we have:

$$g^{-1}(Q(\zeta_k)) = Q(\mu)$$

Lemma A. Let μ and ζ be $(2k+1)$-dimensional vector bundles with structure group $Z(k)$ and let $\chi: E(\mu) \longrightarrow E(\zeta)$ be a map of $Z(k)$-bundles. For the sets $Q(\mu) \subset E(\mu)$ and $Q(\zeta) \subset E(\zeta)$ the following equality holds:

$$\chi^{-1}(Q(\zeta)) = Q(\mu)$$

Proof: Recall that $Q(\zeta)$ is the subbundle of $E(\zeta)$ associated to ζ with fibre Q. If the map χ preserves the $Z(k)$-structure of ζ and μ, then it also preserves the associated subbundles $Q(\zeta)$ and $Q(\mu)$. Lemma A is proved.

Let N be a small neighbourhood of $B\zeta_k$ in $X(k)$ such that $V = g^{-1}(N)$. There exists a retraction

$$\sigma : (X(k) \setminus N) \rightarrow \Gamma_2(k).$$

(The retraction σ is the composition of the projection $(D\zeta_k \setminus N) \longrightarrow \partial D\zeta_k$ along the radii and the glueing map $\rho: \partial D\zeta_k \rightarrow \Gamma_2(k)$. Compare with the definition of τ in §4.)

Consider the composition of maps

$$(S^{n+k} \setminus V) \xrightarrow{g \,|\, S^{n+k} \setminus V} (X(k) \setminus N) \longrightarrow \Gamma_2(k)$$

This defines an immersion into $S^{n+k} \setminus V$ which has no triple points. (See Remark 1 in 2.a.) The image of this immersion is $\tilde{g}^{-1}(Y(k))$, where $\tilde{g}$ is an appropriate approximation of g. The domain of this immersion is a manifold with boundary and the boundary will be immersed into the boundary of $S^{n+k} \setminus V$. The image of this boundary will coincide with the set

$Q(\mu) \cap \partial V$, because

$$\{\sigma \circ (\tilde{g}|\partial V)\}^{-1}(Y(k)) = g^{-1}(Q(\zeta_k)) \cap \partial V = Q(\mu) \cap \partial V$$

The first equality follows from the fact, that

$$\sigma^{-1}(Y(k)) \cap \partial N = Q(\zeta_k) \cap \partial N$$

and the second one follows from Lemma A.

Hence the union of the sets $\tilde{g}^{-1}(Y(k))$ and $Q(\mu)$ is the image of an S-map. The cobordism class of this S-map will correspond to the homotopy class of the map $g: S^{n+k} \longrightarrow X(k)$ under the correspondence φ.

<u>Correctness of $\varphi: \pi_{n+k}(X(k)) \longrightarrow S(n,k)$</u>

For every map $g:S^{n+k} \longrightarrow X(k)$ we constructed an S-map. In the same way, for every homotopy $S^{n+k} \times I \to X(k)$ joining maps g and g' one can construct an S-map $F:W^{n+1} \longrightarrow S^{n+k} \times I$ such that

$\partial W^{n+1} = M^n \cup M'^n$ and $F|_{M^n}$, $F|_{M'^n}$ are the S-maps corresponding to the maps g and g'. Now the correctness is clear.

5.b. Construction of the map $\Psi: S(n,k) \to \pi_{n+k}(X(k))$.

Let $f:M^n \longrightarrow S^{n+k}$ be an arbitrary S-map. Denote $\Sigma^1(f)$ the manifold of singular points, i.e:

$\Sigma^1(f) =: \{ m \in M \mid \text{rank } df(m) = n-1 \}$ and let $f(\Sigma^1(f))$ be its image. Denote V the tubular neighbouhood of $f(\Sigma^1(f))$ in S^{n+k} and μ the fibration of V over $f(\Sigma^1(f))$.

Denote ν the normal bundle of $f(\Sigma^1(f))$ in S^{n+k}.

Lemma B. The normal bundle ν admits the group Z(k) as a structure group. Moreover there exists a natural reduction of the structure group of ν to the group Z(k). This reduction is unique up to isomorphism.

Key Lemma. The fibration μ and its identification with ν can be chosen in such a way that $f(M) \cap V = Q(\nu)$

These lemmas will be proved in §6.

By Lemma B there exists a map χ of the fibration μ into the universal Z(k)-fibration ζ_k

$$\begin{array}{ccc} V & \longrightarrow & D\zeta_k \\ \downarrow & & \downarrow \\ f(\Sigma^1(f)) & \longrightarrow & B\zeta_k \end{array}$$

Denote $\beta: V \longrightarrow X(k)$ the composition of the map χ and the inclusion $D\zeta_k \subset X(k)$. The subset $f(M) \cap (S^{n+k} \setminus V)$ of the manifold $S^{n+k} \setminus V$ is the image of an immersion without triple points, so it defines a map

$(S^{n+k} \setminus V) \to \Gamma_2(k)$. (see Remark 2 in 2.a.)

Denote α the composition of this map and the inclusion $\Gamma_2(k) \subset X(k)$. Now we show that the maps $\alpha|_{\partial V}$ and $\beta|_{\partial V = \partial(S^{n+k} \setminus V)}$ coincide and so they define a map of the whole sphere S^{n+k} into X(k). The homotopy class of this map $S^{n+k} \longrightarrow X(k)$ will correspond to the cobordism class of the S-map f under the map $\psi: S(n,k) \longrightarrow \pi_{n+k}(\ (k))$.

Therefore it remainds to show that $\alpha|_{\partial V} = \beta|_{\partial V}$. Each of these maps can be considered as map into $\Gamma_2(k)$ because $\alpha(\partial V) \subset \Gamma_2(k)$ and $\beta(\partial V) \subset \Gamma_2(k)$. The maps α and β from ∂V to $\Gamma_2(k)$ are defined by the sets $Q(_\nu\alpha) \cap \partial V$ and $f(M) \cap \partial V$, respectively in the sense of Remark 2 in 2.a. . By Key lemma these sets coincide. (Recall that we identified the bundles $_\nu\kappa$ and ν.) The maps $\varphi: \pi_{n+k}(X(k)) \longrightarrow S(n,k)$ and $\psi: S(n,k) \longrightarrow \pi_{n+k}(X(k))$ are two-sided inverses of each other.

§6. Proofs of Lemmas.

6.a. Proof of Lemma B.

We shall need the following results of Haefliger about the structure of an S-map $f: M^n \longrightarrow N^{n+k}$.

1/ There exists a submanifold $\Delta(f)$ of dimension n-k in M^n such that

a/ $\Sigma^1(f) \subset \Delta(f)$ (Recall that $\Sigma^1(f) =: \{m \in M \mid \text{rank } df(m) = n-1\}$ and $\dim \Sigma^1(f) = n-k-1$

b/ $\Delta(f) \setminus \Sigma^1(f) =: \{m \in M \mid \exists!\ m' \in M: m' \neq m \text{ and } f(m) = f(m')\}$

Denote

$\tau: \Delta(f) \longrightarrow \Delta(f)$ the involution defined by

$\tau(m) = m$, if $m \in \Sigma^1(f)$

$\tau(m) = m'$, if $m \in \Delta(f) \setminus \Sigma^1(f)$ and $f(m) = f(m')$, $m \neq m'$.

Introduce the following notations:

L - a small τ-invariant neightbourhood of $\Sigma^1(f)$ in $\Delta(f)$.

∂L - the boundary of L.

ξ - the normal bundle of $\Delta(f)$ in M^n.

Remark 1:

$$\xi|_{\partial L} \approx \tau^*(\xi|_{\partial L})$$

Proof. The inclusion $i: \partial L \subset \Delta(f)$ is homotopic to the composition

$i \circ (\tau \mid \partial L)$ and so the induced bundles
$i^* \xi = \xi|_{\partial L}$ and $\{i \circ (\tau|_{\partial L})\}^*(\xi) = \tau^* \xi|_{\partial L}$ are
isomorphic.

<u>Remark 2.</u>

Let $h = f|_{\partial L} : \partial L \longrightarrow f(\partial L)$. Then h is a 2-fold covering and $h_!(\xi|_{\partial L})$ is the normal bundle of f $(\Delta(f))$ in N^{n+k} restricted to f (∂L).

<u>Proof.</u> In the neighbourhood of ∂L the S-map f is an immersion without triple points. For such an immersion the statement is trivial.

Lemma B follows from Remarks 1 and 2.

6.b. <u>Proof of Key lemma</u>

At this point we need two deep results of Haefliger about the S-maps. The notations below are the same as in 6.a.

<u>Lemma H1.</u> (The local description of an S-map)
For any $x \in \Sigma^1(f)$ there exist

1/ a coordinate system (V, φ) in M^n (where $V \subset M^n$ is a neighbourhood of x and $\varphi : V \longrightarrow R^n = \{x_1, \ldots, x_n\}$ is a homomorphism

2/ a coordinate system (W, ψ) in N^{n+k} where $W \subset N^{n+k}$ is a neighbourhood of f(x) and $\psi : W \longrightarrow R^{n+k} = \{y_1, \ldots, y_{n+k}\}$ is a homomorphism

such that

$$\psi \circ f \circ \varphi^{-1}(x_1, \ldots, x_n) = (y_1, \ldots, y_{n+k})$$

where

$$\begin{array}{lll} y_1 = x^2 & y_2 = x_2 & y_{n+1} = x_1 x_2 \\ & y_3 = x_3 & y_{n+2} = x_1 x_2 \\ & \vdots & \vdots \\ & y_n = x_n & y_{n+k} = x_1 x_{k+1} \end{array}$$

Proof in [Haefliger].

It will be more convenient to introuduce other coordinates

$z_1,\dots,z_{n+k}$ in R^{n+k} by the formulas

$$
\begin{array}{llll}
z_1=y_1\ ; & z_2=y_2+y_{n+1}\ ; & z_{k+2}=y_2-y_{n+1}\ ; & z_{2k+2}=y_{2k+2} \\
 & z_3=y_3+y_{n+2}\ ; & z_{k+3}=y_3-y_{n+2}\ ; & z_{2k+3}=y_{2k+3} \\
 & \vdots & \vdots & \vdots \\
 & z_{k+1}=y_{k+1}+y_{n+k}\ ; & z_{2k+1}=y_{k+1}-y_{n+k}\ ; & z_{n+k}=y_{n+k}
\end{array}
$$

i.e put

$$
\begin{pmatrix} z_1 \\ \vdots \\ z_{n+k} \end{pmatrix} = C \begin{pmatrix} y_1 \\ \vdots \\ y_{n+k} \end{pmatrix}
$$

where C is the following $(n+k) \times (n+k)$ matrix:

$$
\begin{pmatrix}
1 & 0 & \cdots & 0 \\
0 & E_k & 0 & E_k \\
\vdots & E_k & 0 & -E_k \\
0 & \multicolumn{2}{c}{0} & E_{n-2k-1}
\end{pmatrix}
$$

where E_i is the $i \times i$ identity matrix

Denote $\bar{\psi}$ the map $C \circ \psi$ where $C: R^{n+k} \longrightarrow R^{n+k}$ is the linear map corresponding to matrix C.

The map $\bar{\psi} \circ f \circ \varphi^{-1}$ is given by the formulas:

$$
\begin{array}{lll}
z_1 = x_1^2 & & \\
z_i = x_i\,(1+x_1) & \text{for } i=2,\dots,k+1 & \\
z_i = x_i\,(1-x_1) & \text{for } i=k+2,\dots,2k+1 & (!) \\
z_i = x_i & \text{for } i=2k+2,\dots,n+k &
\end{array}
$$

<u>Definition 3.</u> In the situation described in Lemma H1 we shall say, that the coordinate systems (V,φ) and (W,ψ) define <u>canonical coordinates</u> for the S-map f.

<u>Remark.</u> Decompose the euclidean space $R^{n+k} = \{z_1,\dots,z_{n+k}\}$ into the product of $R^{2k+1} = \{z_1,\dots,z_{2k+1}\}$ and

$R^{n-k-1} = \{z_{2k+2},\dots,z_{n+k}\}$. Recall that we defined the subset Q in R^{2k+1} (see 2.c.).
Then

$$\mathrm{Im}\left(\bar{\psi} \circ f \circ \varphi^{-1}\right) = Q \times R^{n-k-1} \subset R^{2k+1} \times R^{n-k-1} = R^{n+k}$$

Lemma H2. Let $f_1 : M^n \longrightarrow N^{n+k}$ and $f_2 : M^n \longrightarrow N^{n+k}$ be two S-maps, such that

$$\Sigma^1(f_1) = \Sigma^1(f_2) \text{ and } \mathrm{Ker}\, d\, f_1(x) = \mathrm{Ker}\, d\, f_2(x)$$

Then there exists a diffeomoprhism H of a neighbourhood V of $\Sigma^1(f_1)$ onto itself such that

$$H \circ f_1 \mid V = f_2 \mid V$$

Proof in [Haefliger].

Definition 4. Given an S-map $f : M^n \longrightarrow N^{n+k}$, we say that the coordinate systems $(V_1, \varphi_1), \dots, (V_r, \varphi_r)$ agree linearly if they define a fibre bundle structure with structure group $O(1) \oplus SO(k)$ in a tubular neighbourhood of $\Sigma^1(f)$ in M^n. (Notice that the normal bundle of $\Sigma^1(f)$ in M^n is the sum of the 1-dimensional normal bundle of $\Sigma^1 f$ in $\Delta(f)$ and the k-dimensional normal bundle ξ of $\Delta(f)$ in M^n restricted to $\Sigma^1(f)$.)

More precisly, the coordinate systems in M^n agree linearly if:

1/ $\Sigma^1(f) \subset \bigcup_{i=1}^{r} V_i$

2/ For any i $1 \le i \le r$ the homomorphism $\varphi_i : V_i \longrightarrow R^n$ has the form $\varphi_i = (\varphi'_i, \varphi''_i, \varphi'''_i)$, where

$\varphi'_i : V_i \longrightarrow R^{n-k-1}$

$\varphi''_i : V_i \longrightarrow R^1$

$\varphi'''_i : V_i \longrightarrow R^k$

i.e. $\varphi_i(v) = (\varphi'_i(v), \varphi''_i(v), \varphi'''_i(v))$ for every $v \in V_i$

3/ $v \in V_i \cap \Sigma^1(f) \iff \varphi''_i(v) = 0$ and $\varphi'''_i(v) = 0$

and $v \in V_1 \cap \Delta(f) \iff \varphi_1''(v) = 0$.

4/ For any i and j, with $V_i \cap V_j \neq \emptyset$ the map
$\varphi_j \circ \varphi_i^{-1} : R^n \longrightarrow R^n$ has the form
$\varphi_j \circ \varphi_i^{-1} = (h^{ij}, \varepsilon^{ij}, A)$

where:

$h^{ij} : R^{n-k-1} \longrightarrow R^{n-k-1}$

$\varepsilon^{ij} : R^1 \longrightarrow R^1$

$A = A(\alpha) : R^k \longrightarrow R^k$

i.e. for

$\forall \alpha \in R^{n-k-1}, \forall \beta \in R^1$ and $\forall \gamma \in R^k$

$$\varphi_j \circ \varphi_i^{-1}(\alpha, \beta, \gamma) = (h^{ij}(\alpha), \varepsilon^{ij}(\beta), A(\alpha)\gamma)$$

the matrix $A(\alpha)$ acts on the vector $\gamma \in R^k$

Here: h^{ij} is a diffeomorphism of R^{n-k-1} onto itself,

$\varepsilon^{ij}(\beta) = \beta$ or $-\beta$

$A^{ij}(\alpha) : R^k \longrightarrow R^k$ is a family of oriented orthogonal operators smoothly depending on the point $\alpha \in R^{n-k-1}$

<u>Definition 5.</u> Given an S-map $f : M^n \longrightarrow N^{n+k}$, we say that the coordinate systems $(W_1, \psi_1), \ldots, (W_r, \psi_2)$ in N^{n+k} <u>agree linearly.</u> if they define a fibre bundle structure with group $SZ(k)$ in a tubular neighbourhood of $f(\Sigma^1(f))$ in N^{n+k}. (Notice that the normal bundle of $f(\Sigma^1(f))$ is the sum of a trivial 1-dimensional bundle (the normal bundle of $f(\Sigma^1(f))$ in $f(\Delta(f))$) and a 2k-dimensional bundle (the normal bundle of $f\Delta(f)$ in N^{n+k}, restricted to $f(\Sigma^1(f))$.)) By the Remarks 1 and 2 of §6 the second bundle admits $Z(k)$ as a structure group.
More precisely, the coordinate systems $(W_1, \psi_1), \ldots, (W_r, \psi_r)$ <u>agree linearly</u> if:

1/ $f(\Sigma^1(f)) \subset \bigcup_{i=1}^{r} W_i$

2/ For any i and j the map $\psi_i : W_i \to R^n$ has the form

$\psi_i = (\psi_i', \psi_i'', \psi_i''')$, where

$\psi_i' : W_i \longrightarrow R^{n-k-1}$

$\psi_i'' : W_i \dashrightarrow R^1$

$\psi_i''' : W_i \longrightarrow R^{2k}$

i.e.: $\psi_i(w) = (\psi_i'(w), \psi_i''(w), \psi_i'''(w))$ for every $w \in W_i$

3/ $w \in W_i \cap f(\Sigma^1(f)) \Longleftrightarrow \{\psi_i''(w) = 0$ and $\psi_i'''(w) = 0\}$

$w \in W_i \cap f(\Delta(f)) \Longleftrightarrow \psi_i''(w) = 0$

4/ For any i and j with $W_i \cap W_j \neq \emptyset$ the map

$\psi_j \circ \psi_i^{-1} : R^{n+k} \longrightarrow R^{n+k}$

has the form $\psi_j \circ \psi_i^{-1} = (\bar{h}^{ij}, \mathrm{id}, B^{ij})$

i.e. for $\forall \alpha \in R^{n-k-1}$, $\forall \beta \in R^1$, $\forall \delta \in R^{2k}$

$\psi_j \circ \psi_i^{-1}(\alpha, \beta, \gamma) = (\bar{h}^{ij}(\alpha), \mathrm{id}(\beta), B^{ij}(\alpha)\cdot\delta)$ the matrix $B^{ij}(\alpha)$ acts on the vector

Here: $\bar{h}^{ij} : R^{n-k-1} \longrightarrow R^{n-k-1}$ is a diffeomorphism

$\mathrm{id}: R^1 \longrightarrow R^1$ is the identity map

$B^{ij} = B^{ij}(\alpha): R^{2k} \longrightarrow R^{2k}$ is a family of linear operators smoothly depending on the point $\alpha \in R^{n-k-1}$ such that $B^{ij}(\alpha) \in Z(k)$

<u>Definition 6</u>: Given an S-map $f: M^n \longrightarrow N^{n+k}$ and coordinate systems (V_1, φ_1), (V_2, φ_2) in M^n and (W_1, ψ_1), (W_2, ψ_2) in N^{n+k} which agree linearly, we say that the transformation

$\psi_1 \circ \psi_2^{-1} : R^{n+k} \longrightarrow R^{n+k}$ is <u>induced from the transformation</u> $\varphi_1 \circ \varphi_2^{-1} : R^n \longrightarrow R^n$ <u>by the map f</u> if

1/ $f(V_i) \subset W_i \quad i=1,2$

2/ $\bar{h}^{ij} = h^{ij}$ (The notations are the same as in the previous definition.)

3/

$$B^{ij}(\alpha) = \begin{cases} \begin{pmatrix} A^{ij}(\alpha) & 0 \\ 0 & A^{ij}(\alpha) \end{pmatrix} & \text{if} \quad \varepsilon^{ij}(\beta) = \beta \\ \begin{pmatrix} 0 & A^{ij}(\alpha) \\ A^{ij}(\alpha) & 0 \end{pmatrix} & \text{if} \quad \varepsilon^{ij}(\beta) = -\beta \end{cases}$$

The Key lemma is an easy consequence of the following

Basic lemma

There exist coordinate systems $(V_1, \varphi_1), \dots, (V_r, \varphi_r)$ in the manifold M^n and coordinate systems $(W_1, \psi_1), \dots, (W_r, \psi_r)$ in N^{n+k} satisfying the following conditions:

A/ the systems $(V_1, \varphi_1), \dots, (V_r, \varphi_r)$ agree linearly (see def.4),

B/ the systems $(W_1, \psi_1), \dots, (W_r, \psi_r)$ agree linearly (see def.5),

C/ the transformations $\psi_i^{-1} \circ \psi_j$ are induced from the transformations $\varphi_i^{-1} \circ \varphi_j$ by the map f for any i and j (see def.6.)

D/ for any i the systems (V_i, φ_i) and (W_i, ψ_i) define canonical coordinate systems for the map f (see def.3.)

Proof:

I. Coordinate systems $(V_1, \varphi_1), \dots, (V_r, \varphi_r)$ satisfying condition A can be chosen because the fibration μ - the tubular neighbourhood of $\Sigma^1(f)$ in M^n - can be identified with the vector bundle ν -the normal bundle of $\Sigma^1(f)$ in M^n.

II. Analogously, there are systems $(\bar{W}_1, \bar{\psi}_1), \dots, (\bar{W}_r, \bar{\psi}_r)$ satisfying condition B (replacing W_i by $\bar{W}_i$ and ψ_i by $\bar{\psi}_i$

because the tubular neighbourhood of $f(\Sigma^1(f))$ in N^{n+k} can be identified with its normal bundle in N^{n+k}.

III. By Remark 2 in 6.a. the systems $(\bar{W}_1, \bar{\Psi}_1), \dots, (\bar{W}_r, \bar{\Psi}_r)$ can be chosen in such a way that condition C will be satisfied too.

IV. Notice, that if the systems $(V_1, \varphi_1), \dots, (V_r, \varphi_r)$ and $(\bar{W}_1, \bar{\Psi}_1), \dots, (\bar{W}_r, \bar{\Psi}_r)$ satisfy conditions A,B,C and $H: U \longrightarrow U$ is an arbitrary diffeomorphism of an arbitrary neighbourhood U of $f(\Sigma^1(f))$ onto itself, then the coordinate systems $(V_1, \varphi_1), \dots, (V_r, \varphi_r)$ and $(H(\bar{W}_1), \bar{\Psi}_1 \circ H^{-1}), \dots, (H(\bar{W}_r), \bar{\Psi}_r \circ H^{-1})$ satisfy them, too.

Indeed, condition A is obviously satisfied and the conditions B and C involve the transformation

$\bar{\Psi}_j \circ \bar{\Psi}_i^{-1}$

which does not change

$$\bar{\Psi}_j \circ \bar{\Psi}_i^{-1} = (\bar{\Psi}_j \circ H^{-1}) \circ (\bar{\Psi}_i \circ H^{-1})^{-1}$$

V. Let us define a map $g: \bigcup_{i=1}^{r} V_i \longrightarrow N^{n+k}$ by the formulas(!) after Lemma H1. The map g is correctly defined. This follows from the following

1/ the conditions A,B,C, hold for the systems
$(V_1, \varphi_1), \dots, (V_r, \varphi_r)$ and $(\bar{W}_1, \bar{\Psi}_1), \dots, (\bar{W}_r, \bar{\Psi}_r)$

2/ $\operatorname{Im} \bar{\Psi}_i \circ g \circ \varphi_i^{-1} = Q \times R^{n-k-1}$

3/ Q is SZ(k)-invariant.

The sets $\Sigma^1(f)$ and $\Sigma^1(g)$ coincide. The kernels Ker df(x) and Ker dg(x) also coincide for any $x \in \Sigma^1(f)$. Hence by Lemma H2 there exists a diffeomorphism $H: U \longrightarrow U$ of a neighbourhood U of $f(\Sigma^1(f))$ in N^{n+k} such that

$$H \circ g|_U = f|_U$$

Let us define the systems (W_i, Ψ_i) $i=1,2,\dots,r$ as follows:

$$W_i \overset{def}{=\!=} H(\overline{W}_i) \quad \text{and} \quad \Psi_i \overset{def}{=\!=} \overline{\Psi}_i \circ H^{-1}$$

Then the systems $(V_1, \varphi_1), \dots, (V_r, \varphi_r)$ and $(W_1, \Psi_1), \dots, (W_r, \Psi_r)$ satisfy all the conditions A,B,C,D. The Basic lemma is proved.

To prove the Key lemma we have only to identify both the fibrations μ and ν with the fibre bundle over $f(\Sigma^1(f))$ defined by the coordinate systems $(W_1, \Psi_1), \dots, (W_r, \Psi_r)$

Bibliography

1/ Barratt-Eccles: Γ^+-structures. I. Topology 13. N°-1 1974

2/ A. Haefliger: Plongements differentiables de veriétés dans veriétés, Comm. Math. Helv. 37. 1962/63 155-17o.

3/ M.W. Hirsch: Immersions of manifolds. Transactions AMS 93. 1959. 242-276

4/ R. Moscher-M. Tangora: Cohomology operations and applications in homotopy theory.

5/ A. Szűcs: Cobordism groups of ℓ-immersions I. in Russian Acta Math. Acad. scient. Hung. 27. N°-3-4 1976 343-358

6/ R. Wells: Cobordism of immersions. Topology 5, N°-3 1966 281-293.

Author's adress: Hungary 672o. Szeged University
Aradi vértanuk tere 1-3.

A SPECTRAL SEQUENCE CONVERGENT TO EQUIVARIANT K-THEORY

Agnieszka Bojanowska and Stefan Jackowski
Instytut Matematyki, Uniwersytet Warszawski,
PKiN IXp., PL-00-901 Warszawa , Poland.

In this work we continue the investigation of the role played by cyclic subgroups in equivariant K-theory. We consider actions of finite groups on equivariant CW-complexes. For an arbitrary finite G-CW-complex X we prove that there is a convergent spectral sequence

$$E_2^{pq} = H_G^*(X \times EC : K_G^*) \Rightarrow K_G^*(X)$$

where H_G^* is the Bredon equivariant singular cohomology and EC denotes a classifying space for the family of cyclic subgroups of G . It means that the E_2-term depends on the G-space which has only cyclic isotropy subgroups and the projection $X \times EC \longrightarrow X$ is a S-homotopy equivalence for all cyclic subgroups S of G .

Combining the spectral sequence with Rubinsztein's result on restriction of equivariant K-theory to cyclic subgroups we obtain a description of the variety of geometric points of equivariant K-theory ring. Such description was suggested by Quillen [1971].

§1. Equivariant K-theory of classifying spaces for families of subgroups.

This note is a continuation of earlier work on equivariant K-theory. Therefore we recall some notions and results.

Let G be a finite group and let F be a family of subgroups of the group G . A classifying space for a family F is a G-space EF such that the fixed point set $(EF)^H$ is contractible for $H \in F$ and empty for $H \notin F$. A classifying space EF can be realized as a G-CW-complex constructed as an infinite join of orbits with isotropy subgroups belonging to F equipped with the compactly generated topology (cf. tom Dieck [1972]) .

We are interested in equivariant K-theory of such classifying spaces. For that purpose we consider an additive extension of equivariant K-theory to the category of all G-spaces (cf. Atiyah, Segal [1969]) .

1.1 THEOREM. Let F be a family of subgroups of a finite group G containing all its cyclic subgroups. Then for an arbitrary finite G-CW-cpmplex X the projection $X \times EF \longrightarrow X$ induces an isomorphism $K_G^*(X \times EF) \cong K_G^*(X)$.

Proof. The theorem follows from Theorem 4.2. of Jackowski [1977].

The proof of Theorem 1.1 mentioned above does not give any information about the filtration of $K_G^*(X)$ EF induced by the skeleton filtration of $X \times EF$. To investigate it we will use the following theorem proved by Rubinsztein [1979].

1.2 THEOREM. Let G be a compact Lie group and let X be a finite G-CW-complex. Then the restriction homomorphism

$$K_G^*(X) \otimes Q \longrightarrow \prod_{S\text{-cyclic}} K_S^*(X) \otimes Q$$

is a monomorphism.

1.3 COROLLARY. Let G be a finite group. For every finite G-CW--complex the kernel of the restriction homomorphism

$$K_G^q(X) \longrightarrow \prod_{S\text{-cyclic}} K_S^q(X)$$

is finite for all q.

Proof. For a finite group G and a finite G-CW-complex X, $K_G^q(X)$ is a finitely generated abelian group. Hence the corollary follows from the last theorem.

The next theorem describes the first derived functor of the inverse limit for the skeleton filtration on $K_G^*(X \times EF)$.

1.4 THEOREM. Let F be a family of subgroups of a finite group G containing all its cyclic subgroups. Then for an arbitrary finite G-CW-complex X the derived functor $\text{inv.lim}^1 \{ K_G^*(X \times (EF)^n) \} = 0$.

Proof. The proof is based on Milnor's lemma which holds also for equivariant K-theory (cf. Atiyah, Segal [1969] Prop.4.1., Remark (b)). Thus we have an exact sequence

$$\text{inv.lim}^1 \{K_G^*(X \times (EF)^n)\} \longrightarrow K_G^*(X \times EF) \xrightarrow{j^*} \text{inv.lim}\{K_G^*(X \times (EF)^n)\}$$

For each n the product $X \times (EF)^n$ is finite CW-complex. Hence the inverse system $K_G^*(X \times (EF)^n)$ is a system of finitely generated abelian groups. The derived functor inv.lim^1 of such system is a divisible abelian group (cf. Bröcker, tom Dieck [1970], Satz V.2.10.) We will show that $\text{inv.lim}^1 K_G^*(X \times (EF)^n)$ is also finite. Consider the restriction homomorphism

$$j_o^* : K_G^*(X \times EF) \longrightarrow K_G^*(X \times (EF)^o) .$$

The zero-skeleton $(EF)^o$ is a disjoint union of orbits G/S for all $S \in F$. Hence we have a commutative diagram

$$\begin{array}{ccc} K_G^*(X \times EF) & \xrightarrow{j_o} & K_G^*(X \times (EF)^o) \\ \| \uparrow & & \| \\ K_G^*(X) & \xrightarrow{\text{res}} & \prod_{S \in F} K_S^*(X) \end{array}$$

in which the left vertical arrow is an isomorphism by Theorem 1.1. From the Corollary 1.3 it follows that $\ker j_o^* = \ker \text{res}$ is finite. Therefore $\ker j^* \subset \ker j_o$ also is a finite group. As a finite divisible group must be trivial the proof of the theorem is completed.

<u>1.5 COROLLARY.</u> Under the assumption of the last theorem the skeleton filtration $F_p := \ker K_G^*(X \times EF) \quad K_G^*((X \times EF)^p)$ is finite and $\bigcap_{p=1}^{\infty} F_p = 0$.

<u>Proof.</u> Theorem 1.4 implies that $\bigcap_{p=1}^{\infty} F_p = 0$. From Theorem 1.3 we infer that for $p > \dim X$ the groups F_p are finite. Hence the filtration F_p is finite.

In the next section we will consder convergence of the Atiyah-Hirzebruch spectral sequence for equivariant K-theory and the complex $X \times EF$.

§2. A criterion for convergence of the Atiyah-Hirzebruch spectral sequence.

We recall first an equivariant version of the Atiyah-Hirzebruch spectral sequence (cf. Bredon [1967], Matumoto [1973]). Let h_G^* be an equivariant cohomology theory defined on the category of G-spaces. Then for every finite G-CW-complex Y there is a convergent spectral sequence E_r^{pq} such that

$$E_2^{pq} = H_G^p(Y:h_G^q)$$

where H_G^* denotes equivariant singular cohomology theory and

$$E_\infty^{pq} = \frac{\ker\{h_G^{p+q}(Y) \longrightarrow h_G^{p+q}(Y^{p-1})\}}{\ker\{h_G^{p+q}(Y) \longrightarrow h_G^{p+q}(Y^p)\}}$$

Some additional functorial and multiplicative properties of the equivariant Atiyah-Hirzebruch spectral sequence will be important in the next section. Assume that a cohomology theory is defined on the category of all equivariant complexes. Every morphism $(\Theta, f) : (G,X) \longrightarrow (G',X')$ of spaces with actions induces a homomorphism of corresponding spectral sequences. This homomorphism coincides with the induced homomorphism on E_2 and E_∞ terms. Moreover, if the cohomology theory h_G^* admits transfer (the Mackey structure) then for an arbitrary G-CW-complex X and a subgroup $H \subset G$ the inclusion $(H,X) \hookrightarrow (G,X)$ defines a homomorphism of spectral sequences $E_*^{**}(H,X) \longrightarrow E_*^{**}(G,X)$ which coincides with transfers on E_2 and E_∞ terms (cf. Illman [1975]). If the cohomology theory is multiplicative then the equivariant Atiyah-Hirzebruch spectral sequence has the usual multiplicative properties.

The Atiyah-Hirzebruch spectral sequence can be constructed also for an infinite complex however it does not have to converge to a graded group associated with cohomology of the complex. In this section we prove a criterion for convergence of such a spectral sequence. This criterion holds for the non-equivariant and equivariant cases and therefore for simplicity of notation we will formulate it for non-equivariant cohomology theories.

2.1 THEOREM. Let h^* be an additive cohomology theory whose coefficients h^q are finitely generated groups for all q. Let X be a countable CW-complex. If $\text{inv.lim}^1\{h^*(X^n)\} = 0$ then the

Atiyah-Hirzebruch spectral sequence of X is convergent to $h^*(X)$ i.e.

i) for every p,q there is r such that $E_r^{pq} = \ldots = E_\infty^{pq}$

ii) $$E_\infty^{pq} = \frac{\ker\{h^{p+q}(X) \longrightarrow h^{p+q}(X^{p-1})\}}{\ker\{h^{p+q}(X) \longrightarrow h^{p+q}(X^{p})\}}$$

<u>Proof.</u> From the assumption about coefficients it follows that $\{h(X^n)\}$ is an inverse system of countable abelian groups for which inv.lim^1 vanishes. According to the lemma of B. Gray (cf. Gray [1966], Bröcker, tom Dieck [1970], Satz V.2.5.) such an inverse system satisfies the Mittag-Leffler condition. The assertion i) follows now from Brocker, tom Dieck [1970] Satz V.3.9.

To prove the second assertion we follow the arguments of A. Jankowski [1972]. From the definition of the Atiyah-Hirzebruch spectral sequence there exists a natural monomorphism

$$\frac{\ker\{h^{p+q}(X) \longrightarrow h^{p+q}(X^{p-1})\}}{\ker\{h^{p+q}(X) \longrightarrow h^{p+q}(X^{p})\}} \longrightarrow E_\infty^{pq}$$

We have to prove that it is also an epimorphism. For every p,q,r there is a natural exact sequence

$$h^{p+q}(X^{p+q},X^p) \xrightarrow{j_r^{pq}} h^{p+q}(X^{p+r},X^{p-1}) \longrightarrow Z_r^{pq} \longrightarrow 0$$

It is easy to prove that for fixed p,q the inverse system $\{\text{im } j_r^{pq}\}$ satisfies the Mittag-Leffler condition and thus $\text{inv.lim}^1\{\text{im } j_r^{pq}\}=0$. Therefore the limit homomorphism

$$\text{inv.lim } h^{p+q}(X^{p+r},X^p) \longrightarrow \text{inv.lim } Z_r^{pq} = \bigcap_{r=1}^{\infty} Z_r^{pq} = Z_\infty^{pq}$$

is an epimorphism. The following commutative diagram

$$\begin{array}{ccccc}
h^{p+q}(X,X^{p-1}) & \longrightarrow & \ker h^{p+q}(X) & \longrightarrow & h^{p+q}(X^{p-1}) \\
\downarrow & & & & \searrow \\
 & & & & E^{pq} \\
 & & & & \nearrow \\
\text{inv.lim } h^{p+q}(X^{p+r},X^{p-1}) & \longrightarrow & Z_\infty^{pq} & &
\end{array}$$

shows that the homomorphism $\ker\{h^{p+q}(X) \longrightarrow h^{p+q}(X^{p-1})\} \longrightarrow E_\infty^{pq}$ is an epimorphism.

A different proof of the last theorem can be found in Meier [1975].

§3. Equivariant K-theory, cyclic subgroups and a spectral sequence.

We will show that there is a spectral sequence relating equivariant singular cohomology $H_G^*(X \times EF:K_G^*)$ to $K_G^*(X)$ where the family F contains all cyclic subgroups of G .

The construction of the spectral sequence is motivated by the Atiyah-Segal Completion Theorem (cf. Atiyah [1961] , Atiyah and Segal [1969]) and the resulting spectral sequence. The Atiyah-Segal Theorem implies that for a finite G-space X there is a convergent spectral sequence

$$E_2^{pq} = H^p(EG \times_G X:K_G^q) \Rightarrow K_G^*(X)$$

where denotes the completion with respect to the augumentation topology which is the same as the topology defined by the filtration of $K_G^*(X)$ associated with the spectral sequence.

The appropriate filtration of $K_G^*(X)$ in our spectral sequence is finite, however E_2-term is more difficult to compute. We consider the Atiyah-Hirzebruch spectral sequence for the G-space $X \times EF$ and then apply Theorem 1.1.

3.1 THEOREM. Let F be a family of subgroups of a finite group G which contains all its cyclic subgroups. For every finite G-CW-complex X there is a multiplicative spectral sequence E_r^{pq} such that

a) $E_2^{pq} = H_G^p(X \times EF:K_G^q)$.

b) $E_2^{pq} = E_r^{pq} = 0$ for odd q .

c) For all p,q,r $E_r^{pq} = E_r^{p,q+2}$ and $E_\infty^{pq} = E_\infty^{p,q+2}$

d) All the even differentials d_{2r} are zero.

e) For every p,q there exists r such that $E_r^{pq} = E_\infty^{pq}$.

f) There is a finite filtration $K_G^*(X) \supset K_G^*(X)_0 \supset \ldots \supset K_G^*(X)_n = 0$ such that $E_\infty^{pq} = K_G^{p+q}(X)_{p-1}/K_G^{p+q}(X)_p$.

g) The multiplicative structure is compatible with the products in $H_G^*(X \times EF:K_G^*)$ and in $K_G^*(X)$.

Proof. We consider the Atiyah-Hirzebruch spectral sequence for the G-CW-complex $X \times EF$. From Theorem 1.4 and the properties of the representation ring it follows that the assumptions of Theorem 2.1

are fulfiled. Moreover Proposition 1.5 tells us that the appropriate filtration is finite. According to Theorem 1.1 $K_G^*(X)$ is isomorphic to $K_G^*(X \times EF)$ and thus assertion e) and f) are proved. Assertions b)-d) follow from the Periodicity Theorem in equivariant K-theory.

The construction of the spectral sequence is natural in the following sense:

3.2 PROPOSITION. Let (G,F,X) and (G',F',X') satisfy the assumptions of Theorem 3.1.

a) A morphism $(\Theta, f) : (G,F,X) \longrightarrow (G',F',X')$ consisting of a homomorphism $\Theta : G \longrightarrow G'$ such that $\Theta(F) \subset F'$ and of the Θ-equivariant map $f : X \longrightarrow X'$ induces a natural homomorphism of spectral sequences $(\Theta, f)^* : E_*^{**}(G',F',X') \longrightarrow E_*^{**}(G,F,X)$.

b) Let $\Theta : G \longrightarrow G'$ be an inclusion and $F = F' \cap G$. The morphism $(\Theta, id) : (G,F,X) \longrightarrow (G',F',X')$ defines a transfer homomorphism of spectral sequences $E_*^{**}(G,F,X) \longrightarrow E_*^{**}(G',F',X')$.

Proof. a) It is easy to see that Θ induces a Θ-equivariant map $E(\Theta) : EF \longrightarrow EF'$. The induced homomorphism of spectral sequences is defined by the map $E\Theta \times f$ (cf. §2).

b) From the definition of classifying spaces for families it follows easily that the space EF' treated as a G-space is classifying for the family F. Then the proof follows from the fact that the Atiyah-Hirzebruch spectral sequence preserves the Mackey structure (cf. Illman [1975]).

We will investigate the edge homomorphism of our spectral sequence

$$\alpha(X) : K_G^*(X) \longrightarrow H_G^o(X \times EF : K_G^*).$$

First we describe the zero cohomology group in terms of fixed point sets of subgroups belonging to the family F. For a G-space X and an arbitrary family of subgroups F we define the category $F(G,X)$. Its objects are pairs (H,c) such that c is a nonempty connected component of X^H. A morphism $(H,c) \longrightarrow (H',c')$ is an equivariant map $[g] : G/H \longrightarrow G/H'$ such that $g(c) \subset c'$. The category $F(G,pt)$ is a category of canonical orbits and equivariant maps. The following property can be easily deduced from the definition of the Bredon cohomology.

3.3 PROPOSITION. Let M be a contravariant coefficient system defined on the category of canonical orbits. Then for an arbitrary family of subgroups F and a G-space X there is a natural isomorphism

$$H^0_G(X \times EF:M) \longrightarrow \operatorname{inv.lim}_{F(G,X)} M$$

where the functor $M:F(G,X) \longrightarrow \underline{Ab}$ is defined by $M(H,c) := M(G/H)$.

Under the above isomorphism the edge homomorphism $\alpha(X)$ corresponds to the homomorphism

$$\alpha(X):K^*_G(X) \longrightarrow \operatorname{inv.lim}_{F(G,X)} K^*_G$$

defined by inclusions $G/H \hookrightarrow X$ such that $G/H \subset c$ for the object (H,c) of $F(G,X)$.

Let us consider the case $X = pt$. Then $K^*_G(X)$ is a complex representation ring and the edge homomorphism

$$\alpha : R(G) \longrightarrow \operatorname{inv.lim}_{F} R(\cdot)$$

is a restriction homomorphism from the representation ring to subgroups belonging to F. The family F contains all cyclic subgroups and therefore α is a monomorphism. Hence we obtain the following purely algebraic consequence of Theorem 3.1.

3.4 COROLLARY. For an arbitrary finite group G and a family of subgroups F containing all cyclic subgroups of G the restriction homomorphism $\alpha : R(G) \rightarrow \operatorname{inv.lim}_F R(\cdot)$ is a monomorphism and its image is equal $\bigcap_{r=1}^{\infty} \ker d_r$ where $d_r : H^0_G(EF:K_G) \longrightarrow H^r_G(EF:K_G)$ are derivations.

Let us observe that the Brauer Induction Theorem (cf. Serre [1967]) can be formulated in terms of the differentials d^0_r. It says that $d_r = 0$ for all $r \geq 2$ iff the family F contains all elementary subgroups.

§4. The spectrum of equivariant K-theory.

This section is devoted to the first application of the spectral sequence 3.2. We describe the variety of geometric points of the equivariant K-theory ring. This description is suggested by Quillen's results on the spectrum of the equivariant Borel cohomology (cf. Quillen [1971]) .

Recall that for a commutative ring A and a field Ω we denote by $A(\Omega)$ the set of ring homomorphisms $A \longrightarrow \Omega$ endowed with the Zariski topology. We denote by $K_G^{\cdot}(X)$ the ring $K_G^{*}X$ divided by nilpotent elements. This is a commutative ring.

<u>4.1 THEOREM.</u> Assume that the assumptions of Theorem 3.1 are satisfied. Then for an arbitrary algebraically closed field the edge homomorphism $\alpha(X)$ induces a homeomorphism of varieties

$$H_G^0(X \times EF : K_G^{\cdot})(\Omega) \longrightarrow K_G^{\cdot}(X)(\Omega)$$

<u>4.2 COROLLARY.</u> The edge homomorphism $\alpha(X)$ induces a bijection

$$\operatorname*{dir.lim}_{F\ G,X} \Omega^{*} \otimes H \longrightarrow K_G^{\cdot}(X)(\Omega)$$

where Ω^{*} denotes the multiplicative group of Ω and $(H,c) \in F(G,X)$.

To prove Theorem 4.1 we first apply the λ-ring structure on $K_G^{\cdot}(X)$ to prove that all elements of $\ker \alpha(X)$ are nilpotent. Then the theorem will follow from the spectral sequence 3.2 and a lemma of Quillen from commutative algebra.

Recall that $K_G^{*}(X)$ carries a λ-ring structure defined by the exterior power operations. We need the follwing elementary lemma due to G. Segal (unpublished).

<u>4.3 LEMMA.</u> Let A be an arbitrary λ-ring. Then any torsion element of A is nilpotent.

<u>4.4 COROLLARY.</u> Every element of $\ker \alpha(X)$ is nilpotent.

<u>Proof.</u> Let us consider the following commutative diagram:

$$\begin{array}{ccccc}
 & \overset{X}{\nearrow} & \operatorname*{inv.lim}_{F(G,X)} K_G^{*} & \hookrightarrow & \prod_{(H,c)\in F(G,X)} K_G^{*}(G/H) \\
K_G^{*}(X) & & & & \downarrow \\
 & \underset{r_1}{\searrow} & \prod_{S\in F} K_S^{*}(X) & \xrightarrow{r_2} & \prod_{S\in F}\ \prod_{(H,c)\in(F\cap S)(S,X)} K_S^{*}(S/H)
\end{array}$$

where r_1 is the restriction homomorphism and $r_2 := \prod_{S\in F} \alpha(S,X)$.

From this diagram it follows that $\ker \alpha(X) = \ker r_1 r_2$.

For every subgroup $S \in F$ all elements of $\ker \alpha(S,X)$ are nilpotent. It follows easily from the Mayer-Vietoris sequence (cf. Quillen [1971] Proposition 3.2). From Corollary 1.3 it follows that elements of $\ker\ r_1$ are torsion hence by Lemma 4.3 they are also nilpotent.
To conclude the proof of Theorem 4.1 we will need the following lemma from commutative algebra proved by Quillen [1971] Corollary B.6.

4.5 LEMMA. Let A be a subring of elements of a ring B killed by a derivation $d : B \longrightarrow M$, where M is a B-module . If A is noetherian and B is a finitely generated A-module, then $A(\Omega) = B(\Omega)$ for all algebraically closed fields Ω.

Proof of Theorem 4.1 . Corollary 4.4 implies that the natural projection $K_G^{\cdot}(X) \longrightarrow K_G^{\cdot}(X)/\ker \alpha(X)$ induces a bijection of the corresponding Ω-varieties. Therefore it is enough to show that the inclusion $K_G^{\cdot}(X)/\ker \alpha(X) \hookrightarrow H_G^{0}(X \times EF : K_G^{\cdot})$ induces a bijection of Ω-varieties. From the spectral sequence 3.1 we know that $\operatorname{im} \alpha(X) = \bigcap_{r=1}^{n} \ker d_r^{0}$ where $d_r^{0} : H_G^{0}(X \times EF : K_G^{\cdot}) \longrightarrow H_G^{r}(X \times EF : K_G^{\cdot})$ are derivations.
Since $\alpha(X)$ defines on $H_G^{0}(X \times EF : K_G)$ a structure of a finite module over a noetherian ring $K_G^{\cdot}(X)$ the assumptions of Lemma 4.5 are satisfied.

4.6 REMARK . Corollary 4.2 holds for an arbitrary compact Lie group G and a finite G-CW-complex X . The proof based on a different approach than the one given here, can be found in Bojanowska [1979].

References.

M.F.Atiyah

[1961] Characters and cohomology of finite groups. Inst.Hautes Etudes Sci.Publ.Math. No.9 (1961) 23-64 .

M.F.Atiyah G.B.Segal

[1969] Equivariant K-theory and completion. J.Diff.Geometry 3 (1969) 1-18.

A.Bojanowska

[1979] The spectrum of equivariant K-theory. Thesis, Warszawa, 1979.

G.Bredon

[1967] Equivariant cohomology theories. Lecture Notes in Mathematics 34, Springer-Verlag 1967.

T.Bröcker, T.tom Dieck

[1970] Kobordismentheorie. Lecture Notes in Mathematics 178, Springer-Verlag 1970

T.tom Dieck

[1972] Orbittypen und aquivariante Homologie . Arch.Math. vol.XXIII (1972) 307-317 .

B.Gray

[1966] Spaces of the same n-type, for all n. Topology 6 No.3 (1966) 241- .

S.Illman

[1975] Equivariant singular homology and cohomology I. Memoirs of the American Mathematical Society No.156 March 1975.

S.Jackowski

[1977] Equivariant K-theory and cyclic subgroups . in Transformation Groups ed.C.Kosniowski, Lecture Note Series, London Mathematical Society No.26 pp.76-92. Cambridge University Press 1977.

A.Jankowski

[1972] Элементы бесконечной фильтрации спектров Мат. зам. Т. 11 (1972), 699-704.

T.Matumoto

[1973] Equivariant cohomology theories on G-CW-complexes. Osaka J.Math. 10 (1973) 51-68.

W.Meier

[1975] Phantomabbildungen und klassifizierende Raume. Thesis,

Zurich, 1975 .

D.Quillen

[1971] The spectrum of an equivariant cohomology ring I,II . Ann.Math. 94 (1971) 549-602.

R.Rubinsztein

[1979] Restriction of equivariant K-theory to cyclic subgroups. to appear.

J.P.Serre

[1967] Representations lineaires des groupes finis. Hermann, Paris 1967 .

AN EQUIVARIANT SURGERY SEQUENCE AND EQUIVARIANT DIFFEOMORPHISM AND HOMEOMORPHISM CLASSIFICATION
(A Survey)

Karl Heinz Dovermann*) and Melvin Rothenberg**)

University of Chicago

University of Chicago
and
University of California
Berkeley

Introduction

The classification of reasonable broad classes of finite group actions on manifolds has become a viable project with the development of modern geometric topology. The central, and by now classical, tool is surgery theory, specifically, the Sullivan-Wall exact sequence. The initial, and in many ways still best, results are found in Wall's book [W1]. The main limitation of the classical method is that the theory, and particularly the Sullivan-Wall exact sequence apply directly only to covering spaces - that is, to free group actions.

While there has been successful applications of surgery techniques to the classification of some simple cases of non free actions, (see Jones [J], Rothenberg [R2], Browder-Petrie [BP] for the initial work in this direction) it was recognized that more comprehensive results depend on the extension of the theory, and in particular the Sullivan-Wall exact sequence, to non free actions. This has turned out to be a rather formidable undertaking. The first attempt to systematically generalize the theory was the paper of Browder-Quinn [BQ]. Although stimulating and provocative, their machinery was hemmed-in by too many restrictive hypotheses (isovariant maps, transverse linearity, etc.). It is valuable more for pointing to the various difficulties than for solving them.

*Partially supported by NSF grant NSF MCS-7905036.

**Partially supported by NSF grant NSF MCS-7701623.

Rather more successful is the theory developed by Petrie and Doverman Petrie [DP]. This is our point of departure. Our results extend and elaborate this theory, partly by relating it to techniques developed by Lashof-Rothenberg [LR] and Rothenberg [R1].

Let G be a finite group. The necessary ingredients for developing a useful surgery theory for G manifolds are

(1) a π - π theorem ,

(2) a G transversality theory ,

(3) a good bundle theory .

(1) has been established in the differentiable category for Petrie's theory in [DP]. (2) is to some extent understood in the smooth category due to the work of Petrie [P]. Part of our work develops a usable transversality theory in a restricted topological category (locally smoothable), at least for $G = \mathbb{Z}/n\mathbb{Z}$, n odd. Further, using the smoothing theory of [LR] we can extend (1) to the topological category. Finally, we have in [LR] a usable G bundle theory for our topological category.

Our work then consists of two parts. The first is to extend the Petrie theory to yield a full generalization of the Sullivan-Wall exact sequence in the smooth category. The second is to extend this whole machinery to the locally smooth category. This extension is not only intrinsically interesting but is valuable since the homotopy theory in making calculations is much more manageable in the locally smooth category than in the smooth category. Finally, we make some calculations and show how they lead to classification results of finite group actions on compact manifolds.

1. Statement of Results.

We consider two categories $\mathcal{C}_{Diff}$ and $\mathcal{C}_{Top}$. Here $\mathcal{C}_{Diff}$ is the category of compact smooth G manifolds and $\mathcal{C}_{Top}$ is the category of compact G manifolds with locally smooth (linear) action [LR]. Furthermore, we assume all our manifolds M are G oriented, i.e., for each subgroup $H \subseteq G$, M^H is oriented. In these categories we consider the following equivalences. If X and Y are objects in $\mathcal{C}_{cat}$, cat = Diff or Top, and $f: X \to Y$ is a G homotopy equivalence, we say that f is an equivalence in the category (cat, ht). If cat = Diff we further consider the case when f is a simple G homotopy equivalence [R1]. Then we say that f is an equivalence in the category (Diff, s). Thus we consider categories (cat, c), where cat = Diff or Top, c = ht or s if cat = Diff, and c = ht if cat = Top.

A G Poset will describe the relevant dimension and local data for an equivariant surgery situation (this generalizes the parameter of a Wall group, namely a group with orientation homomorphism and a dimension). In 2.1 we define G Poset pairs and the G Poset pair of a map. They are denoted by λ and $\lambda(f)$. If $f: X \to Y$, $\lambda(f)$ determines, among other things, dimension X^H. We shall assume for our important results the

Gap Hypothesis on λ: For $H \subset G$ and $X^H \neq \emptyset$, $\dim X^H \geq 5$ and if $H' \subset H$ and $X^H \neq X^{H'}$, $\dim X^{H'} \geq 2 \dim X^H + 1$.

Theorem A. There exists an exact sequence

$$\to N_G^{cat}(Y \times I, \partial(Y \times I), \lambda^+) \xrightarrow{\sigma} I^{cat,c}(G, \lambda^+) \xrightarrow{\partial} hS_G^{cat,c}(Y, \partial Y, \lambda)$$

$$\xrightarrow{d} N_G^{cat}(Y, \partial Y, \lambda) \xrightarrow{\sigma} I^{cat,c}(G, \lambda).$$

Here λ^+ is obtained from λ by raising dimension by 1, λ^+ also satisfies the Gap Hypothesis, $Y \in \mathcal{C}_{cat}$, c = ht and cat = Diff or Top or c = s and cat = Diff. This sequence is a generalization of the Sullivan-Wall surgery sequence [W1]. The terms N, I, and hS denote normal maps, surgery obstructions, and (cat, c) equivalences. They will be defined in section 2. This generalizes section 9 [W1] and is a special case of section 3 [DP].

The term hS contains the geometric information we need to classify group actions; hS is what we seek to calculate. The terms N are normal maps and in favorable cases are calculable using cohomology functors on Y, and thus depend only on the homotopy properties of Y. The terms I are algebraic, depending only on the group G and certain parameters associated to them. The usefulness of the sequence, of course, depends on our ability to calculate I and N, which we can do in simple and favorable cases.

The fact that the functor N is a homotopy functor already has implications for the classification of actions as the following result shows. First we need some notation.

Definition. Let M be a smooth G manifold, $x \in M^G$ and $T_xM = \sum M_\lambda$, where M_λ is the submodule of T_xM belonging to the real irreducible representation λ, and the sum ranges over the different irreducible representations. We say M is G stable if $\dim(M_\lambda) \geq \dim M^G + 2$ for every λ, λ not the trivial representation.

For $G = \mathbb{Z}/2\mathbb{Z}$ or $\mathbb{Z}/3\mathbb{Z}$ the condition is equivalent to $\dim M \geq 2 \dim M^G + 2$.

Suppose $G = \mathbb{Z}/q\mathbb{Z}$ where q is a power of an odd prime. Let D be a disk with semifree G action. Then the tangent bundle of D^G has a stable complex structure, which is unique up to complex conjugation. Such a structure is called an almost complex structure. The Reidemester torsion is denoted by τ [R1].

Theorem B. Let $G = \mathbb{Z}/q\mathbb{Z}$, where q is a prime power. Let D_1 and D_2 be n-dimensional disks on which G acts semifreely and stably. Then D_1 is G diffeomorphic to D_2 if and only if

(i) $\tau(D_1) = \tau(D_2)$ and $\tau(\partial D_1) = \tau(\partial D_2)$,

(ii) there exists a diffeomorphism $f: D_1^G \to D_2^G$ such that if $q \neq 2$ the differential of f preserves the almost complex structure,

(iii) an invariant $\alpha(f) \in W(D_1^G, G)$ (specified below) vanishes.

Indeed, the diffeomorphism from D_1 to D_2 will extend f. We describe the group $W(D^G, G)$ as follows. Let $\mathrm{Irred}(G) = \mathrm{Irred}^+(G)$

$$\mathrm{Irred}(G) = \mathrm{Irred}^+(G) \amalg \mathrm{Irred}^-(G),$$

where $\mathrm{Irred}^+(G)$ are those representations which come from complex representations, $\amalg$ denotes the disjoint union. Consider the map

$$\mu: \sum_{\lambda \in \mathrm{Irred}^+(G)} \widetilde{KU}(D^G)_\lambda + \sum_{\substack{\lambda \in \mathrm{Irred}^-(G) \\ \lambda \neq \mathrm{trivial}}} \widetilde{KO}(D^G)_\lambda \longrightarrow \widetilde{KO}(D^G).$$

Here $KU(D^G)_\lambda$ denotes equivariant K-theory where fibers of bundles are multiples of λ, $\mu = \sum r + \sum \mathrm{id}$, r is realification, and we furthermore forget the action. We define the group $W(D^G, G)$ to be the kernel of μ.

<u>Lemma.</u> $W(D^G, \mathbb{Z}_2) = 0$ and $W(D^G, \mathbb{Z}_3) = 0$.

The invariant $\alpha(f)$ is given as the class of

$$\nu(D_1^G, D_1) - f^* \nu(D_2^G, D_2) \quad \text{in} \quad W(D_1^G, G).$$

The necessity of the conditions in Theorem B is obvious. The proof that the conditions are also sufficient uses Theorems A, D, and 2.10 (π-π Theorem).

In general, for G of odd order, let $I(G) = \ker \varepsilon : R(G) \to \mathbb{Z}$, where ε is the augmentation of $R(G)$ as real representations. Then the above μ can be identified with $\hat{\mu}: I(G) \otimes \widetilde{KU}(D^G) \to \widetilde{KO}(D^G)$ and $W(G)$ is just the kernel.

Remark. The theorem above is a refinement of one of Lowel Jones [J]. Our proof is different from his and is based on the π - π theorem. It does not use the full strength of the machinery developed in Theorem A. However, if we insist on classifying actions on spheres rather than on disks, more delicate considerations come in. Just before completing this manuscript, we received a preprint of A. Assadi [A] also announcing an improvement of Jones's theorem.

If S is a homotopy sphere on which G acts smoothly such that S^H is a homotopy sphere for all $H \subset G$, S is called semilinear. Assume $S^G \neq \emptyset$ and is connected. Then removing a linear disk about $x \in S^G$ turns S into a semilinear disk. For such a semilinear G disk we have a generalized Whitehead torsion invariant [R1] denoted by $\tau(S)$ defined.

If G acts semifreely on S, $\tau(S) \in Wh(G)$. (Here $\dim S - \dim S^G \geq 3$.) The family of G oriented diffeomorphism classes of n-dimensional semilinear G spheres S, where $\dim S^G \geq 1$, with fixed tangent representation (S_x for $x \in S^G$ is a G-module denoted by α called the tangent representation of G) form an abelian group C_α under connected sum. Those for which τ vanishes form a subgroup C_α^+. If A_α is the set of G diffeomorphism classes of G spheres of dimension n with connected non empty fixed point sets and tangent representation α, then C_α^+ acts on A_α via connected sum, and in fact $x, y \in A_\alpha$ are in the same orbit if and only if $\bar{x}$ is G diffeomorphic to $\bar{y}$. Here $\bar{x}$ ($\bar{y}$) denotes x (y) without the interior of a linear disk. Theorem B gives us a criterion for $\bar{x}$ to be G diffeomorphic to $\bar{y}$ in the stable, semifree case. Thus we can classify semifree G spheres which satisfy the Gap Hypothesis if we can classify semilinear G spheres. To do this we need the full strength of Theorem A.

First of all, under some mild dimensionality and gap restrictions, $hS_G^{Diff, s}(S_\alpha, \lambda) = C_\alpha^+$, where $S_\alpha \in C_\alpha^+$ and λ is chosen naturally. To calculate it we must have information about N and I.

To calculate $I^{cat, c}(G, \lambda)$, which we ambiguously denote by $I(G, \lambda)$, we set d_G = dimension of the fixed point set and d_1 = ambient dimension, both specified by λ. We have a homomorphism $\sigma_\lambda : I(G, \lambda) \to L_{d_G}(1)$, where L denotes the Wall groups. Let $\widetilde{K} = \ker \sigma_\lambda$. The following sums up some of our knowledge of I in the semifree case.

Theorem C. Assume the Gap Hypothesis for λ and the action of G is semifree. Then there exists a subgroup $L(\lambda) \subseteq L_{d_1}(G, w) = L$, w depends on λ, such that

a) The obstruction for the free part induces a monomorphism $\sigma_{fr}: \widetilde{K} \to L/L(\lambda)$. If $I(G, \lambda^-) \neq \emptyset$, compared to λ the dimensions λ^- are decreased by 1, the map σ_{fr} is an isomorphism.

b) $L(\lambda) \otimes \mathbb{Q} = 0$

c) For G of odd order, $L(\lambda) = 0$

d) If $\lambda = (\pi, \pi)$, σ_λ is an epimorphism

e) If G is of odd order, $\lambda = (\pi, \pi)$, and $I(G, \lambda^-) \neq \emptyset$, we have an exact sequence

$$0 \to L \to I(G, \lambda) \xrightarrow{\sigma_\lambda} L_{d_G}(1) \to 0$$

The following sums up our knowledge of N.

Theorem D. a) Let M be smooth, semifree, and stable with $\dim M^G \geq 5$. Then for the natural choice of λ,

$$N_G^{Diff}(M, \partial M, \lambda) = [[M, \partial M, F/O, *]].$$

b) Let $G = \mathbb{Z}/q\mathbb{Z}$, q odd, M locally smooth, semifree with $2 \dim M^G + 2 \leq \dim M$. Then for the natural value of λ, $N_G^{Top}(M, \partial M, \lambda)$ sits in an exact sequence

$$H^n(\mathbb{Z}_2, \widetilde{K}_0(\mathbb{Z}(G))) \to N_G^{Top}(M, \partial M, \lambda) \to [[M, \partial M, F/Top, *]]$$

$$\to H^{n-1}(\mathbb{Z}_2, \widetilde{K}_0(\mathbb{Z}(G))) .$$

Here F/O, F/Top have certain universal G action, and [[ ]] denotes G homotopy classes of G maps. The group $\widetilde{K}_0(\mathbb{Z}(G))$ is the reduced projective class group with $\mathbb{Z}_2$ acting by the usual involution. For $G = \mathbb{Z}/q\mathbb{Z}$, $\widetilde{K}_0(\mathbb{Z}(G))$ is finite and so $\widetilde{K}_0(\mathbb{Z}(G)) \otimes \mathbb{Z}_2 = 0$ implies

$$N_G^{Top}(M, \partial M, \lambda) = [[M, \partial M, F/Top, *]].$$

The G spaces F/O, F/Top are equivariant infinite loop spaces, and therefore the maps into them are to some extent calculable.

To show that our machinery is actually practical , here is a simple application of Theorems B, C, D:

<u>Theorem E.</u> For every finite abelian group T having no 2-torsion, there exists up to diffeomorphism and orientation exactly one 12-dimensional $\mathbb{Z}_2$ homotopy sphere Σ with 1-connected 5-dimensional fixed point set $\Sigma^{\mathbb{Z}_2}$ and

$$H_2(\Sigma^{\mathbb{Z}_2}, \mathbb{Z}) \cong T \oplus T.$$

The proof uses Smale's classification of 5-dimensional 1-connected mod 2 homology spheres [Sm] and a computation of semilinear $\mathbb{Z}_2$ homotopy spheres.

The limitations of our machinery lie in the rather complicated homotopy structure of F/O. However F/Top has a simple homotopy structure and one might hope this is true also in the category of G spaces. Infact this is so.

Theorem F. For S_α a linear sphere which satisfies the Gap Hypothesis, and the natural λ

$$N_G^{Top}(S_{\alpha+1}, \lambda^+) \approx I^{Top}(G, \lambda^+) .$$

Proof. This follows from Theorem A and the fact that $hS_G^{Top}(S_\alpha, \lambda)$ and $hS_G^{Top}(S_\alpha \times I, \partial(S_\alpha \times I), \lambda)$ are zero by engulfing arguments. Using Theorems D b) and C, we calculate $[[S_{\alpha+1}, F/Top]]$ for $G = \mathbb{Z}/q\mathbb{Z}$, q odd. In particular, if $\widetilde{K}_0(\mathbb{Z}(G)) \otimes \mathbb{Z}_2 = 0$

$$[[S_{\alpha+1}, F/Top]] \approx I^{Top}(G, \lambda),$$

which was calculated by Theorem C.

We can say a lot more:

Theorem G. Let $G = \mathbb{Z}/p\mathbb{Z}$, p odd, $\widetilde{K}_0(\mathbb{Z}(G)) \otimes \mathbb{Z}_2 = 0$, M a semifree locally linear G manifold, $2 \dim M^G + 1 \leq \dim M$, M^G connected. Let V be a G tubular neighborhood of M^G in M, $W = M - \overset{o}{V}$, with $\partial W = W \cap V$. Then from the G Puppe sequence

$$[[M^G, F/Top]] \longleftarrow [[M, F/Top]] \longleftarrow [[M, M^G, F/Top, *]]$$

we have

a) $[[M, M^G, F/Top, *]] = [W/G, \partial W/G, F/Top, *]$

b) $[[M^G, F/Top]] = [M^G, (F/Top)^G]$.

Let $(F/Top)_\alpha^G$ be the component of $(F/Top)^G$ defined by the tangent representation of M^G.

c) There exists an infinite loop space $F_\alpha(G)$ defined by the fibration

$F_\alpha(G) \to F/Top^L \to \mathcal{L}_d(G)$ and a fibration $F_\alpha(G) \to (F/Top)^G_\alpha \to F/Top$. Here L is a lens space $L = S/G$, S the normal sphere fiber of M^G in M. $\mathcal{L}_d(G)$ are the Wall-Quinn surgery spaces with $\pi_k(\mathcal{L}_d(G)) = L^h_{k+d}(G, e)$ [BLR], and $F/Top^L \to \mathcal{L}_d(G)$ the usual map.

Remark on proof: b) is immediate and a) is elementary using the infinite loop structure of F/Top. c) follows from surgery once we have the control over $[[S_\alpha, F/Top]]$ that our earlier results give.

The homotopy structure of F/Top^L is completely known from Sullivan's work and is simple; i.e., it is calculable in terms of cohomology and K-theory. Further, by L. Jones or Taylor-Williams, the homotopy structure of $\mathcal{L}_d(G)$ is completely known and also simple. See [J2] and [TW]. It follows, in particular, that $a \in [[M, F/Top]]$ is determined completely by cohomology and K-theory invariants of M^G and W/G. It would be nice to have an explicit formula.

The work in the locally smooth category depends fundamentally on a G topological transversality theorem which is interesting in its own right.

For any G manifold M we let $cd(M) = \dim M - \dim M^G$.

Theorem H. Let $G = \mathbb{Z}/q\mathbb{Z}$, q odd. Let M, N, K be semifree locally smooth G manifolds with M^G, N^G, K^G connected and $K \hookrightarrow N$. Suppose that

1) For $x \in M^G$, $y \in K^G$ we have $N_y / N_y^G + K_y \subseteq M_x$ as G-modules,

2) $2 \dim M^G + 2 \leq cd(M) - cd(N) + cd(K)$,

3) $\dim M - \dim N + \dim K \neq 4$, $\dim M^G - \dim N^G + \dim K^G \neq 4$.

Then any continuous G map $f: M \to N$ has a G ε-approximation $\hat{f}$ and $\hat{f}$ is transverse to K. Further, if f is already transverse to K on a neighborhood of a closed invariant subspace $Y \subset M$, then $\hat{f}$ can be chosen so that $\hat{f} = f$ on a neighborhood of Y.

2. The terms of the G surgery sequence.

In this section we define the basic combinatorial concepts, normal maps, and the terms of the surgery sequence. Then we give the assumptions under which Theorem A holds. To simplify the arguments we consider only manifolds M such that for all $H \subseteq G$ the manifold M^H is (simply) connected.

Let G be a finite group and $\mathscr{S}(G)$ the partially ordered G set of subgroups of G. Here G acts by conjugation and the partial ordering is given by $H \leq K$ if and only if $K \subseteq H$ (caution!). A partially ordered G set is a G poset.

Definition 2.1. A G Poset is a quadruple (S, s, d, w) where

(i) $S \subset \mathscr{S}(G)$ is a sub G poset,

(ii) $s: S \to R$, where R is the set of equivalence classes of real representations of subgroups of G. R is a G set under conjugation and s is required to be a G map such that $s(H)$ is a representation of H.

(iii) $d: S \to \mathbb{N}$, $\mathbb{N}$ non negative integers.

(iv) w is a function on S such that for each $H \in S$,

$$w(H): NH/H \to \mathbb{Z}_2 = \{\pm 1\}$$

is a homomorphism (NH is the normalizer of H in G).

<u>The key example.</u> Let M be a G manifold and $\pi(M) = (\mathrm{Iso}(M), s, d, w)$ where $\mathrm{Iso}(M)$ is the set of isotropy groups of points of M, $d(H) = \dim M^H$, $s(H)$ is the slice representation of M^H, and $w(H)$ measures the action of NH on the fundamental class of M^H.

A <u>G Poset pair</u>, denoted by $\lambda = (\pi_1, \pi_2)$, is a pair of G Posets. A <u>G Poset map</u> from π_1 to π_2, $\pi_1 = (S_1, s_1, d_1, w_1)$, is an order preserving map of the underlying sets. It is a G Poset isomorphism if the G Poset map is a bijection and commutes with s_1 and w_1. It is a G Poset equivalence if it furthermore commutes with d_1. Our connectedness assumption as well as the particular types of equivalences we consider allow us to simplify considerably the notation and machinery of [DP] as we have done it in 2.1.

In the category $\mathcal{C}_{\mathrm{Diff}}$ the relevant bundles are real G vector bundles. In the category $\mathcal{C}_{\mathrm{Top}}$ the relevant bundles are locally linear G - $\mathbb{R}^n$ bundles [LR]. By "bundle" we shall mean one of these types, depending on which category we are in.

We now wish to define normal maps of G manifolds. This involves considerations of bundle data which is perhaps the most subtle, troublesome and least understood part of the theory. One must have bundle conditions sufficiently rigid to do modifications and smoothing yet sufficiently flabby to be invariant under surgery. In the case $G = \{e\}$ we have stability results on bundles which allow us to be rather sloppy in dealing with bundle conditions but in the case $G \neq \{e\}$ we do not have these stability conditions and thus are forced to be more careful. In the G case we are forced to assume unstable bundle data, but these are invariant under surgery only when the fixed point sets are non compact, therefore we use the following device. A manifold W satisfies condition *, if for each $H \subset G$ and for each non empty component W_0^H of W^H, $W_0^H \cap \partial W \neq \emptyset$. For such manifolds each component of $(W - \partial W)^H$ is non compact. Unfortunately no closed manifold satisfies this condition, so that we are forced to consider manifold triples $(W, \partial_F W, \bar{\partial} W)$ where $\partial W = \partial_F W \cup \bar{\partial} W$, $\bar{\partial} W \cap \partial_F W = \partial\bar{\partial} W = \partial\partial_F W$, where W and $\bar{\partial}$ are assumed to satisfy *. Then $\bar{\partial} W$ again satisfies * with $\partial_F(\partial W) = \partial\bar{\partial} W$. From now on we restrict ourselves to such manifolds. Presumably the most general case can be reduced to this by removing a finite number of linear disks, however the details can be messy.

2.2. By a G bundle pair over W, $(\varepsilon_1, \varepsilon_2)$, we mean a bundle ε_1 over W, ε_2 over $\bar{\partial} W$ and an isomorphism $\varepsilon_2 \oplus \mathbb{R} \cong \varepsilon_1|_{\bar{\partial} W}$. For example, $(TW, T(\bar{\partial} W))$ is such a bundle.

<u>Definition 2.3.</u> A G normal map is a triad $\mathscr{W} = (X, f, b)$ of the following data.

i) A pair of G manifolds and a G map $f: (X, \partial_F X, \bar{\partial} X) \to (Y, \partial_F Y, \bar{\partial} Y)$ with f^H of degree 1 for all $H \subseteq G$ and $f: \partial_F X \cong \partial_F Y$ is an isomorphism.

ii) A G bundle pair (η_1, η_2) over Y and an isomorphism

$$b: (TX, T\bar{\partial}X) \cong (f^* \eta_1, f^* \eta_2) .$$

As usual b is determined up to bundle isomorphisms of (η_1, η_2).

iii) For cat = Top we need a further condition. For any $X \in \mathcal{C}_{Top}$ we can define a finiteness obstruction $\chi(X) \in \prod_{H \in Iso(X)} \widetilde{K}_0(\mathbb{Z}[NH/H])$. This is the obstruction to X being of the same G homotopy type as a finite G CW-complex and it generalizes the construction of [W2]. For $f: X \to Y$ we let $\chi(f) = \chi(Y) - \chi(X) = \chi(\text{mapping cone of } f)$. For a normal map $\mathscr{W} = (X, f, b)$ we insist that $\chi(f)$ and $\chi(f, \partial F)$ be zero.

<u>Remark.</u> This definition of a G normal map is appropriate for our purpose, particularly when cat = Diff, and when cat = Top and $G = \mathbb{Z}_p$ or if the actions are semi free. The reader can find weaker conditions on bundle data in [DP], and in particular the non compactness assumption is superfluous. For normal maps we use the notation ∂W to denote $(\partial X, \partial f, b|)$. Here $\partial f: \partial X \to \partial Y$ and $b|$ is obtained by restricting b.

Observe that a normal map $\mathscr{N} = (X, f, c)$ with $f: X \to Y$ has a well-defined G Poset attached to it. It is denoted by $\lambda(\mathscr{N}) = \lambda(f) = (\pi(X), \pi(Y))$. In this case, the dimension function and the orientation map coincide for $\pi(X)$ and $\pi(Y)$.

An element in $\mathscr{I}^{cat,c}(G, \lambda)$ is a normal map $\mathscr{N}$ such that $\lambda(\mathscr{N}) = \lambda$ and $\partial\mathscr{N}$ is an equivalence in (cat, c). Two elements $\mathscr{N}_0$ and $\mathscr{N}_1$ in $I^{cat,c}$ are equivalent ($\sim$) if there exists a normal map $\mathscr{N}$ in cat such that $\partial\mathscr{N} = (\mathscr{N}_0 \cup -\mathscr{N}_1) \cup \mathscr{N}_2$, the inclusions induce G Poset isomorphisms $\lambda(\mathscr{N}_0) = \lambda(\mathscr{N}_1) \xrightarrow{\cong} \lambda(\mathscr{N})$ and $\mathscr{N}_2$ is an equivalence in (cat, c). Here "-" denotes that all orientations are reversed, and the union of the $\mathscr{N}_i$ is taken over their boundary components.

<u>Definition 2.4.</u> $\mathscr{I}^{cat,c}(G, \lambda)/\sim \; = I^{cat,c}(G, \lambda)$.

It is shown in [D] that $I^{cat,c}(G, \lambda)$ is a group. For cat = Top we use the remarks on surgery in this category, which the reader finds below. Zero in $I^{cat,c}(G, \lambda)$ is represented by any normal map $\mathscr{N}_0$ in $I^{cat,c}(G, \lambda)$, for which there exists a normal map $\mathscr{N}$ in cat, such that $\partial\mathscr{N} = \mathscr{N}_0 \cup \mathscr{N}_2$ where the inclusion induces an isomorphism $\lambda(\mathscr{N}_0) \xrightarrow{\cong} \lambda(\mathscr{N})$ and $\mathscr{N}_2$ is an equivalence in (cat, c).

Suppose Y is an object in cat. An element in $\eta_G^{cat,c}(Y, \lambda)$ is a normal map $\mathscr{N} = (X, f, c)$ for which $f: X \to Y$, $\lambda(\mathscr{N}) = \lambda$.

Two elements $\mathscr{V}_0$ and $\mathscr{V}_1$ in $\mathcal{N}_G^{cat, c}(Y, \lambda)$ are equivalent ($\sim$) if there exists a normal cobordism $\mathscr{V}$ between $\mathscr{V}_0$ and $\mathscr{V}_1$. This is a normal map $\mathscr{V} = (W, F, C)$, W is an object in cat, $F: W \to Y \times I$, and $\mathscr{V}|_{F^{-1}(Y \times \partial I)} = \mathscr{V}_0 \cup \mathscr{V}_1$. Note: it is implied that the inclusions induce isomorphisms $\lambda(f_i) \xrightarrow{\cong} \lambda(F)$. An element in $\mathcal{N}_G^{cat, c}(Y, \partial Y, \lambda)$ satisfies the further condition that $\partial f: \partial X \to \partial Y$ is an isomorphism in cat (i.e., a G diffeomorphism or homeomorphism). The equivalence relation is the obvious modification of the one given for $\mathcal{N}_G^{cat, c}(Y, \lambda)$.

<u>Definition 2.5.</u> $\mathcal{N}_G^{cat, c}(Y, \lambda)/\sim \; = N_G^{cat, c}(Y, \lambda)$ and

$$\mathcal{N}_G^{cat, c}(Y, \partial Y, \lambda)/\sim \; = N_G^{cat, c}(Y, \partial Y, \lambda).$$

An element in $h\mathscr{S}_G^{cat, c}(Y, \lambda)$ (in $h\mathscr{S}_G^{cat, c}(Y, \partial Y, \lambda)$) is a normal map $\mathscr{V}$ in $\mathcal{N}_G^{cat, c}(Y, \lambda)$ (in $\mathcal{N}_G^{cat, c}(Y, \partial Y, \lambda)$) which is an equivalence in (cat, c), but we do <u>not</u> require bundle data. Two such maps are equivalent ($\sim$) if they are equivalent in $\mathcal{N}_G^{cat, c}(Y, \lambda)$ (in $\mathcal{N}_G^{cat, c}(Y, \partial Y, \lambda)$), again not requiring bundle data, and the cobordism between them is an equivalence in (cat, c).

<u>Definition 2.6.</u> $h\mathscr{S}_G^{cat, c}(Y, \lambda(/\sim \; = hS_G^{cat, c}(Y, \lambda)$ and

$$h\mathscr{S}_G^{cat, c}(Y, \partial Y, \lambda)/\sim \; = hS_G^{cat, c}(Y, \partial Y, \lambda).$$

The sets defined in 2.4-2.6 will be referred to as <u>terms.</u> We shall write the terms without superscripts when the statements are true for any considered choice of superscripts The <u>natural λ for a given Y</u> on N and hS is $(\pi(Y), \pi(Y))$.

Let $\lambda = (\pi_1, \pi_2)$ be a G Poset pair, $\pi_1 = (S, s, d, w)$. Then we define $\lambda^+ = (\pi_1^+, \pi_1^+)$ with $\pi_1^+ = (S, s, d^+, w)$, and $d^+ = d + 1$.

Our aim is to establish, with appropriate assumptions, an exact sequence as stated in Theorem A. The definition of σ is obvious. The definition of d is almost obvious, since as in the classical case the assumption of a (cat, c) equivalence forces an essentially unique choice of bundle data. Therefore, a (cat, c) equivalence defines uniquely a normal map. To define ∂ and show it has nice properties requires the rather stringent Gap Hypothesis on λ

Gap Hypothesis 2.7. For $\lambda = (\pi_1, \pi_2)$, $\pi_1 = (S, s, d, w)$, we require that for $H, K \in S$ with $H \subsetneq K$, $d^+(H) > 2d^+(K)$. Furthermore, assume $d(H) \geq 5$ for all $H \in S$.

The existence and the required properties of ∂ follow from the next two theorems.

Theorem 2.8. Suppose $\lambda = (\pi_1, \pi_2)$ is a G Poset pair, λ satisfies the Gap Hypothesis, and $\lambda' = (\pi_1, \pi_1)$. Let $f: X \to Y$ represent an element in $hS_G(Y, \partial Y, \lambda)$. Then every element in $I(G, \lambda^+)$ is representable by a normal map $\mathscr{W} = (W, F, B)$ such that

(i) $F: W \to X \times I$,

(ii) $\partial W = X \cup \partial X \times I \cup X'$,

(iii) F and B restricted to $X \cup \partial X \times I$ are the identity,

(iv) $F_{|X'}: X' \to X \times 1$ represents an element in $hS_G(Y, \partial Y, \lambda)$.

In the usual way [W1], an element as in the conclusion of 2.8 acts on $f: X \to Y$ as in the assumption of 2.8 by composition.

<u>Theorem 2.9.</u> The construction of 2.8 defines an action of $I(G, \lambda^+)$ on $hS(Y, \partial Y, \lambda)$. This is a group action of the group $I(G, \lambda^+)$.

<u>Theorem A.</u> Suppose λ satisfies the Gap Hypothesis. Then the sequence

$$N_G(Y \times I, \partial(Y \times I), \lambda^+) \xrightarrow{\sigma} I(G, \lambda^+) \xrightarrow{\partial} hS_G(Y, \partial Y, \lambda)$$
$$\xrightarrow{d} N_G(Y, \partial Y, \lambda) \xrightarrow{\sigma} I(G, \lambda)$$

is exact.

If $\lambda = (\pi, \pi)$ and there exists Y in $\mathcal{C}_{cat}$ with $\pi(Y) = \pi$, we have a distinguished class (zero) in $hS_G(Y, \partial Y, \lambda)$ and in $N_G(Y, \partial Y, \lambda)$, namely, $Id: Y \to Y$. In this situation, we define $\partial(a) = a(0)$, where $a \in I(G, \lambda^+)$.

We only want to give some hints for the proofs of the above theorems. Equivariant surgery on relative homotopy classes which are represented by imbedded spheres is possible. This follows in $\mathcal{C}_{Diff}$ as our bundle assumptions imply the ones in [DP]. Working in $\mathcal{C}_{Top}$, we first find an equivariant smoothing of a neighborhood of the imbedded sphere and its translates under the action of G. Then we proceed as in $\mathcal{C}_{Diff}$. Using sections 4 and 5 of [DP] and the full strength of section 3 of [W1], we prove the π-π theorem. The proofs of Theorems 2.8, 2.9, and A are consequences of the π-π theorem and the addition in $I(G, \lambda^+)$. The proofs of the corresponding statements in [W1] do not generalize to our setting.

For our setting, the π - π theorem can be stated as follows.

Theorem 2.10 (π - π Theorem): For λ satisfying the Gap Hypothesis and with Y^H, $\overline{\partial Y}^H \neq \emptyset$ simply connected for all H G we have $hS_G(Y, \lambda) \xrightarrow{d} N_G(Y, \lambda)$ is bijective.

Finally, a note on topological and smooth h and s cobordisms. The question here is: Does the class of an object (X, f) in hS_G determine the G diffeomorphism or homeomorphism type of X ? In the case of $hS_G^{Diff, s}$ we can, under mild gap and dimension hypotheses apply the G - s cobordism theorem of [R1] to argue that for (X_1, f_1) and (X_2, f_2) equal in $hS_G^{Diff, s}$ there exists a G diffeomorphism $\psi: X_1 \to X_2$ with $f_2\psi \simeq f_1$.

As observed by Browder-Hsiang [BH] any smooth G h-cobordism W between semilinear G spheres is topologically a product, provided

(**) $\dim W^H - \dim W^{H'} \geq 3$ when $W^H \neq W^{H'}$, $H \subset H'$ and $\dim W^G \geq 5$.

It then follows easily by the additive properties of generalized Whitehead torsion [R1] that by splicing-in an h cobordism of spheres into an arbitrary smooth h cobordism of G manifolds we can kill its generalized torsion. Thus we have

Theorem 2.11. A smooth h cobordism of G manifolds W satisfying (**) is homeomorphic to a product.

Thus the group $hS^{Diff, h}(Y, \partial Y, \lambda)$ really does detect the G homeomorphism types.

We conjecture that 2.11 is true for locally smooth h cobordisms. We cannot prove that but we can prove a weaker version.

Theorem 2.12. Let W be a locally smooth semifree h cobordism of G manifolds $\partial_0 W$, $\partial_1 W$ which satisfy (**). If $\partial_0 W$ is smoothable, then $\partial_0 W \cong \partial_1 W$.

Thus under the assumptions of 2.12, $hS_G^{Top}(Y, \partial Y, \lambda)$ detects smoothable G manifolds up to G homeomorphism.

References

[A] A. Assadi, Extensions of finite group actions from submanifolds of a disk, preprint (1979).

[BH] W. Browder and W.C. Hsiang, Some problems on homotopy theory, manifolds and transformation groups, Proceedings of Symposia in Pure Mathematics, Vol. 32 (II), (1978), 251-267.

[BP] W. Browder and T. Petrie, Diffeomorphisms of manifolds and semifree actions on homotopy spheres, Bull. of the A.M.S. Vol. 77 (1971), 160-163.

[BQ] W. Browder and F. Quinn, A surgery theory for G manifolds and stratified sets, Manifolds-Tokyo, University of Tokyo Press, (1973).

[BLR] D. Burghelea, R. Lashof, and M. Rothenberg, Groups of automorphisms of manifolds, L.N.M. Vol. 473, (1975).

[D] K.H. Dovermann, Addition of equivariant surgery obstructions, to appear in proceedings of the topology conference in Waterloo, Ontario, Springer LNM (1979).

[DP] K.H. Dovermann and T. Petrie, G Surgery II, to appear.

[DR] K.H. Dovermann and M. Rothenberg, An equivariant surgery sequence and equivariant diffeomorphism classification, to appear.

[J] L. Jones, The converse to the fixed point theorem of P.A. Smith: I, Ann. of Math. 14(1971), 52-68.

[J2] L. Jones, The non simply connected characteristic variety theorem, Proceedings of Symposia in Pure Mathematics, Vol. 32, (1978), 131-140.

[LR] R. Lashof and M. Rothenberg, G smoothing theory, Proceedings of Symposia in Pure Mathematics, Vol. 32, (1978), 211-266.

[P] T. Petrie, Pseudo equivalences of G manifolds, Proceedings of Symposia in Pure Mathematics, Vol. 32, (1978), 169-210.

[R1] M. Rothenberg, Torsion invariants and finite transformation groups, Proceedings of Symposia in Pure Mathematics, Vol. 32, (1978), 267-311.

[R2] M. Rothenberg, Differentiable group actions on spheres, Proc. Adv. Study Inst. Alg. Top. (1970).

[Sm] S. Smale, On the structure of 5 manifolds, Ann. of Math., 75(1962), 38-46.

[W1] C.T.C. Wall, Surgery on compact manfolds, Academic Press (1970).

[W2] C.T.C. Wall, Finiteness conditions for CW-complexes, Ann. of Math. 81(1965), 56-69.

[W3] C.T.C. Wall, Classification of hermitian forms. VI Group rings, Ann. of Math. 103(1976), 1-80.

[TW] L. Taylor and B. Williams, Surgery spaces, formulae and structure, preprint.

Äquivariante Konfigurationsräume und Abbildungsräume

H. Hauschild

I. Übersicht:

Die Konfigurationsräume und ihre Beziehung zu gewissen Abbildungsräumen sind für die Theorie der Schleifenräume und die Homotopietheorie von zentraler Bedeutung. In dieser Note sollen die Ergebnisse aus [4] auf den äquivarianten Fall verallgemeinert werden. Mit Hilfe der Approximation der Abbildungsräume durch Konfigurationsräume ergeben sich neue Beweise und neue Tatsachen über G-Abbildungsräume (Zerspaltungssätze). Hierzu sei G im Folgenden eine endliche Gruppe. Für nicht endliche Gruppen werden alle Ergebnisse falsch. In diesem Fall reichen die Konfigurationen endlicher Mengen nicht aus und man muß allgemeinere Konstruktionen suchen.

Für die G-Mannigfaltigkeit M sei $C(M)_k$ der Raum der k-Konfigurationen in M mit der von M geerbten G-Operation. Weiter setzt man $C(M) = \coprod_k C(M)_k$. Ist TM das Tangentialbündel von M mit dem Einheitsscheibenbündel DTM und dem Einheitskreisbündel STM, so erhält man ein Sphärenbündel $p : E_M \longrightarrow M$, indem man in jeder Faser von DTM die Randsphäre zu einem Punkt identifiziert. Weiter sei $\Gamma(M)$ der Raum der Schnitte (mit kompaktem Träger) von p. Die G-Operation auf M und E_M liefert eine G-Operation auf $\Gamma(M)$. Ist $L \subset M$ ein G-Unterraum, so sei $\Gamma(M,L)$ der Raum der Schnitte, welche auf L verschwinden. Ist X ein G-Raum und $x \in X$, so sei $G_x < G$ die Standuntergruppe an der Stelle x. Weiter setze ich

$$X^H = \{x \in X \mid H < G_x\} = \text{H-Fixpunktmenge}$$

$$X_H = \{x \in X \mid H = G_x\} = \text{H-Orbitbundel.}$$

Bezeichnet NH den Normalisator von H in G und WH den Quotienten NH/H, so sind X^H und X_H offensichtlich WH-Raume.

Weiter sei O(X) die Menge der Konjugationsklassen von auf X auftretenden Standuntergruppen. Der Transfer liefert eine G-Abbildung

$$\Theta : C(M) \longrightarrow \Gamma(M)$$

<u>Theorem</u> (II.14) Ist M eine kompakte G-Mannigfaltigkeit mit Rand ∂M und trifft fur alle $H < G$ jede Komponente von M_H den Rand, so induziert Θ fur alle $K < G$ einen Isomorphismus

$$\Theta_*^K : \underset{k}{\underrightarrow{\lim}} \left\{H_*(C(M)^K_{(k)})\right\} \longrightarrow \underset{k}{\underrightarrow{\lim}} \left\{H_*(\Gamma(M, \partial M)^K_{(k)})\right\} .$$

(Die Definition des Limes wird in § II angegeben; der Index (k) ist ein Multiindex und beschreibt eine Komponente der betreffenden Fixpunktmenge $C(M)^K$ bez. $\Gamma(M, \partial M)^K$. Die G-Abbildung Θ ergibt sich aus der auf Seite 6 definierten G-Abbildung Θ_ε und der G-Homotopieaquivalenz p aus Satz (II.1).)
Der Beweis benutzt eine äquivariante Henkelzerlegung von M und die üblichen Schlüsse von "Gromov-Typ". Da ich nur Anwendungen von Theorem (II.14) angebe, habe ich darauf verzichtet, das Analogon von Theorem (II.14) für geschlossene kompakte G-Mannigfaltigkeiten und offene G-Mannigfaltigkeiten zu diskutieren.

Fur die G-Darstellung V sei S^V die Einpunktkompaktifizierung von V und $\Omega^V S^V = \{S^V | S^V\}_o$ der G-Raum der stetigen Selbstabbildungen von S^V.

Theorem (III.4) Ist V eine G-Darstellung mit einem eindimensionalen trivialen Summanden, so ist die G-Abbildung

$$\Theta : C(V) \longrightarrow \Omega^V S^V$$

eine G-Gruppenkomplettierung, d. h. fur alle $H < G$ ist

$$\Theta^H : C(V)^H \longrightarrow (\Omega^V S^V)^H$$

eine Gruppenkomplettierung.

Korollar (III.8) Sei V eine G-Darstellung mit trivialem Summanden.

a) Man hat eine Gruppenkomplettierung

$$\Theta : \prod_{(H) \in O(V)} C(V_H/WH) \longrightarrow (\Omega^V S^V)^G$$

b) Man hat eine (schwache) Homotopieaquivalenz

$$Z : \prod_{(H) \in O(V)} \Gamma(V_H/WH) \longrightarrow (\Omega^V S^V)^G$$

c) Bildet man in b) den Limes uber "alle" G-Darstellungen V, so erhalt man eine (schwache) Homotopieaquivalenz

$$Z : \prod_{(H) < G} \varinjlim_n \Omega^n(S^n B(WH)^+) \longrightarrow \varinjlim_V (\Omega^V S^V)^G$$

Hierbei ist B(WH) der klassifizierende Raum der Gruppe WH und $(\ )^+$ bezeichnet die Hinzunahme eines disjunkten Grundpunktes.

Alle Resultate dieser Note sind eine Zusammenstellung der Ergebnisse aus den Kapiteln IX und X meiner Habilitationsschrift [7] . Fur weitere Anwendungen insbesondere auf G-Schleifenraume verweise ich auf [7] und [8] . Insbesondere habe ich viele beweistechnische Einzelheiten und Erganzungen, welche in [7] ausgefuhrt sind, weggelassen. Ahnliche etwas speziellere Resultate wurden unabhangig kurzlich von G. Segal [12] bewiesen.

II. G-Konfigurations- und Schnittraume:

Ist $\Delta \subset M^k$ die verallgemeinerte Diagonale und Σ_k die symmetrische Gruppe, so war $C(M)_k = (M^k \setminus \Delta)/\Sigma_k$. Im Folgenden benotigt man relative Konfigurationsraume. Hierzu sei $L \subset \partial M$ ein abgeschlossener G-Unterraum. Fur $s \in C(M)$ und $t \in C(M)$ setzt man $s \sim t$, falls $s \cap (M \setminus L) = t \cap (M \setminus L)$ gilt. Den Raum der Aquivalenzklassen $C(M)/\sim$ bezeichnet man mit $C(M,L)$; fur $C(M, \partial M)$ schreibt man oft auch $\tilde{C}(M)$. Die G-Mannigfaltigkeit M ist ohne Einschrankung mit einer G-invarianten Metrik $d(-,-)$ versehen. Man setzt fur $\varepsilon > 0$, nun $M_\varepsilon = \{m \in M \mid d(m, \partial M) \geq \varepsilon\}$ und $\tilde{C}(M, \varepsilon) = \{s \in \tilde{C}(M) \mid d(s_i, s_j) \geq 2\varepsilon$ fur $i \neq j$ und $s_i \in s,\ s_j \in s\}$. Weiter sei $\tilde{C}_\varepsilon(M) = \bigcup_{0 < \varepsilon' \leq \varepsilon} \tilde{C}(M, \varepsilon') \times \varepsilon' \subset \tilde{C}(M) \times \mathbb{R}$. Man hat eine offensichtliche Projektion $p : \tilde{C}_\varepsilon(M) \longrightarrow \tilde{C}(M)$.

<u>Satz</u> (II.1) a) $C(M)_k$ und $C(M)$ sind vom Homotopietyp eines G-CW-Komplexes.

b) $\tilde{C}(M)$ ist vom Homotopietyp eine G-CW-Komplexes.

c) Die Projektion $p : \tilde{C}_\varepsilon(M) \longrightarrow \tilde{C}(M)$ ist eine G-Homotopieaquivalenz.

<u>Beweis:</u>

a) Es ist $C(M)_k = (M^k \setminus \Delta)/\Sigma_k$ eine differenzierbare G-Mannigfaltigkeit und daher nach [9], Theorem (1) sogar G-triangulierbar.

b) und c) folgen aus a) vollig analog zu [4]. Lemma (2.1) und Lemma (2.3).

∎

Es soll nun der G-Raum $\tilde{C}(M)$ vermoge der Pontrjagin-Thom-Konstruktion mit den Schnitten gewisser G-Bundel identifiziert werden.

Ist TM das Tangentialbundel von M mit dem Diskbundel DTM und dem Spharenbundel STM, so sei $p : E_M \longrightarrow M$ das G-Bundel, dessen Faser uber $m \in M$ der Raum (DTM_m/STM_m) ist. Fur $L \subset N \subset M$ sei $\Gamma_M(N,L)$ der Raum der Schnitte von $p : E_M|N \longrightarrow N$, welche auf allen Punkten $l \in L$ gleich dem Grundpunkt $(STM_l/STM_l) \in (E_M)_l$ sind. Auf $\Gamma_M(N,L)$ liefert die "Abbildungsoperation" definiert durch $g\gamma(m) = g(\gamma(g^{-1}m))$ fur $\gamma \in \Gamma_M(N,L)$ und $m \in N$ eine wohldefinierte G-Operation. Sei nun $\varepsilon > 0$ so gewahlt, daß $M \setminus M_\varepsilon$ ein Kragen von ∂M ist und daß fur jedes $m \in M_\varepsilon$ die Exponentialabbildung eines Diffeomorphismus $Exp_m : \varepsilon DTM_m \longrightarrow \{\varepsilon\text{-Umgebung von } m \text{ in } M\}$ liefert. Man definiert dann eine Abbildung

$$\Theta_\varepsilon \ : \tilde{C}_\varepsilon (M) \longrightarrow \Gamma_M(M_\varepsilon),$$

indem man für $s \in \mathcal{C}(M, \varepsilon')$ und $m \in M_\varepsilon$ setzt:

$$\Theta_\varepsilon (s)(m) = \begin{cases} * & ; \text{ falls nicht } d(m,s_1) < \varepsilon' \text{ für ein } s_1 \in s, \\ d(m,s_1)/\varepsilon' \cdot t(s_1,m); & \text{falls } d(m,s_1) < \varepsilon' \end{cases} .$$

Hierbei ist $t(s_1,m)$ der Tangentialvektor der Länge 1 in TM_m, welcher in Richtung s_1 zeigt. Man rechnet sofort nach, daß Θ_ε G-äquivariant ist, da die Exponentialabbildung G-äquivariant ist.

<u>Definition</u> (II.2) a) Eine G-Abbildung $f: X \longrightarrow Y$ heißt schwache G-Homotopieäquivalenz, falls für alle $H < G$ die Einschränkung $f^H: X^H \longrightarrow Y^H$ eine schwache Homotopieäquivalenz ist.

b) Eine G-Abbildung $p: E \longrightarrow B$ heißt G-Quasifaserung, falls für alle $H < G$ die Einschränkung $p^H: E^H \longrightarrow B^H$ eine Quasifaserung ist.

c) Eine G-Abbildung $p: E \longrightarrow B$ heißt G-Faserung, falls p die Homotopiehochhebungseigenschaft in der Kategorie der G-Räume besitzt. ∎

<u>Lemma</u> (II.3) Ist $r : E \longrightarrow B$ eine G-Abbildung, $B_1 \subset B_2 \subset \dots \subset B_k$ eine aufsteigende Filtrierung von B durch abgeschlossene G-Unterräume, so ist r eine G-Quasifaserung, falls für jedes k gilt:

(i) $r^{-1}(B_k \setminus B_{k-1}) \longrightarrow B_k \setminus B_{k-1}$

ist eine G-Faserung mit der Faser F.

(ii) Es gibt offene Umgebungen U_k von B_{k-1} in B_k und G-äquivariante Homotopien

$h_t : U_k \longrightarrow U_k$ und $H_t : r^{-1}(U_k) \longrightarrow r^{-1}(U_k)$

mit den folgenden Eigenschaften:

a) $h_o = id$; $h_t(B_{k-1}) \subset B_{k-1}$; $h_1(U_k) \subset B_{k-1}$.

b) $H_o = id$; $r \circ H_t = h_t \circ r$

c) $H_1 : r^{-1}(x) \longrightarrow r^{-1}h_1(x)$ ist eine G_x-Homotopieäquivalenz für alle $x \in U_k$.

<u>Beweis:</u>

Es ist zu zeigen, daß $r^H : E^H \longrightarrow B^H$ eine Quasifaserung ist. Dies folgt aus den Bedingungen (i), (ii) und [4], Lemma (3.3). ∎

<u>Definition</u> (II.4) Sind N und N' zwei abgeschlossene Untermannigfaltigkeiten mit Rand von M mit $M = N \cup N'$, für die die Dimensionsgleichung $|M| = |N| = |N'|$ und die Beziehung $N \cap N' = \partial N \cap \partial N' =: B$ gilt,

so heißt die Inklusion $N \hookrightarrow M$ gut, falls B eine G-Untermannigfaltigkeit von ∂N und $\partial N'$ ist und falls jede Zusammenhangskomponente von B_H nicht leeren Durchschnitt mit ∂M_H hat. ∎

Für $N \subset M$ hat man die Restriktionsabbildung

$$r : \tilde{C}(M) \longrightarrow \tilde{C}(N).$$

Lemma (II.5) Ist $N \hookrightarrow M$ eine gute Inklusion von G-Mannigfaltigkeiten mit Rand mit $M = N' \cup N$, dann ist

$$r : \tilde{C}(M) \longrightarrow \tilde{C}(N)$$

eine G-Quasifaserung mit der Faser $C(N', \overline{\partial N' \setminus B})$.

Beweis:

Man filtert $\tilde{C}(N)$ durch die G-Räume

$\tilde{C}^k(N) = \{[s] \in \tilde{C}(N) \mid |s| \leq k\}$. Sei $V_k = \tilde{C}^k(N) \setminus \tilde{C}^{k-1}(N) =$

$C(N \setminus \partial N)_k$ und $F = C(N', \overline{\partial N' \setminus B})$, dann ist F kanonisch in $\tilde{C}(M)$ eingebettet. Nach [4], Seite 98 hat man einen Homöomorphismus

$$t : V_k \times F \longrightarrow r^{-1}(V_k)$$

definiert durch $t([s], [v]) = [s \cup v]$. Man rechnet nach, daß t G-äquivariant ist. Da die Inklusion $N \subset M$ gut ist, hat die Menge $\{y \in M \mid d(y,B) < 2\varepsilon\}$ die Gestalt $B \times (-2\varepsilon, 2\varepsilon)$ und $N \setminus N_{2\varepsilon}$ ist gleich $N \times [0, 2\varepsilon)$. Es gibt weiter eine injektive G-äquivariante Homotopie $f_t: (M, \partial M) \longrightarrow (M, \partial M)$ mit $f_o = id$ und $f_1(N_\varepsilon) = N$.

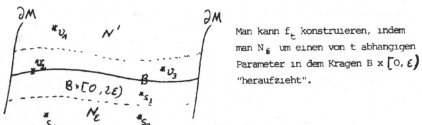

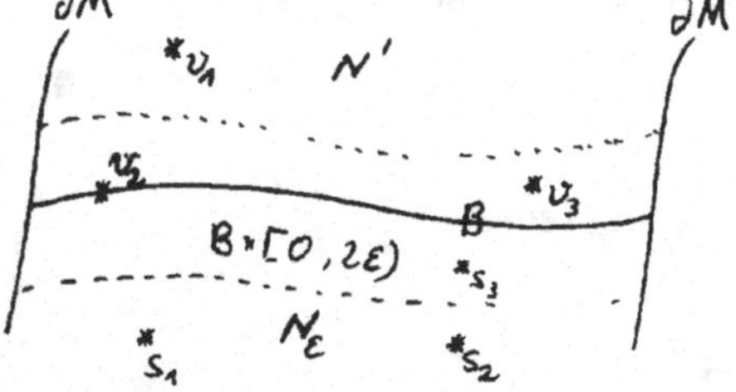

Man kann f_t konstruieren, indem man N_ε um einen von t abhängigen Parameter in dem Kragen $B \times [0, \varepsilon)$ "heraufzieht".

Wegen der Injektivität induziert f_t eine G-Homotopie

$$\hat{f}_t : \tilde{C}(M) \longrightarrow \tilde{C}(M).$$

Man setzt $U_k = \{ [s] \in \tilde{C}^k(N) \,\big|\, |s \cap N_\varepsilon| \leq k-1 \}$ und erhält so eine G-invariante Umgebung von $\tilde{C}^{k-1}(N)$ in $\tilde{C}^k(N)$. Die Homotopie f_t induziert vermöge Einschränkung die in Satz (II.3) gesuchten Homotopien h_t und H_t. Wegen Satz (II.3) genügt es daher zu zeigen, daß für alle $[s] \in U_k$ die Abbildung

$$\hat{f}_1 : r^{-1}([s]) \longrightarrow r^{-1}(\hat{f}_1([s]))$$

eine $G_{[s]}$-Homotopieäquivalenz ist. Hierzu betrachtet man das folgende Diagramm:

$$\begin{array}{ccccccc}
[v \cup s] & \in & r^{-1}[s] & \xrightarrow{\hat{f}_1} & r^{-1}(\hat{f}_1([s])) & \ni & [\hat{f}_1[s] \cup v] \\
\uparrow & & \updownarrow\cong & & \updownarrow\cong & & \uparrow \\
[v] & \in & F & \overset{\tilde{f}_1}{\dashrightarrow} & F & \ni & [v]
\end{array}$$

Es gilt $\tilde{f}_1([v]) = [\hat{f}_1([v]) \cup w]$ mit $w = f_1([s]) \cap N'$; diese Vereinigung ist sinnvoll, da $f_1([v])$ außerhalb der Umgebung $W = f_1(N) \cap N'$ von B in N' liegt.

Da $\tilde{f}_1$ $G_{[s]}$-äquivariant ist, muß w eine $G_{[s]}$-invariante Konfiguration sein. Da mit f_1 auch $\hat{f}_1$ homotop zur Identität ist, genügt es zum Beweis der Behauptung zu zeigen, daß w in

W durch einen Weg $G_{[s]}$-invarianter Konfigurationen mit dem Rand von M verbindbar ist. Nun zerfallt w als $G_{[s]}$-Menge in die Bahnen $w = \coprod_i w_i \cdot G_{[s]}/G_{w_i}$. Da die Inklusion $N \subset M$ gut ist, kann man die Punkte w_i in $W_{G_{w_i}}$ mit dem Rande verbinden durch disjunkte Wege. Die $G_{[s]}$-Operation liefert durch Translation einen"Weg" $G_{[s]}$-invarianter Konfigurationen zwischen w und einer Konfiguration in ∂M. ∎

<u>Satz</u> (II.6) a) Ist M eine kompakte G-Mannigfaltigkeit mit Rand ∂M und hat fur alle $H < G$ jede Komponente M_H^i des Orbitbundels M_H nicht leeren Schnitt mit ∂M, so besitzt M eine aquivariante Henkelkorperzerlegung mit Henkeln $Gx_H(DV \times DW)$, bei denen W immer einen trivialen Summanden enthalt. [Nach ublicher Konvention wird der Henkel langs $Gx_H(SV \times DW)$ angeklebt; sein Index ist $|V|$].

b) Die Standuntergruppen von M seien durch Inklusion partiell geordnet. Es gilt $M_H^i \cap \partial M \neq \emptyset$ fur alle H und alle Komponenten M_H^i von M_H, falls $M_L^i \cap \partial M \neq 0$ gilt fur alle maximalen auf M auftretenden Standuntergruppen L und alle Komponenten M_L^i von M_L.

<u>Beweis</u> [1], Theorem (3.1).
Eine Henkelkorperzerlegung in Henkel $Gx_H(DV \times DW)$, bei der W immer einen trivialen Summanden enthalt, heißt in [1] gut. ∎

<u>Satz</u> (II.7) Ist M eine kompakte G-Mannigfaltigkeit und hat fur alle $H < G$ jede Komponente M_H^i des Orbitbundels M_H nicht leeren Durchschnitt mit ∂M, dann ist

$$\Theta_{\varepsilon} \ : \tilde{C}_{\varepsilon}\ (M) \longrightarrow \Gamma_M(M_{\varepsilon}) \cong \Gamma_M(M)$$

eine (schwache) G-Homotopieäquivalenz.

<u>Beweis:</u>

(1) $M = Gx_H DW$ mit der H-Darstellung W und einem trivialen Summanden $\mathbb{R}^1 \subset W$. Man betrachtet das folgende homotopiekommutative Diagramm:

$$\begin{array}{ccc}
[s] \in \tilde{C}(Gx_H DW,\ \varepsilon') & \xrightarrow{\ \Theta_{\varepsilon}\ } & \Gamma_{Gx_H DW}\ (Gx_H DW_{\varepsilon}) \\
\Big\downarrow \qquad \cong \Big\downarrow \text{radiale Expansion} & & \cong \Big\downarrow \text{Restriktion auf den Nullabschnitt } G/H \\
[(2/\varepsilon')\cdot s] \in \tilde{C}(Gx_H DW, 2) & \xrightarrow{\ \varphi\ } & \{\text{Schnitte von } Gx_H S^W \longrightarrow G/H\}
\end{array}$$

Nun ist φ eine G-Homotopieäquivalenz; denn in jeder Faser DW von $Gx_H DW$ gibt es in $\tilde{C}(Gx_H DW, 2)$ höchstens ein Partikel, da der Abstand untereinander größer als 2 sein muß. Eine Konfiguration in $C(Gx_H DW, 2)$ ist also eine Zuordnung, die jedem Element aus G/H eine Element der darüberliegenden Faser zuordnet. Ebenso läßt sich der Raum der Schnitte von $Gx_H S^W \longrightarrow G/H$ beschreiben.

(2) Der Beweis der Behauptung durch Induktion nach dem Index der in einer äquivarianten Henkelkörperzerlegung auftretenden Henkel. Nach Satz (II.6) gibt es eine Henkelkörperzerlegung in Henkel $Gx_H DVxDW$, bei der W immer einen trivialen Summanden enthält.

(i) Fur einen Henkel $G\times_H DW$ vom Index Null wurde die Behauptung in (1) bewiesen.

(ii) Die Behauptung sei richtig fur alle G-Mannigfaltigkeiten, die durch Ankleben von Henkeln $G\times_H DV\times DW$ entstehen mit $|V|<\lambda$ und $\mathbb{R}^1\subset W$.

(iii) Es entstehe M aus N durch Ankleben des Henkels $G\times_H DV\times DW$ mit $|V|=\lambda$ und $\mathbb{R}^1\subset W$ und die Behauptung sei richtig fur die Mannigfaltigkeit N. Es gilt dann $M = N \underset{G\times_H SV\times DW}{\cup} G\times_H DV\times DW$ und die Restriktionsabbildung liefert das folgende Diagramm:

$$\begin{array}{ccc}
F & \overset{\varphi}{\dashrightarrow} & \bar{F} \\
\downarrow & & \downarrow \\
\tilde{C}_\varepsilon(M) & \xrightarrow{\Theta_\varepsilon} & \Gamma_M(M_\varepsilon) \\
\downarrow r & & \downarrow \bar{r} \\
\tilde{C}_\varepsilon(N) & \xrightarrow{\Theta_\varepsilon|} & \Gamma_N(N_\varepsilon)
\end{array}$$

Da W einen trivialen Summanden enthalt, trifft jede Komponente von $(\partial N \cap \partial(G\times_H DV\times DW))_K = (G\times_H SV\times DW)_K$ die Mannigfaltigkeit $G\times_H SV\times SW$ und damit auch ∂M.

Damit ist r nach Satz (II.5) eine G-Quasifaserung; ebenso ist $\bar{r}$ eine G-Quasifaserung, da die Inklusion $N\subset M$ eine G-Kofaserung ist. Da $\Theta_\varepsilon|$ eine schwache G-Homotopieaquivalenz ist, genugt es zu zeigen, daß auch die auf den Fasern induzierte Abbildung φ eine schwache G-Homotopieaquivalenz ist. Fur die Betrachtung der Fasern kann man sich auf den Fall $M = G\times_H DV\times DW$

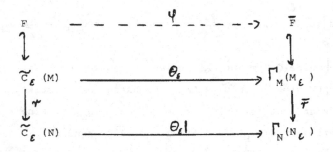

und $N = Gx_H SVx\,[1,a]\,xDW$ beschranken und erhalt hiermit die folgenden G-Quasifaserungen:

$$\begin{array}{ccc}
F & \xrightarrow{\quad\varphi\quad} & \overline{F} \\
\downarrow & & \downarrow \\
\widetilde{C}_{\varepsilon}\,(Gx_H DVxDW) & \xrightarrow{\quad\Theta_{\varepsilon}\quad} & \Gamma_M(Gx_H DVxDW) \\
\downarrow{\scriptstyle r} & & \downarrow{\scriptstyle \mathcal{F}} \\
\widetilde{C}_{\varepsilon}\,(Gx_H SVx\,[1,a]\,xDW) & \xrightarrow{\quad\Theta_{\varepsilon}|\quad} & \Gamma_N(Gx_H SVx\,[1,a]\,xDW)
\end{array}$$

Hier ist Θ_{ε} nach (1) eine schwache G-Homotopieaquivalenz. Da in $Gx_H SVx\,[1,a]\,xDW$ nur Henkel vom Index echt kleiner als λ auftreten, ist auch $\Theta_{\varepsilon}|$ eine schwache G-Homotopieaquivalenz nach Induktionsannahme. Damit ist auch die auf den Fasern induzierte G-Abbildung φ eine schwache G-Homotopieaquivalenz.

∎

Es soll nun noch eine relative Version von Satz (II.7) bewiesen werden. Hierzu sei M eine kompakte G-Mannigfaltigkeit mit Rand ∂M und $\partial M = L \cup \overline{L}$ mit $|L| = |\partial M|$, $|\overline{L}| = |\partial M|$ und $\partial L = \partial\overline{L} = L \cap \overline{L}$.

<u>Satz</u> (II.8) Sei das Tripel $(M,\overline{L},L)$ wie im Vorangehenden gegeben. Die Abbildung

$$\Theta : C(M,L) \longrightarrow \Gamma_M(M,\overline{L})$$

ist eine schwache G-Homotopieaquivalenz, falls fur alle $H < G$ und jede Komponente M_H^1 von M_H gilt $M_H^1 \cap L \neq \emptyset$.

Beweis:

(1) Man betrachtet zunachst das Tripel $(\bar{L}\times[0,1], \bar{L}\times 0, \bar{L}\times 1)$ und beweist die Behauptung fur diesen Spezialfall. (Hierzu lasse ich die unwesentlichen Randstucke $\partial\bar{L}\times[0,1]$ zur Vereinfachung weg). Es sind nun $C(\bar{L}\times[0,1], \bar{L}\times 1)$ und $\Gamma(\bar{L}\times[0,1], \bar{L}\times 0)$ G-zusammenziehbar; denn man kann sowohl die Partikel als auch die Schnitte langs des Einheitsintervalls in den Rand hineindeformieren, wo sie verschwinden. Somit ist Θ in diesem Spezialfall eine schwache G-Homotopieaquivalenz.

(2) Es gibt eine Sequenz von Tripeln $(M_o, \bar{L}_o, L_o) = (\bar{L}\times[0,1], \bar{L}\times 0, \bar{L}\times 1) \subset (M_1, \bar{L}_1, L_1) \subset \ldots \subset (M, \bar{L}, L)$ wobei M_i aus M_{i-1} durch Ankleben eines Henkels $G\times_H DV\times DW$ mit $\mathbb{R}^1 \subset W$ an L_{i-1} entsteht. Man beweist diese Tatsache vollig analog zu Satz (II.6) oder [1] Theorem (3.1) durch Induktion uber die Anzahl der auf M auftretenden Orbittypen. Fur den Induktionsschluß benutzt man Lemma (4.3) aus [1].

(3) Mit Hilfe der Henkelkorperzerlegung aus (2) beweist man die Behauptung wieder durch Induktion nach dem Index der in dieser Zerlegung auftretenden Henkel. Die Ausfuhrung ist wortlich dieselbe wie im Beweis von Satz (II.7). ∎

Es soll nun der eigentlich interessierende Konfigurationsraum $C(M)$ mit dem Schnittraum $\Gamma_M(M, \partial M)$ in Verbindung gebracht werden. Dies geschieht vollig analog zu [4], §4 mit Hilfe der beiden folgenden Satze:

Satz (II.9) Man hat natürliche Zerspaltungen:

(a) $$(C(M))^G \xrightarrow{\cong} \prod_{(H_i)\in O(M)} \prod_{j\in J_i} C(M^j_{H_i}/WH_i)$$

Hierbei durchläuft $(H_i)\in O(M)$ die auf M auftretenden Orbittypen und $j\in J_i$ die Komponenten von M_{H_i}/WH_i.

(b) $$(C(M,L))^G \xrightarrow{\cong} \prod_{(H_i)\in O(M)} \prod_{j\in J_i} C(M^j_{H_i}/WH_i,\ L^j_{H_i}/WH_i)$$

Beweis:

(a) Eine Konfiguration $s = (s_1,\ldots,s_n)$ liegt genau dann in der G-Fixpunktmenge $(C(M))^G$, falls $(s_1,\ldots,s_n)\subset M$ eine G-Untermenge ist. Man zerlegt nun diese G-Menge nach ihren Orbittypen, d. h. $s = \coprod_i \sigma_i\cdot G/H_i$ und unterscheidet für $\sigma_i\cdot G/H_i$ noch, in welche Komponente $M^j_{(H_i)}$ die G-Menge $\sigma_i\cdot G/H_i$ eingebettet ist. Weiter entspricht einer Konfiguration $\{\sigma_i\cdot G/H_i\}\subset M^j_{(H_i)}$ eine Konfiguration $\{\bar{\sigma}_i\}\subset M^j_{H_i}/WH_i$. Man erhält durch diese Zerlegung der G-invarianten Konfiguration s die gewünschte Zerspaltung

$$(C(M))^G \cong \prod_{(H_i)\in O(M)} \prod_{j\in J_i} C(M^j_{H_i}/WH_i).$$

(b) folgt sofort aus a) durch Übergang zu den betreffenden Quotienten. ∎

Korollar (II.10) Es gilt

$$\pi_0(C(M)^G) \cong \prod_{(H_i)\in O(M)} \prod_{j\in J_i} \mathbb{N} \quad .$$

Beweis:

Man benutzt Satz (II.9) und die Beziehung $\pi_o(C(X)) \cong \mathbb{N}$ für eine zusammenhängende Mannigfaltigkeit X mit einer Dimension größer als Null. ∎

Ist (n) ein Tupel in $\pi_o(C(M)^G)$, so sei $C(M)^G_{(n)}$ die zugehörige Komponente.

Sei nun M eine G-Mannigfaltigkeit mit Rand ∂M, so daß jede Komponente $M^j_{(H)}$ von $M_{(H)}$ den Rand in genau einer Komponente $\partial M^j_{(H)}$ von $\partial M_{(H)}$ trifft. Es wird dann für alle $H \in O(M)$ in jeder Komponente $\partial M^j_{(H)}$ von $\partial M_{(H)}$ ein fester Orbit G/H^j gewählt. Der Punkt $(G/H^j/G) \in M_{(H)}/G$ werde mit n^j_H bezeichnet. Man setzt weiter $N = \coprod_{\substack{(H) \in O(M) \\ j \in J_H}} G/H^j$. Wie in [4], §4 heftet man nun an ∂M einen "Kragen mit Löchern" an, d. h. man betrachtet $A = (\partial M \setminus N) \times I$ mit

$\partial A = A_o \sqcup A_1 = (\partial M \setminus N) \times 0 \sqcup (\partial M \setminus N) \times 1$ und $X = M \sqcup_{A_o} A$.

Lemma (II.11) Die Einschränkung

$$r : \tilde{C}(X)^G \longrightarrow \tilde{C}(A)^G$$

ist eine Homologiefaserung mit der Faser $C(M,N)^G$.

Beweis:

Man betrachtet das folgende Diagramm:

$$
\begin{array}{ccccc}
C(M,N)^G & \overset{\cong}{\longleftrightarrow} & \prod\limits_{(H)\in O(M)} & \prod\limits_{j\in J_H} & C(M_H^j/WH, n_H^j) \\
\downarrow & & & & \downarrow \\
C(X)^G & \overset{\cong}{\longleftrightarrow} & \prod\limits_{(H)\in O(M)} & \prod\limits_{j\in J_H} & C(X_H^j/WH) \\
\downarrow r & & & & \downarrow r_H^j \\
C(A)^G & \overset{\cong}{\longleftrightarrow} & \prod\limits_{(H)\in O(M)} & \prod\limits_{j\in J_H} & C(A_H^j/WH)
\end{array}
$$

Nach [4], Lemma (4.1) ist r_H^j eine Homologiefaserung.

∎

Es gibt eine injektive Homotopie $f_t : M \longrightarrow M$, welche außerhalb einer Tubenumgebung von N die Identitat ist und fur die $f_1(N) \cap N = \emptyset$ gilt. Man konstruiert so eine Abbildung

$$\gamma : C(M)^G_{(n)} \longrightarrow C(M)^G_{(n+1)}$$

indem man $\gamma(s) = f_1(s) \sqcup N$ setzt. Das Teleskop der unendlich oft iterierten Abbildung γ

$$C(M)^G_{(0)} \xrightarrow{\ \gamma\ } C(M)^G_{(1)} \xrightarrow{\ \gamma\ } \cdots\cdot$$

werde mit $\mathrm{Tel}\left\{ C(M)^G_{(k)} \right\}$ bezeichnet.

<u>Lemma</u> (II.12) Die offensichtliche Projektion

$$\pi : \mathrm{Tel}\left\{ C(M)^G_{(k)} \right\} \longrightarrow C(M,N)^G$$

ist eine Homotopieaquivalenz.

Beweis:

Man betrachtet das folgende Diagramm:

$$\begin{array}{ccc} \mathrm{Tel}\left\{C(M)^G_{(k)}\right\} & \overset{\simeq}{\longleftrightarrow} & \prod_{(H)\in O(M)} \prod_{j\in J_H} \mathrm{Tel}\left\{C(M^j_H/WH)_k\right\} \\ \Big\downarrow \pi & & \Big\downarrow \pi^j_H \\ C(M,N)^G & \overset{\simeq}{\longleftrightarrow} & \prod_{(H)\in O(M)} \prod_{j\in J_H} C(M^j_H/WH, n^j_H) \end{array}$$

Nach [4], Lemma (4.2) ist π^j_H eine Homotopieaquivalenz.

Nach Theorem (II.8) hat man eine Homotopieaquivalenz

$$\Theta^G : \tilde{C}(A)^G \longrightarrow \Gamma_X(\partial M\times I, N\times I)^G \cong \Gamma_X(\partial M, N)^G.$$

(Um Theorem (II.8) anwenden zu konnen, hat man die Punkte von N naturlich zu kleinen Scheiben zu verdicken). Die Identifikation von $C(X)^G$ mit einem Schnittraum ist etwas muhsamer. Man benutzt hierzu die in [4], §4 zu $\tilde{C}(X^j_H/WH_j)$ und $\tilde{C}(A^j_H/WH_j)$ konstruierten Uberlagerungen $\bar{C}(X^j_H/WH_j)$ und $\bar{C}(A^j_H/WH_j)$ und setzt $\bar{C}(X)^G = \prod_{(H)\in O(M)} \prod_{j\in J_H} \bar{C}(X^j_H/WH)$ und $\bar{C}(A)^G = \prod_{(H)\in O(M)} \prod_{j\in J_H} \bar{C}(A^j_H/WH)$.

Lemma (II.13) a) Man hat eine Homotopieaquivalenz $\bar{C}(X)^G \simeq \Gamma_M(M,N)^G$

b) $\bar{C}(A)^G$ ist eine universelle Uberlagerung von $\tilde{C}(A)^G$ und wegen $\tilde{C}(A)^G \simeq \Gamma_M(\partial M,N)^G$ auch eine universelle Uberlagerung

$\bar{\Gamma}_M(\partial M, N)^G$ von $\Gamma_M(\partial M, N)^G$.

Beweis:

Analog zu [4], Lemma (4.3) bez. (4.4). ∎

Eine Ubungsaufgabe in aquivarianter Obstruktionstheorie [vergl. [6], §4] zeigt, daß die G-Homotopieklassen von Schnitten $\sigma \in \Gamma_M(M, \partial M)^G$ durch einen relativen Multigrad $d(\sigma) \in \prod_{(H) \in O(M)} \prod_{j \in J_H} \mathbb{Z}$ klassifiziert werden. Hierbei ist d(f) so definiert, daß fur $s \in C(M)^G_{(k)}$ gilt $d(\Theta(s)) = (k)$. Man beachte, daß $d(\sigma)$ nicht gleich dem Tupel $\{\delta(\sigma^H)\}$ ist mit $\delta(\sigma^H)$ = nicht aquivarianter Abbildungsgrad des Schnittes σ^H. Ist M_H/WH zusammenhangend fur alle $H \in O(M)$, so laßt sich aber $d(\sigma)$ aus $\{\delta(\sigma^H)\}$ berechnen. Die zu dem Tupel (z) gehorige Komponente von $\Gamma_M(M, \partial M)^G$ werde mit $\Gamma_M(M, \partial M)^G_{(z)}$ bezeichnet. Analog zu $\gamma: C(M)^G_{(k)} \longrightarrow C(M)^G_{(k+1)}$ erhalt man eine Abbildung $\hat{\gamma}: \Gamma_M(M, \partial M)^G_{(k)} \longrightarrow \Gamma_M(M, \partial M)^G_{(k+1)}$ so daß das folgende Diagramm kommutiert bis auf Homotopie:

$$(*)\qquad \begin{array}{ccccc}
C(M,\varepsilon)^G_{(k)} & \xrightarrow{\Theta_\varepsilon} & \Gamma_M(M_\varepsilon, \partial M_\varepsilon)^G_{(k)} & \xrightarrow[\cong]{a} & \Gamma_M(M, \partial M)^G_{(k)} \\
 & & \xrightarrow{\qquad\Theta\qquad} & & \\
\Big\downarrow \gamma & & & & \Big\downarrow \hat{\gamma} \\
C(M,\varepsilon)^G_{(k+1)} & \xrightarrow{\Theta_\varepsilon} & \Gamma_M(M_\varepsilon, \partial M_\varepsilon)^G_{(k+1)} & \xrightarrow[\cong]{a} & \Gamma_M(M, \partial M)^G_{(k+1)} \\
 & & \xrightarrow{\qquad\Theta\qquad} & &
\end{array}$$

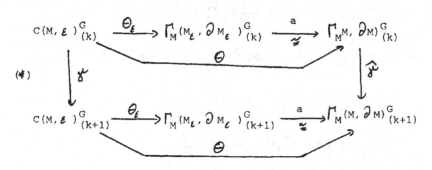

Ist $\sigma \in \Gamma_M(M, \partial M)^G_{(k)}$ gegeben, so betrachtet man $a(\sigma)$ und beachtet, daß $a(\sigma)$ auf einen ε-Kragen von ∂M konstant gleich dem Grundpunkt der Faser ist.

Die injektive Homotopie $f_t : M \longrightarrow M$ schob die Konfiguration N in das Innere von M. Man betrachtet $f_1(N)$ und wahlt f_t ohne Einschrankung so, daß $f_1(N) \subset M \setminus M_\varepsilon$ gilt. Weiter findet man ein δ, so daß $\Theta_\delta(f_1(N))$ ein Schnitt ist, welcher nur auf einer δ Umgebung von $f_1(N)$ lebt, welche ganz im Inneren von $M \setminus M_\varepsilon$ liegt. Man kann daher fur jeden Schnitt σ den Schnitt $\hat{\delta}(\sigma) = a(\sigma) \cup \Theta_\delta(f_1(N))$ definieren. Weiter kann man ebenso einen auf einer δ-Umgebung von $f_1(N)$ lebenden Schnitt wieder "abziehen" und erkennt so, daß $\hat{\delta}$ eine Homotopieaquivalenz ist.

<u>Theorem</u> (II.14) Ist M eine kompakte G-Mannigfaltigkeit mit dem Rand ∂M und gilt fur alle $(H) \in O(M)$ und alle $j \in J_H$ noch $\partial M \cap M^j_{(H)} \neq \emptyset$, so gibt es eine G-Abbildung

$$\bar{\Theta} : C(M) \longrightarrow \Gamma_M(M, \partial M),$$

welche $C(M)^G_{(k)}$ nach $\Gamma_M(M, \partial M)^G_{(k)}$ abbildet und fur welche

$$\bar{\Theta}_* : \varinjlim_{k \in \mathbb{N}} \left\{ H_*(C(M)^G_{(o)}) \xrightarrow{\delta_*} H_*(C(M)^G_{(1)}) \xrightarrow{\delta_*} \dots \right\} \longrightarrow$$

$$\varinjlim_{k \in \mathbb{N}} \left\{ H_*(\Gamma_M(M, \partial M)^G_{(o)}) \xrightarrow{\hat{\delta}_*} \dots\dots \right\}$$

isomorph ist. Hierbei ist fur $k \in \mathbb{N}$ das Tupel (k) gegeben durch $(k) = (k, k, \dots k)$.

Beweis:

a) Sei zunachst $M^{j}_{(H)}/G \cap \partial M_{(H)}/G$ zusammenhangend fur alle $H \in O(M)$ und $j \in J_H$. Man erhalt dann mit Lemma (II.13) das folgende Diagramm:

$$
\begin{array}{ccccl}
 & C(M,N)^G & & \Gamma_M(M,\partial M)^G_{(\alpha)} & \\
 & \downarrow & & \downarrow & \\
C(M,N)^G & \bar{C}(X)^G & \xrightarrow{\cong} & \Gamma_M(M,N)^G & \text{(zusammenhangend)} \\
\downarrow & \swarrow \quad \downarrow \bar{r} & & \downarrow \hat{r} & \\
(**)\ \tilde{C}(X)^G & \bar{C}(A)^G & \xrightarrow{\cong} & \bar{\Gamma}_M(\partial M,N)^G & \text{(einfach zusammenhangend)} \\
\downarrow r & \swarrow & & \swarrow p & \\
\tilde{C}(A)^G & & \xrightarrow{\cong} & \Gamma_M(\partial M,N)^G & \text{(zusammenhangend)}
\end{array}
$$

(In the top row, $C(M,N)^G$ on the left of the second row is joined to the upper $C(M,N)^G$ by a double line.)

Die Faser von $p \circ \hat{r}$ ist $\Gamma_M(M,\partial M)^G$. Analog zu

$$\pi_0(\Gamma_M(M,\partial M)^G) \cong \prod_{(H)\in O(M)} \prod_{j\in J_H} \mathbb{Z} \quad \text{gilt}$$

$$\pi_1(\Gamma_M(\partial M,N)^G) \cong \prod_{(H)\in O(M)} \prod_{j\in J_H} \mathbb{Z}.$$

(Der degenerierte Fall von nulldimensionalen Orbitbundeln ist naturlich auszuschließen und gesondert zu behandeln.)

Die Faser von $\hat{r}$ ist daher genau eine Zusammenhangskomponente $\Gamma_M(M,\partial M)^G_{(\alpha)}$ von $\Gamma_M(M,\partial M)^G$. Da r und $\bar{r}$ nach Lemma (II.11) Homologiefaserungen sind und $\hat{r}$ eine Faserung ist, induzieren die horizontalen Abbildungen des Diagramms (**) eine Homologieaquivalenz

$$\psi_*: H_*(C(M,N)^G) \longrightarrow H_*(\Gamma_M(M,\partial M)^G_{(\alpha)})$$

Mit Hilfe von Lemma (II.12) erhalt man die Homologieaquivalenz

$$\overline{\Psi} = \psi_* \circ \pi_* : \varinjlim_{k \in \mathbb{N}} \{H_*(C(M)^G_{(k)})\} \longrightarrow H_*(C(M,N)^G) \longrightarrow H_*(\Gamma_M(M,\partial M)^G_{(\alpha)})$$

b) Man hat zu zeigen, wie $\overline{\Psi}$ mit der Abbildung $\overline{\Theta}^G$ zu identifizieren ist.

Hierzu betrachtet man das Diagramm (*) und bildet fur $k \in \mathbb{N}$ den Homotopielimes uber die vertikalen Abbildungen $\hat{\gamma}$. Da die $\hat{\gamma}$ Homotopieaquivalenzen sind, ist die Inklusion $i: \Gamma_M(M,\partial M)^G_{(\alpha)} \longrightarrow \underset{k \in \mathbb{N}}{\text{H-}\varinjlim} \{\Gamma_M(M,\partial M)^G_{(k)}\}$ eine Homotopieaquivalenz. Man rechnet explizit nach, daß $i_* \circ \overline{\Psi} = \overline{\Theta}^G_*$ gilt. Hierbei ist $\overline{\Theta}^G_*$ die von Θ auf den Homotopielimites in Diagramm (*) induzierte Abbildung. Mit i_* und $\overline{\Psi}$ ist also auch $\overline{\Theta}^G_*$ isomorph.

c) Ist $(M^j_H/WH \cap \partial M_H/WH)$ nicht zusammenhangend, so heftet man an ∂M einen etwas modifizierten Kragen an. Hierzu wahlt man fur alle $H \in O(M)$ und $j \in J_H$ in $(M^j_{(H)} \cap \partial M)$ eine Komponente $\overline{\partial M^j_{(H)}}$ aus und in dieser Komponente den homogenen Raum G/H^j.

Man setzt $C = \bigcup_{\substack{(H) \in O(M) \\ j \in J_H}} \overline{\partial M^j_H}$ und $N = \bigcup_{\substack{(H) \in O(M) \\ j \in J_H}} G/H^j$.

Sodann verdickt man das Komplement von C in ∂M zu einer Mannigfaltigkeit U und heftet an $\overline{\partial M} = \partial M \setminus U$ den Kragen mit Löchern $(\bar{\partial}M \setminus N) \times I$ an. Ersetzt man in den Beweisschritten aus a) und b) und den zugehörigen Lemmata ∂M durch $\bar{\partial}M$, so bleibt der Beweis richtig, auch falls $M_H^{\delta}/WH \cap \partial M_H/WH$ nicht zusammenhängend ist. ∎

III. Anwendungen:

Definition (III.1) a) Ein H-Raum X mit G-Operation ist ein H-Raum $X \in$ G-Top, für den die Komposition $m: X \times X \longrightarrow X$ äquivariant ist.

b) Ein G-äquivarianter H-Raum X heißt G-gruppenartig, falls für alle $H < G$ die Menge der Komponenten $\pi_0(X^H)$ eine Gruppe ist.

c) Eine G-Abbildung $f : X \longrightarrow Y$ zwischen G-äquivarianten H-Räumen heißt G-Gruppenkomplettierung, falls Y G-gruppenartig ist und falls alle $K < G$ $f_*^K : H_*(X^K)[\pi_0(X^K)^{-1}] \longrightarrow H_*(Y^K)$ isomorph ist. Hierbei bezeichnet $H_*(X^K)[-]$ die betreffende Lokalisierung. ∎

Für ein topologisches Monoid M sei BM der klassifizierende Raum.

Satz (III.2) Ist $f : X \longrightarrow Y$ eine G-Gruppenkomplettierung zwischen G-gruppenartigen G-äquivarianten H-Räumen, ist f eine schwache G-Homotopieäquivalenz.

Beweis:

$f^K : X^K \longrightarrow Y^K$ ist eine Gruppenkomplettierung zwischen gruppenartigen H-Raumen und damit nach [10] , Remark (1.5) eine schwache Homotopieaquivalenz. ∎

<u>Satz</u> (III.3) Ist M ein topologisches G-zulassiges Monoid mit G-Operation, so ist die naturliche G-Abbildung

$$\xi : M \longrightarrow \Omega BM$$

eine G-Gruppenkomplettierung. (Hierbei heißt M G-zulassig, falls fur alle $H < G$ noch $\pi_o(M^H)$ im Zentrum von $H_*(M^H;\mathbb{Z})$ liegt.)

<u>Beweis:</u>
Man verwendet [5] , Proposition(1)fur alle Fixpunktmengen. ∎

Die G-Darstellung V enthalte einen trivialen Summanden, d. h. $V = \mathbb{R}^1 \times \bar{V}$. Fur zwei reelle Zahlen $a < b$ setzt man $V_{(a,b)} = (a,b) \times \bar{V} \subset V$. Man definiert nun einen modifizierten Konfigurationsraum

$$C'(V)_n = \{ (s,t) \in C(V)_n \times \mathbb{R} \mid s \subset V_{(o,t)} \} \text{ und setzt } C'(V) = \bigcup_n C(V)_n .$$

Nach [11] , Seite 213 ist C'(V) G-homotopieaquivalent zu C(V). Bezeichnet $T_\varepsilon : \mathbb{R} \times \bar{V} \longrightarrow \mathbb{R} \times \bar{V}$ die durch $T_\varepsilon(t,v) = (t+\varepsilon, v)$ gegebene Translation, so erhalt man auf C'(V) eine Monoidstruktur, indem man $(s,t) + (s',t') = (s \cup T_t(s'), t+t')$ setzt; und in [11] , Seite 213 ist ausgefuhrt, wie die G-Abbildung

$$\Theta : C'(V) \longrightarrow \Omega^V S^V$$

als H-Raumabbildung aufzufassen ist.

In [8] wird C'(V) durch die zu der G-Operade $\mathcal{C}_V(-)$ der kleinen normierten Kugeln in V gehorige G-Monade C_V ersetzt und man erhalt eine G-Abbildung von G-Monaden

$$\alpha : C_V \longrightarrow \Omega^V S^V.$$

<u>Theorem</u> (III.4) Die G-Abbildung

$$\Theta : C'(V) \longrightarrow \Omega^V S^V$$

ist eine G-Gruppenkomplettierung.

<u>Beweis:</u>

1) $\Omega^V S^V$ ist G-gruppenartig, da V einen trivialen Summanden enthalt.

2) Es ist zu zeigen, daß fur alle $H < G$ die Abbildung

$$\Theta^H : C'(V)^H \longrightarrow (\Omega^V S^V)^H$$

die Homologie lokalisiert. Ohne Einschrankung setze ich G = H Man erhalt

$$H_*(C'(V)^G)[(\pi_0(C'(V)^G)^{-1}] \cong \varinjlim_{\delta} \{H_*(C(DV)^G)\} \xrightarrow{\Theta_*} H_*(\Omega^V S^V)^G)$$

Hierbei entsteht der direkte Limes wie in §II beschrieben wurde, indem man eine feste G-invariante Konfiguration N, die fur jedes $(H) \in O(V)$ und jede Komponente $M^j_{(H)}$ von $M_{(H)}$

eine Bahn $G/H^j \subset M^j_{(H)}$ enthält, unendlich oft zu $C(DV)^G$ hinzuaddiert.

Hiermit folgt die Behauptung, da Θ_* nach Theorem (II.14) isomorph ist. ∎

Sei V eine G-Darstellung, welche alle irreduziblen G-Darstellungen mindestens einmal enthält, und V^n die n-fache Summe. In der offensichtlichen Weise kann man die direkten Limiten $\varinjlim_n C'(V^n)$ und $\varinjlim_n \Omega^{V^n} S^{V^n}$ definieren.

<u>Korollar</u> (III.5) Die G-Abbildung

$$\Theta : \varinjlim_n C'(V^n) \longrightarrow \varinjlim_n \Omega^{V^n} S^{V^n}$$

ist eine G-Gruppenkomplettierung.

<u>Beweis:</u>
Die Behauptung folgt sofort aus Theorem (III.4) durch Limesbildung. ∎

Man beachte, daß $\varinjlim_n \Omega^{V^n} S^{V^n}$ das darstellende Objekt der stabilen äquivarianten Homotopie $\pi^0_G(-)$ ist.

Bezeichnet $\hat{C}(V)_k$ die geordneten k-Konfigurationen in V, so operiert Σ_k frei und G-äquivariant auf $\hat{C}(V)_k$ und man erhält das G-äquivariante Σ_k-Prinzipalbündel $p_V : \hat{C}(V)_k \longrightarrow \hat{C}(V)_k/\Sigma_k = C(V)_k$.

<u>Satz</u> (III.6) Die Projektion

$$P_\infty : \varinjlim_n \widehat{C}(V^n)_k \longrightarrow \varinjlim_n C(V^n)_k$$

ist ein universelles G-aquivariantes Σ_k-Prinzipalbundel,

d. h. $\varinjlim_n C(V^n)_k \cong B(\Sigma_k, G)$ = klassifizierender Raum fur G-aquivariante Σ_k-Prinzipalbundel.

<u>Beweis:</u>

Fur die Limiten $\varinjlim_n C(V^n)$ bez. $\varinjlim_n \Omega^{V^n} S^{V^n}$ schreibe ich der Einfachheit halber $C(V^\infty)$ bez. $\Omega^{V^\infty} S^{V^\infty}$.

a) Man hat eine klassifizierende Abbildung

$$f : C(V^\infty)_k \longrightarrow B(\Sigma_k, G)$$

und es genugt zu zeigen, daß

$$f^H : C(V^\infty)_k^H \longrightarrow B(\Sigma_k, G)^H$$

eine Homotopieaquivalenz ist. Ohne Einschrankung sei H = G.

b) $B(\Sigma_k, G)^G = \coprod_{\substack{N = \text{k-punktige} \\ \text{G-Mengen}}} B(\text{Aut}_G(N))$.

Hierbei ist $\text{Aut}_G(N)$ die Gruppe der G-aquivarianten Automorphismen von N.

c) Nach Satz (II.9) gilt

$$(C(V^\infty)_k)^G = \coprod_{\substack{N = k\text{-punktige} \\ G\text{-Menge}}} \text{Einb}_G(N;V^\infty) / \text{Aut}_G(N) \,.$$

Hierbei ist $\text{Einb}_G(N;V)$ der Raum der G-aquivarianten Einbettungen von N in V.

Es genugt also zu zeigen, daß $\text{Einb}_G(N,V^\infty)$ zusammenziehbar ist, denn dann ist $\text{Einb}_G(N,V^\infty)/\text{Aut}_G(N) \cong B(\text{Aut}_G(N))$.

d) Es genugt zu zeigen, daß fur passend große n

$\pi_1(\text{Einb}_G(N,V^n)) = 0$ gilt. Seien also

α bez. $\beta \in \pi_i(\text{Einb}_G(N,V^n))$ gegeben. Man betrachtet die adjungierten Abbildungen $\bar{\alpha}$ bez. $\bar{\beta} : S^1 \times N \longrightarrow V^n$, welche nach dem G-Einbettungssatz [13] G-homotop zu G-Einbettungen sind. Ebenso sind $\bar{\alpha}$ und $\bar{\beta}$ G-homotop durch G-Einbettungen, falls nur n hinreichend groß ist. Es folgt, daß auch α und β G-homotop sind, d. h. $\pi_1(\text{Einb}_G(N,V^n)) = 0$ fur alle hinreichend großen n. ∎

Mit Hilfe von Satz (III.6) kann man die folgende oft nutzliche Umformulierung von Korollar (III.5) geben. Hierzu bezeichne $A_G(Y)$ die Menge der Isomorphieklassen von G-aquivarianten endlichen Uberlagerungen uber Y. Die disjunkte Vereinigung macht $A_G(-)$ zu einem halbgruppenwertigen kontravarianten halbexakten Homotopiefunktor, welcher nach Satz (III.6) dargestellt wird durch den G-Raum $\coprod_k B(\Sigma_k,G) \cong C(V^\infty)$.

<u>Korollar</u> (III.7) Die durch die G-Abbildung Θ aus Korollar (III.5) definierte naturliche Transformation

$$\Theta_* : A_G(-) \cong [- \mid C(V^\infty)]^G \longrightarrow [- \mid \Omega^{V^\infty} S^{V^\infty}]^G = \pi_G^0(-)$$

ist universell in dem folgenden Sinn:

Ist $h : A_G(-) \longrightarrow T_G(-)$ eine naturliche Transformation von halbexakten halbgruppenwertigen Homotopiefunktoren und ist $T_G(-)$ gruppenwertig, so gibt es genau eine naturliche Transformation

$$\hat{h} : \pi_G^o(-) \longrightarrow T_G(-) \text{ mit } \hat{h} = h \circ \Theta_4 \quad .$$

Beweis:

a) Nach dem Darstellungssatz fur halbexakte Homotopiefunktoren ist $T_G(-)$ darstellbar durch einen G-Raum $\tilde{T}$ und die Transformation h ist darstellbar durch eine G-Abbildung $\tilde{h} : C(V^\infty) \longrightarrow \tilde{T}$.

b) Der Limes in Theorem (III.4) entstand durch unendlich oft iterierte Addition γ einer festen G-invarianten Konfiguration N. Fur den dargestellten Funktor $A_G(-)$ bedeutet diese Addition, daß man zu jeder Uberlagerung die triviale G-Uberlagerung $(N \times -)$ hinzuaddiert. Die Addition von $h(N \times -)$ in $T_G(-)$ ist darstellbar durch eine G-Abbildung $\tilde{\alpha}_{h(N)} : \tilde{T} \longrightarrow \tilde{T}$ und $\tilde{\alpha}_{h(N)}$ ist eine G-Homotopieaquivalenz, da $T_G(-)$ gruppenwertig ist.

Man erhalt also das folgende homotopie-kommutative Diagramm:

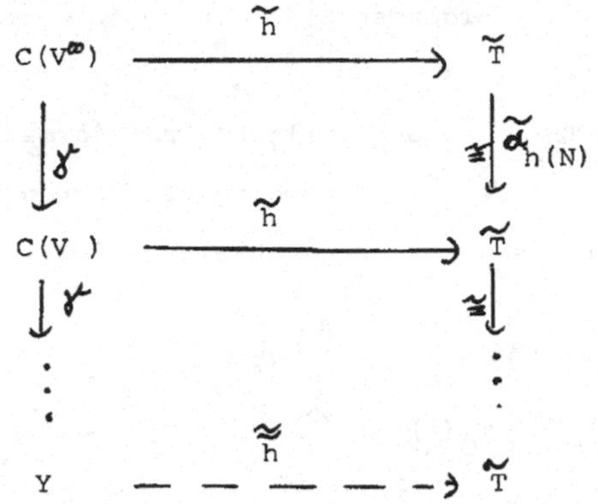

Hierbei ist $\tilde{\tilde{h}}$ die von $\tilde{h}$ auf den betreffenden Teleskopen induzierte G-Abbildung. Mit dieser Uberlegung ergibt sich das folgende homotopiekommutative Diagramm:

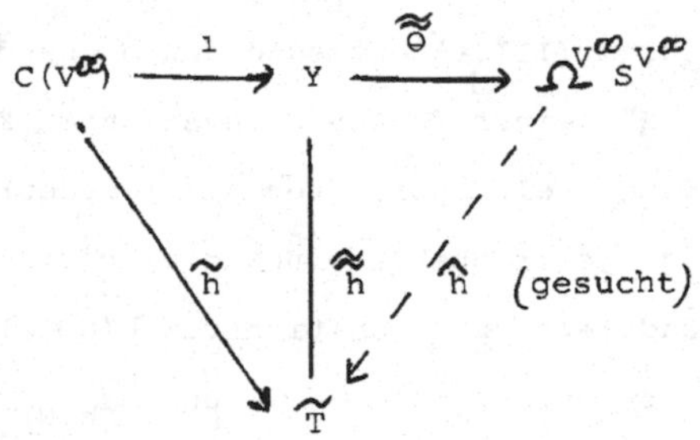

c) Da sowohl ι als auch Θ die Homologie aller Fixpunktmengen lokalisieren, ist $\tilde{\tilde{\Theta}}$ eine Homologieaquivalenz auf allen Fixpunktmengen.

Mit Hilfe der Spektralsequenz (I, 10.4) aus [2] erkennt man, daß $\tilde{\tilde{\Theta}}$ fur die Bredonkohomologie $H^*_G(-,\underline{A})$ mit beliebigem Koeffizienten $\underline{A}$ einen Isomorphismus liefert. Da $T_G(-)$ gruppenwertig ist, ist fur alle $H < G$ der Raum $\tilde{T}^H$ einfach; denn er stellt den Funktor $\mathcal{T}_G(G\times_K -)$ dar auf der Kategorie der trivialen K-Räume. Mit Hilfe äquivarianter Obstruktionstheorie ([2], Theorem (2.5)) gibt es daher die gesuchte Erweiterung $\hat{h}$, welche bis auf G-Homotopie eindeutig ist. ∎

Beweis von (III.8) aus der Einleitung:

a) Nach Theorem (III.4) ist

$$\Theta^G : (C'(V))^G \longrightarrow (\Omega^V S^V)^G$$

eine Gruppenkomplettierung. Nach Satz (II.9) gilt $C'(V)^G = \prod_{(H)\in O(V)} C'(V_H/WH)$. Damit erhält man eine Gruppenkomplettierung

$$\Theta : \prod_{(H)\in O(V)} C'(V_H/WH) \longrightarrow (\Omega^V S^V)^G .$$

b) Das Tangentialbündel von V_H/WH ist gleich $V_H \times_{WH} V^H \longrightarrow V_H/WH$. Nach Theorem (II.14) und da $C'(V_H/WH)$ ein Monoid ist, hat analog zu Theorem (III.4) die Gruppenkomplettierung von $C'(V_H/WH)$ die Gestalt $\Gamma_{V_H/WH}(V_H/WH)$. Damit folgt aus (a) die Behauptung $\prod_{(H)\in O(V)} \Gamma_{V_H/WH}(V_H/WH) \xrightarrow{\simeq} (\Omega^V S^V)^G$

c) Nach Satz (III.6) gilt

$$C(V^\infty)^G \simeq \coprod_{\substack{N = \text{endliche}\\ \text{G-Mengen}}} B(\mathrm{Aut}_G(N)) .$$

Ist $N = \coprod_i n_i\, G/H$ eine Zerlegung von N in Bahnen, so gilt $\mathrm{Aut}_G(N) = \prod_i \left(\Sigma_{n_i} \int WH_i\right)$; hierbei bedeutet

$(\Sigma_{n_1} \int WH_1)$ das semidirekte Produkt von $(WH_1)^{n_1}$ mit Σ_{n_1}.

Weiter ist $B(\Sigma_{n_1} \int WH_1) \cong E\Sigma_{n_1} \times_{\Sigma_{n_1}} (BWH_1)^{n_1}$; denn auf $E\Sigma_{n_1} \times E(WH_1)^{n_1}$ operiert $(\Sigma_{n_1} \int WH_1)$ frei und der Quotient ist $E\Sigma_{n_1} \times_{\Sigma_{n_1}} B(WH_1)^{n_1}$. Es folgt

$C(V^\infty)^G \cong \prod_{(H)<G} \coprod_n E\Sigma_n \times_{\Sigma_n} B(WH)^n$. Nach [11] ist die Gruppenkomplettierung von $(\coprod_n E\Sigma_n \times_{\Sigma_n} B(WH)^n)$ homotopieäquivalent zu $\varinjlim_n \Omega^n(S^n \wedge B(WH)^+)$. Es folgt die Behauptung

$$\prod_{(H)<G} \varinjlim_n \Omega^n(S^n \wedge B(WH)^+) \xrightarrow{\cong} (\Omega^{V^\infty} S^{V^\infty})^G.$$

■

<u>Bemerkung</u> (III.9) a) Die Zerspaltung aus Korollar (III.8) wurde in [6] auf rein homotopietheoretischem Wege erhalten. Zusammen mit nicht äquivarianten Approximationssätzen liefert diese Zerspaltung natürlich umgekehrt, daß $\Theta : C'(V) \longrightarrow \Omega^V S^V$ eine G-Gruppenkomplettierung ist.

b) Ist X ein punktierter G-Raum, so kann man die Konfigurationen in der G-Mannigfaltigkeit M durch Werte aus X parametrisieren und erhält in der üblichen Weise die durch X parametrisierten Konfigurationsräume $C(M)[X]$. Ist E_M das zu TM assoziierte G-Sphärenbündel aus §II, so kann man in jeder Faser das Smash-Produkt mit X bilden und erhält das Bündel $E_M[X] \longrightarrow M$. Der G-Raum der Schnitte von $E_M[X]$ sei $\Gamma_M(M)[X]$. Alle unsere G-Approximationssätze bleiben richtig, falls man die Konfigurations- und Schnitträume durch einen zusätzlichen G-Raum parametrisiert. Die Beweise sind wörtlich dieselben wie im unparamatrisierten Fall. Ist der Parameterraum X G-zusammenhangend, d. h. ist X^H fur alle $H < G$ zusammenhangend, so sind die G-Gruppenkomplettierungen sogar schon G-Homotopieaquivalenzen. ∎

Literaturverzeichnis

1) Bierstone, E.: Equivariant Gromov theory. Topology 13, 327 - 345 (1974)

2) Bredon, G.: Equivariant cohomology theories. Lecture notes in mathematics 34, Springer 1973.

3) tom Dieck, T.: Faserbündel mit Gruppenoperation. Archiv der Mathematik 20, 136 - 143 (1969)

4) Mc Duff, D.: Configuration spaces of positive and negative particles. Topology 14, 91 - 110 (1975)

5) Mc Duff, D.; Segal, G.: Homology fibrations and the group completion theorem. Inventiones 31, 279 - 285 (1976)

6) Hauschild, H.: Zerspaltung aquivarianter Homotopiemengen. Math. Annalen 230, 279 - 292 (1977).

7) Hauschild, H.: Klassifizierende Raume, G-Konfigurationsraume und aquivariante Schleifenraume. Habilitationsschrift Gottingen 1978.

8) Hauschild, H.; May, P.; Waner, S.: Equivariant loop spaces. (to appear).

9) Illman, S.: Smooth equivariant triangulations of G-manifolds. Math. Annalen 233, 199-220 (1978).

10) May, P.: E_∞-spaces, group completions and permutative categories. New developments in Topology. Lond. Math. Soc. Lecture Notes 11 (1974).

11) Segal, G.: Configuration spaces and iterated loop spaces. Inventiones 21, 213 - 221 (1873).

12) Segal, G.: Some results in equivariant topology. preprint.

13) Wasserman, A.: Equivariant differentiable topology. Topology 8, 128 - 144 (1969).

On Equivariant Homotopy Theory

I.M. James and G.B. Segal

Introduction

Many of the ideas of homotopy theory seem to belong most naturally to the category G-Top_B of G-spaces over a given G-space B, where G is a topological group. Both Top_B (the case when G is trivial) and G-Top (the case when B is trivial) have been extensively studied but G-Top_B itself appears to have been somewhat neglected. In this note we discuss some aspects of the homotopy theory of G-Top_B and show how certain fundamental results in G-Top and Top_B can be better understood from this point of view.

In the combined theory appropriate versions of the notions of absolute retract (AR) and absolute neighbourhood retract (ANR) play an important role. The definitions we adopt are based on Fox's characterization although one can, of course, make definitions more in the spirit of Borsuk and Čech.[1] As we shall see in §2, the familiar properties of AR/ANR theory in Top generalise easily to G-Top_B. In §3 we give some applications, including a version of the well-known theorem of Dold [4] on the fibre homotopy equivalences. The remaining sections are devoted to the proof of a theorem about G-AR's, where G is a compact Lie group, which implies a recent result of our own on equivariant homotopy theory. Indeed it was the desire to place this result on a more satisfactory footing that led to the development of ideas in the present note.

2. AR/ANR Theory in G-Top_B

Let B be a G-space, where G is a topological group. We say that a G-space E over B is a G-AR over B if it has the following extension property[2]. Let (Y,X) be a paracompact G-pair, ie. a pair in which Y is a paracompact G-space and X is a closed invariant subspace. Let θ, ϕ be G-maps such that the diagram shown below is commutative.

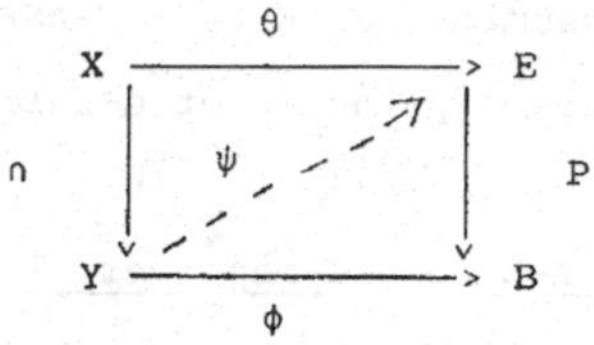

By an extension of θ over ϕ we mean a G-map ψ: $Y \to E$ such that $\psi|X = \theta$ and $p\psi = \phi$. If such an extension always exists we say that E is a G-AR over B.

Note that in making this definition it is not necessary to describe X,Y as G-spaces over B and θ, ϕ, ψ as G-maps over B, since they aquire this character automatically from the structure of the diagram.

When B is paracompact we see, taking $(Y,X) = (B,\emptyset)$, that if E is a G-AR over B then E admits an equivariant section $s: B \to E$. When E is a G-AR over B then E is G-contractible over B: the composition sp: $E \to E$ is G-homotopic to the identity over B.

We say that a G-space E over B is a G-ANR over B if, given any paracompact G-pair (Y,X) and G-maps θ, ϕ over B as before, there exists an open invariant neighbourhood U of X in Y and an extension ψ: U→E of θ over $\phi|U$.

Note that if E is a G-AR (resp. G-ANR) over B then E|A is a G-AR (resp. G-ANR) over A for all invariant subspaces A of B; indeed f^*E is a G-AR (resp. G-ANR) over B' for any G-map f: B'→B.

Our first two results are straightforward generalisations of well known results in ordinary ANR theory and we therefore omit the proofs. In both results it is assumed that E is a G-ANR over a given G-space B and that (Y,X) is a paracompact G-pair.

Proposition (2.1)

Let ϕ : Y→B be a G-map. Let ψ: Y→E be a G-map over ϕ and let θ_t: X→E be a G-homotopy of $\psi|X$ over $\phi|X$. Then θ_t can be extended to a G-homotopy ψ_t: Y→E of ψ over ϕ .

Proposition (2.2)

Let ϕ : Y→B be a G-map. Let ψ_0, ψ_1: Y→E be G-maps over ϕ , and let θ_t: X→E be a G-homotopy of $\psi_0|X$ into $\psi_1|X$ over $\phi|X$. Then there exists an open invariant neighbourhood U of X in Y and an extension ξ_t: U→E of θ_t to a G-homotopy of $\psi_0|U$ into $\psi_1|U$ over $\phi|U$.

Thus (2.1) shows that a G-ANR over B has the G-homotopy extension property over B, for paracompact G-pairs. We deduce

Corollary (2.3)

Let E be a G-ANR over B. If E is G-contractible over B then E is a G-AR over B.

For let s: B→E be an equivariant section and let η_t: E→E be a G-homotopy over B of sp into the identity. Given θ:X→E, ϕ : Y→B as before, we see that $\eta_t\theta$ is a G-homotopy over $\phi|X$ of spθ into θ. Since spθ = s $\phi|X$ has an extension to the G-map sϕ: Y→E over ϕ it follows that θ also has such an extension.

Let E, F be G-spaces over B and let $h: E\to F$ be a G-map over B. If h has the G-homotopy lifting property[3] for G-maps then we describe h as a <u>G-fibration over B.</u> Again the well-known results of the ordinary theory generalise. In particular one has the equivariant version of (6.1) of [3] which asserts that if h is a G-fibration over B and if h is also a G-homotopy equivalence over B then E is G-contractible <u>over F.</u>

Any G-map $h: E\to F$ over B can be replaced by a G-fibration $\rho: W\to F$, over B in the following sense. Take $W = W_B(h)$ to be the mapping path-space in the category $G\text{-Top}_B$, ie. the appropriate subspace[4] of the fibre product $E\times F^I$ with the diagonal action of G. The G-map ρ is defined by evaluation, in the usual way, and a G-homotopy equivalence σ : $E\to W$ over B is given by the stationary path, so that $\rho\sigma = h$. Note that if f, and hence ρ, is a G-homotopy equivalence over B then W is G-contractible over F.
We use (2.1) and (2.2) to prove

<u>Proposition</u> (2.4)

<u>Let $h: E\to F$ be a G-map over B, where E,F are G-ANR's over B. Then $W = W_B(h)$ is a G-ANR over F.</u>

For let (Y,X) be a paracompact G-pair. Maps into W are described, in the usual way, by giving the components in E and F^I; the latter we rewrite by taking the adjoint.

Thus let

$\alpha: X\times 0\to E$, $\beta: X\times I\to F$, $\gamma: Y\times 1\to F$ be G-maps over B such that
$\beta|X\times 0 = f\alpha$, $\beta|X\times 1 = \gamma|X\times 1$.
Since E is a G-ANR over B there exists a closed invariant neighbourhood $\bar{U}_1$ of X in Y and a G-map $\alpha_1 : \bar{U}_1\times 0\to E$ over B

extending α. Now β constitutes a G-homotopy over B of $f\alpha_1|X\times 0$ into $\gamma|X\times 1$. Since F is a C-ANR over B there exists, by (2.2), an open invariant neighbourhood U of X in Y, contained in U_1, and a G-homotopy β': $U\times I\to F$ over B, extending β, of $f\alpha_1|U\times 0$ into $\gamma|U\times 1$. This proves (2.4).

3. Some Results In Overhomotopy Theory

Proposition (3.1)

Let E be a space over the paracompact space B. If E|U is an AR (resp ANR) over U for every member U of an open covering of B then E is an AR (resp ANR) over B.

We will give the proof in the AR case: the ANR case is very similar. The first step is to replace the given covering by a countable covering, using the well-known Milnor procedure. Each member of the new covering is a disjoint union of open sets, each contained in a member of the old covering, and so the AR hypothesis carries over. The next step is to use the Dieudonné shrinking theorem to find a closed set in each member of the new covering such that the interiors of these closed sets still cover B.

Without real loss of generality, therefore, we may start with a countable sequence $A_1,\dots,A_n,\dots$ of closed subsets of B, such that the interiors of the A_n cover B, and such that $E|A_n$ is an AR over A_n for all n. Let θ: $X\to E$, ϕ : $Y\to B$ as before.
Consider the expanding sequence
$X = Y_0\subset Y_1\subset\dots\subset Y_n\subset\dots\subset Y$ of subsets of Y defined by
$Y_n = Y_{n-1}\cup\phi^{-1}A_n$. The filler ψ: $Y\to E$ will be defined by
$\psi|Y_n = \psi_n$ (n = 0,1,...), where ψ_n: $Y_n\to E$ is the map over $\phi|Y_n$ defined inductively as follows.

Define $\psi_0 = \theta$. Make the inductive hypothesis that ψ_{n-1} is defined, where $n \geq 1$. Since $E|A_n$ is an AR over A_n, we can extend $\psi_{n-1}|(Y_{n-1} \cap \phi^{-1}A_n)$ to a map $\psi'_{n-1}: \psi^{-1}A_n \to E$ over A_n, and then glue ψ'_{n-1} and ψ_{n-1} together to define ψ_n. Since ψ_n is an extension of ψ_{n-1} over $\phi|Y_n$ this establishes the inductive step and completes the proof of (3.1).

Proposition (3.2).

Let E be an ANR over the paracompact space B. Let B' be a closed subspace of B, and write B" = B-B'. Suppose that E|B' is an AR over B' and that E|B" is an AR over B". Then E is an AR over B.

Let $\theta: X \to E$, $\phi: Y \to B$ be as before, and write $E' = E|B'$, $E'' = E|B''$, $Y' = Y|B'$, $Y'' = Y|B''$. Since E' is an AR over B' we can extend $\theta|(X \cap Y')$ to a map $\theta': Y' \to E'$ over $\phi|Y'$. Then θ and θ' can be glued together to construct an extension of θ to a map $\psi': X \cup Y' \to E$ over $\phi|(X \cup Y')$. Since E is an ANR over B there exists an open neighbourhood U of $X \cup Y'$ in Y and an extension $\sigma: U \to E$ of ψ' over $\phi|U$. Let $W \subset U$ be a closed neighbourhood of $X \cup Y'$ in Y. Since E" is an AR over B" we can extend $\sigma|(W - Y'')$ to a map $s: Y'' \to E''$ over $\phi|Y''$. Finally we can glue s and $\sigma|W$ together to obtain a filler $Y \to E$ as required.

As an application of (3.1) we prove

Proposition (3.3)

Let E,F be paracompact ANR's over the space B. Let h: E→F be a map over B such that h|U is a homotopy equivalence over U for every member U of some open covering of B. Then h is a homotopy equivalence over B.

Of course this is a special case of the well-known theorem of Dold [4] but with the (admittedly unnecessary) ANR restriction the following simple proof can be given. First replace h: E→F by the fibration p: W→F. Here W is an ANR over F, by (2.4), hence W|U is an ANR over F|U. Since h|U is a homotopy equivalence over U it follows at once that ρ|U is a homotopy equivalence over U and hence that W|U is contractible over F|U, by (6.1) of [4], since ρ|U is a fibration. Therefore W|U is an AR over F|U, by

(2.3). The sets F|U, for all members of the given covering of B, form an open covering of F. By (2.4), therefore, W is an AR over F. Hence[5] ρ admits a section and so h admits a homotopy right inverse k: F→E over B. Now apply the argument to k, showing that k admits a homotopy right inverse over B as well as the homotopy left inverse h. We conclude that k, and hence h, is a homotopy equivalence over B.

By the same argument, but using (3.2) in place of (3.1) we obtain

Proposition (3.4)

Let E,F be paracompact ANR's over the space B. Let h: E→F be a map over B such that h|B' is a homotopy equivalence over B', where B' is a closed subspace of B, and such that h|(B-B') is a homotopy equivalence over B-B'. Then h is a homotopy equivalence over B.

4. Some Results in the Equivariant Theory

Let B be a G-space, where G is a topological group. Clearly (see §1 of [7]) if E is a G-AR (resp. G-ANR) over B then E^H is an AR (resp. ANR) over B^H for all closed subgroups H of G. The remainder of this note is mainly devoted to the proof of

Proposition (4.1)

Let G be a compact Lie group. Let B be a paracompact G-space and let E be a G-ANR over B. Suppose that E^H is an AR over B^H for all closed subgroups H of G. Then E is a G-AR over B.

Corollary (4.2)

Let G be a compact Lie group. Let f: A→B be a G-map, where A and B are paracompact G-ANR's. Suppose that f^H→B^H is a homotopy equivalence for all closed subgroups H of G. Then f is a G-homotopy equivalence.

The proof of (4.1) will be given in the next section. Here we show how (4.2) will follow. We apply (4.1) to the mapping path-space W of f, with G-action as in §2. Then W is a G-ANR over B, by (2.4), and W^H is an ANR over B^H for all H. Now W^H is the mapping path-space of f^H: $A^H \to B^H$, which is a homotopy equivalence by hypothesis. Therefore W^H is contractible over B^H and so is an AR over B^H, by (2.4). Hence W is a G-AR over B, by (4.1), and so admits an equivariant section. Hence f admits a G-homotopy right inverse f: B→A, say. Now g^H: $B^H \to A^H$ is a homotopy right inverse of f^H and so is a homotopy equivalence. We may therefore repeat the argument, with g in place of f, and obtain that g admits a G-homotopy inverse on the right, as well as on the left. Thus g is a G-homotopy equivalence, and so f is a G-homotopy equivalence, as asserted.

The proof of (4.2) given in [7] is essentially the same as the one we are giving here but makes unnecessary restrictions on the topology of A and B. The corresponding theorem for G-complexes has been proved by Illman [6], Matumoto [8] and others.

An example purporting to show that (4.2) breaks down when G is no longer compact was given at the end of [7]. Unfortunately the example is incorrect, and should be replaced by the following. Let f be the inclusion of $A = S^1 \times \{o\}$ in $B = S^1 \times \mathbb{R}$. Let $G = \mathbb{R}$, acting trivially on A, and acting on B by

$$x.(y,z) = (y + xz, z)$$
$$x.(\infty.z) = (\infty, z)$$

where S^1 is regarded as $\mathbb{R} \cup \{\infty\}$. The spaces A and B are not G-homotopy equivalent. To see that we need only show that any G-map $B \to A$ is nullhomotopic as a map. But when $z \neq o$ the subspace $S^1 \times \{z\}$ of B consists of just two orbits, the line $\mathbb{R} \times \{z\}$ and the point $\{\infty\} \times \{z\}$, and so any G-map $B \to A$ factors through the second projection $B \to \mathbb{R}$. Now $A^H = A$ is a deformation retract of $B^H = A \cup (\{\infty\} \times \mathbb{R})$ when H is a non-trivial subgroup of G, while $A^H = A$ is a deformation retract of $B^H = B$ when H is trivial. Thus the inclusion $f^H: A^H \to B^H$ is a homotopy equivalence for all $H \subset G$.

This example might still be thought to be defective because the preceding argument clearly shows also that A is not a G-ANR. But really the argument shows two things:

(a) it is essentially impossible for a space to be a G-ANR when $G = \mathbb{R}$, and

(b) there is no reasonable class of G-spaces which one can use instead of G-ANR's to obtain an analogue of (4.2) in the non-compact case.

For any reasonable class of G-spaces ought surely to contain all smooth manifolds with smooth G-action, and both G-spaces A and B in the example are of this type.

5. Proof of (4.1)

In the proof of (4.1) we require the following technical device. Here it is sufficient for G to be any compact topological group. For any G-space X and Hausdorff G-space Y let (X:Y) denote the orbit space W/G, where W is the subspace of X×Y consisting of pairs (x,y) such that $G_y \subset G_x$. We regard (X:Y) as a space over Y/G, in the obvious way, noting that the fibre over an orbit Gy is the fixed point get X^{G_y}. Any G-map f: Y→X determines a section f': Y/G→(X:Y), where f' (Gy) = G(fy,y), and it is easy to see that the correspondence is bijective. In fact W is a pull-back of the projections.

$$Y \to Y/G \leftarrow W/G = (X:Y) .$$

Using this observation we obtain

Proposition (5.1)

Let B be a G-space and let E be a G-space over B. Then E is a G-AR (G-ANR) over B if and only if (E:Z) is an AR (ANR) over (B:Z) for all paracompact G-spaces.

To prove "if", let (Y,X) be a paracompact G-pair and let θ,φ be G-maps such that the diagram on the left commutes

$$\begin{array}{ccc} X & \xrightarrow{\theta} & E \\ \downarrow & & \downarrow p \\ Y & \xrightarrow[\phi]{} & B \end{array} \qquad\qquad \begin{array}{ccc} X/G & \xrightarrow{\tilde\theta} & (E:Y) \\ \cap & & \downarrow \tilde p \\ Y/G & \xrightarrow[\tilde\phi]{} & (B:Y) \end{array}$$

Then the diagram on the right also commutes, where $\tilde\theta, \tilde\phi, \tilde p$ are given by θ, ϕ, p. If $(E:Y)$ is an AR over $(B:Y)$ then there exists an extension $\tilde\psi$: $Y/G \to (E:Y)$ of $\tilde\theta$ over $\tilde\phi$. The corresponding G-map ψ : $Y \to E$ makes the diagram on the left commute. The proof in the ANR case is similar.

To prove "only if", let (Q,P) be a paracompact pair and let θ', ϕ' be maps such that the diagram shown below is commutative.

$$\begin{array}{ccc} P & \xrightarrow{\theta'} & (E:Z) \\ \cap & & \downarrow \tilde p \\ Q & \xrightarrow[\phi']{} & (B:Z) \end{array}$$

Define $P^* \subset P \times Z$ to be the pull-back of the G-space Z with respect to

$$P \to (E:Z) \to Z/G \quad ,$$

and define $Q^* \subset Q \times Z$ similarly with respect to

$$Q \to (B:Z) \to Z/G$$

We identify $(Q^*/G), P^*/G)$ with (Q,P) in the obvious way. Now θ', ϕ' determine sections $\tilde\theta: P \to (E:P^*)$, $\tilde\phi: Q \to (B:Q^*)$ and hence G-maps θ, ϕ such that the following diagram is commutative.

$$\begin{array}{ccc} P^* & \xrightarrow{\theta} & E \\ {\scriptstyle \eta}\downarrow & & \downarrow{\scriptstyle p} \\ Q^* & \xrightarrow[\phi]{} & B \end{array}$$

When E is a G-AR over B there exists an extension ψ: $Q^* \to E$ of θ over ψ. Combining this with the canonical map $Q^* \to Z$ determines an extension $Q \to (E:Z)$ of θ' over ϕ' in the original diagram. The ANR case is similar.

In view of (5.1) it is clear that (4.1) will be an immediate consequence of

Proposition (5.2).

Under the same hypotheses as (4.1), the space (E:Z) is an AR over the space (B:Z), for all paracompact G-spaces Z.

We shall proceed in four stages of increasing generality, depending on the orbit structure of Z.

As a preliminary to the proof of (5.2), consider a closed subgroup H of the compact Lie group G. Suppose that $Z = (G/H) \times S$, where G acts trivially on S. Then (E:Z) can be identified with $E^H \times S$, and (B:Z) with $B^H \times S$, so that

$$\tilde{f}: (E:Z) \to (B:Z)$$

is identified with

$$f^H \times 1: E^H \times S \to B^H \times S,$$

as a map over S. Since E^H is an AR over B^H by hypothesis, it follows at once that (E:Z) is an AR over (B:Z) in this special case.

In the proof of (4.2) we first suppose (case I) that only one orbit type $[H]$ occurs in Z. Then (see 1.6 of [2], for example) we can cover Z by open sets of the form $(G/H) \times S$ as above. Hence (5.2) in this case follows at once from (3.1).

Next suppose (case II) that only a finite number of orbit types occur in Z, say $[H_1],\ldots,[H_n]$. We partially order in the usual way, so that $i \geq j$ when H_i is conjugate to a subgroup of H_j. Let $Z_j \subset Z$ ($j = 1,\ldots,n$) denote the subset of points where the stabilizer is conjugate to H_i for some $i \leq j$. Then Z_j is closed and

$$\emptyset = Z_o \subset Z_1 \subset \ldots \subset Z_n = Z .$$

Also only one orbit type occurs in each of $Z_1 - Z_o$, $Z_2 - Z_1$, ..., $Z_n - Z_{n-1}$. Hence $(E: Z_j - Z_{j-1})$ is an AR over $(B: Z_j - Z_{j-1})$ $(j=1,\ldots,n)$, by case I. Now (5.2) in case II follows by applying (3.2) successively over the closed pairs

$$(Z_1/G,\ Z_o/G),\ldots,(Z_n/G,\ Z_{n-1}/G).$$

Next suppose (case III) that the orbit structure of Z is locally finite. This means that there exists an open covering of Z/G such that only a finite number of conjugacy classes of isotropy subgroups occur in each member U of the covering. By case II, therefore, $(E:U)$ is an AR over $(B:U)$, for all such U, and so (5.2) in case III follows at once from (3.1).

Finally we turn to the general case (case IV) where no assumption is made about the orbit structure of Z. Write

$$Z = Z_o \cup Z_1 \cup \ldots \cup Z_d ,$$

where $d = \dim G$ and

$$Z_i = \{z \in Z \mid \dim G_z = i\}.$$

Then $Z_1 \cup \ldots \cup Z_d$ is closed, for $i = 1,\ldots,d$. Also the orbit structure in each Z_i is locally finite. For, as is well-known, there exists a neighbourhood of every point $z \in Z$ such that the stabilizers of points in that neighbourhood are all conjugate to subgroups of G_z, while no compact Lie group contains more than a finite number of closed subgroups with the same dimension as itself. Hence $(E:Z_i)$ is an AR over $(B:Z_i)$, by case III. Hence and from (2.5) it follows, using descending induction, that $(E:Z_i \cup \ldots \cup Z_d)$ is an AR over $(B:Z_i \cup \ldots \cup Z_d)$ for $i = d,\ldots,0)$. This proves (5.2) in the general case.

Footnotes

1. The notion of euclidean neighbourhood retract (ENR) over B has been considered by Dold [5]. It is not hard to show that an ENR over B is an ANR over B in our sense.

2. Note that we make no metrizability assumption.

3. The usual form of words but with G-spaces etc., in place of spaces etc.

4. That is, the subspace of pairs (e,λ), where $e \in E_b$ and $\lambda: I \to F_b$ $(b \in B)$ are related by $h(e) = \lambda(o)$.

5. If we were prepared to assume W paracompact we could, of course,complete the proof straightaway at this stage.

References

1. G. Bredon, Equivariant cohomology theories, Springer Lecture Notes in Mathematics, no.34 (1967).

2. G. Bredon, Introduction to compact transformation groups, Academic Press (1972).

3. T. tom Dieck, K.H. Kamps and D. Puppe, Homotopietheorie, Springer Lecture Notes in Mathematics, no.157 (1970).

4. A. Dold, Partitions of unity in the theory of fibrations, Ann. of Math. 78 (1963), 223-255.

5. A.Dold, The fixed-point index of fibre-preserving maps, Inv. Math. 25 (1974), 281-297.

6. S. Illman, Equivariant algebraic topology, Ph.D. thesis, Princeton 1972.

7. I.M. James and G.B. Segal, On equivariant homotopy type, Topology 17 (1978), 267-272.

8. T. Matumoto, On G-CW complexes and a theorem of J.H.C. Whitehead, J. Fac. Sci. Univ. Tokyo 18 (1971), 363-374.

SOME FORMULAE AND CONJECTURES ASSOCIATED WITH CIRCLE ACTIONS

Czes Kosniowski

University of Newcastle upon Tyne.

Let n be a positive integer and let $P_o = \{p_1, p_2, \ldots, p_n\}$ be any set consisting of n distinct non-zero integers. Form n more sets $P_1, P_2, \ldots, P_n$ from P_o as follows:

$$P_i = \{p_1 - p_i, \ldots, p_{i-1} - p_i, -p_i, p_{i+1} - p_i, \ldots, p_n - p_i\}$$

for $i = 1, 2, \ldots, n$. It is quite well known that

$$(1) \qquad \sum_{i=o}^{n} \frac{s_\omega(P_i)}{e(P_i)} = 0 \text{ if } |\omega| < n$$

where ω is an unordered n-tuple of non-negative integers, $\omega = (\omega(1), \omega(2), \ldots, \omega(n))$ and $|\omega| = \omega(1) + \omega(2) + \ldots + \omega(n)$, if $\{y_1, y_2, \ldots, y_n\}$ is a set of variables then $s_\omega(\{y_1, y_2, \ldots, y_n\})$ is the smallest symmetric polynomial that contains $y_1^{\omega(1)} y_2^{\omega(2)} \ldots y_n^{\omega(n)}$ and $e(\{y_1, y_2, \ldots, y_n\}) = y_1 y_2 \ldots y_n$, the meaning of $s_\omega(P_i)$ and $e(P_i)$ should then be clear. Furthermore if $|\omega| = n$ then the expression on the left hand side of (1) is an integer which, for n fixed, is independent of the original set P_o (it is a multinomial coefficient depending only on n and ω).

We shall remind the reader of how formula (1) arises shortly. It turns out that there are other formulae relating the sets $P_o, P_1, \ldots, P_n$. Let m be a positive integer. If $Q = \{q_1, q_2, \ldots, q_n\}$ is a set containing n distinct integers then reducing modulo m divides Q into m subsets. We say that the m-division of Q, abbreviated to m-div(Q), is $D = (d_o, d_1, \ldots, d_{m-1})$ if the number of elements in Q which are congruent to k modulo m is precisely d_k, $k = 0, 1, \ldots, m-1$. Returning to our sets $P_o, P_1, \ldots, P_n$, we have the following formulae: let m be a positive integer and let $D = (d_o, d_1, \ldots, d_{m-1})$ be a sequence of non-negative integers whose sum is n. Then

$$(2) \qquad \sum_{\{i;\, m\text{-div}(P_i) = D\}} \frac{s_{\omega_o}(\{p \in P_i ; p \equiv 0(m)\})}{e(\{p \in P_i ; p \equiv 0(m)\})} \prod_{k=1}^{m-1} s_{\omega_k}(\{p \in P_i ; p \equiv k(m)\}) = 0$$

for all $\omega_o, \omega_1, \ldots, \omega_{m-1}$ with $\sum_{i=o}^{m-1} |\omega_i| < d_o$, where each ω_i is an unordered d_i-tuple of non-negative integers. Furthermore, if $\Sigma|\omega_i| = d_o$ then the left hand side of (2) is an integer (which is not always independent of P_o).

Notice that if $m = 1$ and $D = (n)$ then formula (2) becomes formula (1).

The above formulae arise as a simple case of some results about S^1 actions on unitary manifolds. Let M^{2n} be a 2n dimensional unitary S^1 manifold. This means that M has a unitary structure (weakly almost complex) and the group S^1 acts on M preserving this structure. The fixed point set Fix(M) under the S^1 action consists of a number of unitary submanifolds of even codimensions. If F is a component of $Fix(M)$ then the normal bundle $\nu(F,M)$ of F in M is a complex vector bundle which decomposes into a sum of complex subbundles

$$\nu(F,M) = \bigoplus_{i \in \mathbb{Z}-\{0\}} N_i$$

according to how S^1 acts on $\nu(F,M)$. In each fibre of N_i an element $t \in S^1$ acts by multiplication by t^i - locally this is obvious - globally it is relatively easy result in equivariant K-theory, see for example Proposition 1.6.2 of M.F. Atiyah's book [1]. The component F is given an orientation so that together with the natural orientation of $\nu(F,M)$ determined by the complex structure on it we get the orientation on $TM|F$ as given by the unitary structure. The bundles N_i have characteristic classes

$$Y_i = \{y_{i,1}, y_{i,2}, \dots, y_{i,k}\}, \quad k = \dim_{\mathbb{C}} N_i$$

where $\sigma_j(Y_i) = c_j(N_i)$, i.e. the j-th elementary symmetric function of Y_i gives the j-th Chern class of N_i. The tangent bundle of the submanifold F has classes

$$Y_0 = \{y_{0,1}, y_{0,2}, \dots, y_{0,\ell}\}, \quad 2\ell = \dim F$$

where $\sigma_j(X) = c_j(TF)$.

Note that although we have written Y_i $(i \in \mathbb{Z})$ as a set, repeated elements of Y_i may not be discarded. In particular Y_i $(i \in \mathbb{Z} - \{0\})$ consists of precisely $\dim_{\mathbb{C}} N_i$ elements while Y_0 consists of precisely $\frac{1}{2}\dim F$ elements, not necessarily distinct. Of course the Y_i depend upon the component F, but with this notation we have the following well known formula

$$\sum_{F \in Fix(M)} \frac{s_\omega(\bigcup_i 1+Y_i)}{e(\bigcup_{i \neq 0} 1+Y_i)}[F] = 0 \text{ if } |\omega| < n$$

where $1+Y_i = \{1+y, \ y \in Y_i\}$ and the left hand side of formula (3) is an evaluation of a cohomology class on the fundamental homology class of F. Furthermore, if $|\omega| = n$ then the left hand side of (3) is equal to the integer $s_\omega[M]$.

Formula (3) has been around since the late 60's following the Atiyah-Singer Index Theorem (see for example [5]), although it follows essentially from the Atiyah-Segal localization Theorem [3,4]. Formula (3) has been the source of much fruitful investigation. There are many ways of proving (3) and many papers contain the formula implicitly, also generalizations have been made to $\mathbb{Z}/p$ actions and to $\mathbb{Z}/2$ or $(\mathbb{Z}/2)^k$ actions on unoriented manifolds. It is beyond the object of this paper to go into details but some of the papers that deserve mention are [6, 7, 9, 10, 12, 13, 14, 16, 17, 18, 19].

Formula (1) is obtained from formula (3) simply by looking at the S^1 action on complex projective n space $\mathbb{C}P^n$ given by

$$[z_0, z_1, \dots ; z_n] \to [z_0 ; t^{p_1} z_1, \dots ; t^{p_n} z_n]$$

for $t \in S^1$, where $[z_0, z_1, \dots, z_n]$ denotes, as usual, homogeneous coordinates.

To state the generalization of (3) that leads to (2) recall that if F is a component of the fixed point set then

$$\nu(F,M) = \bigoplus_{i \in \mathbb{Z}-\{0\}} N_i$$

where the N_i are complex vector bundles over F. Define the <u>m-division</u> of F, abbreviated to m-div(F), to be $(d_0, d_1, \ldots, d_{m-1})$ where

$$2d_0 = \dim F + 2 \dim_{\mathbb{C}} \Big(\bigoplus_{i \equiv 0(m)} N_i \Big)$$

$$d_k = \dim_{\mathbb{C}} \Big(\bigoplus_{i \equiv k(m)} N_i \Big) \qquad \text{for } k = 1, 2, \ldots, m-1.$$

<u>Theorem 1</u>. Let m be a positive integer and let $D = (d_0, d_1, \ldots, d_{m-1})$ be a sequence of non-negative integers whose sum is n. Then, using the notation above for fixed point set Fix(M) of the S^1 manifold M^{2n}.

$$\sum_{\substack{F \\ m\text{-div}(F)=D}} \frac{s_{\omega_0}\big(\bigcup_{i\equiv 0(m)} 1+Y_i\big)}{e\big(\bigcup_{i\equiv 0(m),\, i\neq 0} 1+Y_i\big)} \prod_{k-1}^{m-1} s_{\omega_k}\big(\bigcup_{i\equiv k(m)} 1+Y_i\big)[F] = 0$$

for all $\omega_0, \omega_1, \ldots, \omega_{m-1}$ with $\sum_{i=0}^{m-1} |\omega_i| < d_0$, where each ω_i is an unordered d_i-tuple of non-negative integers. Furthermore, if $\Sigma|\omega| = d_0$ then the left hand side is an integer.

To prove Theorem 1 we need another result which generalises formula (3) to the case of S^1 actions on vector bundles. Let $E_1, E_2, \ldots, E_{m-1}$ be complex vector bundles over the unitary S^1 manifold M. Suppose that S^1 acts on each E_i covering the action on M. Now if $F \in \text{Fix}(M)$ then

$$\nu(F,M) = \bigoplus_{i \in \mathbb{Z}-\{0\}} N_i$$

and N_i has characteristic classes Y_i while F has characteristic classes Y_0. Furthermore

$$E_k|F = \bigoplus_{i \in \mathbb{Z}-\{0\}} E_{k,i}$$

and $E_{k,i}$ has characteristic classes $Y_{k,i}$. With this notation we have.

<u>Theorem 2</u>. Suppose that $\dim M = 2d_0$ and $\dim_{\mathbb{C}} E_k = d_k$, $k = 1, 2, \ldots, m-1$, then

$$\sum_{F \in \text{Fix}(M)} \frac{s_{\omega_0}\big(\bigcup_i 1+Y_i\big)}{e\big(\bigcup_{i\neq 0} 1+Y_i\big)} \prod_{k=1}^{m-1} s_{\omega_k}\big(\bigcup_{i\neq 0} 1+Y_{k,i}\big)[F] = 0$$

for all $\omega_0, \omega_1, \ldots, \omega_{m-1}$ with $\sum_{i=0}^{m-1} |\omega_i| < n$ where each ω_i is an unordered d_i-tuple of non-negative integers.

<u>Proof</u>. Take your favourite proof of formula (3) and modify it to take care of the vector bundles present - no problems should arise.

<u>Proof of Theorem 1</u>. Let m be a positive integer, then $M^{\mathbb{Z}/m}$, the fixed point set of M under the $\mathbb{Z}/m \subset S^1$ action in an S^1 manifold. Because $\mathbb{Z}/m$ acts trivially on each component L of $M^{\mathbb{Z}/m}$ the bundle $\nu(L,M)$ splits

$$\nu(L,M) = \bigoplus_{i=1}^{m-1} E_i$$

where $g \in \mathbb{Z}/m$ acts on E_i by multiplication by g^i in each fibre of E_i. Each bundle E_i

is a vector bundle over L with an S^1 action covering the S^1 action on L. Note that if $F \in Fix(M)$ then F is a submanifold of $M^{\mathbb{Z}/m}$ and if $F \subseteq L$ then in the decomposition

$$E_k|F = \bigoplus_{i \in \mathbb{Z}-\{0\}} E_{k,i}$$

$E_{k,i} = \emptyset$ unless $i \equiv k(m)$.

For a fixed $D = (d_o, d_1, \ldots, d_{m-1})$, look at the union of the components L of $M^{\mathbb{Z}/m}$ with

$$\dim L = 2d_o$$

$$\dim_{\mathbb{C}} E_k = d_k, \quad k = 1, 2, \ldots, m-1,$$

applying Theorem 2 immediately gives Theorem 1.

There is a converse to Theorem 1 which tells us that a collection of vector bundles with S^1 action is the fixed point data of some S^1 manifold if the collection satisfies the equations of Theorem 1. Let

$$s_\Omega^{m,D}(Fix(M^{2n}), \circ\, Fix(M^{2n}))$$

denote the expression on the left hand side of the equation in Theorem 1. Here m is a positive integer, $D = (d_o, d_1, \ldots, d_{m-1})$ is an m-tuple of non-negative integers with $|D| = d_o + d_1 + \ldots + d_{m-1} = n$ and $\Omega = (\omega_o, \omega_1, \ldots, \omega_{m-1})$ where each ω_i is an unordered d_i-tuple of non-negative integers. Let $|\Omega| = \sum_{i=o}^{m-1} |\omega_i|$.

<u>Theorem 3</u>. Let $\{(F, E_F); F \in \mathcal{F}\}$ be a collection of complex vector bundles over unitary manifolds $F \in \mathcal{F}$ in which $\dim F + 2\dim_{\mathbb{C}} E_F = 2n$. Suppose each bundle E_F, $F \in \mathcal{F}$, comes equipped with an S^1 action in which $Fix(E_F) = F$. Then, the collection $\{(F, E_F); F \in \mathcal{F}\}$ with the specified S^1 action is the fixed point data of some (closed) unitary S^1 manifold if and only if

$$s_\Omega^{m,d}\Big(\bigcup_{F \in \mathcal{F}} (F, E_F)\Big) = 0$$

for all positive integers m, all m-tuples of non-negative integers $D = (d_o, d_1, \ldots, d_{m-1})$ with $|D| = n$ and all $\Omega = (\omega_o, \omega_1, \ldots, \omega_{m-1})$ with $|\omega| < d_o$ where each ω_i is an unordered d_i-tuple of non-negative integers.

<u>Proof</u>. Consider $N = \bigcup_{F \in \mathcal{F}} S(E_F)$ where $S(E_F)$ denotes the sphere bundle of the vector bundle E_F. The S^1 action on E_F gives us an S^1 action on N which is fixed point free (i.e. $Fix(N) = \emptyset$).

Suppose we could prove that N is the boundary of an S^1 manifold W in which S^1 acts on W with no fixed points. Let $D(E_F)$ denote the disc bundle of E_F. By glueing equivariantly $\bigcup D(E_F)$ and -W along their common boundary we get a closed unitary manifold with S^1 action having the fixed point data in question. It remains therefore to prove that N is indeed the boundary of a fixed point free S^1 manifold. This follows from the next result, Theorem 4.

Let N^{2n-1} be a fixed point free unitary S^1 manifold. It is a fact that N is the boundary of an S^1 manifold W^{2n} (with fixed points) - this was first proved by E. Ossa

[20], see also [11] and the preamble following Theorem 4. Since the fixed point set of W is away from the boundary it makes sense to define

$$s_{\Omega}^{m,d}(\mathrm{Fix}(W), \nu\,\mathrm{Fix}(W)),$$

by Theorem 1 this depends only upon N and not upon the choice of W.

<u>Theorem 4.</u> Let N^{2n-1} be a fixed point free S^1 manifold and let $N=\partial W$. Then N is the boundary of a fixed point free S^1 manifold if and only if

$$s_{\Omega}^{m,D}(\mathrm{Fix}(W), \nu\,\mathrm{Fix}(W)) = 0$$

for all positive integers m, all m-tuples of non-negative integers $D = (d_o, d_1, \ldots, d_{m-1})$ with $|D| = n$ and all $\Omega = (\omega_o, \omega_1, \ldots, \omega_{m-1})$ with $|\Omega| < d_o$, where each ω_i is an unordered d_i-tuple of non-negative integers.

The proof of Theorem 4 requires some preamble and two lemmas. Let $\mathfrak{A}_*^G[F_\infty]$ denote the bordism ring arising from bordism of fixed point free $G = S^1$ unitary manifolds. From [20] we see that

$$\mathfrak{A}_{2n-1}^G[F_\infty] \cong \bigoplus_{m\geq 1} \bigoplus_{\substack{D \\ |D|=n}} \mathfrak{A}_{2d_o-2}(BS^1 \times BU(d_1)\times\ldots\times BU(d_{m-1}))$$

where $D = (d_o, d_1, \ldots, d_{m-1})$ runs through all m-tuples of non-negative integers (with sum n). We abbreviate

$$\mathfrak{A}_{2d_o-2}(BS^1 \times BU(d_1)\times\ldots\times BU(d_{m-1}))$$

by $B_{2n}(m, D)$. An element of $B_{2n}(m, D)$ is a complex vector bundle E over some unitary manifold F of dimension $2d_o-2$. This bundle splits as

$$E = E_o \oplus E_1 \oplus \ldots \oplus E_{m-1}$$

where E_o is a complex line bundle and $\dim_{\mathbb{C}} E_i = d_i$ for $i = 1, 2, \ldots, m-1$. Note that dim $F + 2\dim_{\mathbb{C}} E = 2n$. We give E an S^1 action by letting $t \in S^1$ act by multiplication by t^m in the fibre of E_o and by multiplication by t^i in the fibres of E_i, $i = 1, 2, \ldots, m-1$. Note that S^1 then acts trivially on F and that $\mathrm{Fix}(E) = F$. The isomorphism

$$\bigoplus_{m\geq 1} \bigoplus_{\substack{D \\ |D|=n}} B_{2n}(m, D) \to \mathfrak{A}_{2n-1}^G(F_\infty)$$

is given by $E \to S(E)$ the sphere bundle of E. Note then that D(E), the disc bundle of E, is a unitary S^1 manifold with boundary S(E).

<u>Lemma.</u> If $(F, E) \in B_{2n}(k, D')$ and either $m > k$ or $m = k$ and $D \neq D'$ then

$$s_{\Omega}^{m,D}(F, D(E)) = 0$$

for all Ω with $|\Omega| < d_o$.

<u>Proof.</u> The latter case follows by definition since the m-division of F is D'. On the other hand if $m > k$ then the m-division of F is

$$D'' = (d_o'-1, d_1', \ldots, d_k', 1, 0, 0, \ldots, 0)$$

where $(d'_o, d'_1, \ldots, d'_k) = D'$. If $D \neq D''$ then we have nothing to show. However, if $D = D''$ then

$$s_\Omega^{m,D}(F, D(E))$$

is of the form

$$s_{\omega_o}(Y_o) \prod_{i=1}^{m-1} s_{\omega_i}(1+Y_i)[F]$$

since $2(d'_o-1) = \dim F$. But $|\Omega| = d_o = d'_o-1$ and so the cohomology class evaluates to zero.

<u>Lemma</u>. If $(F, E) \in B_{2n}(m, D)$ and if

$$s_\Omega^{m,D}(F, D(E)) = 0$$

for all Ω with $|\Omega| < d_o$ then (F, E) is the zero element in $B_{2n}(m, D)$.

<u>Proof</u>. Recall that $\dim F = 2(d_o-1)$ so that

$$s_\Omega^{m,D}(F, D(E))$$

takes the form

$$\frac{s_{\omega_o}(Y_o \cup \{m+y\})}{m+y} \prod_{i=1}^{m-1} s_{\omega_i}(1+Y_i)[F].$$

which we may rewrite as

$$\frac{s_{\omega_o}(Y_o)}{m+y} \Pi s_{\omega_i}(1+Y_i)[F] + \sum_{j>0} s_{\omega_o-j}(Y_o)(m+y)^{j-1} \Pi s_{\omega_i}(1+Y_i)[F].$$

But the degrees of the latter terms are less than $\dim F$ and hence evaluate to zero so that the expression above becomes

(*)
$$\frac{s_{\omega_o}(Y_o)}{m+y} \Pi s_{\omega_i}(1+Y_i)[F].$$

If $\Omega = (\omega_o, 0, 0, \ldots, 0)$ then the term (*) which does not vanish for dimensional reasons is

$$\frac{s_{\omega_o}(Y_o)}{(m)^{\ell+1}} y^\ell [F]$$

where $\ell = d_o - |\Omega| - 1 = d_o - |\omega_o| - 1$. It follows, from the assumption in the lemma, that

$$s_{\omega_o}(Y_o) y^\ell [F] = 0 \qquad (\ell = d_o - |\Omega| - 1)$$

for all ω_o with $2|\omega_o| \leq \dim F$. Assume inductively for $p \geq 1$ that

$$s_{\omega_o}(Y_o) y^\ell \; \Pi s_{\omega_i}(Y_i)[F] = 0 \qquad (\ell = d_o - |\Omega| - 1)$$

for all Ω with $|\Omega| < d_o$ and $|\Omega| - |\omega| < p$. Let $|\Omega| - |\omega_o| = p$, the terms in (*) which do not vanish for purely dimensional reasons are

$$\frac{s_{\omega_o}(Y_o)y^{\ell}}{(m)^{\ell+1}} \Pi s_{\omega_1}(Y_1)[F] \qquad (\ell = d_o - |\Omega| - 1)$$

and terms involving expressions of the form

$$s_{\omega_o}(Y_o)y^{\ell'} \Pi s_{\omega_1'}(Y_1)[F]$$

where $|\omega_o| + \Sigma|\omega_1'| < |\Omega|$. By induction these latter terms vanish and so we conclude that

$$s_{\omega_o}(Y_o)y^{\ell} \Pi s_{\omega_1}(Y_1)[F] = 0 \qquad (\ell = d_o - |\Omega| - 1)$$

for all Ω with $2|\Omega| \leq \dim F$. In other words all characteristic numbers of (F, E) vanish and hence (F, E) is the zero element of $B_{2n}(m, D)$.

Proof of Theorem 4. Given N there is an integer m_o so that for all $m > m_o$ the subgroup $\mathbb{Z}/m$ of S^1 acts without fixed points. This means that N comes from

$$\bigoplus_{m_o \geq m \geq 1} \bigoplus_{\substack{D \\ |D|=n}} B_{2n}(m, D)$$

Using the two lemmas above inductively starting with $m = m_o$ and decreasing, the result follows easily.

The formulae of Theorem 1 lead to some interesting results about S^1 manifolds with isolated fixed points. The most tractable cases are when the number of fixed points is small or when the dimension of the manifold is small. We mention some of these results, details of proofs being left for another occasion.

If P is an isolated fixed point of an S^1 manifold M^{2n} then the normal bundle to P is of the form

$$\nu(P, M) = \varepsilon(P)\, V_{m(1)} V_{m(2)} \cdots V_{m(n)}$$

where $\varepsilon(P) = \pm 1$ is the orientation of P determined by the convention described earlier on, $m(1), m(2), \ldots, m(n) \in \mathbb{Z}$ and $V_{m(j)}$, $j = 1, 2, \ldots, n$ denotes the complex numbers with $t \in S^1$ acting by multiplication by $\exp(2\pi i t m(j))$. Thus $\nu(P, M)$ is an element of

$$\mathbb{Z}[\{V_j,\ j \in \mathbb{Z} - \{0\}\}].$$

Let $\rho(M)$ be the sum of the $\nu(P, M)$ as P runs through the fixed points of M. If M is an S^1 boundary then (clearly) $\rho(M) = 0$. Indeed it is evident that two unitary S^1 manifolds M, M' with isolated fixed points are S^1 bordant if and only if $\rho(M) = \rho(M')$.

Theorem 5. If M is a unitary S^1 manifold with 2 fixed points then $\rho(M)$ is either 0, $V_k + V_{-k}$ or $V_k V_\ell V_{-k-\ell} + V_{-k} V_{-\ell} V_{k+\ell}$ for some $k, \ell \in \mathbb{Z} - \{0\}$.

Corollary. If M is a unitary S^1 manifold with 2 fixed points then M is either a boundary or dim M is 2 or 6.

In fact if $\dim M = 2$ and M is not a boundary then M is S^1 bordant to $\mathbb{C}P^1$ with S^1 action given in homogeneous coordinates by

$$t[z_o, z_1] = [z_o, t^k z_1] \qquad t \in S^1$$

for some $k \in \mathbb{Z}-\{0\}$. If dim M=6 and M is not a boundary then M is S^1 bordant to S^6 with a suitable S^1 action. Here S^6 has the standard almost complex structure (so that as such it is not a boundary) determined for example by writing it as the quotient G_2/SU_3 of the Lie group G_2 by SU_3. A suitable S^1 action is then not difficult to describe

Theorem 5 and its Corollary generalise results of the author in [15].

There are formulae, analogous to those in Theorem 1, for S^1 actions on oriented manifolds. In this case if P is an isolated fixed point then $\nu(P,M)$ is of the form

$$\nu(P,M) = \varepsilon(P)\, V_{m(1)} V_{m(2)} \cdots V_{m(n)}$$

with $m(1), m(2), \ldots, m(n) \in \mathbb{N}$ the set of positive integers, and we can prove the following result.

<u>Theorem 6.</u> If M is an oriented S^1 manifold with two fixed points P, Q then $\nu(P,M) = -\nu(Q,M)$.

This result generalises, to a certain extent, previous results of this nature [2] where the additional assumption that M is a homology sphere is required. Combining Theorem 6 with results of [20] we deduce

<u>Corollary</u> If M is an oriented S^1 manifold with two fixed points then $2^m M$ is an S^1 boundary for some integer m.

Returning to unitary S^1 manifolds, if dim M = 4 and M has isolated fixed points then the polynomials $\rho(M)$ that can arise are quite easy to find by using the formulae of Theorem 1. Indeed, we can find a geometric basis for the subgroup of $\mathfrak{U}_4^G$ $(G=S^1)$ generated by S^1 manifolds with isolated fixed points. We shall not write down either of these results here but it is worth mentioning that this subgroup is strictly larger than the subgroup generated by complex surfaces having a holomorphic S^1 action with isolated fixed points (a basis of this subgroup is described in [8]).

We end by giving two conjectures. The first relates the number of fixed points to the dimension of the manifold and roughly suggests that the number of fixed points is large if the dimension is large and if the manifold is not a boundary.

<u>Conjecture A.</u> Suppose that M is a unitary S^1 manifold with isolated fixed points. If M is not a boundary then the number of fixed points is greater than f(dim M) where f is some (linear) function.

The most likely function is $f(x) = x/4$, which certainly works in low dimensions. One could also formulate a conjecture in the case of non-isolated fixed points. This would involve the euler characteristic of M and the codimension of the fixed point set. (Note that if M has isolated fixed points then the euler characteristic is equal to the number of fixed points.) If we remove the condition that M is not a boundary then the conjecture is clearly false.

The second conjecture is concerned with the roots of Lie groups. Consider the root system of a complex Lie algebra and let W be the Weyl group (which of course acts on the set of roots). Let $S = \{x_1, x_2, \ldots, x_n\}$ be a subset of the set of roots such that if $x \in S$ then $-x \notin S$. Let W_S be a subgroup of W that fixes S, i.e. $W_S(S) = S$.

Finally note that if for each fundamental root y we choose a positive integer I(y) then we get, in an obvious way, a non-zero integer I(x) for each root x.

Conjecture B. Given S and W_S as above and a positive integer I(y) for each fundamental root y then there exists a unitary S^1 manifold M^{2n} with isolated fixed points so that

$$\rho(M) = \sum_{h \in W/W_S} \prod_{x \in S} V_{Ih(x)}.$$

There is some very strong evidence for this conjecture based upon some unpublished work of the author. As for examples the reader can easily furnish these by looking, for example, at the root systems of the Lie algebras of type A_n and G_2.

Much of the above research was done while the author was visiting the I.H.E.S. and the author is indebted to the members of the I.H.E.S. for their generous hospitality. The author would also like to thank U. Koschorke and W. Neumann for the well organised conference at Siegen.

References

1. M.F. Atiyah, K-theory, Benjamin 1967.
2. M.F. Atiyah and R. Bott, "A Lefshetz fixed point formula for elliptic complexes II. Applications.", Ann. of Math. 88 (1968) 451-491.
3. M.F. Atiyah and G.B. Segal, "Equivariant K-theory." University of Warwick Notes. 1965.
4. M.F. Atiyah and G.B. Segal, "The index of elliptic operators II." Ann. of Math. 87 (1968) 531-545.
5. M.F. Atiyah and I.M. Singer, "The index of elliptic operators III." Ann. of Math. 87 (1968) 546-604.
6. R. Bott, "Vector fields and characteristic numbers." Michigan Math. J. 14 (1967) 231-244.
7. R. Bott, "A residue formula for holomorphic vectorfields." J. Diff. Geometry 1 (1967) 311-330.
8. J. Carrell, A. Howard and C. Kosniowski, "Holomorphic vector fields on complex surfaces." Math. Ann. 204 (1973) 73-81.
9. T. tom-Dieck, "Characteristic numbers on G manifolds I." Inv. Math. 13 (1971) 213-224.
10. T. tom-Dieck, "Periodische Abbildungen unitarer Mannigfaltigkeiten." Math. Z. 126 (1972) 275-295.
11. G. Hamrick and E. Ossa, "Unitary bordism of monogenic groups and isometries." Springer Lecture Notes in Math. 298 (1972) 172-182.
12. A. Hattori, "Equivariant characteristic numbers and integrality theorem for unitary T^n-manifolds." Tôhuku Math. J. 26 (1974) 461-482.
13. L. Illusie, "Nombres de Chern et groupes finis." Topology 7 (1968) 255-269.
14. K. Kawakubo, "Global and local equivariant characteristic numbers of G-manifolds." To appear.
15. C. Kosniowski, "Holomorphic vector fields with simple isolated zeros." Math. Ann. 208 (1974) 171-173.
16. C. Kosniowski, "Characteristic numbers of $\mathbb{Z}/p$ manifolds." J. Lond. Math. Soc. 14 (1976) 283-295.
17. C. Kosniowski, "$\mathbb{Z}/p$ manifolds with low dimensional fixed point set." Transformation Groups. Cambridge University Press. L.M.S. Lecture Note Series 26 (1977) 92-120.
18. C. Kosniowski and R.E. Stong, "Innvolutions and characteristic numbers." Topology 17 (1978) 309-330.
19. C. Kosniowski and R.E. Stong, "$(\mathbb{Z}/2)^k$ actions and characteristic numbers." To appear in Indiana Math. J. Sept.-Oct. 1979.
20. E. Ossa, "Fixpunktfreie S^1-Aktionen." Math. Ann. 186 (1970) 45-52.

Equivariant K-theory and homotopy rigidity

Arunas Liulevicius*)

The aim of this paper is to present a proof of a theorem of Snaith on equivariant K-theory of homogeneous spaces with linear actions and to show how it leads to a new method for proving homotopy rigidity of these actions. We illustrate by proving that linear actions of arbitrary compact groups on complex Grassman manifolds are homotopy rigid.

The paper is organized as follows: §1 gives a statement of the results, §2 proves Snaith's theorem, §3 examines the Steinberg basis, §4 proves the homotopy rigidity of linear actions on complex Grassman manifolds.

---

*) Research partially supported by NSF grant MCS 77-01623.

1. Statement of results

Let U be a compact connected Lie group with $\pi_1(U)$ free abelian, let $\imath: H \hookrightarrow U$ be a closed connected subgroup of U of maximal rank with $T \subset H$ a maximal torus. We let $W = N_U(T)/T$ be the Weyl group of U and $W' = N_H(T)/T$ be the Weyl group of H.

Multiplication $m: U \times U \to U$ induces a left action $\mu: U \times U/H \to U/H$. If $\gamma: G \to U$ is a continuous homomorphism, then we denote by $\gamma^* U/H$ the G-space structure on U/H given by $\mu(\gamma \times 1)$. Snaith's theorem will give us complete information about $K_G(\gamma^* U/H)$ and will present it in a functorial fashion.

Pittie [13] proved that under our hypotheses above the complex representation ring R(H) is a free R(U)-module under the restriction homomorphism $i^*: R(H) \to R(U)$. Steinberg [16] gave an algorithm to construct a basis $\{f_1, \dots, f_n\}$ of R(H) as a R(U)-module, where $n = [W: W']$. We have the structure equations

$$f_i f_j = \sum_k u_{ij}^k f_k ,$$

$$\psi^i f_j = \sum_k v_j^{ik} f_k ,$$

where u_{ij}^k, v_j^{ik} are elements of R(U).

If $\gamma: G \to U$ is a continuous homomorphism, then the induced homomorphism $\gamma^*: R(U) \to R(G)$ makes $R(G)$ into a $R(U)$-module. There is a homomorphism

$$A(\gamma): R(G) \otimes_{R(U)} R(H) \to K_G(\gamma^* U/H)$$

defined by a bilinear function

$$\alpha: \mathrm{Rep}(G) \times \mathrm{Rep}(H) \to \mathrm{Vect}_G(\gamma^* U/H)$$

which is given by $\alpha(V, W) = V \times (U \times_H W)$ with diagonal G-action. The homomorphism for $G = E$ the trivial group was examined by Atiyah and Hirzebruch [1].

Theorem 1 (Snaith): The homomorphism

$$A: R(G) \otimes_{R(U)} R(H) \to K_G(\gamma^* U/H)$$

is an isomorphism of $R(G)$-modules.

The next question to ask is: what is the $R(G)$-algebra structure of the left-hand side corresponding to the tensor product of vector bundles in $K_G(\gamma^* U/H)$? What about the Adams operations ψ^1? The answer turns out to be easy:

Corollary 2. The product and the action of ψ^1 on $R(G) \otimes_{R(U)} R(H)$ is given in terms of the structure equations by

$$1 \otimes f_i \cup 1 \otimes f_j = \sum_k \gamma^*(u_{ij}^k) \otimes f_k ,$$

$$\psi^1(1 \otimes f_j) = \sum_k \gamma^*(v_j^{1k}) \otimes f_k .$$

If $\alpha, \beta : G \to U$ are continuous homomorphisms such that $\beta = \chi\alpha$ where $\chi : G \to \text{Center } U$ is a homomorphism (a linear character of G), then $\alpha^* U/H = \beta^* U/H$, since Center $U \subset T \subset H$. Now let U be a unitary group, $c: U \to U$ conjugation and $H \subset U$ a subgroup stable under c, so c induces a homeomorphism $\underline{c}: U/H \to U/H$. If $\beta: G \to U$ is a representation of G, then $\underline{c}: \beta^* U/H \to \bar{\beta}^* U/H$ is a G-homotopy equivalence.

Definition. We say that linear actions of G on U/H are homotopy rigid if given two representations $\alpha, \beta: G \to U$ there exists a G-homotopy equivalence $f: \alpha^* U/H \to \beta^* U/H$ if and only if there is a linear character χ such that $\beta = \chi\alpha$ or $\bar{\beta} = \chi\alpha$ as elements of the representation ring R(G).

Theorem 3. Linear actions of a compact group G on the complex Grassman manifold $G_m(\mathbb{C}^{m+n}) = SU(m+n)/SU(m,n)$ are homotopy rigid.

Back [2] has shown homotopy rigidity of linear actions of a compact connected Lie group G on complex Grassmannians. His technique is to exploit the Borel rational equivariant cohomology studied by Wu-Yi Hsiang [8] and to use the results of O'Neill [12], Glover and Homer [5], and Brewster [3] on automorphisms of $H^*(G_m(\mathbb{C}^{m+n}); Q)$. Our technique is to use equivariant K-theory, Snaith's theorem and Brewster's thesis [3].

Conjecture 4. If H is a closed connected subgroup of U of maximal rank then linear actions of any compact group G on U/H are homotopy rigid.

A special case of this conjecture has been proved by Ewing and Liulevicius [4] for $U = U(N)$ and H a subgroup fixing a line in C^N under the birth certificate action of U. Theorem 3 is of course supporting evidence for the conjecture, and a key ingredient in one method of attacking it.

We wish to thank V. Snaith for all sorts of useful information about K-theory. Thanks also go to A. Back, H. Dovermann, H. Glover and T. Petrie for numerous conversations about homotopy rigidity.

2. Proof of Snaith's theorem

Our main tool will be the Kunneth spectral sequence in equivariant K-theory which was first studied by Hodgkin [6], [7], then by Snaith [14] and McLeod [10], [11]. If X and Y are locally contractible U-spaces, then there is a spectral sequence with

$$E_2^{*,*} = \mathrm{Tor}^{*,*}_{R(U)}(K_U^*(X), K_U^*(Y))$$

such that if $\pi_1(U)$ is free abelian then the spectral sequence converges to $K_U(X \times Y)$.

Let $\gamma: G \to U$ be a continuous homomorphism, X a G-space, Y a U-space. We will use the construction $U \times_G X$ (topological induction) which has the property $\mathrm{Map}_U(U \times_G X, Y) = \mathrm{Map}_G(X, \gamma^* Y)$. In particular, $K_U(U \times_G X) = K_G(X)$. For example, if $*$ is a point then $U \times_G * = U/\gamma$, the orbit space of the right action on U via γ, and $K_U(U/\gamma) = K_U(U \times_G *) = K_G(*) = R(G)$.

<u>Lemma 4.</u> Let $\gamma: G \to U$ be a homomorphism, $\pi: U \to U/\gamma$ the orbit map. If Y is a U-space, then the canonical U-map $f: U \times_G \gamma^* Y \to U/\gamma \times Y$ given by $f[u, y] = (\pi(u), uy)$ is a U-equivalence, where U acts diagonally on $U/\gamma \times Y$.

<u>Proof.</u> Let $\tilde{g}: U \times Y \to U \times_G \gamma^* Y$ be the map $\tilde{g}(u, y) = [u, u^{-1}y]$. If $x \in G$ then $\tilde{g}(u\gamma(x), y) = [u\gamma(x), \gamma(x)^{-1}u^{-1}y] = [u, u^{-1}y] = \tilde{g}(u, y)$, so $\tilde{g}$ induces a map $g: U/\gamma \times Y \to U \times_G \gamma^* Y$. The map g is a two-sided

inverse to f (hence a U-equivalence), because $fg(\pi(u), y) = f[u, u^{-1}y]$ $= (\pi(u), y)$, $gf[u, y] = g(\pi(u), uy) = [u, u^{-1}uy] = [u, y]$.

We are now ready to prove Theorem 1. Let $\gamma: G \to U$ be a homomorphism, $X = U/\gamma$, $Y = U/H$, then

$$K_U(U/\gamma \times U/H) = K_U(U \times_G \gamma^* U/H)$$
$$= K_G(\gamma^* U/H) .$$

The E_2-term of the Künneth spectral sequence is $\operatorname{Tor}^{*,*}_{R(U)}(K_U(X), K_U(Y))$. But $K_U(U/\gamma) = K_U(U \times_G *) = K_G(*) = R(G)$, an $R(U)$-module via the ring homomorphism $\gamma^*: R(U) \to R(G)$. Similarly, $K_U(U/H) = K_U(U \times_H *)$ $= K_H(*) = R(H)$, an $R(U)$-module via the ring homomorphism $i^*: R(U) \to R(H)$.

Now $R(H)$ is $R(U)$-free, so $\operatorname{Tor}^{i,*}_{R(U)}(R(G), R(H)) = 0$ for $i > 0$ and $\operatorname{Tor}^{0,*}_{R(U)}(R(G), R(H),) = R(G) \otimes_{R(U)} R(H)$. This means that the spectral sequence collapses and the edge homomorphism

$$A: R(G) \otimes_{R(U)} R(H) \to K_G(\gamma^* U/H)$$

is an isomorphism.

We stop to record a useful remark:

<u>Lemma 5.</u> Let $\varphi : F \to G$ be a continuous homomorphism, $\gamma : G \to U$ as before. The following diagram commutes.

$$\begin{array}{ccc}
R(G) \otimes_{R(U)} R(H) & \xrightarrow{\quad A \quad} & K_G(\gamma^* U/H) \\
\Big\downarrow \varphi^* \otimes 1 & & \Big\downarrow \varphi^! \\
R(F) \otimes_{R(U)} R(H) & \xrightarrow{\quad A \quad} & K_R((\gamma\varphi)^* U/H)
\end{array}$$

We shall use this result for the special case $F = E$, the trivial subgroup of G. Here $\varphi^* = \dim$, the dimension function and $\varphi^! : K_G(\gamma^* U/H) \to K(U/H)$ is the forgetful homomorphism.

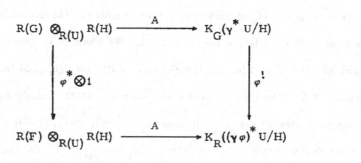

3. RSU(m,n) as a module over RSU(m+n)

If U is a simply connected Lie group and H is a closed connected subgroup of maximal rank, then Steinberg [16] has shown how to construct a $R(U)$-free basis of $R(H)$. We shall be interested in the special case $U = SU(m+n)$, $H = SU(m,n) = U(m) \times U(n) \cap SU(m+n)$. We let $T \subset H$ be the standard maximal torus of U consisting of diagonal matrices with coordinates $x_i : T \to S^1$, $i = 1, \ldots, m+n$. These linear characters satisfy the relation $x_1 \cdots x_{m+n} = 1$, and the group $\hat{T}$ of all linear characters $T \to S^1$ is free abelian with $x_1, \ldots, x_{m+n-1}$ as a free basis. The representation ring $R(T) = Z[\hat{T}]$ is the group ring of $\hat{T}$ (where the group operation in $\hat{T}$ is written multiplicatively). The Weyl group $W = N_U(T)/T$ is the symmetric group S_{m+n} on $m+n$ letters and its action on $\hat{T}$ is by permutation of subscripts. The Weyl group $N_H(T)/T = W'$ is precisely the subgroup $S_m \times S_n$. We let $x_i x_j^{-1}$ for $i < j$ be the positive roots of U and let

$$\Pi = \{x_1 x_2^{-1}, x_2 x_3^{-1}, \ldots, x_{m+n-1} x_{m+n}^{-1}\}$$

be the system of simple roots of U. The fundamental weights associated to this system of simple roots are:

$$\{x_1, x_1 x_2, x_1 x_2 x_3, \ldots, x_1 \cdots x_{m+n-1}\}$$

and form a basis for $\hat{T}$ since U is simply connected. Let W'' be the subset in W mapping the positive roots of H into the positive roots of U. That is, $w \in W''$ if and only if for each $1 \leq i < j \leq m$, we have $w(i) < w(j)$ and for each $m+1 \leq p < q \leq m+n$ we have $w(p) < w(q)$. Thus W'' consists precisely of all (m, n)-shuffles. Given a (m, n)-shuffle v, we let λ_v be the product in $\hat{T}$ of all those fundamental weights for which the corresponding simple root is mapped into a negative root under v^{-1}. The element $E(v)$ is the sum in $Z[\hat{T}] = R(T)$ of elements in the orbit of $v^{-1}\lambda_v$ in $\hat{T}$ under W'. The set $\{E(v) \mid v \in W''\}$ is the Steinberg basis of $RSU(m, n) = R(H)$ over $RSU(m+n) = R(U)$.

An example will make this clearer. Let $m = 3$, $n = 2$ and v be the shuffle 1 4 2 5 3, that is

$$v = \begin{pmatrix} 1 & 2 & 3 & 4 & 5 \\ 1 & 3 & 5 & 2 & 4 \end{pmatrix}$$

$$v^{-1} = \begin{pmatrix} 1 & 2 & 3 & 4 & 5 \\ 1 & 4 & 2 & 5 & 3 \end{pmatrix}$$

The only simple roots which are made negative by v^{-1} are $x_2x_3^{-1}$ and $x_4x_5^{-1}$, so $\lambda_v = x_1^2x_2^2x_3x_4$, $v^{-1}\lambda_v = x_1^2x_4^2x_2x_5 = x_1^2x_2x_4^2x_5$, and $E(v) = s(2,1)\cdot t(2,1)$, where

$$s(2,1) = x_1^2x_2 + x_1^2x_3 + x_1x_2^2 + x_1x_3^2 + x_2^2x_3 + x_2x_3^2,$$

$$t(2,1) = x_4^2x_5 + x_4x_5^2.$$

We will also use a similar notation for the general case. Let $y_i = x_{m+i}$, $i = 1, \ldots, n$. If $(e_1, \ldots, e_r) = E$ is a sequence of natural numbers with $e_1 \geq e_2 \geq \cdots \geq e_r$, then we shall denote by $s(E) = s(e_1, \ldots, e_r)$ the sum of the elements in the orbit of $x_1^{e_1} x_2^{e_2} \cdots x_r^{e_r}$ under S_m and by $t(e_1, \ldots, e_r)$ the sum of the elements in the orbit of $y_1^{e_1} \cdots y_r^{e_r}$ under S_n. If $r > m$ we set $s(E) = 0$ and if $r > n$ we let $t(E) = 0$.

Proposition 6. Let v be the shuffle given schematically by

$$\underset{m_1}{\underline{\qquad}} \;\; \underset{n_1}{\underline{\qquad}} \;\; \underset{m_2}{\underline{\qquad}} \;\; \underset{n_2}{\underline{\qquad}} \cdots \underset{m_A}{\underline{\qquad}} \;\; \underset{n_A}{\underline{\qquad}}$$

where $m_1 + \ldots + m_A = m$, $n_1 + \ldots + n_A = n$. Then

$$E(v) = s(X(v)) \cdot t(Y(v)),$$

$$X(v) = (m_i \text{ times } A-i),$$

$$Y(v) = (n_i \text{ times } A-i), \quad i = 1, \ldots, A-1.$$

Example. Let $m = 5$, $n = 3$, $A = 4$, v the shuffle 6 1 2 7 3 4 8 5, that is

i	1	2	3	4
m_i	0	2	2	1
n_i	1	1	1	0

then $E(v) = s(2, 2, 1, 1) \cdot t(3, 2, 1)$.

We shall let $s(k \text{ times } 1) = g_k$, $t(k \text{ times } 1) = h_k$, the k-th elementary symmetric functions in $x_1, \ldots, x_m$ and $y_1, \ldots, y_n$, respectively. Let v_k be the shuffle $m+1, \ldots, m+k, 1, 2, \ldots, m, m+k+1, \ldots, m+n$. That is, v_k is given by $A = 2$, $m_1 = 0$, $m_2 = m$, $n_1 = k$, $n_2 = n-k$ (where, of course, $k \leq n$). Then $E(v_k) = t(k \text{ times } 1) = h_k$. Similarly, define w_k by $m_1 = 1$, $m_2 = m-1$, $n_1 = k$, $n_2 = n-k$, then $E(w_k) = s(1)t(k \text{ times } 1) = g_1 h_k$. We record this for future reference.

Corollary 7. Let g_k be the k-th elementary symmetric function of $x_1, \ldots, x_m$, h_k the k-th elementary symmetric function of $y_1, \ldots, y_n$. We have $g_1 h_k$ and h_k for $k \leq n$ part of the Steinberg basis of $RSU(m,n) = Z[g_1, \ldots, g_{m-1}, h_1, \ldots, h_n, h_n^{-1}]$ over $RSU(m+n) = Z[b_1, \ldots, b_{m+n-1}]$, where b_k is the k-th elementary symmetric function of $x_1, \ldots, x_m$, $y_1, \ldots, y_n$.

Example. We have $b_1 \cdot 1 = g_1 + h_1$. More generally, $b_k \cdot 1 = \sum_{i+j=k} g_i h_j$. Here are some expansions in terms of the Steinberg basis:

$$g_1 = b_1 \cdot 1 - h_1 ,$$

$$h_1^2 = b_1 \cdot h_1 - g_1 h_1 ,$$

$$t(2) = b_1 \cdot h_1 - g_1 h_1 - 2h_2 .$$

$$t(3) = (b_1^2 - b_2) \cdot h_1 - 2b_1 \cdot h_2 - b_1 \cdot g_1 h_1 + g_2 h_1 + 2g_1 h_2 + 3h_3 .$$

Has the reader noticed a pattern emerging ?

Lemma 8. The expansion of $t(k+1)$ in terms of the Steinberg basis of $RSU(m,n)$ over $RSU(m+n)$ is

$$t(k+1) = x + (-1)^k k g_1 h_k + (-1)^k (k+1) h_{k+1} ,$$

where x is an $RSU(m+n)$-linear combination of the remaining generators in the Steinberg basis.

The hint is to use the Newton relations between $t(r)$ and h_s and induction. It is also convenient for the argument to use a basis consisting of monomials in g_i and h_j instead of the Steinberg $s(E)t(F)$. This is also convenient later (in the proof of Lemma 12).

The main point to remember in Lemma 8 is that if $2 \leq k+1 \leq n$ then $t(k+1) \neq 0$ and the coefficients of $g_1 h_k$ and h_{k+1} are non-zero integers.

3. Proof of Theorem 3.

Let $U = SU(m+n)$, $H = SU(m,n)$ and let $Y = U/H = G_m(\mathbb{C}^{m+n})$ be the Grassman manifold of complex m-planes in $\mathbb{C}^{m+n}$. The group $N_U(H)/H$ is $Z/2Z$ if $m = n$, and is the trivial group if $m \neq n$. Let $\underline{c} : Y \to Y$ be the map induced by complex conjugation in U.

Theorem 9. If φ is an automorphism of the algebra $H\ (Y; Q)$ then there exists a $w \in N_U(H)/H$ such that $\varphi = w^*$ or $\varphi = w^* \underline{c}^*$.

This is a theorem of Brewster [3]. It had been proven earlier for $n \geq 2m^2 - m - 1$ by Glover and Homer [5] (they study endomorphisms of $H^*(Y; Q)$, not just automorphisms).

Suppose now that $\alpha, \beta: G \to U$ are representations and $f: \alpha^*Y \to \beta^*Y$ is a G-homotopy equivalence. According to Brewster's theorem there exists a G-equivalence $g: \beta^*Y \to \beta^*Y$ or $g: \beta^*Y \to \bar{\beta}^*Y$ such that $(gf)^* =$ identity map on $H^*(Y; Q)$. We will replace f by gf, so we assume that f^* is the identity. Now $K(Y)$ is torsion-free and $ch: K(Y) \to H^{**}(Y; Q)$ is an injection. It follows that $f^*: K(Y) \to K(Y)$ is the identity. Let $i: E \to G$ be the inclusion of the trivial subgroup, then we have the commutative diagrams

$$\begin{array}{ccc}
K_G(\beta^*Y) & \xrightarrow{\;f^!\;} & K_G(\alpha^*Y) \\
\Big\downarrow i^! & & \Big\downarrow i^! \\
K(Y) & \xrightarrow{\;1\;} & K(Y)
\end{array}$$

and

$$\begin{array}{ccc}
R(G) \otimes_{R(U)} R(H) & \xrightarrow{\;A(\gamma)\;} & K_G(\gamma^*Y) \\
\Big\downarrow \dim \otimes 1 & & \Big\downarrow i^! \\
Z \otimes_{R(U)} R(H) & \xrightarrow{\;A\;} & K(Y)
\end{array}$$

(see Lemma 5) since $i^* = \dim : R(G) \to R(E) = Z$. If $y \in R(H)$ we shall write $y(\gamma) = A(\gamma)(1 \otimes y)$.

Proposition 10. For $i = 1, \ldots, n$, $f^! h_i(\beta) = \nu_i h_i(\alpha)$, $f^! g_1(\beta) = \mu g_1(\alpha)$, $\mu, \nu_i \in R(G)$, $\dim \mu = \dim \nu_i = 1$.

Sketch proof. Use the equivariant Chern character of Slominska [15] to prove that $f^! h_1(\beta) = \nu_1 h_1(\alpha)$, $f^! g_1(\beta) = \mu g_1(\alpha)$. Now notice that $h_k = \Lambda^k h_1$ and use the expansion of $\Lambda^k(\nu_1 h_1(\alpha))$ together with the assumption that $f^* : K(Y) \to K(Y)$ is the identity.

Our aim is to prove that μ is a linear character and $\beta = \mu\alpha$. The second is immediate:

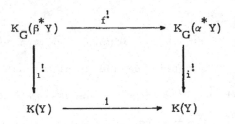

Lemma 11. $\beta = \mu\alpha$, $\nu_1 = \mu$.

Proof. The identity $b_1 \cdot 1 = g_1 + h_1$ yields $\beta \cdot 1 = g_1(\beta) + h_1(\beta)$, $\alpha \cdot 1 = g_1(\alpha) + h_1(\alpha)$. We apply $f^!$ to the first identity and obtain $\beta \cdot 1 = \mu g_1(\alpha) + \nu_1 h_1(\alpha)$. However $\mu\alpha \cdot 1 = \mu g_1(\alpha) + \mu h_1(\alpha)$. Since 1 and $h_1(\alpha)$ are part of a $R(G)$-basis of $K_G(\alpha^* Y)$, we obtain the result from the equation $\beta \cdot 1 - \nu_1 h_1(\alpha) = \mu\alpha \cdot 1 - \mu h_1(\alpha)$.

Lemma 12. For all k, $\nu_k = \nu_1^k$.

Proof. We do this by induction on k. Assume $k \geq 1$ and $\nu_k = \nu_1^k$. Notice that $\psi^{k+1} h_1 = t(k+1)$. Using $f^! \psi^{k+1} h_1(\beta) = \psi^{k+1} f^! h_1(\beta)$ and evaluating the coefficient of $g_1(\alpha) h_k(\alpha)$ we obtain (using Lemma 8):

$$(-1)^k k \psi^{k+1}(\nu_1) = (-1)^k k \mu \nu_k ,$$

and using the coefficient of $h_{k+1}(\alpha)$:

$$(-1)^k (k+1) \psi^{k+1}(\nu_1) = (-1)^k (k+1) \nu_{k+1} .$$

Thus $\nu_{k+1} = \mu \nu_k = \nu_1 \nu_1^k = \nu_1^{k+1}$ by Lemma 11 and the inductive hypothesis. Lemma 12 now follows by mathematical induction.

The proof of Theorem 3 will be completed if we can show that μ is a linear character. Recall the exact sequence (see [9]) of Picard groups (of complex line bundles under tensor product)

$$\mathrm{Hom}(G, S^1) \to \mathrm{Pic}_G(\alpha^* Y) \xrightarrow{i^!} \mathrm{Pic}_E(Y) .$$

Notice that $f^! h_n(\beta)$ and $h_n(\alpha)$ determine the same element in $\mathrm{Pic}_E(Y)$, so $f^! h_n(\beta) = \nu_n h_n(\alpha)$ where ν_n is a linear character of G. We claim: the equation $\mu^n = \nu_1^n = \nu_n$ implies that μ is a linear character of G. Think of μ as a class function on G (i.e., as a virtual character). Then for each $g \in G$ we have $|\mu(g)| = 1$ since $|\mu(g)|^n = |\nu_n(g)| = 1$. Now calculate the Schur inner product

$$(\mu, \mu) = \int_G |\mu(g)|^2 \, dg$$

$$= \int_G 1 \, dg = 1,$$

so μ is an irreducible character or its negative. However, $1 = \dim \mu = \mu(e)$, so μ is a linear character, as claimed.

REFERENCES

1. M.F. Atiyah and F. Hirzebruch, Vector bundles and homogeneous spaces, Proc. Symp. Pure Math. AMS 3 (1961), 7-38.

2. A. Back, Homotopy rigidity for Grassmannians (to appear).

3. S. Brewster, Automorphisms of the cohomology ring of finite Grassman manifolds. Ph.D. Dissertation, Ohio State Univ., 1978.

4. J. Ewing and A. Liulevicius, Homotopy rigidity of linear actions on friendly homogeneous spaces (to appear).

5. H. Glover and W. Homer, Endomorphisms of the cohomology ring of finite Grassman manifolds, Springer Lecture Notes in Math. 657 (1978), 170-193.

6. L. Hodgkin, An equivariant Kunneth formula in K-theory, Univ. of Warwick, preprint, 1968.

7. ________, The equivariant Kunneth theorem in K-theory, Springer Lecture Notes in Math. 496 (1975), 1-100.

8. W.-Y. Hsiang, Cohomology Theory of Topological Transformation Groups, Springer, 1975.

9. A. Liulevicius, Homotopy rigidity of linear actions: characters tell all, Bulletin AMS, 84 (1978), 213-221.

10. J. McLeod, Thesis, Cambridge University, 1975.

11. ________, The Künneth formula in equivariant K-theory, Proceedings of the Waterloo Topology Conference June 1978, Springer Lecture Notes in Math. (to appear).

12. L. O'Neill, The fixed point property for Grassman manifolds, Ph.D. Dissertation, Ohio State Univ., 1974.

13. H. V. Pittie, Homogeneous vector bundles on homogeneous spaces, Topology 11 (1972), 199-203.

14. V. P. Snaith, On the Künneth formula spectral sequence in equivariant K-theory, Proc. Camb. Phil. Soc. 72 (1972), 167-77.

15. J. Slominska, On the equivariant Chern homomorphism Bulletin de l'Académie Polonaise des Sciences (Ser. math., astr. et phys.) 24 (1976), 909-913.

16. R. Steinberg, On a theorem of Pittie, Topology 14 (1975), 173-177.

The University of Chicago
Chicago, Illinois 60637

August 1979

Homotopielineare Involutionen auf Sphären

Peter Löffler

Wir nennen eine Involution auf einer Homotopiesphäre (kurz: Sphäre) M^{n+k} homotopielinear, falls die Fixpunktmenge F^k wieder eine Sphäre ist.

Ziel dieser Note ist ein Beweis des folgenden Satzes:

<u>Satz A:</u> π_n^s bezeichne den stabilen n-Stamm. Es repräsentiere $x \in \pi_{2t}^s$ eine Sphäre. Dann gibt es auf der durch $x^2 \in \pi_{4t}^s$ dargestellten Sphäre eine homotopielineare Involution mit Fixpunktmenge $x (t \geq 5)$.

Zuerst einige Bezeichnungen: Es sei $\mathbb{R}^{n,k}$ der $\mathbb{R}^{n+k}$ mit nichttrivialer Involution auf den ersten n Koordinaten. Mit $S^{n,k-1}$ bezeichnen wir die Einheitssphäre im $\mathbb{R}^{n,k}$. Schließlich sei $\varepsilon^{n,k}$ das triviale Z_2-Vektorraumbündel mit Faser $\mathbb{R}^{n,k}$.

<u>Definition 1:</u> Es sei M eine Mannigfaltigkeit mit Involution und tM ihr Tangentialbündel. M heiße speziell (n,k)-rahmbar, falls es einen Z_2-Vektorraumbündelisomorphismus $\bar{\varphi}$ gibt, mit

$$\bar{\varphi} : tM \oplus \varepsilon^{o,s} \xrightarrow{\simeq} \varepsilon^{n,k+s}$$

Mit ${}^{sp}\Omega_{n,k}$ bezeichnen wir die Bordismengruppe der speziell (n,k)-gerahmten Mannigfaltigkeiten. Wie üblich gibt es eine Conner-Floyd-Folge

$$^{sp}\Omega_{n,k}[\text{frei}] \xrightarrow{i} {}^{sp}\Omega_{n,k} \xrightarrow{j} \pi^{s}_{k} \xrightarrow{\delta} {}^{sp}\Omega_{n,k-1}[\text{frei}]$$

($^{sp}\Omega_{n,k}[\text{frei}]$ ist die Bordismengruppe freier speziell (n,k)-gerahmter Mannigfaltigkeiten, i, δ die offensichtlichen Abbildungen, j ist die Beschrankung auf die Fixpunktmenge.)

Lemma 2: Die Abbildung $^{sp}\Omega_{k,k} \xrightarrow{j} \pi^{s}_{k}$ ist surjektiv.

Beweis: Wir setzen $\pi_{n,k} = \lim\limits_{r,s} [S^{n+r,k+s}, S^{r,s}]^{o}_{Z_2}$ wobei $[\ ,\ ]^{o}_{Z_2}$ die Menge der basispunkterhaltenden Homotopieklassen von Z_2-aquivarianten Abbildungen bezeichnet. Wegen [H] Satz IV 2 definiert die Pontrjagin-Thom-Konstruktion $\tau : {}^{sp}\Omega_{n,k} \longrightarrow \pi_{n,k}$ einen Isomorphismus abelscher Gruppen und man erhalt ein kommutatives Diagramm

$$\begin{array}{ccc} {}^{sp}\Omega_{n,k} & \xrightarrow{j} & \\ \downarrow \tau & & \pi^{s}_{k} \\ \pi_{n,k} & \xrightarrow{j} & \end{array}$$

Wird $x \in \pi^{s}_{k}$ durch $f : S^{k+t} \longrightarrow S^{t}$ reprasentiert, so stellt $f \wedge f : S^{k+t} \wedge S^{k+t} \longrightarrow S^{t} \wedge S^{t}$ ein Element in $\pi_{k,k}$ dar, das $j(f \wedge f) = x$ erfullt.

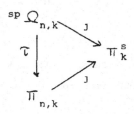

Definition 3: Es sei ${}^{sp}\sigma_{n,k}$ die folgende Bordismengruppe: Die Elemente seien speziell (n,k)-gerahmte Mannigfaltigkeiten, deren Fixpunktmenge eine (k-dimensionale) Sphäre ist. Die Bordismen seien h-Kobordismen auf der Fixpunktmenge. Die Summe sei gegeben durch zusammenhängende Summe an einem Fixpunkt ($k \geq 1$).

Satz 4: Es gibt für $k \geq 5$ eine exakte Folge

$$\to L_{k+1} \longrightarrow {}^{sp}\sigma_{n,k} \xrightarrow{\varphi} {}^{sp}\Omega_{n,k} \longrightarrow L_k \to$$

wobei die L_k die L-Gruppen aus [KM] [W] sind.

Beweis: Völlig analog zur bekannten Kervaire-Milnor-Folge.

Im folgenden sei von nun an $k \geq 5$ und $n \geq 3$.

Definition 5: Es sei ${}^{sp}\theta_{n,k}$ die Gruppe der speziell (n,k)-gerahmten Diffeomorphieklassen (= äquivarianten h-Kobordismenklassen) von homotopielinearen Involutionen.

Bemerkung: Aus [L1] folgt, daß für $k \leq n$ jede homotopielineare Involution speziell (n,k)-rahmbar ist.

Satz 6: Es gibt exakte Folgen

$$\to {}^{sp}\theta_{n,k} \xrightarrow{\psi} {}^{sp}\sigma_{n,k} \longrightarrow A_{n,k} \longrightarrow {}^{sp}\theta_{n,k-1} \to$$

hierbei ist 1) $A_{n,k} \cong L_{n+k}(Z_2, (-1)^n)$ für $k < n$

sowie 2) $A_{2t,2t}$ wird entdeckt durch beide Signaturen.

Beweis: Wir definieren eine Bordismengruppe $A_{n,k}$ wie folgt: Objekte sind speziell (n,k)-gerahmte Mannigfaltigkeiten $(M,\partial M)$ mit 1) M^{Z_2} ist ein k-Ball 2) ∂M ist äquivariant homotopieäquivalent zu $S^{n,k-1}$. Bordismen sind h-Kobordismen auf dem Rand und auf der Fixpunktmenge. Wir setzen $(M,\partial M)\sim 0$, falls M azyklisch ist. Die Summe wird durch zusammenhängende Summe an einem Fixpunkt auf dem Rande gegeben.

Die Abbildung von ${}^{sp}\sigma_{n,k} \longrightarrow A_{n,k}$ ist wie zu erwarten durch Herausschneiden eines Balles gegeben.

Die Exaktheit der langen Folge ist eine einfache Übungsaufgabe. Der Isomorphismus 1 wird wie in [W] Kapitel 5, 6 gezeigt. Der Isomorphismus 2 wird z. B. in [Do] bewiesen.

Beweis von Satz A: Wegen Lemma 2 gibt es $y \in {}^{sp}\Omega_{2t,2t}$ mit $j(y) = x$. Nach Vergessen der Gruppenoperation stellt y gerade x^2 dar. Wegen Satz 4 gibt es $y_1 \in {}^{sp}\sigma_{2t,2t}$ mit $\varphi(y_1) = y$. Da der gewöhnliche und der äquivariante Index von y_1 verschwinden, gibt es $y_2 \in {}^{sp}\Theta_{2t,2t}$ mit $\psi(y_2) = y_1$. Dann ist y_2 das gesuchte Element.

Bemerkung: 1) Eine breitere Darstellung dieses Falles und Verallgemeinerungen auf andere Gruppen werden in $[L_2]$ gegeben.

2) Die Idee des Beweises von Satz A ist einfach die folgende: Man nehme M x M $\#(-S^{2t}\times S^{2t})$ und mache Chirurgie auf dem Komplement der Fixpunktmenge. Ist man einmal überzeugt, daß 6.2 richtig ist, so ist der Beweis eine Frage des Sitzfleisches.

Literatur

[Do] K. H. Dovermann: The equivariant surgery problem for involutions, preprint.

[H] H. Hauschild: Bordismentheorie stabil gerahmter G-Mannigfaltigkeiten, Math. Z. 139 (1974), 165 - 172.

[KM] M. A. Kervaire - J. W. Milnor: Groups of homotopy spheres, Ann. of Math. 77 (1963), 504 - 537.

[L_1] P. Loffler: Equivariant framability of involutions on homotopy spheres, manuscr. math. 23 (1978), 161 - 171.

[L_2] P. Loffler: Homotopielineare Z_p-Operationen auf Spharen, in Vorbereitung.

[W] C. T. C. Wall: Surgery on compact manifolds, Academic Press 1970.

GROUP ACTIONS ON HYPERTORAL MANIFOLDS. I

Reinhard Schultz
Mathematics Department, Purdue University
West Lafayette, Indiana 47907 (USA)

Given a closed oriented manifold, it is natural to ask the extent to which its cohomology ring structure restricts the sorts of group action that the manifold can admit. Some early results in this direction - which apply immediately to surfaces - were obtained by P. A. Smith [13,14]. He also speculated that, in general, more elaborate cohomology structures imply more restrictive conditions on possible group actions. Unfortunately there does not appear to be a comprehensive summary of further studies in this direction, and it is really beyond our scope to attempt one here. However, the following result does provide one example of the existing support for Smith's speculation:

THEOREM. Suppose M^n is a closed oriented manifold with integral (equivalently) rational cohomological cuplength n and suppose a compact connected Lie group G acts effectively on M. Then G is a torus and all isotopy subgroups of the action are finite.

Many proofs have been given for this theorem in varying degrees of generality; a proof in the topological category was given by D. Burghelea and the author [3,Thm. A].

Working independently, S. Schwartzman had obtained a somewhat similar result by quite different methods [8,16]. Assuming that the generator of $H^n(M^n;\mathbb{Z})$ was a product of n one-dimensional classes, Schwartzman proved that G must act freely and the resulting principal G-bundle $M \to M^n/G$ must be a product bundle.

For the sake of uniform simplicity we shall say that manifolds that satisfy the hypothesis of Schwartzman's Theorem are hypertoral; Since $H^1(X;\mathbb{Z})$ is isomorphic to the cohomotopy group $[X,S^1]$ this is equivalent to assuming that a manifold M^n admits a degree one map onto a torus T^n, a fact which should explain our choice of terminology.

Partially supported by NSF Grants MPS74-03609, MCS76-08794, and MCS78-02913A1.

We shall begin this paper by giving a proof of Schwartzman's Theorem in the topological category using the methods of [3] and the fibering theorem for injective toral actions due to P. Conner and F. Raymond [4]; this is done in Section 1. This approach is basically converse to Schwartzman's, for he observes that the splitting theorems of [16] imply the Conner-Raymond theorem. Neither his work nor ours properly contains the other.

A mapping torus construction shows that Schwartzman's Theorem also has very strong implications about finite group actions on hypertoral manifolds, and we develop some of these implications in Section 2; special cases were first observed at the end of [16]. The implications clearly resemble some results of P. A. Smith [13,14] and also some results of A. Borel developed further by Conner and Raymond [5,6]. However, our results do not require that the relevant manifolds be aspherical as in [5,6].

An important feature of the Conner-Raymond work is that it led to the first known manifolds admitting no finite group actions except trivial ones [1,6]. Since the methods of [6] involve excrutiatingly specific computations, this could only be verified for finitely many dimensions in [6]. Other infinite classes in dimension 4 were given by E. Bloomberg [1], and a remark at the end of [1] indicates how one might use constructions from [6] to find manifolds without group actions in all higher dimensions. In dimension 3 examples were given by F. Raymond and J. Tollefson [12]; L. Siebenmann has also informed me about similar families of hyperbolic manifolds. We shall conclude this paper by adding to this list of examples, showing in particular that <u>every oriented bordism class with odd signature contains infinitely many manifolds without group actions</u> (except maybe dimension 4 in the topological case, see Proposition 4.3. The key feature of our approach is that we need not limit ourselves to aspherical manifolds; because of this, we can and do consider manifolds that are glued together from many different sorts of pieces. This at least superfically resembles Bloomberg's approach [1] in which connected sums of two aspherical manifolds with little symmetry are considered.

In a subsequent paper the methods presented here will be developed further to find new families of manifolds without group actions in still other dimensions, and some other consequences of Schwartzman's Theorem will also be discussed.

Acknowledgments. This paper could not have been written if Ed Grove had not informed me of Schwartzman's unpublished work [8], and accordingly I am deeply grateful to him for doing so. I would also like to thank Sol Schwartzman for showing me the details of [16], which had not appeared in print when this was written. The relationship between [3] and [4] was first pointed out to me explicitly by Frank Raymond, to whom I am also grateful for this and other comments on the problems treated in this paper.

1. Schwartzman's Theorem

We shall now prove the topological version of Schwartzman's Theorem (specifically, the fourth theorem in [16]). Accordingly, our standing hypothesis in this section is that M^n is an oriented closed n-manifold with a given S^1 action, and there exist cohomology classes $w_1,\ldots,w_n \in H^1(M;\mathbb{Z})$ so that $w_1\ldots w_n$ generates $H_n(M^n;\mathbb{Z}) = \mathbb{Z}$. All cohomology groups are Čech groups unless stated otherwise.

The first step is essentially a simple application of an important result due to P. Conner and F. Raymond[4]:

(1.1) <u>The manifold M^n as above is equivariantly isomorphic to $S^1 \times_\pi X^{n-1}$ where $\pi \subseteq S^1$ is a finite cyclic group and X^{n-1} is a suitable space with π-action.</u>

<u>Proof</u>. The Leray spectral sequence argument of [3] shows that the map $H^1(M^n;\mathbb{Q}) \to H^1(S^1;\mathbb{Q}) = \mathbb{Q}$ induced by restriction to an orbit is non zero and hence onto. Therefore one can apply [4, Theorem 4.2]∎

<u>Remark</u>. Without loss of generality, we may assume that X is connected. For π permutes the components of X, and each component of N corresponds to an equivalence class of components of X. Since N is assumed connected, π must permute the components of X transitively. But if π_o is the stability subgroup of a component, it follows that N is also isomorphic to $S^1 \times_{\pi_o} X$.

For the next step in the analysis of M^n, it is useful to consider a slight weakening of our standing hypothesis, replacing $\mathbb{Z}$ by the localization $\mathbb{Z}_{(p)}$ for some arbitrary prime p. The proof of (1.1) goes through without change in this case.

The following lemma, although stated in a very specialized setting, is technically the key step in proving that torus actions on hypertoral manifolds are free.

LEMMA 1.2. <u>Let N^n be a closed oriented n-manifold satisfying the p-local weakening of our standing hypothesis, and suppose $N^n = S^1 \times_\pi Y^{n-1}$ where π is a cyclic p-group. Then π acts freely on Y^{n-1}, and hence S^1 acts freely on N^n.</u>

<u>Remark</u>. In the smooth category, Y is a smooth manifold with group action. On the other hand, in the topological category all we know is that Y is a cohomology manifold (being a manifold factor). Fortunately, the cohomological machinery of [2] allows us to handle topological actions with the same ease as smooth ones.

PROOF. Some preliminary information on the set of nonfree orbits is necessary. In the first place, the cohomology manifold Y is orientable, and π acts orientation preservingly on Y; for otherwise N could not be an orientable manifold. This follows from the Wang sequence for the twisted product $N = S^1 \times_\pi Y$, which is the mapping torus for the generator of π acting on Y (See [10,p.67], for example). It follows that the sets of nonprincipal orbits in N and Y

have $\mathbb{Z}_{(p)}$ cohomological codimension at least two (compare [2,V.4.4 and V.4.7]), and consequently the set of principal orbits in N/S^1 is a connected open dense set.

The $E_2^{0,1}$ term of the Leray spectral sequence for $N \to N/S^1$ is the group of sections of the sheaf whose stalk at $[x] \in N/S^1$ is just the first cohomology group of the orbit $S^1 \cdot x$. It follows from the slice theorem, the connectedness of the set of principal orbits, and the finiteness of all isotropy subgroups that $E_2^{0,1}$ is isomorphic to $\mathbb{Z}_{(p)}$. Moreover, if $\pi' \subseteq \pi$ is the minimal isotropy subgroup of the action (this exists because the subgroups lattice of π is linearly ordered), then the value of a section on the stalk of a principal orbit $H^1(S^1 \cdot x^*) = \mathbb{Z}_{(p)}$ is divisible by the p-power $|\pi/\pi'| = p^s$. For convenience, let $p^r = |\pi'|$.

Let $u_1, \ldots, u_n$ be the classes which exist by the standing hypothesis that M is hypertoral. Let m be the least integer such that some u_j belongs to $p^m E_2^{0,1}$; without loss of generality we may assume that the last class u_n lies in $p^m E_2^{0,1}$ (reorder if necessary). Then by subtracting off suitable multiples of u_n from $u_1, \ldots, u_{n-1}$ we get new classes $v_1, \ldots, v_{n-1}$ such that $u_1 \ldots u_{n-1} u_n = v_1 \ldots v_{n-1} u_n$ but the v_j's have positive filtration in the Leray spectral sequence. In other words, the v_j's lie in the image of the edge homomorphism $H^1(N/S^1) \to H^1(N)$.

Recall that we also have the equivariant fibering $N \to S^1/\pi$, and in fact we have a map $N \to N/S^1 \times S^1/\pi$ satisfying the following conditions:

(i) The composite $S^1 \to S^1 \cdot x^* \subseteq N \to S^1/\pi$ for x^* on a principal orbit is just the usual quotient map.

(ii) The $n-1$ dimensional cohomology of N/S^1 is $\mathbb{Z}_{(p)}$, the n-dimensional cohomology of $N/S^1 \times S^1/\pi$ is also $\mathbb{Z}_{(p)}$, and the composite

$$H^n(N/S^1 \times S^1/\pi) \to H^n(N)$$

is multiplication by $|\pi| = p^{r+s}$.

The only nontrivial points in the above claims involve the cohomology of N/S^1 and the statement regarding degrees. First, recall that the (generalized) manifold Y^{n-1} is orientable and π acts orientation preservingly, as noted in the first paragraph of this proof. The compact T_2 space $Y/\pi \cong N/S^1$ contains an open dense set V that is a connected generalized $(n-1)$-manifold by the principal orbit theorem and our previous remarks; to prove the assertions regarding

$H^{n-1}(Y/\pi;Z)$, it suffices to recall that the closed set $Y/\pi - V$ has cohomological dimension $\leq n-3$ by the remarks in the first paragraph of this proof. Next, consider the effective action of $\pi \times \pi$ on $S^1 \times Y$. By construction $N = S^1 \times Y/\Delta$ where Δ is the diagonal subgroup of $\pi \times \pi$; therefore N inherits a natural orientation preserving, effective action of $\pi \cong \pi \times \pi/\Delta$ whose quotient is just $Y/\pi \times S^1/\pi = N/S^1 \times S^1/\pi$ (Since $Y/\pi \cong N/S^1$). Clearly the degree of $N \to N/S^1 \times S^1/\pi$ must be $|\pi|$ under these circumstances.

Now let $W_n \in H^1(N)$ be the image of a generator in $H^1(S^1/\pi)$. By (1) it follows that the image of W_n in $H^1(S^1 x^*)$ is divisible by $|\pi| = p^{r+s}$. On the other hand, the image of u_n in this group is divisible by p^{s+t} for some $t \leq r$. Therefore $w_n = \alpha\, p^{r-t}\, u_n + v_n$ where α is a p-local unit and v_n lies in the image of $H^1(N/S^1)$. Since $v_1 \ldots v_{n-1}\, v_n = 0$ for dimensional reasons, it follows that $v_1 \ldots v_{n-1}\, W_n$ is p^{r-t} times a generator of $H^n(N)$. On the other hand, $v_1, \ldots, v_{n-1}$ lie in the image of $H^1(N/S^1)$ while w_n lies in the image of $H^1(S^1/\pi)$, and therefore $v_1 \ldots v_{n-1}\, w_n$ is divisible by p^{r+s} in $H^n(N)$. This can only happen if $r-t = r+s$ and $t = s = 0$. But $s = 0$ implies π' is the trivial group, and therefore π must act freely on Y∎

We can now patch these local results together and complete the crucial step of the argument.

LEMMA 1.3. <u>Under the standing hypotheses, the action of π on X as in (1.1) is free.</u>.

PROOF. If π does not act freely, then the restricted action of some Sylow p-subgroup π' is not free. But consider the finite covering $N \downarrow M = S^1 \times_{\pi'} X \longrightarrow S^1 \times_{\pi} X$ for an arbitrary Sylow p-subgroup π'. It is immediate that N^n satisfies the p-local analog because the index of the covering $= |\pi/\pi'|$ is prime to p. By Lemma 1.2 we know that π' acts freely on X, and thus by the first sentence π itself acts freely∎

LEMMA 1.4. <u>Under the standing hypotheses, the principal S^1 bundle $S^1 \to S^1 \times_{\pi} X \to X/\pi$ is totally nonhomologous to zero.</u>

PROOF. Consider the Serre spectral sequence; we need to verify that the transgression

$$d_2^{0,1}\colon\ H^1(S^1) \to H^2(X/\pi)$$

is zero. In the first place the image of d_2 cannot have infinite order, for then $S^1 \times_{\pi} X$ would not have rational cuplength n by [3, Thm.A]. On the other hand, if d_2 on a generator of $\mathbb{Z} = H^1(S^1)$ had order μ, then the multiplicative properties of the Serre spectral sequence would imply that $u_1 \ldots u_n$ was divisible by μ in $H^n(N^n = S^1 \times_{\pi} X)$∎

It is now easy to prove the main result of this section:

(1.5) TOPOLOGICAL VERSION OF SCHWARTZMAN'S THEOREM.

Let M^n be a closed hypertoral manifold with an effective TOP or DIFF action of a compact connected Lie group G. Then G is a torus, the action of G on M is free, and the principal G-bundle $M \to M/G$ is a product bundle.

PROOF: We already know that G must be a torus by [3, Thm.A], and by Lemma 1.3 the restriction to every circle subgroup is free. But if the total G-action were not free, then there would be a circle subgroup of G that also did not act freely.

Now suppose that torus G has rank k and is the direct product of circle subgroups $C_1, \ldots, C_k$. Of course, each C_j acts freely, and by Lemma 1.4 we know that the principal circle bundles $M \to M/C_j$ all have zero Euler classes X_j. On the other hand, the bundle $M \to M/G$ is completely determined by the sequence of Euler classes $(X_1, \ldots, X_k)$, and accordingly the principal bundle $M \to M/G$ must be trivial. Finally if the action of G is smooth, then the isomorphism $M \cong M/G \times G$ may also be chosen to be smooth∎

2. Finite group actions

Since Schwartzman's Theorem deals so pointedly with actions of connected groups, it is mildly surprising to realize that the theorem has implications for finite groups also. Here is the basic result we want for further use:

THEOREM 2.1. Suppose that M^n is a hypertoral manifold, and suppose $\pi = \mathbb{Z}_p$ (p prime) acts effectively but nonfreely on M^n. Then the functorially induced antihomomorphism

$$H^{1*}: \pi \to \mathrm{Aut}\, H^1(M^n; \mathbb{Z})$$

sending g to $H^1(g; \mathbb{Z})$ is injective.

PROOF: If H^{1*} is not injective, then it is constant. Consider the manifold

$$V^{n+1} = S^1 \times_{\pi} M^n;$$

as in [4], this fibers over S^1/π and it is in fact the mapping torus of the homomorphism of M^n determined by a generator of π. Furthermore, it has an obvious effective S^1 action whose isotropy subgroups are exactly $\{1\}$ and $\mathbb{Z}_p$ (recall that π acts nonfreely but effectively). By the Wang sequence for the cohomology of a mapping torus (e.g., see [10] again), it follows that the restriction map $H^1(V;\mathbb{Z}) \to H^1(N;\mathbb{Z})$ is onto. From this and Poincaré duality it is immediate that V^{n+1} is also hypertoral. However, Schwartzman's Theorem implies that a circle action on V^{n+1} must be free - contradicting our assumption on π. Therefore either π acts freely or H^{1*} is injective ∎

Of course, if M^n is the torus T^n, then the numerous free actions induced by subgroups of order p show that the nonfreeness hypothesis is crucial. On the other hand, there are some circumstances under which the freeness hypothesis can be relaxed. Here are two examples:

COROLLARY 2.2. Suppose that M^n is given as in 2.1, and in addition assume that $H^*(M^n;\mathbb{Q})$ is generated multiplicatively by $H^1(M^n;\mathbb{Q})$ and the Euler characteristic $\chi(M^n)$ is nonzero. Then H^{1*} is always injective.

Example: A connected sum of $g > 1$ copies of T^{2n} satisfies this. For $n = 1$ this result was initially proved by P.A. Smith [13,14].

PROOF: Suppose H^{1*} is trivial. Then the assumption on $H^*(M^n;\mathbb{Q})$'s multiplicative generators implies π acts trivially on rational cohomology. Therefore π must act freely. But now the Lefschetz fixed point theorem (the version in [7] suffices here) implies that the generator of π must have a fixed point, for the generator's Lefschetz number equals the Euler characteristic of M^n - which is nonzero. The contradiction implies H^{1*} is injective ∎

COROLLARY 2.3. Under the assumptions on M^n in 2.1, assume that $\chi(M^n) = \pm 1$. Then H^{1*} is always injective.

PROOF: As in Corollary 2.2, we must exclude the possibility of free actions. But if π acts freely on M^n, then the orbit manifold M^n/π has the homotopy type of a finte complex [9] and accordingly $\chi(M^n)$ is divisible by the order of π. But $\chi(M^n) = \pm 1$ implies that no such divisibility condition is possible ∎

Results resembling Theorem 2.1 have proved to be useful tools in restricting the orders of periodic maps on manifolds; the faithfulness of a representation of π induced by a homotopy functor obviously allows one to reduce the question to studying the appropriate faithful representations of π. For example, such results were applied in the work of P.A. Smith and Conner and Raymond.

3. Rigid hypertoral manifolds

Obviously the results of Section 2 place some strong restrictions on the possible nonfree (sometimes even free) maps of prime period. In particular, if $p-1 > \text{rank } H^1(M^n,\mathbb{Z})$, then $\mathbb{Z}_p$ cannot act unless it does so freely, and the assumptions of 2.2 or 2.3 serve to exclude the latter possibility too in those cases. On the other hand, if M^n is hypertoral it is obvious that rank $H^1(M^n,\mathbb{Z}) \geq n$, and thus there is an *a priori* possibility for many different prime periods. In order to eliminate such primes, we have to look a little deepter into the structure of geometrically realized periodic automorphisms. Since every periodic automorphism of $H^1(T^n;\mathbb{Z})$ is induced by an appropriate periodic diffeomorphism from $GL(n,\mathbb{Z})$, it is clear that such inspection requires an appropriately chosen class of examples. Needless to say, the aim of this section is to produce such exampes. For the sake of brevity, we shall restrict attention manifolds of dimension $4n$, where $n > 1$; further examples in other dimensions will appear in a sequel to this paper.

THEOREM 3.1. *Let $n > 1$ be given and let $(p_1,\ldots,p_{4n})$ be an arbitrary sequence of $4n$ primes congruent to 3 mod 4. Then there is a manifold $B^{4n}(p_1,\ldots,p_{4n})$ with the following properties:*

(i) *Each $B^{4n}(p_1,\ldots,p_{4n})$ is hypertoral and has the rational cohomology of T^{4n}.*

(ii) *The torsion in the integral cohomology of $B^{4n}(p_1,\ldots,p_{4n})$ has exponent exactly $p_1\cdots p_{4n}$.*

(iii) *Every periodic self-map of $B^{4n}(p_1,\ldots,p_{4n})$ induces the identity on $H^1(B^{4n}(p_1,\ldots,p_{4n});\mathbb{Z})$.*

Remark. There are infinitely many such primes (by Dirichlet's Theorem or an elementary undergraduate exercise in many textbooks), and hence there are infinitely many such sequences.

PROOF: Let $L^{4n-1}(p)$ denote the simple lens space of dimension $(4n-1)$ for the prime p, and let $L_o^{4n-1}(p)$ be the manifold with the interior of a closed disk removed.

Let $\Gamma_1, \ldots, \Gamma_{4n} \subseteq T^{4n}$ be smoothly embedded circles corresponding to the standard generators e_j of $H_1(T^{4n};\mathbb{Z})$, and isotop them into pairwise disjoint smoothly embedded circles C_j. Extend these embeddings to pairwise disjoint closed tubular neighborhoods $C_j \times D^{4n-1}$, and form the manifold $B^{4n}(p_1,\ldots,p_n)$ by cutting out the interiors of these closed tubes and replacing them with copies of $C_j \times L_o^{4n-1}(p_j)$. It is immediate that (i) and (ii) are satisfied.

To prove (iii), first denote the Poincaré duality isomorphism $H_1 \cong H^{4n-1}$ by D, and let $f:B \to B$ be a self-map of prime period q.

Consider the cohomology of B with coefficients in $\mathbb{Z}_{p_j}$; it is immediate that B has the same cohomology as $[T^{4n-1} \# L^{4n-1}(p_j)] \times S^1$. Let $\rho_j:\mathbb{Z} \to \mathbb{Z}_{p_j}$ be the usual surjection with induced cohomology operation $\rho_j{}^*$, and let β_j be the Bockstein operation arising from the short exact sequence

$$0 \to \mathbb{Z}_{p_j} \to \mathbb{Z}_{p_j^2} \to \mathbb{Z}_{p_j} \to 0 .$$

Then $\rho_{j*} De_j = y_j(\beta_j y_j)^{2k-1}$, where $y_j \in H^1(B;\mathbb{Z}_{p_j})$ lies in the complement of Image ρ_{j*}. Since the latter and y_j generate the cohomology group, it follows that $f^*y_j = d_j y_j + \rho_{j*} z_j$ for suitable z_j. But $\beta_j \rho_{j*} = 0$, and therefore

$$f^*De_j = d_j^{2k} De_j \mod p_j \text{ and torison.}$$

Next let $e^j \in H^1(B;\mathbb{Z})$ $1 \le j \le 4n$ be a dual basis to the e_j. We claim that $f^* e^j = \pm e^j$. First, f is a homotopy equivalence, and therefore the infinite cyclic coverings of B determined by f^*e^j and e^j are homotopy equivalent.

We now need the following computational result:

SUBLEMMA 3.2. <u>In the above notation, suppose that</u> $\sum n_j e^j \in H^1(B;\mathbb{Z})$ <u>is a nonzero class with the nonzero</u> n_j <u>all relatively prime. Let</u> $\tilde{B}$ <u>be the associated infinite cyclic covering.</u> (i) If $u_k = 0$ <u>for some</u> k, <u>then the</u> p_k<u>-torsion in</u> $H_1(\tilde{B};\mathbb{Z})$ <u>is countably infinite.</u> (ii) <u>If</u> $n_k \neq 0$ <u>for some</u> k, <u>then the</u> p_k<u>-torsion in</u> $H_1(\tilde{B};\mathbb{Z})$ <u>is a mod</u> p_k <u>vector space of dimension</u> $|n_k|$.

REMARKS ON THE PROOF. It is easy to construct a degree one map from $B^{4n}(p_1,\ldots,p_{4n})$ to $C^{4n-1} \times S^1 = [T^{4n-1} \# L^{4n-1}(p_k)] \times S^1$ that is an isomorphism in p_k-local cohomology. Using this, one can reduce the proof to the corresponding assertion for $C^{4n-1} \times S^1$. After an allowable change of coordinates, the corresponding infinite cyclic covering on $C^{4n-1} \times S^1$ may be assumed to have a classifying map of the form

$$C^{4n-1} \times S^1 \xrightarrow{g \times id} S^1 \times S^1 \xrightarrow{\varphi} S^1,$$

where $\varphi(x,y) = x^a y^b$ for some relatively prime integers a and b (in fact, $b = n_k$ by construction), and g is given by collapsing C^{4n-1} onto T^{4n-1} and projecting onto a circle factor. If we denote the induced infinite cyclic covering of C^{4n-1} by $\hat{C}$ and its monodromy by S, then the desired infinite cyclic covering of $C \times S^1$ may be recovered by factoring out the $\mathbb{Z}$ action on $\hat{C} \times R$ generated by sending (x,y) to $(S^b x, y-a)$. But the homological and monodromy structure of C is well understood, and the sublemma now follows from some elementary Wang sequence calculations ∎

Using the sublemma, we obtain a good hold on the homological torsion of the infinite cyclic coverings associated to e^j and $f^* e^j$ for all j. In particular, if we do this for all primes p_k, the sublemma tells us that $f^* e^j$ must equal $\pm e^j$; the sign need not be the same for all j, at least by what we now know.

Of course, we want $f^* e^j = e^j$. Let $\omega = \neq 1$ be chosen so that f^* in dimension 4n is multiplication by ω. It then follows that $f^*De_j \cdot f^* e^j = \omega \cdot$ generator. But we know that $f^*De_j \equiv d_j^{2k} De_j$ mod p_j and torsion, and a little arithmetic now shows that $f^* e^j \equiv d_j^{-2k} \omega e^j$ mod p_j for <u>all</u> j. But we have chosen p_j congruent to 3 and 4, so that -1 is not a square mod p_j, and from this it follows that $f^* e^j = \omega e^j$ must hold for all j. Since $H^{4n-1}(B)/\text{Torsion}$ is generated by monomials in H^1, it follows that $f^* = \omega$ also on $H^{4n-1}(B)/\text{Torsion}$. Now apply $f^*De_j \equiv d_j^{2k} De_j$ mod p_j plus $d_j^{2k} \not\equiv -1$ mod p_j (we need $p_j \equiv 3$ mod 4 again) to deduce that $\omega = 1$. Therefore we have shown that f^* is the identity on $H^1(B;\mathbb{Z})$, as required ∎

COROLLARY 3.3. <u>The third conclusion of Theorem</u> 3.1 <u>remains true if we replace</u> B <u>by</u> $B\#\Sigma^{4n}$, <u>where</u> Σ^{4n} <u>is an arbitrary</u> 1-<u>connected manifold with no torsion of order</u> $p_1,\ldots,p_{4n}$.

This follows immediately upon retracing the proof ∎

4. Construction of examples

To give the desired examples, we need only find simply connected manifolds Σ^{4n} such that

(4.1) Σ^{4n} <u>has no odd torsion on its homology</u>.

(4.2) <u>The Euler characteristic of</u> Σ^{4n} <u>is</u> 3.

For suppose we have take such manifolds and form their connected sums with the manifolds $B^{4n}(p_1,\ldots,p_{4n})$. Then every periodic self-map of the connected sum $B\#\Sigma$ induces the identity in one-dimensional integral cohomology by 3.3. On the other hand, the Euler characteristic of $B\#\Sigma$ is 1 (recall that $\chi(A\#B) = \chi(A) + \chi(B)-2$ in even dimensions) and therefore by 2.3 very periodic map induces a faithful representation into $\mathrm{Aut}\, H^1(B\#\Sigma;\mathbb{Z})$. Therefore $B\#\Sigma$ admits no periodic maps.

There are many ways of finding such manifolds Σ^{4n}. Perhaps the easiest is to take the connected sum of CP^{2n} with (n-1) copies of $S^{2n+1} \times S^{2n-1}$ (notice that zero copies are used if n=1). Here is another elaboration which illustrates the ubiquity of manifolds without group actions:

PROPOSITION 4.3. Let M^{4n} be a closed oriented CAT manifold (= TOP, PL, or DIFF) with odd signature, and assume CAT $\neq$ TOP or $n \geq 2$. Then M^{4n} is CAT orientably bordant to infinitely many homologically distinct hypertoral manifolds that admit no group actions.

PROOF. We know that M^{4n} is orientably bordant to a simply connected manifold N^{4n} [15]; in the smooth case this manifold may be assumed to have only 2-primary torsion, while in the other cases the torsion involves only a finite list of primes depending only on n [11]. When necessary we may assume that $(p_1,\ldots,p_{4n})$ are none of these primes.

Since the index and Euler characteristic are congruent mod 2, the number $\chi(N^{4n})$ is also odd. Construct a simply connected π-manifold P^{4n} so that $\chi(P^{4n}) = 5-\chi(N)$; one can take P to be a connected sum of $S^{2n} \times S^{2n}$'s and $S^{2n-1} \times S^{2n+1}$'s. It follows that $\chi(N\#P\#B)=1$, and we merely need to check that B is an oriented smooth boundary (P is already because it is a π-manifold). Of course, it suffices to check that B has no Stiefel-Whitney or rational Pontrjagin numbers. But B is assembled from manifolds with boundary that have neither Stiefel-Whitney nor rational Pontrjagin classes, and therefore the only way characteristic classes might arise is from the boundary identifications. Thus all relevant characteristic classes are induced by the collapsing map

$$q:B \longrightarrow \bigvee_{i=1}^{4n} \Sigma(S^1 \times S^{4n-2}),$$

followed by a map from the codomain of q into $BSO_{(2)}$. The only possible problem might come from a second Stiefel-Whitney class (for example, the Hirzebruch signature theorem implies there is no top dimensional Pontrjagin class). However, one can always make $w_2 = 0$

by choosing the tubular neighborhoods $C_j \times D^{4n-1}$ with sufficient care ■

Final Remark. As for other examples of manifolds with group actions, ours depend crucially on the use of information about the fundamental group. Accordingly, the question of whether one can find *simply connected* closed manifolds without group actions - even involutions - still must be regarded as completely open. In fact, it even may be premature to try and make an educated guess about the answer.

REFERENCES

1. E. M. Bloomberg, Manifolds with no periodic homeorphisms, Trans. Amer. Math. Soc. 202 (1975), 67-78.

2. A. Borel (ed), Seminar on Transformation Groups Ann. of Math. Studies No. 46. Princeton University Press, Princeton, 1960.

3. D. Burghelea and R. Schultz, On the semisimple degree of symmetry, Bull. Soc. Math. France 103 (1975), 433-440.

4. P. Conner and F. Raymond, Injective operations of the toral groups I, Topology 10(1971), 283-296.

5. ―――――― and ――――――, Manifolds with few periodic homeomorphisms Proc. Second Conf. on Compact Transf. Gps. (U. of Mass, Amherst, 1971) Part II, Lecture Notes in Mathematics Vol 299, 1-75. Springer, New York, 1972.

6. ――――――, ――――――, and P. Weinberger, Manifolds with no periodic maps, Proc. Second. Conf. on Compact Transf. Gps. (U. of Mass., Amherst, 1971) Part II, Lecture Notes in Mathematics Vol. 299, 81-108. Springer, New York, 1972.

7. M. Greenberg, Lectures on Algebraic Topology. Benjamin, New York, 1967.

8. E. A. Grove, letter to the author (University of Rhode Island, Kingston RI, dated January 24, 1975).

9. R. C. Kirby and L. C. Siebenmann, On the triangulation of manifolds and the Hauptvermuting, Bull. Amer. Math. Soc. 75(1969), 742-749.

10. J. Milnor, Singular Points on Complex Surfaces, Ann. of Math Studies No. 61. Princeton University Press, Princeton, 1968.

11. S. Papastavridis, A note on killing torsion of manifolds by surgery, Proc. Amer. Math. Soc. 69(1978), 181-182.

12. F. Raymond and J. Tollefson, Closed 3-manifolds with no periodic maps, Trans. Amer. Math. Soc. 221(1976), 403-418.

13. P. A. Smith, The topology of transformation groups, Bull. Amer. Math. Soc. 44(1938), 497-514.

14. ――――――, Periodic and nearly periodic transformations Lectures in Topology (Univ. of Michigan Conference, 1940), 159-190. University of Michigan Press, Ann Arbor, 1941.

15. R. Stong, Notes on Cobordism Theory Mathematical Notes No. 7. Princeton University Press, Princeton, 1968.

16. S. Schwartzman, A split action associated with a compact transformation group, preprint, University of Rhode Island, 1978.

Graeme Segal's Burnside Ring Conjecture

J.F. Adams

§1. In my lecture, I spent a little time explaining Graeme Segal's Burnside Ring Conjecture. Here I shall assume that those who read a written account are more likely to have some background; if not they can be referred to accounts such as [3]. I shall simply say that the starting-point for Segal's conjecture seems to have been the theorem of Atiyah [2] which says that a certain map $R(G)^{\wedge} \longrightarrow K(BG)$ is iso. For present purposes G is a finite group. Segal replaced the representation ring R(G) by the Burnside ring A(G), which is defined like R(G), but using finite G-sets instead of representations. For the moment it is enough to know that A(G) is of an elementary and computable nature. Segal also replaced K-theory by stable cohomotopy $\pi_S^*(\ )$; this is the generalised cohomology theory corresponding to the sphere spectrum. (Like homotopy, it is hard to compute, but it contains valuable topological information if you can compute it.) Segal then defined a natural transformation

$$A(G)^{\wedge} \longrightarrow \pi_S^0(BG),$$

and conjectured that it is iso.

In the trivial case G = 1 the conjecture is trivially true. The next case is $G = Z_2$; in this case the space BG is real projective space RP^{∞}, so that the problem is to compute the stable cohomotopy group $\pi_S^0(RP^{\infty})$; and the suggested answer $A(Z_2)^{\wedge}$ is $Z \oplus Z_2^{\wedge}$, the direct sum of the integer and the 2-adic integers. This case presents a definite problem, which has been giving trouble for nine or ten years.

However, this case $G = Z_2$ has recently been settled (in the affirmative) by W.H. Lin.

Theorem 1.1. (Lin) (i) If $n > 0$, then the stable cohomotopy group $\pi_S^n(RP^\infty)$ is zero.

(ii) In the case $n = 0$, Graeme Segal's map

$$A(Z_2)^\wedge \longrightarrow \pi_S^0(RP^\infty)$$

is iso.

Here I should explain that part (i) is not one of those trivial results which hold for dimensional reasons. If you attack this problem by the methods of obstruction-theory, you find (for each $n > 0$) infinitely many non-zero cohomology groups $H^m(RP^\infty; \pi_S^{n-m}(pt))$. The content of part (i) is that in some way all these groups must cancel out.

Lin's manuscript is 57 pages long and involves substantial calculation. A simplification of the algebraic part of Lin's proof has been found by Davis and Mahowald. I shall present a version of their work which I hope is simpler yet.

Lin's methods prove a result a little more general that (1.1); one can replace the space RP^∞ by a spectrum X which has the cohomological behaviour of RP^∞/RP^{k-1}; and we shall see that this formulation makes sense even for $n \leq 0$.

To be more precise, let M be a given module over the mod 2 Steenrod algebra A; I suppose further that M is bounded below, and finitely generated over Z_2 in each degree. I shall say that a spectrum X is "of type M" if it satisfies the following conditions. (i) Its mod 2 cohomology $H^*(X; Z_2)$ is isomorphic (as an A-module) to M. (ii) Its homotopy groups are bounded below, so that X is m-connected for some m. (iii) $H_*(X; Z)$ is finitely-generated in each degree, and has no odd torsion.

The modules M which I propose to use are as in [1]. Take the ring of finite Laurent polynomials $P = Z_2[x, x^{-1}]$ in one variable x of degree 1, and make it into an A-module by setting

$$Sq^i(x^j) = \frac{j(j-1)\dots(j-i+1)}{1.\ 2\ \dots\dots\ i}\ x^{i+j}.$$

Let P_k^∞ be the submodule of P which has as a Z_2-base the powers x^j with $j \geq k$. It now makes sense to speak of a spectrum "of type P_k^∞". In particular, the suspension spectrum of the space RP^∞/RP^{k-1} qualifies as a spectrum of type P_k^∞ if $k \geq 1$; and the suspension spectrum of $RP^\infty \cup pt$ qualifies as a spectrum of type P_0^∞.

<u>Theorem 1.2 (after Lin)</u>. Let X be a spectrum of type P_k^∞.

(i) If $n > 0$ and $k < n$ then $[X, S^n] = 0$.

(ii) If $n = 0$ and $k < n$ then the group $[X, S^n]$ (with its filtration topology) is $Z_2^\wedge$ (with its 2-adic topology); and a map $f: X \longrightarrow S^0$ is zero or non-zero mod $2[X, S^0]$ according as $f^*: H^0(X; Z_2) \longleftarrow H^0(S^0; Z_2)$ is zero or non-zero.

(iii) If $n < 0$ and $k < 0$, let $f: X \longrightarrow S^0$ be non-zero mod 2, and consider the map $f^*: {}_2[S^0, S^n] \longrightarrow [X, S^n]$, where ${}_2G$ means the 2-component of G. This map f^* is iso if $k < n-1$, epi if $k = n-1$.

In this theorem, [X, Y] means homotopy classes of maps from X to Y in a suitable category of spectra. In line with this, the symbol S^n means the n-fold suspension of the sphere spectrum; and $H^0(S^0; Z_2)$ means the 0th cohomology of the sphere spectrum, that is, Z_2.

The idea of the restriction $k < n$ in parts (i), (ii) is that $[X, S^n]$ does not depend on the structure of X in degrees less than (n-1), but does depend on its structure in higher degrees. The idea of the restriction $k < n-1$ in part (iii) is that the proof unfortunately loses one dimension.

Theorem 1.1(i) follows immediately from Theorem 1.2(i), by taking X to be the suspension spectrum of $RP^{\infty} \cup pt$. The deduction of Theorem 1.1(ii) is not quite so immediate, because to apply Theorem 1.2(ii) we need to relate $RP^{\infty} \cup pt$ to a spectrum X which goes down and has a cell in degree -1; it is easy to supply such a cell and keep track of the difference it makes.

Theorem 1.2 can be S-dualised into a statement about stable homotopy. The number 1.3 is reserved for this statement, but in this account I shall omit it.

To prove Theorem 1.2 one can use the Adams spectral sequence

$$\mathrm{Ext}_A^{**}(Z_2, H^*(X; Z_2)) \Longrightarrow [X, S^0]_*.$$

To use this spectral sequence - apart from overcoming technical difficulties with its convergence - one must begin by computing the Ext groups. Now on the face of it, these are some of the most difficult Ext groups which have ever been computed in this line of business; and the only thing which gives one hope of a simple proof is that the answer turns out to be simple.

In order to state it, let $P = Z_2[x, x^{-1}]$ be as above. Z_2, as a graded module, will mean Z_2 in degree zero; then we have a monomorphism

$$\phi : Z_2 \longrightarrow P$$

defined by $\phi(\lambda) = \lambda x^0$. (The algebraic map ϕ corresponds to f in (1.2).) The suspension $\Sigma^j M$ of a graded module M is defined by $(\Sigma^j M)_{i+j} = M_i$; then we have an epimorphism

$$\gamma : P \longrightarrow \Sigma^{-1} Z_2$$

defined by $\gamma(\sum_i \lambda_i x^i) = \lambda_{-1}$.

<u>Theorem 1.4</u> (Lin).

(i) The induced map

$$\gamma_*: \mathrm{Tor}^A_{s,t}(Z_2, P) \longrightarrow \mathrm{Tor}^A_{s,t}(Z_2, \Sigma^{-1} Z_2)$$

is iso.

(ii) The induced map

$$\gamma^*: \mathrm{Ext}_A^{s,t}(\Sigma^{-1} Z_2, Z_2) \longrightarrow \mathrm{Ext}_A^{s,t}(P, Z_2)$$

is iso.

(iii) The induced map

$$\phi_*: \mathrm{Ext}_A^{s,t}(Z_2, Z_2) \longrightarrow \mathrm{Ext}_A^{s,t}(Z_2, P)$$

is iso.

The deduction of the topological results from Theorem 1.4 is due to Lin; Davis and Mahowald do not seem to suggest any change in that part of the argument.

I will now move on to discuss the contribution of Davis and Mahowald, who give a new proof of Theorem 1.4 by studying the structure of P. In fact, I will sketch a series of reductions of (1.4). Let A_r be the subalgebra of A generated by the Sq^{2^i} with $i \leq r$; then we wish to compute (for example) $\mathrm{Tor}^A_{s,t}(Z_2, P)$; but it is sufficient to compute $\mathrm{Tor}^{A_r}_{s,t}(Z_2, P)$, because we can pass to a (direct) limit over r.

<u>Lemma 1.5</u>. As a module over A_r, P is generated by the powers x^j with $j \equiv -1 \bmod 2^{r+1}$.

Proof. If $j \equiv -1 \bmod 2^{r+1}$ and $0 \le i < 2^{r+1}$, then $Sq^i \in A_r$ and $Sq^i x^j = x^{i+j}$.

Let $F_{\ell,r}$ be the A_r-submodule of P generated by the x^j with $j < \ell$. By Lemma 1.5 it is actually sufficient to consider the $F_{\ell,r}$ with $\ell \equiv -1 \bmod 2^{r+1}$; but for (1.8), (1.9) it will be convenient to index the $F_{\ell,r}$ as above.

We wish to compute (say) $\mathrm{Tor}^{A_r}_{s,t}(Z_2, P)$, but it is sufficient to compute $\mathrm{Tor}^{A_r}_{s,t}(Z_2, P/F_{\ell,r})$, because we can pass to an (attained) limit over ℓ. It is now sufficient to compute $\mathrm{Tor}^{A}_{s,t}(Z_2, A \otimes_{A_r} (P/F_{\ell,r}))$, by a change-of-rings theorem.

Lemma 1.6 (Davis and Mahowald). There is an isomorphism of A-modules

$$A \otimes_{A_r} (P/F_{\ell,r}) \cong \bigoplus_j \Sigma^j (A \otimes_{A_{r-1}} Z_2)$$

where j runs over the set $j \equiv -1 \bmod 2^{r+1}$, $j \ge \ell$.

This lemma answers the purpose of computing $\mathrm{Tor}^{A}_{s,t}(Z_2, A \otimes_{A_r} (P/F_{\ell,r}))$.

I remark that $P/F_{\ell,r}$ does not split as a sum of cyclic modules over A_r; it is essential to pass to $A \otimes_{A_r} (P/F_{\ell,r})$. (For all homological purposes, to look at $A \otimes_{A_r} (P/F_{\ell,r})$ over A is the same as looking at $P/F_{\ell,r}$ over A_r; but structure-theory is not part of homology.) Moreover, it is no use to go to a limit over ℓ and try to state a similar result for $A \otimes_{A_r} P$; it is essential to pass to Tor before taking the limit over ℓ.

In order to make good the steps of the reduction, one must consider the behaviour of the isomorphism in (1.6) as ℓ and r vary, so that one can pass to the limits in question; and one must also consider the relation of the isomorphism in (1.6) to the map γ which occurs in (1.4). In the former direction, we have to consider the following diagrams (1.7), (1.8).

$$
(1.7) \qquad \begin{array}{ccc}
A \otimes_{A_r} P/F_{\ell,r} & \stackrel{\cong}{\longleftrightarrow} & \bigoplus_j \Sigma^j (A \otimes_{A_{r-1}} Z_2) \\
\big\downarrow & & \big\downarrow \theta \\
A \otimes_{A_r} P/F_{m,r} & \stackrel{\cong}{\longleftrightarrow} & \bigoplus_k \Sigma^k (A \otimes_{A_{r-1}} Z_2)
\end{array}
$$

Here the left-hand vertical arrow is the obvious quotient map, which exists when $\ell \le m$; and the horizontal arrows are as in Lemma 1.6. The index j runs over the set $j \equiv -1 \bmod 2^{r+1}$, $j \ge \ell$, and the index k runs over the set $k \equiv -1 \bmod 2^{r+1}$, $k \ge m$. The map θ has the obvious components, namely 0 if $j < m$, and the identity map if $j = k \ge m$.

$$
(1.8) \qquad \begin{array}{ccc}
A \otimes_{A_r} P/F_{\ell,r} & \stackrel{\cong}{\longleftrightarrow} & \bigoplus_j \Sigma^j (A \otimes_{A_{r-1}} Z_2) \\
\big\downarrow & & \big\downarrow \psi \\
A \otimes_{A_{r+1}} P/F_{\ell,r+1} & \stackrel{\cong}{\longleftrightarrow} & \bigoplus_k \Sigma^k (A \otimes_{A_r} Z_2)
\end{array}
$$

Here the left-hand vertical arrow is the obvious quotient map, and the horizontal arrows are as in Lemma 1.6. The index j runs over the set $j \equiv -1 \bmod 2^{r+1}$, $j \ge \ell$, and the index k runs over the set $k \equiv -1 \bmod 2^{r+2}$, $k \ge \ell$; that is, just half of the values of j correspond to values of k. The map ψ has the obvious components: if $j = k \equiv -1 \bmod 2^{r+2}$ we take the obvious quotient map

$$\Sigma^k(A \otimes_{A_{r-1}} Z_2) \longrightarrow \Sigma^k(A \otimes_{A_r} Z_2),$$

and if $j \equiv 2^{r+1} - 1 \bmod 2^{r+2}$ we take the zero map of $\Sigma^j(A \otimes_{A_{r-1}} Z_2)$.

<u>Lemma 1.9</u> (Davis and Mahowald). The isomorphisms of Lemma 1.6 can be chosen so that Diagrams 1.7 and 1.8 commute, and so that for $\ell \leq -1$ the composite

$$\Sigma^{-1}(A \otimes_{A_{r-1}} Z_2) \longrightarrow \bigoplus_j \Sigma^j(A \otimes_{A_{r-1}} Z_2) \cong A \otimes_{A_r} P/F_{\ell,r} \xrightarrow{1 \otimes \gamma} A \otimes_{A_r} \Sigma^{-1} Z_2$$

is the obvious quotient map.

The proof which I suggest for Lemma 1.6 will be given in §2, and the proof which I suggest for Lemma 1.9 will be given in §3.

It remains to discuss our hopes of further progress.

(i) First one should try the case $G = Z_p$, $p > 2$. There is no visible reason why the same method should not be tried; in fact, I have a student trying it.

(ii) Secondly one should try the case in which G is a p-group. The obvious way is to try to copy Atiyah's argument by induction over the order of G, using an exact sequence $H \longrightarrow G \longrightarrow Z_p$ and assuming the result for H. For various reasons it is clear that this cannot work in quite the same way as in Atiyah's case; however, I am not yet convinced that it cannot work at all. Of course, I have no reason to think that it can work, either.

(iii) Thirdly one should try the general case. The obvious way is to study the relation between G and its Sylow p-subgroups. I have the impression that if steps (i) and (ii) work then one should be able to deduce something for a general G, but I do not know if one could deduce all that is conjectured.

§2. In this section I will prove Lemma 1.6.

The modules $P/F_{\ell,r}$ for different values of ℓ become isomorphic after regrading; so it is sufficient to consider one value of ℓ, say $\ell = -1$. And as we only have to consider one value of r in this section, there is no need to display r either; so for brevity I write

$$F = F_{-1,r}\ ,\ F' = F_{2^{r+1}-1,r}\ ,\ F'' = F_{2.2^{r+1}-1,r}\ .$$

Lemma 2.1. In P we have $Sq^{2^i} x^{-1} \in F = F_{-1,r}$ if $i < r$.

Proof. It is sufficient to display the equation

$$Sq^{2^i} x^{-1} = Sq^{2^r} x^{2^i-1-2^r}\ .$$

Lemma 2.2. We have the following exact sequence of A_r-modules:

$$0 \longrightarrow \Sigma^{-1} A_r \otimes_{A_{r-1}} Z_2 \longrightarrow P/F \longrightarrow P/F' \longrightarrow 0.$$

Proof. It is clear that we have an exact sequence

$$0 \longrightarrow F'/F \longrightarrow P/F \longrightarrow P/F' \longrightarrow 0;$$

moreover, Lemma 2.1 shows that one can define a map

$$\Sigma^{-1} A_r \otimes_{A_{r-1}} Z_2 \longrightarrow F'/F$$

by sending $a \otimes 1$ to ax^{-1}. This map is onto, by (1.5); to show that it is iso, it is sufficient to show that both sides have rank 2^{r+1} over Z_2. This is known for $\Sigma^{-1} A_r \otimes_{A_{r-1}} Z_2$, and I prove it for F'/F.

In fact, choose a residue class ρ mod 2^{r+1}. Then the values of j in ρ for which x^j lies in F' form a descending segment, say $(j_o, j_o - 2^{r+1}, j_o - 2.2^{r+1}, \ldots,)$; and the values of j in ρ for which x^j lies in F form the subsegment $(j_o - 2^{r+1}, j_o - 2.2^{r+1}, \ldots)$.

So F'/F has a Z_2-base consisting of one power x^{j_o} for each residue class mod 2^{r+1}. This proves Lemma 2.2.

Lemma 2.3. We have the following exact sequence of A-modules.

$$0 \longrightarrow \Sigma^{-1}A \otimes_{A_{r-1}} Z_2 \xrightarrow{\alpha} A \otimes_{A_r} P/F \longrightarrow A \otimes_{A_r} P/F' \longrightarrow 0.$$

Proof. This follows by taking (2.2) and applying the functor $A \otimes_{A_r}$, which preserves exactness since A is free as a right module over A_r.

I will next show that the exact sequence in (2.3) splits. For this purpose I recall Milnor's work on the dual of the Steenrod algebra [4]. Let A_* be the dual of A; it is a polynomial algebra $Z_2[\xi_1,\xi_2,\ldots,\xi_k,\ldots]$ on generators ξ_k of degree 2^k-1; and we have

$$\psi\, \xi_k = \sum_{i+j=k} \xi_i^{2^j} \otimes \xi_j$$

where ξ_o is interpreted as 1. The dual of the subalgebra A_r is the quotient

$$A_*/(\xi_1^{2^{r+1}}, \xi_2^{2^r}, \ldots, \xi_r^{2^2}, \xi_{r+1}^2, \xi_{r+2}, \xi_{r+3}, \ldots).$$

I now introduce the quotient of A_* by a smaller ideal namely

$$B_* = A_*/(\xi_2^{2^r}, \ldots, \xi_r^{2^2}, \xi_{r+1}^2, \xi_{r+2}, \xi_{r+3}, \ldots).$$

It is easy to verify that B_* is a left comodule with respect to A_{r_*} and a right comodule with respect to A_{r-1_*}. Let B^* be the dual of B_*; it is a subspace of $A^* = A$; it is a left module over A_r and a right module over A_{r-1}.

Lemma 2.4. There is an isomorphism of A_r-modules

$$\beta : \sum^{-1} B \otimes_{A_{r-1}} Z_2 \longrightarrow P/F$$

which sends $b \otimes 1$ to bx^{-1}.

Proof. The prescription $\beta(b \otimes 1) = bx^{-1}$ gives a well-defined map of $\sum^{-1} B \otimes_{A_{r-1}} Z_2$, by (2.1); and it is a map of A_r-modules. It is onto, for $Sq^i \in B^*$ (each $i \geq 0$), $Sq^i x^{-1} = x^{i-1}$ and the elements x^{i-1} span P/F. In order to prove that β is iso, it is sufficient to note that $\sum^{-1} B^* \otimes_{A_{r-1}} Z_2$ and P/F have the same Poincaré series. In fact, since we know the structure of B^*, and B^* is free as a right module over A_{r-1} by Theorem 4.4 of Milnor-Moore [5], we find that the Poincaré series for $B^* \otimes_{A_{r-1}} Z_2$ is

$$\frac{1}{1-t^{2^r}} (1 + t^{3.2^{r-1}})(1 + t^{7.2^{r-2}})\dots(1 - t^{(2^r-1)2})(1 - t^{2^{r+1}-1}).$$

On the other hand, using Lemma 2.2 one can filter P/F so as to obtain a subquotient $A_r \otimes_{A_{r-1}} Z_2$ every 2^{r+1} dimensions, and we find that the Poincaré series for $\sum(P/F)$ is

$$\frac{1}{(1-t^{2^{r+1}})} (1 + t^{2^r})(1 + t^{3.2^{r-1}})\dots(1 - t^{(2^r-1)2})(1 - t^{2^{r+1}-1}).$$

This proves Lemma 2.4.

Proof of Lemma 1.6. Consider the following diagram.

$$\begin{array}{ccc}
\sum^{-1} A \otimes_{A_{r-1}} Z_2 & \xrightarrow{\ \alpha\ } & A \otimes_{A_r} P/F \\
\uparrow 1 & & \uparrow 1\otimes\beta \ \cong \\
\sum^{-1} A \otimes_{A_{r-1}} Z_2 & \xleftarrow{\ \mu\otimes 1\ } & \sum^{-1} A \otimes_{A_r} B^* \otimes_{A_{r-1}} Z_2
\end{array}$$

Here α and β are as in (2.3), (2.4), while μ is given by the product map for A, that is, $\mu(a \otimes b) = ab$. It is easy to check that the diagram is commutative. Thus the exact sequence in (2.3) splits and gives

$$A \otimes_{A_r} P/F \cong (\Sigma^{-1} A \otimes_{A_{r-1}} Z_2) \oplus (A \otimes_{A_r} P/F').$$

But the same conclusion applies to P/F', so that

$$A \otimes_{A_r} P/F \cong (\Sigma^{-1} A \otimes_{A_{r-1}} Z_2) \oplus (\Sigma^{2^{r+1}} A \otimes_{A_{r-1}} Z_2) \oplus (A \otimes_{A_r} P/F'').$$

Continuing by induction, we obtain Lemma 1.6.

§3. In this section I will prove Lemma 1.9. To this end, I begin by giving more explicit formulae for the splitting which was obtained by induction at the end of §2.

I first introduce the element

$$y_k = \sum_{i+j=k} \chi(Sq^i) \otimes x^j \in A \otimes_{A_r} P/F_{\ell,r}.$$

Here χ is the canonical anti-automorphism of A; and the sum is finite, since we only have to consider the range $i \geq 0$, $j \geq \ell$. Then we have the following more precise form of Lemma 1.6.

Lemma 3.1. The A-module $A \otimes_{A_r} P/F_{\ell,r}$ is a direct sum (over k such that $k \equiv -1 \bmod 2^{r+1}$, $k \geq \ell$) of cyclic submodules $\Sigma^k (A \otimes_{A_{r-1}} Z_2)$ with generators y_k.

Proof. Consider the explicit splitting used in proving Lemma 1.6; it displays P/F as the direct sum of the cyclic submodule

$\Sigma^{-1} A \otimes_{A_{r-1}} Z_2$, on the generator $x^{-1} = y_{-1}$, and a complementary summand, namely the kernel of the splitting map $(\mu \otimes 1)(1 \otimes \beta)^{-1}$. I claim that this kernel contains the remaining elements y_k, that is, those with $k \geq 0$. In fact, we have

$$\beta \, Sq^{j+1} = x^j$$

(where j runs over the range $j \geq -1$, so that $j+1$ runs over the range $j+1 \geq 0$). Thus

$$\begin{aligned} &(\mu \otimes 1)(1 \otimes \beta)^{-1} \Big(\sum_{i+j=k} \chi(Sq^i) \otimes x^j \Big) \\ &= \sum_{i+(j+1)=(k+1)} \chi(Sq^i) Sq^{j+1} \\ &= 0 \quad \text{if} \quad k+1 \geq 1. \end{aligned}$$

On the other hand, the periodicity isomorphisms

$$P/F' \cong \Sigma^{2^{r+1}} P/F$$

$$P/F'' \cong \Sigma^{2^{r+1}} P/F', \text{ etc},$$

clearly carry elements y_k to other elements $y_{k'}$. It is now clear that the inductive process used in proving Lemma 1.6 displays P/F as a direct sum of cyclic submodules $\Sigma^k (A \otimes_{A_{r-1}} Z_2)$ on generators y_k. This proves Lemma 3.1.

It is now clear that Diagram (1.7) commutes; in fact, this has been clear since we constructed the splitting by induction. Moreover, the composite

$$\Sigma^{-1}(A \otimes_{A_{r-1}} Z_2) \longrightarrow \bigoplus_j \Sigma^j (A \otimes_{A_{r-1}} Z_2) \cong A \otimes_{A_r} P/F_{\ell,r} \xrightarrow{1\otimes\gamma} A \otimes_{A_r} \Sigma^{-1} Z_2$$

carries the generator 1, via y_{-1}, to 1. To complete the proof of Lemma 1.9, I have to show that Diagram (1.8) commutes; and this will follow from the first half of the following lemma.

Lemma 3.2. The element $y_k \in A \otimes_{A_r} P/F_{\ell,r}$ is zero unless $k \equiv -1 \bmod 2^{r+1}$; and then it is equal to the sum

$$\sum_{i+j=k} \chi(Sq^i) \otimes x^j$$

where i and j are restricted to the residue classes

$$i \equiv 0 \bmod 2^{r+1}$$

$$j \equiv -1 \bmod 2^{r+1}.$$

The proof of this lemma requires identities in A and P.

Lemma 3.3. There exist a finite number of elements $a_i = a_{i,r} \in A_r$, of degree $2^{r+1} i + 2^r$, such that

(i) $$Sq^{2^{r+1}k+2^r} = \sum_{i+j=k} a_i \, Sq^{2^{r+1}j},$$

(ii) $$\sum_{i+j=k} \chi(a_i) x^{2^{r+1}j-1} = x^{2^{r+1}k+2^r-1},$$

(iii) $$\sum_{i+j=k} \chi(a_i) x^{2^{r+1}j+2^r-1} = 0.$$

The prototype of these identities may be seen for $r = 0$; we have one element Sq^1, and

$$Sq^{2k+1} = Sq^1 \, Sq^{2k},$$

$$Sq^1 x^{2k-1} = x^{2k},$$

$$Sq^1 x^{2k} = 0.$$

Otherwise, the best way to justify these identities is to use them.

Proof of Lemma 3.2, assuming Lemma 3.3. The proof is by induction over r. The result is trivially true for $r = -1$ provided we interpret A_{-1} as Z_2, so we assume it true for $r-1$. Then y_k is zero in $A \otimes_{A_{r-1}} P/F_{\ell,r-1}$ unless $k \equiv -1 \bmod 2^r$, so we have to consider only two cases, $k \equiv -1 \bmod 2^{r+1}$ and $k \equiv 2^r-1 \bmod 2^{r+1}$. In the first case, let $k = 2^{r+1}m - 1$; the inductive hypothesis gives

$$y_k = \sum_{i+j=m} \chi(Sq^{2^{r+1}i}) \otimes x^{2^{r+1}j-1}$$
$$+ \sum_{i+j=m-1} \chi(Sq^{2^{r+1}i+2^r}) \otimes x^{2^{r+1}j+2^r-1} .$$

One can rewrite the second sum using (3.3)(i), and we obtain

$$\sum_{e+h+j=m-1} \chi(Sq^{2^{r+1}e})\chi(a_h) \otimes x^{2^{r+1}j+2^r-1}$$
$$= \sum_{e+h+j=m-1} \chi(Sq^{2^{r+1}e}) \otimes \chi(a_h)x^{2^{r+1}j+2^r-1}$$

(since the tensor product is taken over A_r, and $\chi(a_h) \in A_r$). But this gives zero, by (3.3)(iii).

In the second case, let $k = 2^{r+1}m+2^r-1$; the inductive hypothesis gives

$$y_k = \sum_{i+j=m} \chi(Sq^{2^{r+1}i+2^r}) \otimes x^{2^{r+1}j-1}$$
$$+ \sum_{i+j=m} \chi(Sq^{2^{r+1}i}) \otimes x^{2^{r+1}j+2^r-1}.$$

One can rewrite the first sum using (3.3)(i) as above, and we get

$$\sum_{e+h+j=m} \chi(Sq^{2^{r+1}e})\ \chi(a_h) \otimes x^{2^{r+1}j-1}$$
$$= \sum_{e+h+j=m} \chi(Sq^{2^{r+1}e}) \otimes \chi(a_h)\ x^{2^{r+1}j-1}$$
$$= \sum_{e+n=m} \chi(Sq^{2^{r+1}e}) \otimes x^{2^{r+1}n+2^r-1}$$

(using (3.3)(ii)). So we see that $y_k = 0$ in this case. This proves Lemma 3.2, assuming Lemma 3.3.

Proof of Lemma 3.3. With the notation of §2, B^* is free as a left module over A_r, by Theorem 4.4 of Milnor-Moore [5]; and in fact one can take the elements $Sq^{2^{r+1}j}$ as an A_r-base. Therefore, we have for each k a unique formula

$$Sq^{2^{r+1}k+2^r} = \sum_{i+j=k} a_i(k) Sq^{2^{r+1}j}$$

with coefficients $a_i(k) \in A_r$. Since A_r is a finite algebra and $a_i(k)$ is of degree $2^{r+1}i+2^r$, the sum can be taken over a finite range of i which does not depend on k. Moreover, the coefficient $a_i(k)$ does not depend on k provided that k is sufficiently large; for in the dual, multiplying with $\xi_1^{2^{r+1}}$ gives an isomorphism of everything at issue. Let us write a_i for the common value of $a_i(k)$; then the formula

$$Sq^{2^{r+1}k+2^r} = \sum_{i+j=k} a_i \ Sq^{2^{r+1}j}$$

remains true for small values of k, provided we interpret $Sq^{2^{r+1}j}$ as zero for $j < 0$.

This proves part (i); I turn to parts (ii) and (iii). That is, I want to determine the sum of the operations $\chi(a_i)$ in P which map into a degree congruent mod 2^{r+1} to -1 or 2^r-1 according to the case (where this sum is counted as 0 or 1 in the obvious way). In part (iii) the sum is obviously zero, for every operation $a \in I(A)$ in P is zero out of degree 0 or (S-dually) into degree -1, and every operation $a \in I(A_r)$ is zero into a degree congruent to -1 mod 2^{r+1}. In part (ii), it is equivalent (by S-duality) to determine the sum of the operations a_i mapping out of a degree congruent to 2^r mod 2^{r+1}. But one can take the original formula

$$Sq^{2^{r+1}k+2^r} = \sum_{i+j=k} a_i \, Sq^{2^{r+1}j}$$

and apply it to the class x^{-2^r}; on this class all the operations $Sq^{2^{r+1}k+2^r}$ and $Sq^{2^{r+1}j}$ are 1, so we see that the required sum of the operations a_i is 1. This proves Lemma 3.3, which completes the proof of lemmas 3.2 and 1.9.

References

[1] J.F. Adams, Operations of the nth kind in K-theory, and what we don't know about RP^{∞}, in London Math. Soc. Lecture Note Series no 11, Cambridge U.P. 1974, pp1-9.

[2] M.F. Atiyah, Characters and cohomology of finite groups, Publ. Math. of the I.H.E.S. no 9, 1961.

[3] E. Laitinen, On the Burnside ring and stable cohomotopy of a finite group, Aarhus University publication, 1978.

[4] J. Milnor, The Steenrod algebra and its dual, Ann. Math.(2) 67 (1958) pp150-171.

[5] J. Milnor and J.C. Moore, On the structure of Hopf algebras, Ann. Math.(2) 81 (1965) pp211-264.

[6] V. Snaith, On the stable cohomotopy of RP^{∞}, Proc. Amer. Math. Soc. 69 (1978) pp174-176.

AN ALGEBRAIC PROOF OF A THEOREM OF J. MILNOR

Jean BARGE

In [1], J. Milnor proved by geometric means the following result. Let $\overline{V}$ be an infinite cyclic covering of an n-dimensional, oriented, closed manifold V. Assume that the homology groups $H_*(\overline{V}, \mathbb{Q})$ are finite dimensional, then the covering $\overline{V}$ satisfies Poincaré duality with rational coefficients like an n-1-dimensional, oriented, closed manifold.

The goal of this lecture is to provide a completely algebraic proof of this result. This method allows us to generalize this result elsewhere to a whole family of Galois coverings [2]. Since on one hand, the manifold V satisfies Poincaré duality with coefficients $\mathbb{Q}[\mathbb{Z}]$ and since, on the other hand, we have :

$$H_*(V, \mathbb{Q}[\mathbb{Z}]) = H_*(\overline{V}, \mathbb{Q}),$$

Milnor's theorem reduces to the following equations :

$$H^{i+1}(V, \mathbb{Q}[\mathbb{Z}]) = H^i(\overline{V}, \mathbb{Q}), \quad \text{for all } i.$$

But, in fact, these equations result from the following purely algebraic claim.

THEOREM. Let C_* be a differential complex of projective $\mathbb{Z}[\mathbb{Z}]$-modules and denote by $\overset{\circ}{C}_*$ the same complex considered as complex of modules over $\mathbb{Z}$. Assume that the homology groups $H_*(\overset{\circ}{C}_*; \mathbb{Q}) = H_*(C_*, \mathbb{Q}[\mathbb{Z}])$ are finite dimensional over $\mathbb{Q}$; then the cohomology groups

$$H^{i+1}(C_*; \mathbb{Q}[\mathbb{Z}]) \text{ and } H^i(\overset{\circ}{C}_*; \mathbb{Q})$$

are isomorphic for all i.

In fact, one can in the preceeding result, replace the rationals by any unitary subring, Λ. Therefore, Milnor's theorem is true for coefficients Λ if one assumes of course that the homology groups $H_*(\overline{V}\,;\Lambda)$ are finitely generated over Λ. For the sake of simplicity, we leave to the reader the proof of these generalizations.

I am grateful to J. Lannes and J.J. Sansuc for fruitful conversations.

One identifies the ring $\mathbb{Z}[\mathbb{Z}]$ (resp $\mathbb{Q}[\mathbb{Z}]$) with the ring of Laurent polynomials $\mathbb{Z}[x,x^{-1}]$ (resp $\mathbb{Q}[x,x^{-1}]$). One denotes by $\mathbb{Q}(x)$ their quotient field and by $\mathbb{Q}[[x,x^{-1}]]$ the module of formal series (infinite on both sides)

LEMMA 1. For any differential complex of $\mathbb{Z}[x,x^{-1}]$-modules, C_*, one has :

$$H^1(\mathring{C}_*,\mathbb{Q}) = H^1(C_*;\mathbb{Q}[[x,x^{-1}]]) .$$

Proof. On the category of $\mathbb{Z}[x,x^{-1}]$-modules, the two functors $\mathrm{Hom}_{\mathbb{Z}}(-\,;\mathbb{Q})$ and $\mathrm{Hom}_{\mathbb{Z}[x,x^{-1}]}(-,\mathbb{Q}[[x,x^{-1}]])$ are isomorphic. □

Thus it is now sufficient to compare the coefficients $\mathbb{Q}[x,x^{-1}]$ and $\mathbb{Q}[[x,x^{-1}]]$.

LEMMA 2. The sequence : $0 \to \mathbb{Q}[x,x^{-1}] \xrightarrow{\ i\ } \mathbb{Q}(x) \xrightarrow{\ T\ } \mathbb{Q}[[x,x^{-1}]]$ where i is the canonical injection and T defined by : $T(\frac{\alpha}{\beta})$ = Laurent expansion at 0, - Laurent expansion at ∞, is exact. Moreover, T identifies the quotient $\dfrac{\mathbb{Q}(x)}{\mathbb{Q}[x,x^{-1}]}$ with the torsion of the $\mathbb{Q}[x,x^{-1}]$-module $\mathbb{Q}[[X,x^{-1}]]$.

Proof. The exactness is obvious. The only point to check is the surjectivity of T on the torsion of $\mathbb{Q}[[x,x^{-1}]]$.

So let S be such a torsion element and β be a polynomial such that $\beta S = 0$ Write $S = S_+ + S_-$ with S_+ infinite on the right and S_- infinite on the left The product βS_+ is in fact a polynomial, say α, and thus $\beta S_- = -\alpha$. One checks that $S = T(\alpha/\beta)$. □

LEMMA 3. Over $\mathbb{Q}[x,x^{-1}]$, the modules $\mathbb{Q}(x)$ and coker T are torsionless and injective.

Proof. It is obvious for the field $\mathbb{Q}(X)$. The module coker T is a quotient of the divisible module $\mathbb{Q}[[x,x^{-1}]]$. Coker T is divisible, hence injective, and torsionless according to Lemma 2. □

The theorem will now follow from :

LEMMA 4. Let C_* be a differential complex of projective $\mathbb{Z}[x,x^{-1}]$-modules such that the homology groups $H_*(\overset{\circ}{C}_*; \mathbb{Q})$ are finite dimensional over $\mathbb{Q}$ and let I be an injective torsionless $\mathbb{Q}[x,x^{-1}]$-module, then the colomology groups $H^*(C_*; I)$ are trivial.

Proof. The result follows from the universal coefficients formula for the principal ideal domain $\mathbb{Q}[x,x^{-1}]$, if one remarks that any $\mathbb{Q}[x,x^{-1}]$-module which is finite dimensional over $\mathbb{Q}$, is a torsion $\mathbb{Q}[x,x^{-1}]$-module. □

[1] J.W. MILNOR, Infinite cyclic coverings, Conference on the topology of manifolds, Prindle, Weber and Schmidt (1968).

[2] J. BARGE, Dualité dans les revêtements galoisiens, to appear.

REPRESENTATIONS OF BROWN-GITLER SPECTRA

by
Ralph L. Cohen [(1)]

1. In [2], E. Brown and S. Gitler constructed spectra B(k), $k \geq 0$, that have since been applied to several seemingly diverse areas of topology. In their original form they were used by Brown and Peterson to give evidence supporting the immersion conjecture for compact manifolds [4]. More recently, they were used in [6] to construct universal spaces BO/I_n, for normal bundles of n-manifolds.

Brown and Peterson also showed that the B(k)'s can be represented as certain stable wedge summands of $\Omega^2 S^n$ [5]. As remarked by Mahowald in [10], this leads to an easy proof of his theorem stating that for $i \neq 2$, $h_1 h_i$ is a permanent cycle in the Adams spectral sequence converging to the 2-primary part of the stable homotopy groups of spheres. This representation of Brown-Gitler spectra also led to a classification of braid-oriented manifolds through a range of dimensions [5, 9].

---

(1) This research was partially supported by National Science Foundation grant NSF MCS-7701623.

In [9], the author proved a conjecture of B. Sanderson, stating that the spectra $B(k)$ could also be represented as the Thom spectra of certain stable vector bundles over filtrations of $\Omega^2 S^3$. This led to a classification of primitive Mahowald-oriented manifolds through a range of dimensions, and was the theorem reported on by the author at the Siegen symposium.

Also at the symposium, Peterson reported on a result of his and Brown's that the author's method of proof in [9] can be modified to give a much simpler proof of their theorem about representing Brown-Gitler spectra as wedge summands of $\Omega^2 S^n$ [7].

In this paper we shall show that the methods of [9] in fact generalize to give a complete homotopy characterization of the family of spectra $\{B(k)\}$, by three relatively easy to verify properties.

Before we state our results more precisely, we recall some information about Brown-Gitler spectra.

For an integer $k \geq 0$, let $M(k)$ denote the following cyclic module over the mod 2 Steenrod algebra A.

$$M(k) = A/A\{\chi(Sq^i) : i > k\},$$

where χ is the canonical antiautomorphism.

Theorem 1.1 (Brown and Gitler [2]). For every $k \geq 0$ there exists a 2-local spectrum $B(k)$ satisfying the following properties.

(1) $H^*(B(k); \mathbb{Z}_2) = M(k)$

(2) If $j_k : B(k) \to K(\mathbb{Z}_2)$ generates $H^*(B(k); \mathbb{Z}_2)$ as an A-module, then for any C. W. complex X, the induced map in generalized homology theories

$$j_{k_*} : B(k)_q(X) \to H_q(X; \mathbb{Z}_2)$$

is surjective for $q \leq 2k+1$.

Moreover, Brown and Peterson proved in [5] that properties (1) and (2) above completely characterize the homotopy type of $B(k)$. Clearly, the geometric input in this characterization is embodied in property (2). Although this is a very attractive property about the generalized homology theory that $B(k)$ represents, it is in practice very difficult to verify that a given spectrum satisfies it.

The purpose of this paper is to describe a more readily verifiable set of conditions that will ensure that a given family $\{Y_k\}$ of spectra are homotopy equivalent to Brown-Gitler spectra. That is, we shall prove the following theorem.

Theorem 1.2. Suppose $\{Y_k;\ k \geq 0\}$ is a family of 2-local spectra. Then each $Y_k \simeq B(k)$ if and only if the family satisfies the following properties:

(1) $H^*(Y_k; \mathbb{Z}_2) = M(k)$ generated by a class $u_k \in H^0(Y_k; \mathbb{Z}_2)$.

(2) For every pair of integers $r, s \geq 0$ there exists a pairing

$$\mu_{r,s}: Y_r \wedge Y_s \to Y_{r+s}$$

such that $\mu_{r,s}{}^*(u_{r+s}) = u_r \otimes u_s \in H^0(Y_r \wedge Y_s; \mathbb{Z}_2)$.

(3) For every $i \geq 0$ there exists a "cup-1 product":

$$\zeta_i: S^1 \ltimes_{\mathbb{Z}_2} Y^{(2)}_{2^i} \to Y_{2^{i+1}}$$

such that $\zeta_i^*(u_{2^{i+1}}) = e_0 \otimes_{\mathbb{Z}_2} u_{2^i} \otimes u_{2^i} \in H^0(S^1 \ltimes_{\mathbb{Z}_2} Y^{(2)}_{2^i}; \mathbb{Z}_2)$.

In part (3), $Y^{(2)}_{2^i} = Y_{2^i} \wedge Y_{2^i}$ has the $\mathbb{Z}_2$-action which permutes coordinates, and S^1 has the usual antipodal $\mathbb{Z}_2$-action. A precise definition of the quadratic construction $S^1 \ltimes_{\mathbb{Z}_2} E^{(2)}$ for a spectrum E, as well as other spectrum level extended power constructions, can be found in [12].

Theorem 1.2 will be proved in section 2 using techniques similar to those used in [9]. In section 3 we will demonstrate how the two representations of Brown-Gitler spectra discussed above are easily seen to satisfy the hypotheses of 1.2.

We remark that odd primary analogues of Brown-Gitler spectra have been constructed and studied by the author in [8] and [9]. In this paper, however, we shall be concerned only with the prime 2, and all homology will be taken with $\mathbb{Z}_2$ coefficients.

2. Proof of Theorem 1.2.

Before proceeding directly to the proof of Theorem 1.2, we collect and summarize some of the essential properties of the original construction of the Brown-Gitler spectra $B(k)$ [2].

First of all, let Λ be the algebra over $\mathbb{Z}_2$ generated by elements λ_i, $i \geq 0$, of dimension i, subject to the relations

$$(2.1) \qquad \lambda_i \lambda_j = \sum_s \binom{s-1}{2s-j+2i} \lambda_{i+s} \lambda_{j-s} .$$

This is the algebra originally studied by Bousfield, Curtis, Kan, et. al. [1].

If J is a sequence of integers $J = (j_1, \ldots, j_s)$, then as usual we let $\lambda_J = \lambda_{j_1} \lambda_{j_2} \cdots \lambda_{j_s}$ and define the length of $J = \ell(J)$ to be s. J is defined to be admissible if $2j_i \geq j_{i+1}$ for $1 \leq i \leq s-1$. It was shown in [1] that $\{\lambda_J \mid J \text{ is admissible}\}$ forms a $\mathbb{Z}_2$-vector space basis for Λ.

Let $\Lambda_k \subset \Lambda$ be the subspace generated by $\{\lambda_J \mid J = (j_1, \ldots, j_s)$ is admissible and $j_s \geq k\}$, and grade Λ_k by subspaces $\Lambda_{k,r}$ generated by $\{\lambda_J \mid J \text{ admissible}, \lambda_J \in \Lambda_k, \text{ and } \ell(J) = r\}$.

Let $H_{k,r}$ be a product of suspensions of Eilenberg-MacLane spectra such that

$$\pi_*(H_{k,r}) = \Lambda_{k,r}$$

as graded $\mathbb{Z}_2$ vector spaces, where $\Lambda_{k,r}$ is graded by dimension. Notice that $H_{k,0} = K(\mathbb{Z}_2)$.

The spectra $B(k)$ were originally constructed by Brown and Gitler in [2] by building a Postnikov tower of fibrations

$$\begin{array}{lccccccccccc}
 & & & H_{k,1} & & & & H_{k,q} & & & \\
 & & & \downarrow i_1 & & & & \downarrow i_q & & & \\
(2.2) & K(\mathbb{Z}_2) = H_{k,0} & \xleftarrow{\rho_1} & E_{k,1} & \xleftarrow{\rho_2} & \cdots & \leftarrow E_{k,q-1} & \leftarrow E_{k,q} & \xleftarrow{\rho_{q+1}} & E_{k,q+1} & \leftarrow \cdots \\
 & \downarrow s_1 & & & & & \downarrow s_q & & & & \\
 & \Sigma H_{k,1} & & & & & \Sigma H_{k,q} & & & &
\end{array}$$

and defining $B(k) = \varprojlim_q E_{k,q}$.

This tower was shown to satisfy the following properties:

(2.3) $$H^*(B(k)) = \varinjlim_q H^*(E_{k,q}) \cong M(k) \text{ as } A\text{-modules.}$$

(2.4) If X is any C. W. complex, then the homotopy exact sequence

$$\pi_r(H_{k,q} \wedge X) \xrightarrow{i_{q*}} \pi_r(E_{k,q} \wedge X) \xrightarrow{\rho_{q*}} \pi_r(E_{k,q-1} \wedge X)$$

is

(a) split, short exact if $r < 2k$,

(b) short exact (but not necessarily split) if $r = 2k$, and

(c) ρ_{q*} is surjective if $r = 2k+1$.

Remark. Although properties (2. 4)(a)-(c) are not explicitly stated in [2], they follow immediately by applying the Pontrjagin duality functor "χ" described in section 6 of [2] to Theorem 5.1 of that paper.

Notice that by letting $X = S^0$ in property (2. 4)(a) and using induction on q, we obtain the following result, which was stated in [5].

Corollary 2. 5. Let $(\Lambda_k)_r$ denote the subgroup of Λ_k consisting of elements having dimension r. Then

$$\pi_r(B(k)) = (\Lambda_k)_r$$

for $r < 2k$.

The generator $j_k: B(k) \to K(\mathbb{Z}_2) = H_{k,0}$ of $H^*(B(k))$ clearly factors as the composition of the ρ_q's in tower (2. 2). Thus property (2. 4)(c) above implies that if X is any C. W. complex, $j_{k*}: B(k)_q(X) \to H_q(X)$ is surjective for $q \leq 2k+1$. This is property (2) of Theorem 1.1.

Moreover, if N is a closed n-manifold with stable normal bundle ν_N having Thom spectrum $M(\nu_N)$, then by applying S-duality to this property, we have the following useful corollary (see Cor. 1.2 of [2]).

<u>Corollary 2.6.</u> If $n \leq 2k+1$ and $\tau_N : M(\nu_N) \to K(\mathbb{Z}_2)$ is the Thom class, then there exists a map $\tilde{\tau}_N : M(\nu_N) \to B(k)$ such that the following diagram commutes.

$$\begin{array}{ccc} M(\nu_N) & \xrightarrow{\tau_N} & K(\mathbb{Z}_2) \\ & {\scriptstyle\tilde{\tau}_N}\searrow \quad \nearrow{\scriptstyle j_k} & \\ & B(k) & \end{array}$$

<u>Definition.</u> If N and $\tilde{\tau}_N$ are as in the corollary, we say that N is <u>adapted</u> to the module $M(k) = H^*(B(k))$, if the homomorphism $\tilde{\tau}_N^* : M(k) \to H^*(M(\nu_N))$ is injective.

Notice that this definition is somewhat more restrictive than the definition of adapted manifold given in [5], but this is all we need in this paper.

The next lemma was first proved by Brown and Peterson in [5]. The following considerable simplification of the proof is also due to Brown and Peterson, and is to be found in [7].

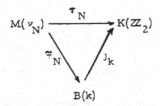

Lemma 2.7. For every $n \geq 0$ there exists a closed, n-dimensional manifold Q_n adapted to $M[\frac{n}{2}]$.

Proof. If N is a closed n-manifold, let J_N be the kernel of

$$\tau_N^* : A = H^*(K(\mathbb{Z}_2)) \to H^*(M(\nu_N)).$$

Let $J_n = \bigcap_N J_N$, where the intersection is taken over all closed n-manifolds N. In [3] it was proven that J_n is the left ideal of A generated by $\{\chi(Sq^i) : i > [\frac{n}{2}]\}$. Thus $A/J_n = M[\frac{n}{2}]$ which is a finite dimensional $\mathbb{Z}_2$-vector space.

For each $\mathbb{Z}_2$-basis element $v \in M[\frac{n}{2}]$, we may therefore find a closed n-manifold N(v) such that $\tilde{\tau}_{N(v)}^*(v) \neq 0$. Let Q_n be the disjoint union of all such N(v)'s, where the v's span a (finite) basis for $M[\frac{n}{2}]$. Clearly $M(\nu_{Q_n}) \simeq \bigvee_v M(\nu_{N(v)})$, and therefore $\tilde{\tau}_{Q_n} = \bigvee_v \tilde{\tau}_{N(v)} : M(\nu_{Q_n}) \to B[\frac{n}{2}]$ induces an injection in cohomology. Thus Q_n is adapted to $M[\frac{n}{2}]$.

The following corollary of the properties of the Postnikov tower 2.2 defining B(k), can be viewed as a strengthening of Corollary 2.6.

Corollary 2.8. Let N be as in 2.5, having dimension $\leq 2k+1$. Suppose $\sigma_q : M(\nu_N) \to E_{k,q}$ is any lifting in tower 2.2 of the Thom class $\tau_N : M(\nu_N) \to K(\mathbb{Z}_2) = H_{k,0} = E_{k,0}$. Then there exists a lifting of σ_q to $E_{k,q+1}$. That is, there exists a map $\sigma_{q+1} : M(\nu_N) \to E_{k,q+1}$ making the following diagram commute.

$$\begin{array}{ccc} & & E_{k,q+1} \\ & \overset{\sigma_{q+1}}{\nearrow} & \\ M(\nu_N) & & \downarrow \rho_{q+1} \\ & \underset{\sigma_q}{\searrow} & \\ & & E_{k,q} \end{array}$$

<u>Proof.</u> This follows from property (2.4)(c) and S-duality.

The next lemma, proved by Brown and Peterson in [5], is crucial to our proof of Theorem 1.2. We include their proof of this lemma for the sake of completeness.

<u>Lemma 2.9.</u> Suppose X_k is a 2-local spectrum such that $H^*(X_k) = M(k)$ as A-modules, generated by a class $w_k \in H^0(X_k)$. Let N_k be a closed manifold of dimension $\leq 2k+1$, which has stable normal bundle ν_k, and which is adapted to $M(k)$. If there exists a map

$$g_k : M(\nu_k) \to X_k$$

such that $g_k^*(w_k)$ is the Thom class τ_k, then $X_k \simeq B(k)$.

<u>Proof.</u> We shall show that the generator $w: X_k \to K(\mathbb{Z}_2) = H_{k,0}$ lifts all the way up the Postnikov tower 2.2. This will yield a map $f: X_k \to B(k)$ which, in cohomology, preserves A-module generators, and hence is an isomorphism.

Inductively, assume there is a lifting $f_q: X_k \to E_{k,q}$ of w_k, where $E_{k,0} = H_{k,0}$. We shall prove that there is a lifting $f_{q+1}: X_k \to E_{k,q+1}$ of f_q.

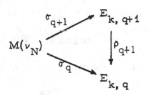

Let $\sigma_q = f_q \circ g_k : M(\nu_k) \to X_k \to E_{k,q}$. Since f_q is a lifting of w_k, σ_q is a lifting of $w_k \circ g_k = \tau_k$. Thus by 2.8, there exists a lifting $\sigma_{q+1} : M(\nu_k) \to E_{k,q+1}$ of σ_q. Since

$$E_{k,q+1} \xrightarrow{\rho_{q+1}} E_{k,q} \xrightarrow{s_{q+1}} \Sigma H_{k,q+1}$$

is a fibre sequence, we have that $s_{q+1} \circ \sigma_q = s_{q+1} \circ f_q \circ g_k$ is null homotopic. But $\Sigma H_{k,q+1}$ is a product of suspensions of Eilenberg-MacLane spectra, and N_k being adapted to M(k) implies that g_k^* is injective in cohomology. Thus $s_{q+1} \circ f_q$ is null homotopic, and hence there exists a lifting $f_{q+1} : X_k \to E_{k,q+1}$ of f_q, as claimed.

We are now ready to prove Theorem 1.2. So let $\{Y_k\}$ be a family of 2-local spectra as in the hypotheses of the theorem. Since $H^*(Y_k) = M(k)$ generated by a class $u_k \in H^0(Y_k)$, it is enough by 2.9 to construct a manifold N_k adapted to M(k), together with a map $g_k : M(\nu_k) \to Y_k$ such that $g_k^*(u_k) = \tau_k$. We do this by induction on k.

For $k = 0$, $Y_0 = B(0)$ = the sphere spectrum S. Let N_0 be a point. Then $M(\nu_0) = S$, and we may let g_0 be the identity.

Now let $k = 1$. Then for purely algebraic reasons, Y_1 and B(1) are both homotopy equivalent to the $\mathbb{Z}_2$-Moore spectrum, $S^0 \cup_2 e^1$, and hence are homotopy equivalent to each other. We may therefore let $N_1 = Q_2$ as in 2.7. Dim $N_1 = 2 \leq 2k+1$. Let $g_1 = \tilde{\tau}_{Q_2} : M(\nu_1) = M(\nu_{Q_2}) \to B(1) = Y_1$, where $\tilde{\tau}_{Q_2}$ is as in 2.6. Clearly, N_1 and g_1 satisfy the required properties.

Inductively, for an integer $k > 1$ we assume that manifolds N_j and maps g_j have been constructed for all $j < k$. By 2.9 this implies that $Y_j \simeq B(j)$ for all $j < k$. We now construct N_k and g_k.

Write k in the form $k = 2^i + r$ where $1 \leq r \leq 2^i$. If $r < 2^i$, let $N_k = Q_{2^{i+1}} \times Q_{2r}$, where the Q_j's are as in 2.7. $\operatorname{Dim} N_k = 2^{i+1} + 2r = 2k$. Define $g_k : M(\nu_k) \to Y_k$ to be the following composition.

$$g_k : M(\nu_k) = M(\nu_{Q_{2^{i+1}}}) \wedge M(\nu_{Q_{2r}}) \xrightarrow{\tilde{\tau}_{Q_{2^{i+1}}} \wedge \tilde{\tau}_{Q_{2r}}}$$

$$B(2^i) \wedge B(r) \simeq Y_{2^i} \wedge Y_r \xrightarrow{\mu_{2^i, r}} Y_k$$

By hypothesis and the definition of the $\tilde{\tau}_{Q_j}$'s, $g_k^*(u_k) = \tau_k$. We therefore need only to check that g_k^* is injective in cohomology. To do this we need the following straightforward algebraic result which was proved in [5].

Let $\Delta : A \to A \otimes A$ be the comultiplication in A. Clearly Δ induces maps

$$\Delta_{s,t} : M(s+t) \to M(s) \otimes M(t) .$$

<u>Lemma 2.10.</u> (a) If $r < 2^i$, $\Delta_{2^i, r} : M(2^i + r) \to M(2^i) \otimes M(r)$ is injective.

(b) The kernel of $\Delta_{2^i, 2^i} : M(2^{i+1}) \to M(2^i) \otimes M(2^i)$ is $\{0, \chi(Sq^{I_{i+1}})\}$ where $I_j = (2^j, 2^{j-1}, \ldots, 2^1, 1)$.

Notice that by part (2) of the hypotheses of Theorem 1.2, the following diagram commutes.

$$\begin{array}{ccc} H^*(Y_k) & \xrightarrow{\mu^*_{2^i,r}} & H^*(Y_{2^i}) \otimes H^*(Y_r) \\ \| & & \| \\ M(k) & \xrightarrow[\Delta_{2^i,r}]{} & M(2^i) \otimes M(r) \end{array}$$

So in the case we are considering, namely $k = 2^i + r$ where $1 \leq r < 2^i$, Lemma 2.10.a implies that $\mu^*_{2^i,r}$ is injective. Since the Q_j's are adapted manifolds, $\mp^*_{Q_{2^{i+1}}}$ and $\mp^*_{Q_{2r}}$ are also injective, and hence so is g^*_k.

We are therefore left with the case $k = 2^i + r$ where $r = 2^i$, that is, $k = 2^{i+1}$. In this case, we let $N_{2^{i+1}} = S^1 \times_{\mathbb{Z}_2} (Q_{2^{i+1}} \times Q_{2^{i+1}})$. Dim $N_{2^{i+1}} = 2^{i+2} + 1 = 2k+1$. It is easy to verify (and was proven in [9]) that

$$M(\nu_{2^{i+1}}) \cong S^1 \ltimes_{\mathbb{Z}_2} (M(\nu_{Q_{2^{i+1}}}))^{(2)} .$$

We define $g_{2^{i+1}} : M(\nu_{2^{i+1}}) \to Y_{2^{i+1}}$ to be the following composition.

$$g_{2^{i+1}} : M(\nu_{2^{i+1}}) = S^1 \ltimes_{\mathbb{Z}_2} M(\nu_{2^{i+1}})^{(2)} \xrightarrow[1 \ltimes \mp^{(2)}_{Q_{2^{i+1}}}]{} S^1 \ltimes_{\mathbb{Z}_2} B(2^i)^{(2)}$$

$$\cong S^1 \ltimes_{\mathbb{Z}_2} Y_{2^i}^{(2)} \xrightarrow[\zeta_1]{} Y_{2^{i+1}}$$

By part (3) of the hypotheses of Theorem 1.2 and the definition of the $\mp_{Q_j}$'s, $g^*_{2^{i+1}}(u_{2^{i+1}}) = \tau_{2^{i+1}}$. Therefore to complete the inductive step, and hence the proof that the family $\{Y_j\}$ is a family of Brown-Gitler spectra, we need only check that $g_{2^{i+1}}^*$ is injective.

Now since $\tilde{\tau}_{Q_{2^i}}^*$ is injective, so is $(1 \ltimes \tilde{\tau}_{Q_{2^i}}^{(2)})^*$. It is therefore sufficient to prove that $\zeta_i^*: H^*(Y_{2^{i+1}}) = M(2^{i+1}) \to H^*(S^1 \ltimes_{\mathbb{Z}_2} Y_{2^i}^{(2)})$ is injective. Now if $j: Y_{2^i}^{(2)} \hookrightarrow S^1 \ltimes_{\mathbb{Z}_2} Y_{2^i}^{(2)}$ is the natural inclusion then properties (2) and (3) of 1.2 imply that

$$(\zeta_i \circ j)^* = \mu_{2^i, 2^i}^* = \Delta_{2^i, 2^i}: H^*(Y_{2^{i+1}}) = M(2^{i+1}) \to H^*(Y_{2^i} \wedge Y_{2^i})$$

$$= M(2^i) \otimes M(2^i) .$$

Hence by 2.10, the kernel of $j^* \circ \zeta_i^*$ is $\{0, \chi(Sq^{I_{i+1}}(u_{2^{i+1}}))\}$. Thus to show ζ_i^* is injective, it suffices to show that

$$\chi(Sq^{I_{i+1}})\zeta_i^*(u_{2^{i+1}}) = \chi(Sq^{I_{i+1}})(e_0 \otimes_{\mathbb{Z}_2} u_{2^i} \otimes u_{2^i}) \in H^*(S^1 \ltimes_{\mathbb{Z}_2} Y_{2^i}^{(2)})$$

is nonzero.

Now $\chi(Sq^{I_{i+1}}) = Sq^1\chi(Sq^{J_{i+1}})$, where $J_{i+1} = (2^{i+1}, 2^i, \ldots, 2^1)$. Since $M(2^i) = H^*(Y_{2^i})$ has as a $\mathbb{Z}_2$-basis $\{\chi(Sq^I) \mid I = (s_1, \ldots, s_m)$ is admissible and $s_1 \leq 2^i\}$ [2], a simple exercise with the Cartan formula and the Adem relations yields

$$\chi(Sq^{J_{i+1}})(e_0 \otimes_{\mathbb{Z}_2} u_{2^i} \otimes u_{2^i}) = e_0 \otimes_{\mathbb{Z}_2} \chi(Sq^{I_i})(u_{2^i}) \otimes \chi(Sq^{I_i})(u_{2^i}) .$$

For ease of notation, let $a_i = \chi(Sq^{I_i})(u_{2^i}) \in H^{2^{i+1}-1}(Y_{2^i})$. Since the dimension of a_i is odd, then a standard result about the quadratic construction yields $Sq^1(e_0 \otimes_{\mathbb{Z}_2} a_i \otimes a_i) = e_1 \otimes_{\mathbb{Z}_2} a_i \otimes a_i$, which is nonzero in $H^{2^{i+2}-1}(S^1 \ltimes_{\mathbb{Z}_2} Y^{(2)}_{2^i})$. Thus $\chi(Sq^{I_{i+1}})(e_0 \otimes_{\mathbb{Z}_2} u_{2^i} \otimes u_{2^i}) = \zeta_i^*(\chi(Sq^{I_{i+1}})(u_{2^{i+1}}))$ is nonzero, which implies that ζ_i^*, and hence $g_{2^{i+1}}{}^*$ are injective.

We have now shown that if $\{Y_k\}$ is a family of 2-local spectra satisfying properties (1)-(3) of Theorem 1.2, then each $Y_k \simeq B(k)$. Notice, however, that it is not a priori clear that the B(k)'s themselves satisfy these properties. To prove this, and therefore complete the proof of Theorem 1.2, it is sufficient by what we have shown so far, to give an example of any family $\{Y_k\}$ satisfying these properties. We shall now give two such examples.

3. Examples

The examples we shall describe are the two representations of Brown-Gitler spectra discussed in section 1. To describe them more precisely, first recall May's representation of $\Omega^2 S^3$ as a filtered complex [11].

$$\Omega^2 S^3 = \Omega^2 \Sigma^2 S^1 \simeq C_2(S^1) = \coprod_{j \geq 0} \mathcal{C}_2(j) \times_{\Sigma_j} (S^1)^j / \sim .$$

$\mathcal{C}_2(j)$ is Boardman and Vogt's space of "j-little 2-cubes", Σ_j is the symmetric group on j letters, and the relations "$\sim$" is generated by

$$(c_1, \ldots, c_j) \times_{\Sigma_j} (t_1, \ldots, t_{j-1}, *) \sim (c_1, \ldots, c_{j-1}) \times_{\Sigma_{j-1}} (t_1, \ldots, t_{j-1}),$$

where $* \in S^1$ is the basepoint. $\Omega^2 S^3 \simeq C_2(S^1)$ is given the obvious filtration

$$F_n(\Omega^2 S^3) = \coprod_{j=0}^{n} \mathcal{C}_2(j) \times_{\Sigma_j} S^{1^j} / \sim .$$

Let X_k be the localization at 2 of the suspension spectrum of $F_{2k}(\Omega^2 S^3)/F_{2k-1}(\Omega^2 S^3) = \mathcal{C}_2(2k) \ltimes_{\Sigma_{2k}} S^{1^{(2k)}}$. Let Y_k be the 2-localization of the Thom spectrum of the stable vector bundle γ_k over $F_k(\Omega^2 S^3)$, which is classified by the composition

$$\gamma_k : F_k(\Omega^2 S^3) \hookrightarrow \Omega^2 S^3 \xrightarrow[\gamma]{} BO .$$

Here γ is the unique 2-fold loop map which, when restricted to $S^1 \subset \Omega^2 S^3$, generates $\pi_1 BO = \mathbb{Z}_2$.

In [10], Mahowald proved that

$$H^*(X_k) = H^*(Y_k) = M(k)$$

as A-modules. X_k was proven to be equivalent to $B(k)$ in [5], and Y_k was proven to be equivalent to $B(k)$ in [9]. By what we have shown so far, in order to verify these results it is sufficient to construct maps

$$\mu_{r,s} : X_r \wedge X_s \to X_{r+s} \quad , \quad \mu'_{r,s} : Y_r \wedge Y_s \to Y_{r+s} \, ,$$

$$\zeta_i : S^1 \ltimes_{\mathbb{Z}_2} X^{(2)}_{2^i} \to X_{2^{i+1}} \quad \text{and} \quad \zeta'_i : S^1 \ltimes_{\mathbb{Z}_2} Y^{(2)}_{2^i} \to Y_{2^{i+1}}$$

that satisfy hypotheses (2) and (3) of 1.2.

The maps $\mu_{r,s}$ and ζ_i are given by the operad structure of $C_2(S^1)$ [11], recalling that S^1 is $\mathbb{Z}_2$-equivariantly homotopy equivalent to $\mathcal{C}_2(2)$. Clearly, these maps satisfy the required properties.

Now $Y_r \wedge Y_s$ is the Thom spectrum of the bundle $\gamma_r \times \gamma_s$ over $F_r(\Omega^2 S^3) \times F_s(\Omega^2 S^3)$, and $S^1 \ltimes_{\mathbb{Z}_2} Y_{2i}^{(2)} \simeq \mathcal{C}_2(2) \ltimes_{\mathbb{Z}_2} Y_{2i}^{(2)}$ is the Thom spectrum of the bundle $\mathcal{C}_2(2) \times_{\mathbb{Z}_2} \gamma_{2i}^2$ over $\mathcal{C}_2(2) \times_{\mathbb{Z}_2} F_{2i}(\Omega^2 S^3)^2$. The maps $\mu'_{r,s}$ and ζ'_i are defined to be the maps of Thom spectra induced by the multiplications

$$m_{r,s} : F_r(\Omega^2 S^3) \times F_s(\Omega^2 S^3) \longrightarrow F_{r+s}(\Omega^2 S^3)$$

and $$\xi_i : \mathcal{C}_2(2) \times_{\mathbb{Z}_2} F_{2i}(\Omega^2 S^3)^2 \longrightarrow F_{2i+1}(\Omega^2 S^3) ,$$

given by the operad structure of $C_2(S^1) \simeq \Omega^2 S^3$. The maps $\mu'_{r,s}$ and ζ'_i also obviously satisfy the hypotheses of Theorem 1.2.

We have thus shown that $\{X_k\}$ and $\{Y_k\}$ are both families of spectra satisfying properties (1)-(3) of Theorem 1.2, and hence are families of Brown-Gitler spectra.

References

1. A. Bousfield, E. Curtis, D. Kan, D. Quillen, D. Rector, and J. Schlessinger: The mod p lower central series and the Adams spectral sequence, Topology 5 (1966), 331-342.

2. E.H. Brown, Jr. and S. Gitler: A spectrum whose cohomology is a certain cyclic module over the Steenrod algebra, Topology 12 (1973), 283-295.

3. E.H. Brown, Jr. and F.P. Peterson: Relations among characteristic classes - I, Topology 3 (1964), 39-52.

4. E.H. Brown, Jr. and F.P. Peterson: On immersions of n-manifolds, Adv. in Math. 24 (1974), 74-77.

5. E.H. Brown, Jr. and F.P. Peterson: On the stable decomposition of $\Omega^2 S^{r+2}$, Trans. of the A.M.S. 243 (1978), 287-298.

6. E.H. Brown, Jr. and F.P. Peterson: A universal space for normal bundles of n-manifolds, to appear.

7. E.H. Brown, Jr. and F.P. Peterson: The Brown-Gitler spectrum and $\Omega^2 S^3$, Proc. of the Moscow symposium on Alg. Top. 1979, to appear.

8. R.L. Cohen: New infinite families in the stable homotopy groups of spheres and of Moore spaces, to appear.

9. R.L. Cohen: The geometry of $\Omega^2 S^3$ and braid orientations, Invent. Math., to appear.

10. M. Mahowald: A new infinite family in ${}_2\pi_*^s$, Topology 16 (1977), 249-254.

11. J.P. May: The Geometry of Iterated Loop Spaces, Springer Lecture Notes in Math. Vol. 271, 1972.

12. R. Bruner, G. Lewis, J. P. May, J. McClure, and M. Steinberger. H_∞ Ring Spectra and their Applications. Springer Lecture Notes in Math., to appear.

University of Chicago
Chicago, Illinois 60637

PSEUDO-ISOTOPY AND INVARIANT THEORY, II: RATIONAL ALGEBRAIC K-THEORY OF A SPACE WITH FINITE FUNDAMENTAL GROUP

by

W. Dwyer[1], W C. Hsiang[2], and R Staffeldt[3]

I. Introduction and statement of results

To any space X Waldhausen [9] has associated another space $A(X)$[4] whose homotopy groups are, in a certain sense, the algebraic K-groups of X. This paper is concerned with the problem of computing the rational homotopy groups $\pi_* A(X) \otimes \mathbb{Q}$ under the assumption that the fundamental group of X is finite.

The approach is as follows: Let X be a path-connected space with finite fundamental group π. In §2 we define and construct algebraic models for the loop space ΩX. In §3 we work out the stability properties of some invariant homology groups associated to the models. Finally in §4 we show that these invariant homology groups determine the rational homology groups, and therefore the rational homotopy groups, of $A(X)$.

In more detail, we formulate in §2 what it means for a connected differential graded algebra K over $\mathbb{Q}$ with a right π-action (a π-DGA) to be a split DGA model for ΩX. In particular, this implies that if $\mathbb{Z}\pi \tilde{\otimes} K$ is the obvious twisted tensor product (i.e., the algebra with multiplication $(g \otimes k)(g' \otimes k') = gg' \otimes (kg')k'$), then the homology of $\mathbb{Z}\pi \tilde{\otimes} K$ is the ring $H_*(\Omega X, \mathbb{Q})$. Now let K be a π-DGA, and let $\overline{K}$, the augmentation ideal, be the kernel of the unique augmentation $K \to \mathbb{Q}$. For each n consider the differential graded Lie algebra (DGL) of $(n \times n)$-matrices

$$L_n = M_n(\mathbb{Z}\pi \widetilde{\otimes} \overline{K})$$

and the differential graded coalgebra

$$C_n = \mathcal{C}(L_n)$$

obtained as in [8] (cf. [5]) by the graded Koszul construction. The left (right) action of $GL_n = GL_n(\mathbb{Z}[\pi])$ on L_n. Define the (twisted) adjoint representation of GL_n on L_n by letting

$$M \circ g = g^{-1}Mg$$

for $g \in GL_n$, $M \in L_n$. By naturality this representation induces a right action of GL_n on the coalgebra C_n. These actions are compatible with the "upper" inclusion $GL_n \to GL_{n+1}$, $L_n \to L_{n+1}$ and the induced inclusion $C_n \to C_{n+1}$.

Inside the chain complex C_n is the subcomplex $C_n^{GL_n}$ of invariant chains.

Theorem 1.1.

(1) The natural map

$$H_i(C_n^{GL_n}) \to H_i(C_n)^{GL_n} \qquad (i \geq 0)$$

is an isomorphism.

(2) The natural chain map

$$C_n^{GL_n} \to C_{n+1}^{GL_{n+1}}$$

is an isomorphism in dimension i for $i << n$, in a range tending to infinity with n.

(3) As a consequence of (1) and (2), the natural map

$$H_i(C_n)^{GL_n} \to H_i(C_{n+1})^{GL_{n+1}}$$

is an isomorphism for $i \ll n$, *in a range tending to infinity with* n.

The proof of Theorem 1.1 will be given in §3. Let $C = \varinjlim C_n$, $GL = \varinjlim GL_n$. It follows from Theorem 1.1 that we may piece together the complexes for various n to obtain a single chain complex

$$\varinjlim C_n^{GL_n}$$

which is just C^{GL}. Moreover, the natural map

$$H_i(C_n)^{GL_n} \to H_i(C)^{GL} = H_i(C^{GL})$$

is an isomorphism for $i \ll n$. This group will be denoted by $Inv_i(K)$ and called the i-th *invariant homology group* associated to K. If K is a split DGA model for ΩX, this is also denoted by $Inv_i(X)$. (It will be shown that it only depends on X if $\pi_1 X$ is finite.) In addition, we may consider the cochain complex of invariant functions and denote the corresponding invariant cohomology group by $Inv^i(X)$. (Cf. [5].)

Note that in §2 we show that split DGA models for ΩX exist if $\pi_1 X$ is finite.

Theorem 1.2. Let $A(X)$ *be the Waldhausen algebraic K-theory space associated to the space* X. *If* $\pi = \pi_1 X$ *is finite, then there is an isomorphism of* $H_n(A(X),\mathbb{Q})$, *respectively* $H^n(A(X);\mathbb{Q})$, *with*

$$\sum_{p+q=n} H_p(GL(\mathbb{Z}\pi),\mathbb{Q}) \otimes Inv_q(X),$$

respectively

$$\sum_{p+q=n} H^p(GL(\mathbb{Z}\pi);\mathbb{Q}) \otimes \mathrm{Inv}^q(X).$$

For $\pi_1 X = \{1\}$, these results are proved in [5]. (See also [2].) Note that since $A(X)$ is an infinite loop space, the rational homology groups of $A(X)$ determine the rational homotopy groups of $A(X)$ in a simple way. We shall apply these theorems for computing some special cases in a future paper.

II. Split DGA models for ΩX

Assume that X is a 1-reduced simplicial set, i.e., a simplicial set with only one zero-simplex. Such a simplicial set can be obtained from any pointed path-connected topological space Y by forming the subcomplex of the singular complex of Y containing all singular simplices with their vertices at the base point. Recall that Kan [6] has constructed a simplicial loop group GX for X.

Call a map $G_1 \to G_2$ between two simplicial groups an h-isomorphism if it induces isomorphisms

$$\pi_0 G_1 \simeq \pi_0 G_2$$

$$(\pi_i G_1) \otimes \mathbb{Q} \simeq (\pi_i G_2) \otimes \mathbb{Q}, \qquad i > 0.$$

Two simplicial groups are said to be h-equivalent if they are related by the equivalence relation generated by h-isomorphisms.

Definition 2.1. A simplicial group is a (rational) group model for the loop space ΩX if it is h-equivalent to the Kan loop group GX.

Let $\pi = \pi_1 X$ be fixed. Define a simplicial π-group to be a connected simplicial group H together with a right action of π on H.

Definition 2.2. The simplicial π-group H is a split group model for ΩX if the semidirect product $\pi \tilde{\times} H$ is a group model for ΩX.

Recall that a π-DGA is a connected (i.e., $K_0 \simeq \mathbb{Q}$) DGA over $\mathbb{Q}$ together with a right π-action. A morphism $f : K_1 \to K_2$ between two π-DGA's is an ordinary DGA map that respects the action of π. Call such a morphism an h-*isomorphism* if it induces isomorphisms

$$H_i(K_1) \simeq H_i(K_2), \qquad i \geq 0.$$

Two π-DGA's are said to be h-*equivalent* if they are related by equivalence relation generated by h-isomorphisms.

Note that if H is a simplicial π-group, the normalized group ring [8] $N\mathbb{Q}H$ is a π-DGA.

Definition 2.3. Suppose that the simplicial π-group H is a split group model for ΩX. Then the π-DGA K is a *split DGA-model* for ΩX if K is h-equivalent to $N\mathbb{Q}H$.

The following lemma shows that split group models and split DGA models exist in the case we are interested in.

Lemma 2.4. *Suppose that* $\pi = \pi_1 X$ *is finite. Then there exists a 1-reduced free simplicial group* H *and a right action of* π *on* H *such that, with this action,* H *is a split group model for* ΩX.

Proof. To construct H, let $p : X \to B\pi$ be the first stage of the Postnikov system of X. By localization theory, find a map $X \to X'$ which induces isomorphisms

$$\pi \simeq \pi_1 X \simeq \pi_1 X'$$

$$(\pi_i X) \otimes \mathbb{Q} \simeq \pi_i X', \qquad i > 1.$$

Since π is finite, the corresponding map

$$p' : X' \to B\pi$$

has a section. For simplicity, assume that p has a section $s : B\pi \to X$.

(We may also assume that p gives an isomorphism on the one-skeleton.) Note that s induces a fundamental group isomorphism. Consider the induced map

$$\tilde{s} : E\pi = B\pi \to \tilde{X}$$

of universal covers. Clearly, π acts freely on $\tilde{X}$ and the image of $\tilde{s}$ is an invariant contractible simplicial subset containing the one-skeleton of $\tilde{X}$, so that the quotient $Y = \tilde{X}/\text{image}(\tilde{s})$ is a 2-reduced simplicial set which is weakly equivalent to $\tilde{X}$ and on which π acts in a basepoint-preserving way. By naturality π acts on the loop group $H = GY$, and it is easy to see that this gives a split group model for ΩX.

III. Proof of Theorem 1.1

Let $\mathbb{E}$ be a splitting field for π, i.e., a finite normal extension of $\mathbb{Q}$ such that

$$\mathbb{E}\pi = \mathbb{E} \otimes_{\mathbb{Q}} \mathbb{Q}\pi$$

is a product of full matrix algebras over $\mathbb{E}$, [3]. Consider the DGA over $\mathbb{E}$

$$\mathbb{E}\pi \tilde{\otimes}_{\mathbb{Q}} K = \mathbb{E} \otimes_{\mathbb{Q}} (\mathbb{Q}\pi \otimes_{\mathbb{Q}} K)$$
$$= \mathbb{E} \otimes_{\mathbb{Q}} (\mathbb{Z}\pi \otimes_{\mathbb{Q}} K)$$

and the DGL over $\mathbb{E}$

$$M_n(\mathbb{E}\pi \tilde{\otimes}_{\mathbb{Q}} \overline{K}) = \mathbb{E} \otimes_{\mathbb{Q}} M_n(\mathbb{Z}\pi \tilde{\otimes}_{\mathbb{Z}} \overline{K}).$$

We can define the DGC over $\mathbb{E}$ from this DGL and denote it by $\mathcal{C}_{\mathbb{E}}(M_n(\mathbb{E}\pi \tilde{\otimes}_{\mathbb{Q}} \overline{K}))$ We have a canonical isomorphism

$$\mathcal{C}_{\mathbb{E}}(M_n(\mathbb{E}\pi \tilde{\otimes}_{\mathbb{Q}} \overline{K})) \simeq \mathbb{E} \otimes_{\mathbb{Q}} \mathcal{C}(M_n(\mathbb{Z}\pi \tilde{\otimes}_{\mathbb{Z}} \overline{K})).$$

$GL_n(\mathbb{Z}\pi)$ acts on $\mathcal{C}_{\mathbb{E}}(M_n(\mathbb{E}\pi \tilde{\otimes}_{\mathbb{Q}} \overline{K}))$ by the extension of scalars. (This action extends to an action of $GL_n(\mathbb{E}\pi)$.) It is compatible with the upper inclusion $GL_n(\mathbb{Z}\pi) \to GL_{n+1}(\mathbb{Z}\pi)$ and the chain map $\mathcal{C}_{\mathbb{E}}(M_n(\mathbb{E}\pi \tilde{\otimes}_{\mathbb{Q}} \overline{K})) \to \mathcal{C}_{\mathbb{E}}(M_{n+1}(\mathbb{E}\pi \tilde{\otimes}_{\mathbb{Q}} \overline{K}))$ induced by the upper inclusion $M_n(\mathbb{E}\pi \tilde{\otimes}_{\mathbb{Q}} \overline{K}) \to M_{n+1}(\mathbb{E}\pi \tilde{\otimes}_{\mathbb{Q}} \overline{K})$.

Recall a few facts concerning a $\mathbb{Q}$-structure on an $\mathbb{E}$-vector space [1, pp. 40-41 and 52-53] In general, let $\mathbb{F}' \supset \mathbb{F}$ be fields, and let V be an $\mathbb{F}'$-vector space An $\mathbb{F}$-subspace $V_{\mathbb{F}}$ of V is called an $\mathbb{F}$-structure on V if the canonical map $\mathbb{F}' \otimes_{\mathbb{F}} V_{\mathbb{F}} \to V$ is an isomorphism. If U is an $\mathbb{F}'$-subspace of V, put $U_{\mathbb{F}} = U \cap V_{\mathbb{F}}$. We say that U is defined over

$\mathbb{F}$ if $U_{\mathbb{F}}$ is an $\mathbb{F}$-structure on U. We say an $\mathbb{F}'$-linear map $f : V \to W$, both V and W having $\mathbb{F}$-structures, is defined over $\mathbb{F}$ if $f(V_{\mathbb{F}}) \subset W_{\mathbb{F}}$. Then the $\mathbb{F}$-subspace $\mathrm{Hom}_{\mathbb{F}'}(V, W)_{\mathbb{F}}$ of $\mathrm{Hom}_{\mathbb{F}'}(V, W)$ consisting of all maps from V to W defined over $\mathbb{F}$ is an $\mathbb{F}$-structure on $\mathrm{Hom}_{\mathbb{F}'}(V, W)$ if W is finite-dimensional. There is also a canonical $\mathbb{F}$-structure $V_{\mathbb{F}} \otimes_{\mathbb{F}} W_{\mathbb{F}}$ on $V \otimes_{\mathbb{F}'} W$, and similar canonical structures on symmetric and exterior powers. The following statement is a modification of the Proposition of [1, p. 54]: Suppose $\mathbb{F}'$ is a finite Galois extension of $\mathbb{F}$ with group Γ. Let W be a subspace of a space V with an $\mathbb{F}$-structure. Then W is defined over $\mathbb{F}$ if and only if W is Γ-stable.

<u>Lemma</u> 3.1.

$$\mathcal{C}_{\mathbb{E}}(M_n(\mathbb{E}\pi \otimes_{\mathbb{Q}} \overline{K}))^{GL_n(\mathbb{Z}\pi)} \simeq \mathbb{E} \otimes_{\mathbb{Q}} \mathcal{C}(M_n(\mathbb{Z}\pi \otimes_{\mathbb{Z}} \overline{K}))^{GL_n(\mathbb{Z}\pi)}$$

<u>and</u>

$$H_i(\mathcal{C}_{\mathbb{E}}(M_n(\mathbb{E}\pi \otimes_{\mathbb{Q}} \overline{K}))^{GL_n(\mathbb{Z}\pi)}) \simeq \mathbb{E} \otimes_{\mathbb{Q}} H_i(\mathcal{C}(M_n(\mathbb{Z}\pi \otimes_{\mathbb{Z}} \mathbb{Q}))^{GL_n(\mathbb{Z}\pi)}).$$

<u>Moreover, these isomorphisms (over $\mathbb{E}$) are compatible with the maps induced by the upper inclusion.</u>

<u>Proof.</u> Since

$$\mathbb{E} \otimes_{\mathbb{Q}} \mathbb{Q}\pi \otimes_{\mathbb{Q}} V \simeq \mathbb{E}\pi \otimes_{\mathbb{E}} (\mathbb{E} \otimes V),$$

it follows from the remarks above that $\mathcal{C}(M_n(\mathbb{Q}\pi \widetilde{\otimes}_{\mathbb{Q}} \overline{K}))$ is a $\mathbb{Q}$-structure on $\mathcal{C}_{\mathbb{E}}(M_n(\mathbb{E}\pi \widetilde{\otimes}_{\mathbb{Q}} \overline{K}))$. By the remarks again and the statement about Γ-stable subspaces, it follows that $\mathcal{C}(M_n(\mathbb{Q}\pi \widetilde{\otimes}_{\mathbb{Q}} \overline{K}))^{GL_n(\mathbb{Z}\pi)}$ and

$H_i(\mathcal{C}(M_n(\mathbb{Q}\pi \widetilde{\otimes}_{\mathbb{Q}} \overline{K}))^{GL_n(\mathbb{Z}\pi)})$ are $\mathbb{Q}$-structures on $\mathcal{C}_{\mathbb{E}}(M_n(\mathbb{E}\pi \widetilde{\otimes}_{\mathbb{Q}} \overline{K}))^{GL_n(\mathbb{Z}\pi)}$ and $H_i(\mathcal{C}_{\mathbb{E}}(M_n(\mathbb{E}\pi \widetilde{\otimes}_{\mathbb{Q}} \overline{K}))^{GL_n(\mathbb{Z}\pi)}$ respectively. This proves the lemma.

Since $\mathbb{E}$ is a splitting field for π, $\mathbb{E}\pi = \mathbb{E} \otimes_{\mathbb{Q}} \mathbb{Q}\pi$ is a product of full matrix algebras over $\mathbb{E}$. Let $\{V_\alpha | \alpha \in S\}$ be the finite collection of irreducible right representations of π on $\mathbb{E}$-vector spaces. ($\mathbb{E}$ acts on V_α from the left, and π acts on V_α from the right.) Form $\mathbb{E}\pi$-bimodules $\{\mathbb{E}\pi \widetilde{\otimes}_{\mathbb{E}} V_\alpha | \alpha \in S\}$ as follows: As an $\mathbb{E}$-vector space, $\mathbb{E}\pi \widetilde{\otimes}_{\mathbb{E}} V_\alpha$ is just $\mathbb{E}\pi \otimes_{\mathbb{E}} V_\alpha$. For $x \in \pi$, $r \in \mathbb{E}\pi$ and $v \in V_\alpha$, the left multiplication is given by $x(r \widetilde{\otimes} v) = xr \widetilde{\otimes} v$, and the right multiplication is given by $(r \widetilde{\otimes} v)x = rx \widetilde{\otimes} vx$. In the obvious way we have an $\mathbb{E}$-space

$$M_n(\mathbb{E}\pi \widetilde{\otimes} V_\alpha)$$

of $(n \times n)$-matrices with entries in $\mathbb{E}\pi \widetilde{\otimes} V_\alpha$, on which $GL_n(\mathbb{E}\pi)$ acts by the formula

$$M \circ g = g^{-1}Mg$$

where the left (right) matrix multiplication is according to the left (right) $\mathbb{E}\pi$-module structure on $\mathbb{E}\pi \widetilde{\otimes} V_\alpha$. Denote this representation by Ad_n^α. We want to describe $\mathbb{E}$-valued, $GL_n(\mathbb{E}\pi)$-invariant multilinear functions of p_α vectors from Ad_n^α, p_β vectors from Ad_n^β, etc. In other words, we will find the $GL_n(\mathbb{E}\pi)$-invariant linear functions on

$$\bigotimes_{\alpha \in S} (Ad_n^\alpha)^{\otimes p_\alpha}$$

(We shall only consider tensor products over $\mathbb{E}$, so we omit the $\mathbb{E}$ from the notation. We also remark that the same question asked when the V_α's are not assumed irreducible is obviously reduced to the one we discuss now.)

We first describe the $\mathbb{E}\pi$-bimodule $\mathbb{E}\pi \tilde{\otimes} V_\alpha$ in more detail. Write $A = \mathbb{E}\pi$ from now on and recall that $A = \prod_{\alpha \in S} A_\alpha$ and each $A_\alpha \simeq M_{d_\alpha}(\mathbb{E})$. The representation V_α is afforded by the right action of A on an ideal $\mathfrak{a}_\alpha$ corresponding to a row of $M_{d_\alpha}(\mathbb{E})$. Then it is clear that the left A-module

$$A \tilde{\otimes} V_\alpha \simeq \prod_{\alpha \in S} M_{d_\beta, d_\beta d_\alpha}(\mathbb{E})$$

($d_\beta \times d_\beta d_\alpha$ rectangular matrices), with the subalgebra A_β of A acting by the left matrix multiplication on the factor $M_{d_\beta, d_\beta d_\alpha}(\mathbb{E})$. Since the left and right A actions commute with each other, the left A-submodule gotten from $M_{d_\beta, d_\beta d_\alpha}(\mathbb{E})$ by the left multiplication of an idempotent is also a right A-submodule. By Wedderburn theory each right submodule is a product of simple right A_γ-modules (for γ running through some not necessarily distinct indices in S). The explicit bimodule structure on $M_{d_\beta, d_\beta d_\alpha}(\mathbb{E})$ is

$$aMa' = a_\beta M_\beta P_\beta^\alpha \begin{pmatrix} a'_{\gamma_1} & & 0 \\ & \ddots & \\ 0 & & a'_{\gamma_r} \end{pmatrix} (P_\beta^\alpha)^{-1}$$

where $M_\beta \in M_{d_\beta, d_\beta d_\alpha}(\mathbb{E})$, $a = (a'_\gamma)$ are in A, the indices $\gamma_i \in S$ are not necessarily distinct, and P_β^α is an invertible $(d_\beta d_\alpha \times d_\beta d_\alpha)$ matrix over $\mathbb{E}$. So we observe that mapping

$$(M_\beta) \to (M_\beta P_\beta^\alpha)$$

(and blocking of the matrices $M_\beta P_\beta^\alpha$) defines an A-bimodule isomorphism of $A \widetilde{\otimes} V_\alpha$ interpreted as matrices to an appropriate product of $M_{d_\beta}(\mathbb{E})$-left, $M_{d_{\beta'}}(\mathbb{E})$-right modules. (The actions on the both sides are given by the usual matrix multiplications.) We also record the fact that the right A-module $A \widetilde{\otimes} V_\alpha$ is isomorphic to the right A-module A^{d_α} (not respecting the given left structure on $A \widetilde{\otimes} V_\alpha$ which is isomorphic to the obvious left A-structure on A^{d_α}!) so that the total multiplicities of the simple left and right $M_{d_\beta}(\mathbb{E})$ modules are the same.

We now apply these remarks to each entry of the $(n \times n)$-matrices of Ad_n^α and permute the coordinates to obtain a description of the representation of $GL_n(A)$. Note first that $GL_n(A) = \prod_{\alpha \in S} GL_n(A_\alpha) \simeq \prod_{\alpha \in S} GL_{nd_\alpha}(\mathbb{E})$.

Ad_n^α is isomorphic to an appropriate product of spaces $M_{nd_\beta, nd_{\beta'}}(\mathbb{E})$ on which $g = (g_\alpha) \in GL_n(A)$ acts by the formula

$$M \cdot g = g_\beta^{-1} M g_{\beta'} .$$

If we replace the set S by the set $\{1, \ldots, r = \operatorname{Card} S\}$, we may describe this decomposition of Ad_n^α as a matrix of representations

$$Ad_n^\alpha \simeq \begin{pmatrix} a_{11}\rho_1' \otimes \rho_1 & \ldots & a_{1r}\rho_1' \otimes \rho_r \\ & & \\ a_{r1}\rho_r' \otimes \rho_1 & \ldots & a_{rr}\rho_r' \otimes \rho_r \end{pmatrix}$$

where each a_{ij} is a nonnegative integer and ρ_i denotes the standard representation of the factor $GL_n(A_i) \simeq GL_{nd_i}(\mathbb{E})$ of $GL_n(A)$ and ρ_i' denotes the contragredient representation of ρ_i (If ρ_i is the representation on row vectors $v \to vg$, then the contragredient representation is the representation on column vectors $\varphi \mapsto g^{-1}\varphi$.) It is clear that $Ad_n^\alpha | GL_n(A_\iota)$ is a sum of

(1) trivial representations,

(2) adjoint representations,

(3) a number of standard representations ρ_ι and an equal number of representations ρ'_ι. The equality is a consequence of the last remark above on the left and right A structure of the A bimodule $A \widetilde{\otimes} V_\alpha$.

Example 3.2. Let V_0, V_1 be respectively the trivial and nontrivial representations of $\mathbb{Z}_2$. Then $\mathbb{Q}\mathbb{Z}_2 \simeq \mathbb{Q} \times \mathbb{Q}$ and

$$GL_n(\mathbb{Q}) \times GL_n(\mathbb{Q}) \simeq GL_n(\mathbb{Q}\mathbb{Z}_2).$$

Representing an element of Ad_n^ι $(\iota = 0, 1)$ by a pair of $n \times n$ matrices (M_0, M_1), we interpret the action of $g = (g_0, g_1)$ as follows:

The Ad_n^0 action $(M_0, M_1)g$

$$= (g_0^{-1} M_0 g_0, g_1^{-1} M_1 g_1),$$

the Ad_n^1 action $(M_0, M_1)g$

$$= (g_0^{-1} M_0 g_1, g_1^{-1} M_1 g_0).$$

The matrices are therefore

$$Ad_n^0 \simeq \begin{pmatrix} \rho'_0 \otimes \rho_0 & 0 \\ 0 & \rho'_1 \otimes \rho_1 \end{pmatrix},$$

$$Ad_n^1 \simeq \begin{pmatrix} 0 & \rho'_0 \otimes \rho_1 \\ \rho'_1 \otimes \rho_0 & 0 \end{pmatrix}.$$

For a slightly more complicated version of this phenomenon, consider the cyclic group $\mathbb{Z}_p$ (p a prime). The splitting field $\mathbb{E} = \mathbb{Q}(\zeta)$, where $\zeta = \zeta_p$ is a primitive p-th root of 1 fixed once and once for all. $A = \mathbb{E}\mathbb{Z}_p = \prod_{i=1}^{p} \mathbb{E}_i$, a product of copies of $\mathbb{E}$, where $\mathbb{E}_i$ corresponds to the representation $t \to \zeta^i$ for some generator t of $\mathbb{Z}_p$. We find that $Ad_n^i = M_n(A \widetilde{\otimes} \mathbb{E}_i)$ may be interpreted as follows Take a p-fold product of $M_n(\mathbb{E})$ and let $GL_n(A) \simeq \prod_{i=1}^{p} GL_n(\mathbb{E}_i)$ act by

$$(M_k)_{1 \le k \le p}\, g = (g_k^{-1} M_k g_{k+i})_{1 \le k \le p}$$

where $k+i$ is calculated mod p. Here the matrix of representations is

$$\begin{pmatrix} 0 & \dots & \rho_1' \otimes \rho_{1+k} & \dots & 0 \\ 0 & & & \rho_2' \otimes \rho_{2+k} & \vdots \\ \vdots & & & & \vdots \\ \rho_{p-k+1}' \otimes \rho_1 & & & & \vdots \\ \vdots & & & & 0 \end{pmatrix}$$

which looks like the k-th power of the permutation matrix

$$\begin{pmatrix} 0 & 1 & & & \\ & 0 & 1 & & \\ & & \ddots & \ddots & \\ 0 & & & \ddots & 1 \\ 1 & 0 & & & 0 \end{pmatrix}.$$

Example 3.3. The most general situation is illustrated by $\pi = S_3$ and $V = V_\alpha$ the irreducible 2-dimensional $\mathbb{Q}$ representation of S_3. Take $A = \mathbb{Q}\pi$ and note that $A = A_1 \times A_2 \times A_3 \simeq \mathbb{Q} \times \mathbb{Q} \times M_2(\mathbb{Q})$. Then $M_n(A \widetilde{\otimes} V)$ is represented by the matrix

$$\begin{pmatrix} 0 & 0 & \rho_1' \otimes \rho_3 \\ 0 & 0 & \rho_2' \otimes \rho_3 \\ \rho_3' \otimes \rho_1 & \rho_3' \otimes \rho_2 & \rho_3' \otimes \rho_3 \end{pmatrix}.$$

In terms of matrices the action is to carry

$$\begin{pmatrix} 0 & 0 & M_{13} \\ 0 & 0 & M_{23} \\ M_{31} & M_{32} & M_{33} \end{pmatrix}$$

to

$$\begin{pmatrix} g_1^{-1} & 0 & 0 \\ 0 & g_2^{-1} & 0 \\ 0 & 0 & g_3^{-1} \end{pmatrix} \begin{pmatrix} 0 & 0 & M_{13} \\ 0 & 0 & M_{23} \\ M_{31} & M_{32} & M_{33} \end{pmatrix} \begin{pmatrix} g_1 & 0 & 0 \\ 0 & g_2 & 0 \\ 0 & 0 & g_3 \end{pmatrix}.$$

Now we observe that the rules of matrix multiplication give a description of the representation $\bigotimes_{\alpha \in S} (Ad_n^{\alpha})^{\otimes p_\alpha}$ when we interpret the product of matrix entries as tensor product and sum of products as direct sum. For example, the second tensor power of the representation of Example 3.2 is represented by

$$\begin{pmatrix} \rho_1' \otimes \rho_3 \otimes \rho_3' \otimes \rho_1 & \rho_1' \otimes \rho_3' \otimes \rho_3 \otimes \rho_2 & \rho_1' \otimes \rho_3' \otimes \rho_3 \otimes \rho_3 \\ \rho_2' \otimes \rho_3 \otimes \rho_3' \otimes \rho_1 & \rho_2' \otimes \rho_3 \otimes \rho_3' \otimes \rho_2 & \rho_2' \otimes \rho_3' \otimes \rho_3 \otimes \rho_3 \\ \rho_3' \otimes \rho_3 \otimes \rho_3' \otimes \rho_1 & \rho_3' \otimes \rho_3 \otimes \rho_3' \otimes \rho_2 & \bigoplus_{i=1}^{3} \rho_3' \otimes \rho_i \otimes \rho_i' \otimes \rho_3 \end{pmatrix}$$

Next we argue for the following claim: The space of the linear $GL_n(A)$ invariant functions on such a representation is the sum of the spaces of invariant functions on the representations appearing along the diagonal of the product matrix. When we have this claim and an interpretation of the invariant functions, it will be easy to see that the stability for invariant linear functions $\bigotimes_{\alpha\in S}(\mathrm{Ad}_n^{\alpha})^{\otimes p_\alpha} \to \mathbb{E}$ as n tends to infinity is a consequence of the stability result for the adjoint representation of $GL_m(\mathbb{E})$: For $m > M$ the spaces of $\mathbb{E}$-valued invariant linear functions on $(\mathrm{Ad}_m)^{\otimes M}$ are all canonically isomorphic to the space obtained when $m = M+1$. We also observe that the invariants are those gotten from the corresponding invariants of $\prod_{\alpha\in S} SL_{nd_\alpha}(\mathbb{E})$. It follows from Zariski density theorem for SL_m that the invariants are exactly those invariants under $GL_n(\mathbb{Z}\pi)$. Let us give some examples to illustrate this claim.

Example 3.4. In the case of our Example 3.3 the linear invariant function

$$M_n(A \widetilde{\otimes} V) \to \mathbb{Q}$$

is

$$M \to \mathrm{Tr}\, M_{33},$$

the six bilinear invariant functions

$$M_n(A \widetilde{\otimes} V) \times M_n(A \widetilde{\otimes} V) \to \mathbb{Q}$$

are

$$M \otimes N \to \begin{cases} \mathrm{Tr}\, M_{13}N_{31}, & \mathrm{Tr}\, M_{31}N_{13}, \\ \mathrm{Tr}\, M_{23}N_{32}, & \mathrm{Tr}\, M_{32}N_{23}, \\ \mathrm{Tr}\, M_{33}N_{33}, & \mathrm{Tr}\, M_{33}\, \mathrm{Tr}\, N_{33}. \end{cases}$$

Returning to Example 3.2, there is clearly no linear invariant functional on Ad_n^i if $i \neq p$. If $i+j = p$, then there is a family of bilinear invariant functions

$$f_\ell : Ad_n^i \times Ad_n^j \to \mathbb{E}$$

$$f_\ell((M_k),(N_k)) = \mathrm{Tr}\, M_\ell N_{\ell+1}$$

for $\ell = 1,\ldots,p$.

The proof of the claim depends on the fact that if M is any $G = G_1 \times \ldots \times G_r$ module, then the invariant submodule $M^G = (\ldots(M^{G_1})\ldots)^{G_r}$. Recall also that, according to [7, pp. 345-349], the invariants of $GL_m(\mathbb{E})$ acting on k $(m\times m)$-matrices, p m-vectors (row vectors), and q m-covectors (column vectors) are as follows:

(1) invariants of the matrices alone, that is, products of traces of monomials in the matrices [7, pp. 312-313] [5].

(2) products of scalar products $<v_{i_1}M_1,\varphi_{j_1}> \ldots <v_{i_t}M_t,\varphi_{j_t}>$ where v_α is a vector, M_α is a monomial in matrices, and φ_α is a covector;

(3) products of these two types.

This was proved by interpreting Ad_m as $\rho'_m \otimes \rho_m$, using the classical description of invariants of vectors and covectors [10], and interpreting the classical description in terms of matrices

Writing the representation $\bigotimes_{\alpha\in S} Ad_n^{\otimes p_\alpha}$ as a matrix, we may see immediately its structure when restricted to the first factor $GL_n(A_1)$. More importantly, we can see the $GL_n(A_1)$-invariant function and the action of the other factors on them. Invariant functions are obtained by contracting a covector and a vector to a scalar. In a summand

$$\rho'_{a_1} \otimes \rho_{a_2} \otimes \rho'_{a_2} \otimes \ldots \otimes \rho_{a_t} \otimes \rho'_{a_t} \otimes \rho_{a_{t+1}}$$

we create $GL(A_1)$ invariant tensors by performing all possible contractions

$<v_{a_i}, \psi_{a_j}>$ where $a_i = a_j = 1$. Interpreting an element of the summand as a tensor product of matrices

$$M_{a_1, a_2} \otimes \ldots \otimes M_{a_t, a_{t+1}}$$

where M_{a_1, a_2} is a matrix on which $GL_n(A_{a_1})$ acts from the left and $GL_n(A_{a_2})$ acts from the right, we are forming all possible products

$$M_{*, a_1} M_{a_1, *}$$

and dropping a $\otimes$ symbol, or we are taking $\mathrm{Tr}\, M_{a_1 a_1}$, etc. On such things we still see the action of the other factors. Now this representation is on the diagonal if and only if $a_{t+1} = a_1$. In this case and only in this case, it is possible to pair off all indices to obtain a $GL_n(A)$ invariant. Interpreting this process in terms of matrices [7, pp. 311-312], we are taking products of matrices $M_{a_1 a_2} M_{a_2 a_3} \ldots M_{a_t, a_1}$, then the trace of the product, and possibly multiplying by traces of other products to obtain the invariants.

To answer the stability question for $\underset{\alpha \in S}{\otimes} (Ad_n^{\alpha})^{\otimes p_\alpha}$, the important thing is to determine for which factor $GL_n(A_i)$ of $GL_n(A)$ the largest number of adjoint representations of $GL(A_i)$ appear as the action of the other factors on a summand $\rho'_{a_1} \otimes \rho_{a_2} \otimes \rho'_{a_2} \otimes \ldots \otimes \rho_{a_t} \otimes \rho'_{a_t} \otimes \rho_{a_1}$ is contracted away. One may observe that the number of such representations is at most $\frac{1}{d_i} \sum_{\alpha \in S} p_\alpha d_\alpha$ $(GL_n(A_i) \simeq GL_{nd_i}(\mathbb{E}))$. Therefore, we will have stability when $n > \sum_{\alpha \in S} p_\alpha d_\alpha$. Let us summarize the discussions into the following lemma.

Lemma 3.5. The natural chain map

$$\mathcal{C}_{\mathbb{E}}(M_n(\mathbb{E}\pi \tilde{\otimes}_{\mathbb{Q}} \overline{K}))^{GL_n(\mathbb{E}\pi)} \to \mathcal{C}_{\mathbb{E}}(M_{n+1}(\mathbb{E}\pi \tilde{\otimes}_{\mathbb{Q}} \overline{K}))^{GL_{n+1}(\mathbb{E}\pi)}$$

is an isomorphism for degree $i << n$, *in a range tending to infinity with* n.

Theorem 1.1 follows immediately from this lemma and the observation that the $GL_n(\mathbb{E}\pi)$-invariant subcomplex of $\mathcal{C}_{\mathbb{E}}(M_n(\mathbb{E}\pi \tilde{\otimes}_{\mathbb{Q}} \overline{K}))$ is equal to

$$\mathcal{C}_{\mathbb{E}}(M_n(\mathbb{E}\pi \tilde{\otimes}_{\mathbb{Q}} \overline{K}))^{GL_n(\mathbb{Z}\pi)}$$

This completes the proof of Theorem 1.1.

IV. Proof of Theorem 1.2

Let G be a simplicial group with component group π. Let $\mathbb{Q}G$ be the simplicial rational group ring of G and let $\mathbb{Q}\pi$ (respectively $\mathbb{Z}\pi$) be the ordinary rational (respectively integral) group ring of π (both can be considered as the constant simplicial rings). There is a simplicial ring homomorphism

$$M_n(\mathbb{Q}G) \to M_n(\mathbb{Q}\pi)$$

induced by the component map $G \to \pi$, and $\widehat{GL}_n(\mathbb{Q}G)$ is defined to be the pullback of the above map over the natural inclusion $GL_n(\mathbb{Z}\pi) \to M_n(\mathbb{Q}\pi)$:

$$\begin{array}{ccc} \widehat{GL}_n(\mathbb{Q}G) & \to & M_n(\mathbb{Q}G) \\ \downarrow & & \downarrow \\ GL_n(\mathbb{Z}\pi) & \to & M_n(\mathbb{Q}\pi) . \end{array}$$

Matrix multiplication induces a simplicial monoid structure on the direct limit $\widehat{GL}(\mathbb{Q}G)$ of the upper inclusion maps $\widehat{GL}_n(\mathbb{Q}G) \to \widehat{GL}_{n+1}(\mathbb{Q}G)$. The plus construction $B\widehat{GL}(\mathbb{Q}G)^+$ with respect to the perfect commutator subgroup of $\pi_1 B\widehat{GL}(\mathbb{Q}G) = GL(\mathbb{Z}\pi)$ is a space which, if G is the loop group GX (see §2), is denoted by $A(X)$. This is rationally equivalent to the $A(X)$ of [9]. (Cf. [5].)

The following two lemmas are obvious.

Lemma 4.1. *If* $f : G_1 \to G_2$ *is an* h-*isomorphism of simplicial groups (see §2), then* f *induces isomorphisms*

$$H_i(B\widehat{GL}(\mathbb{Q}G_1), \mathbb{Q}) \simeq H_i(B\widehat{GL}(\mathbb{Q}G_2); \mathbb{Q}), \qquad i \geq 0.$$

Lemma 4.2. *If* $f : K_1 \to K_2$ *is an h-isomorphism of* π-DGA *(see §2), then* f *induces an isomorphism*

$$\mathrm{Inv}_i(K_1) \simeq \mathrm{Inv}_i(K_2), \qquad i \geq 0.$$

Now suppose that X is a 1-reduced simplicial set with finite fundamental group π, and let H be a 1-reduced simplicial π-group which is a split group model for ΩX (Lemma 2.4). Let G be the semidirect product $\pi \tilde{\times} H$, and let K be the π-DGA given by $N\mathbb{Q}H$. According to 4.1, 4.2, and a simple duality argument, to prove Theorem 1.2 it is enough to show that there are isomorphisms

$$H_n(B\widehat{GL}(\mathbb{Q}G),\mathbb{Q}) \simeq \sum_{p+q=n} H_p(GL(\mathbb{Z}\pi),\mathbb{Q}) \otimes \mathrm{Inv}_q(X).$$

(It follows from 4.1, 4.2 that $\mathrm{Inv}_q(X)$ only depends on X if $\pi = \pi_1 X$ is finite.) Let $S = \mathbb{Q}H$ (so that $K = NS$), and let $\overline{K}$ be the augmentation ideal of K. We may consider $\mathbb{Z}\pi \tilde{\otimes} K$ as the normalization of the simplicial ring $R = \mathbb{Q}G$ and $\mathbb{Z}\pi \tilde{\otimes} \overline{K}$ the augmentation ideal of $\mathbb{Z}\pi \tilde{\otimes} K \to \mathbb{Q}\pi$. Form the DGL $M_n(\mathbb{Z}\pi \tilde{\otimes} \overline{K})$, and let $GL_n(\mathbb{Z}\pi)$ act on it by conjugation. By functoriality this carries over into an action of $GL_n(\mathbb{Z}\pi)$ on $H_*(\mathcal{C}(M_n(\mathbb{Z}\pi \tilde{\otimes} \overline{K}))$ and on $H_*(\mathcal{C}(M_n(\mathbb{Z}\pi \tilde{\otimes} \overline{K})))$

Let $\widehat{GL}{}^0_n(R)$ be the kernel of the map $\widehat{GL}_n(R) \to \widehat{GL}_n(\mathbb{Z}\pi)$ provided above in the definition of $\widehat{GL}_n(R)$. Let

$$\widehat{GL}{}^0(R) = \varinjlim \widehat{GL}{}^0_n(R).$$

The exact sequence of simplicial monoids

$$1 \to \widehat{GL}{}^0(\mathbb{R}) \to \widehat{GL}(R) \to GL(\mathbb{Z}\pi) \to 1$$

induces a fibration

$$B\widehat{GL}^0(R) \to B\widehat{GL}(R) \to BGL(\mathbb{Z}\pi). \tag{4.3}$$

We need the following lemma, which is a generalization of Lemma 3.1 of [5].

<u>Lemma</u> 4.4. $H_*(B\widehat{GL}^0(R);\mathbb{Q})$ <u>is isomorphic to</u> $H_*(\mathcal{C}(M_n(\mathbb{Z}\pi \tilde{\otimes} \overline{K})))$ <u>in a way that respects the natural</u> $GL_n(\mathbb{Z}\pi)$ <u>actions</u>.

The argument of Lemma 3.1 of [5] works for 4.4 if we can complete $R = \mathbb{Q}G$ appropriately. To do this, just define

$$\hat{R} = \mathbb{Z}\pi \tilde{\otimes} \hat{S}$$

where $\hat{S}$ is the completion of S with respect to the augmentation ideal $\overline{S}$ of $QH = S \to \mathbb{Q}$.

Using 4.4 and the argument of [5], we see that the Serre spectral sequence of 4.3 collapses. This gives the result.

Yale University, New Haven, CT, U.S.A.
Princeton University and The Institute for Advanced Study, Princeton, NJ, U.S.A.
The Institute for Advanced Study, Princeton, NJ, and Pennsylvania State University, University Park, PA, U.S.A.

Footnotes

(1) Supported by NSF Grant MCS 78-02977

(2) Partially supported by NSF Grant GP 3432X1.

(3) Partially supported by NSF Grant MCS 77-18723 A01.

(4) See [9] for the relationship of $A(X)$ to the space of pseudo-isotopies.

References

[1] A. Borel, Linear Algebraic Groups, W. A. Benjamin (New York), 1969.

[2] D. Burghelea, The rational homotopy groups of Diff(M) and Homeo(M) in stable range, preprint.

[3] C. Curtis and I. Reiner, Representation Theory of Finite Groups and Associative Algebras, Interscience Publishers (New York), 1962.

[4] A. Dold, Homology of symmetric products and other functors of complexes, Annals of Math., 68 (1958), 54-80.

[5] W. Dwyer, W. C. Hsiang and R. Staffeldt, Pseudo-isotopy and invariant theory I, preprint.

[6] J. P. May, Simplicial Objects in Algebraic Topology, Van Nostrand (Princeton, New Jersey), 1967

[7] C. Procesi, The invariant theory of $n \times n$ matrices, Advances in Math., 19 (1976), 306-381.

[8] D. G. Quillen, Rational homotopy theory, Annals of Math., 90 (1969), 205-295

[9] F. Waldhausen, Algebraic K-theory of topological spaces I, Proc. Symp. Pure Math., 32, Part I (1978), 35-60

[10] H. Weyl, The Classical Groups, Princeton University Press (Princeton, New Jersey), 1946.

Secondary Cohomology Operations applied to the Thom class

by Friedrich Hegenbarth

Introduction

Let A denote the mod 2 Steenrod algebra and $H^*(X)$ the mod 2 cohomology of the space X. A is the graded algebra of stable cohomology operations and acts on $H^*(X)$. If we have a none-trivial relation $r = \Sigma a_i b_i$ among some elements of A, that is $r = 0$, and an element $x \in H^*(X)$ such that $b_i(x) = 0$ for all i, then there is defined an element $\Phi(r) \in H^*(X)/\Sigma a_i H^*(X)$. $\Sigma a_i H^*(X)$ is the indeterminancy. The precise definition is contained in § 1. r is assumed to be homogeneous of degree m Mainly we think of the Adém-relations

$$Sq^a Sq^b + \sum_j \binom{b-1-j}{a-2j} Sq^{a+b-j} Sq^j$$

with $0 < b \leq a < 2b$. The object of this paper is to prove the following

__Theorem:__ Let $r = \Sigma Sq^{l_i} Sq^{k_i}$ be a relation of degree $m \geq 3$. Denote by $\Phi(r)$ the secondary cohomology operation based on r. Assume for all i, $l_i \geq k_i$. Then there exists an (n-1)-spherical fibration ξ (n sufficiently large) such that $\Phi(r)$ is defined on the Thom class $U \in H^n(T\xi)$ with vanishing indeterminancy ($T\xi$ the Thom space of ξ) and $\Phi(r)(U) \neq 0$.

Let us denote by BSF_n the classifying space for (n-1)-spherical fibrations and $q_i \in H^i(BSF_n)$ the ith Wu-class. Further let $K_i = K(i, Z_2)$ be the Eilenberg-Mac Lane space of type (i, Z_2) and set $Z = \prod_{i \geq 2}^{n} K_i$. The Wu classes q_i define a map $q : BSF_n \to Z$. Denote by $\widetilde{BSF}_n \to BSF_n$ the induced path fibration over Z by q. The universal (n-1)-spherical fibration ξ_n over BSF_n lifts to $\tilde{\xi}_n$. Then $q_i(\tilde{\xi}) = 0$ and we may apply $\Phi(r)$ to $\tilde{U} \in H^n(T\tilde{\xi})$, $\tilde{U}$ the Thom class of $\tilde{\xi}$. If we have a relation r as in the Theorem, the indeterminancy of $\Phi(r)(\tilde{U})$ is 0 by Lemma 3.1 below. Hence via the Thom isomorphism we obtain an element $\tilde{e}(r) \in H^*(\widetilde{BSF}_n)$. By our Theorem $\tilde{e}(r) \neq 0$. The question now is . does $\tilde{e}(r)$ lift to a class

$e(r) \in H^* BSF_n)$? For that it is natural to consider twisted secondary operations. They define an element $e(r) \in H^*(BSF_n)$ (see [8] or [4]). However it is not obvious that $e(r)$ is a lifting of $\tilde{e}(r)$.

I would like to thank F. P. Peterson and M. Mahowald for discussions during the Siegen Topology Symposium in June 1979.

§ 1 Some auxiliary Propositions

For an abelian group G we denote by $K(n,G)$ the Eilenberg-Mac Lane space of type (G,n). In particular we abbreviate $K_n = K(n, Z_2)$. Associated to a relation $r = \sum_{i=1}^{s} a_i b_i$ in A, the Steenrod algebra mod 2, is a secondary cohomology operation $\Phi(r)$ of degree $-1 + l_i + k_i$, $i = 1, \ldots, s$, where $l_i = \deg a_i$ and $k_i = \deg b_i$. $\Phi(r)$ is defined on an element $H^n(X; Z_2)$ if $b_i(x) = 0$ for $i = 1, \ldots, s$. If $\Phi(r)$ is defined on x then

$$\Phi(r)(x) \in H^{n+q-1}(X; Z_2) / \sum_{i=1}^{s} a_i H^*(X; Z_2).$$

$\sum_{i=1}^{s} a_i H^*(X; Z_2)$ is called the indeterminancy.

$\Phi(r)$ is natural and there is a universal model. Consider the 2-stage Postnikov system

$$\begin{array}{ccc} \Omega K \xrightarrow{j} E & & \\ \big\downarrow p & & \\ K(n,Z) & \xrightarrow{\prod_{i=1}^{s} Sq^{k_i}} & \prod_{i=1}^{s} K_{n+k_i} = K \end{array}$$

If $\iota_n \in H^n(K(n, Z); Z_2)$ is the generator, then $\Phi(r)(p^*(\iota_n))$ is defined and

$$j^* \Phi(r)(p^*(\iota_n)) = \sum_{i=1}^{s} a_i (\iota_{n+k_i-1}) \in H^{n+q-1}(\Omega K)$$

For details see [1].

Consider now the stable 2-stage Postnikov system

$$\text{(2)} \qquad \begin{array}{ccc} K_{n+k-1} & \rightarrow & E \\ & & \downarrow p \\ & & K(Z,n) \xrightarrow[Sq^k]{} K_{n+k} \end{array}$$

and its various deloopings

$$\text{(3)} \qquad \begin{array}{ccc} K_{r+k-1} & \rightarrow & \Omega^{n-r} E \\ & & \downarrow \\ & & K(Z,r) \xrightarrow[Sq^k]{} K_{k+r} \end{array}$$

in particular

$$\begin{array}{ccc} K_{k-1} & \rightarrow & \Omega^n E \\ & & \downarrow \\ & & K(0,Z) \xrightarrow[Sq^k]{} K_k \end{array}$$

In $H_*(\Omega^n E, Z_2)$ one has the Dyer-Lashof operations

$$Q^i . H_s(\Omega^n E) \rightarrow H_{s+i}(\Omega^n E)$$

having the following properties

(i) $Q^i(x)$ is defined if $i - \deg x < n - 1$

(ii) $Q^i(x) = 0$ for $i < \deg x$

(iii) $Q^i(x) = x^2$ for $i = \deg x$

(iv) Q^i is stable (in the range where it is defined).

For these and other properties we refer to [2] or [5]. Note that $\Omega^n E \simeq Z \times K_{n-1}$ as spaces (but not as n^{th} loop spaces as the next proposition will show). If $I = (i_1, \ldots, i_l)$ is an l-tuple of non-negative integers one puts $Q^I = Q^{i_1} \cdots Q^{i_l}$, $l = l(I)$ is the length of I.

Proposition 2.1.

Let $[j] \in H_0(\Omega^n E) \cong Z$ be the generator of the jth component and $c_{k-1} \in H_{k-1}(K_{k-1})$ the generator, then

(a) $Q^{k-1}[1] = [2] \times c_{k-1}$

(b) $Q^I[1] = 0$ if $l(I) \geq 2$ and $I \neq (0,0,\ldots,0)$

(coefficients are understood to be in Z_2).

Proof of (a). Let σ_*^{k-1} $\tilde{H}_0(\Omega^n E) \to H_{k-1}(\Omega^{n-k+1} E)$ be the iterated ((k-1)-times) homology suspension. It is $\Omega^{n-k-1} E \simeq K_{k-1} \times K_{2k-2}$ as spaces and $\sigma_*^{k-1}[1] = c_{k-1} \times 1$.

Therefore if n is large enough:

$\sigma_*^{k-1}(Q^{k-1}[1]) = Q^{k-1}(\sigma_*^{k-1}[1]) = Q^{k-1}(c_{k-1} \times 1) = (c_{k-1} \times 1)^2$. The following holds

Lemma: $(c_{k-1} \times 1)^2 \neq 0$.

By this lemma we have $Q^{k-1}[1] \neq 0$, hence $Q^{k-1}[1] = [j] \times c_{k-1}$. That $j = 2$ follows as in the case $X = \Omega^{\infty} S^{\infty}$ from the structure map $S^{\infty} \underset{Z_2}{\times} X \times X \to X$ which is used for the definition of the Q^i (see [2]).

Part (b) is proved in [6], proposition 6.2.

It remains to prove the lemma. We look at the Postnikov system (3) with $r = k - 1$

$$\begin{array}{ccc} K_{2k-2} & \xrightarrow{j} & \Omega^{n-k-1} E \\ & & \downarrow p \\ & & K(k-1, Z) \xrightarrow[Sq^k]{} K_{2k-1} \end{array}$$

This is trivial, so

$$\Omega^{n-k+1} E \simeq K(k-1, Z) \times K_{2k-2}$$

as spaces, but not as H-spaces. Let $y = p^*(\iota_{k-1} \times 1)$, there exists an element $x \in H^{2k-2}(\Omega^{n-k+1})$ such that $j^*(x) = \iota_{2k-2}$, the generator of $H^{2k-2}(K_{2k-2})$, and

(4) $\quad \psi^*(x) = 1 \otimes x + y \otimes y + x \otimes 1$

where $\psi^* : H^*(\Omega^{n-k+1} E) \to H^*(\Omega^{n-k+1} E) \otimes H^*(\Omega^{n-k+1} E)$ is the co-multiplication arising from the loop structure. (for a proof see [7], Lemma 3.1.1). The lemma is a direct consequence of formula (3).

2.1 (a) can be improved to

Proposition 2.2:

If $k - 1 \geq r$, then $\Omega^{n-r} E \simeq K(r, Z) \times K_{k+r-1}$ and $Q^{k-1}(c_r \times 1) = 1 \times c_{k+r-1}$. ($c_r \in H_r(K(r,Z); Z_2)$, $c_{k+r-1} \in H_{k+r-1}(K_{k+r-1}; Z_2)$ are the generators).

Proof: The homology suspension

$$\sigma_* : \tilde{H}_*(K_p \times K_q) \to H_{*+1}(K_{p+1} \times K_{q+1})$$

satisfies $\sigma_*(C_p \times 1) = C_{p+1} \times 1$ and $\sigma_*(1 \times C_q) = 1 \times C_{q+1}$.
Therefore

$$Q^{k-1}(C_r \times 1) = Q^{k-1}(\sigma_*^r([1]) = \sigma_*^r(Q^{k-1}[1]) = \sigma_*^r(1 \times C_{k-1}) = 1 \times C_{r+k-1}.$$

We consider now 3-stage Postnikov systems. Assume given

$$\text{(5)} \qquad \begin{array}{ccccc} E'' & \overset{j'}{\hookrightarrow} & E' & & \\ \downarrow p'' & & \downarrow p' & & \\ K_{n+k-1} & \overset{j}{\hookrightarrow} & E & \underset{f}{\to} & K_{n+q-1} \\ & & \downarrow p & & \\ & & K(n;Z) & \underset{Sq^k}{\to} & K_{n+k} \end{array}$$

with $Sq^l(\iota_{n+k-1}) = j^* f^*(\iota_{n+q-1})$, $l + k = q$.

p', p'' are induced fibrations by f and $f \circ j = Sq^l$. E'' is also the fiber of $p \circ p' \cdot E' \to K(n,Z)$. Hence $\Omega^n E' \simeq Z \times \Omega^n E''$ and if $l \geq k$ then $\Omega^n E'' \simeq K_{k-1} \times K_{q-2}$.

Proposition 2.3:

If $l \geq k$, then in $H_*(\Omega^n E')$ one has

(a) $Q^{l-1} Q^{k-1}[1] = [4] \times 1 \times C_{q-2}$

(b) $Q^I[1] = 0$ if $l(I) \geq 3$

Proof: By our assumption $l \geq k$, $\Omega^n E' \simeq Z \times K_{k-1} \times K_{q-2}$ and $\Omega^n p' \cdot \Omega^n E' \to \Omega^n E \simeq Z \times K_{k-1}$ is the projection onto the first two factors. $(\Omega^n p')_* : H_*(\Omega^n E') \to H_*(\Omega^n E)$ maps $Q^{k-1}[1]$ to $Q^{k-1}[1] = 2 \times C_{k-1}$ (by Proposition 2.1). But the only element which maps to $2 \times C_{k-1}$ is $[2] \times C_{k-1} \times 1$ because $q - 2 \geq k - 1$. Hence in $H_*(\Omega^n E')$ we have

$$Q^{k-1}[1] = [2] \times C_{k-2} \times 1.$$

Further it is $Q^{l-1}Q^{k-1}[1] = Q^{l-1}([2]\times C_{k-1}\times 1) = [4]\times Q^{l-1}(C_{k-1}\times 1)$ (we refer to [2] for the properties of the Q^i). To compute $Q^{l-1}(C_{k-1}\times 1)$ in $H_*(\Omega^n E'')$ we can apply Proposition 2.2 with $r = k-1$ and n replaced by $m = n-k+1$ and k replaced by l. Then $l-1 \geq r = k-1$ and we find $Q^{l-1}(C_{k-1}\times 1) = 1\times C_{q-2}$. This proves (a). For the proof of (b) we refer to [6], Proposition 6.2.

Let $r = \Sigma\, Sq^{l_i} Sq^{k_i}$ be a relation in A, $\Phi(r)$ the associated secondary cohomology operation. We are going to apply Proposition 2.3 to the following situation:

$$
\begin{array}{ccccc}
E'' & \xrightarrow{j'} & E' & & \\
\downarrow & & \downarrow p' & & \\
\Omega K & \xrightarrow{j} & E & \xrightarrow{\Phi(r)(p^* \iota_n)} & K_{n+q-1} \\
 & & \downarrow p & & \\
 & & K(n,Z) & \xrightarrow{\Pi Sq^{k_i}} & \Pi K_{n+k_i} = K
\end{array}
\tag{6}
$$

We know that $j^*\Phi(r)(p^*(\iota_n)) = \sum_i Sq^{l_i}(\iota_{n+k_i-1})$.

Proposition 2.4.

In $H_*(\Omega^n E')$ the following holds if $l_i \geq k_i$.

(a) $Q^{l_i-1}Q^{k_i-1}[1] = [4]\times 1\times c_{q-2}$

(b) $Q^I[1] = 0$ for $l(I) \geq 3$.

To prove 2.4 we note that we can use the inclusions $K_{n+k_i} \subset K$ (choosing base points) and the projections $K \to K_{n+k_i}$ to "split" (6) and then apply Proposition 2.3.

Inductively one can obtain similar results on k-stage Postnikov systems. Using the Nishida relations the action of the Dyer-Lashof algebra might be computed.

§ 3 Proof of the Theorem

First we are going to define the spherical fibration ξ mentioned in the Theorem. Let $r = \Sigma a_i \cdot b_i$ be a given relation with $a_i = Sq^{l_i}$, $b_i = Sq^{k_i}$. Let $l_i + k_i = q$.

Consider the diagram

$$
(7) \qquad
\begin{array}{ccccccc}
\Omega K' & \overset{j'}{\to} & E' & & & & \\
 & \overset{x''}{\nearrow} & \downarrow p' & & \Phi(r)p^*(\iota_n) & & \\
 & & E & & \to & K_{n+q-1} & = K' \\
 & \overset{x'}{\nearrow} & \downarrow p & & & & \\
S^n & \underset{x}{\to} & K(n,Z) & \underset{\Pi Sq^{k_i}}{\to} & & \Pi K_{n+k_i} & = K
\end{array}
$$

where $x \in H^n(S^n; Z)$ is a generator, x', x'' liftings of x and $j' : \Omega K' = K_{n+q-2} \to E'$ the inclusion of the fiber of p'.

For any based space X we denote by $\pi : PX \to X$ the path fibration. The fiber of π is ΩX.

Then we define M by the pull-back diagram

$$
(8) \qquad
\begin{array}{ccc}
M & \to & PE \\
\downarrow t & & \downarrow \pi \\
S^n & \overset{x'}{\to} & E
\end{array}
$$

M is n-connected and $\Omega^n t$ maps $\Omega^n M$ in the 0-component of $\Omega^n S^n$, denoted by $(\Omega^n S^n)_0$.

Let us denote the composite map

$$\Omega^n M \to (\Omega^n S^n)_0 \overset{*1}{\to} (\Omega^n S^n)_1 \simeq SF(n)$$

by f. $*1$ is a homotopy equivalence.
($*$ means loop sum in $\Omega^n S^n$, $1 \in \Omega^n S^n$ is the identity $S^n \to S^n$ viewed as element in $\Omega^n S^n$). The adjoint of f

$$\text{ad}\, f \cdot \Sigma\Omega^n M \to BSF(n)$$

defines a n-spherical fibration ξ over $\Sigma\Omega^n M$.

It is easily shown that all Stiefel-Whitney classes of ξ vanish. Hence $\Phi(r)$ is defined on the Thom class $U \in H^n(T\xi)$.

Lemma 3.1:

If $\deg a_i \geq \deg b_i$, that is $l_i \geq k_i$, then the indeterminancy of $\Phi(r)\,(U)$ is zero. Hence we have a homomorphism

$$\Phi(r) : H^n(T\xi) \to H^{n+q-1}(T\xi).$$

Proof: We have to prove that $Sq^{l_i} H^{n-1+k_i}(T\xi) = \{0\}$. Any $x \in H^{n-1+k_i}(T\xi)$ is of the form $x = y \cdot U$ where $y \in H^{k_i-1}(\Sigma\Omega^n M)$. But $Sq^{l_i}(y\,U) = Sq^{l_i}(y)\cdot U$ (since the Stiefel Whitney classes vanish). Since $l_i > k_i - 1$, $Sq^{l_i}(y) = 0$ and the lemma is proved.

We have now a welldefined class

$$e(r) \in H^{q-1}(\Sigma\Omega^n M)$$

such that

$$e(r) \cdot U = \Phi(r)\,(U).$$

We will show that $e(r) \neq 0$.
With $*$ we denote also the loop sum in $H_*(\Omega^n S^n)$.

Proposition 3.2.

If $I = (i_1, \ldots, i_e)$ and $l(I) = l \geq 2$ then $Q^I([1] * [1-2^{l(I)}]$ is in the image of

$$f_* . H_* (\Omega^n M) \to H_*(SF(n)).$$

Before we prove 3.2 let us mention that $Q^I([0]) \in H_*((\Omega^n S^n)_0)$ is in the image of $(\Omega^n x')_*$, since Q^I is natural under n^{th}-loop maps. But 3.2 does not follow from that.

Proof: Of course the statement of 3.2 makes only sense when $Q^I[1]$ is defined which will be assumed. We loop diagram (8) n times and obtain

$$\begin{array}{ccc} \Omega^n M & \to & \Omega^n PE \simeq pt \\ \downarrow \Omega^n t & & \downarrow \\ (\Omega^n S^n)_0 & \overset{\Omega^n x'}{\to} & (\Omega^n E)_0 \end{array}$$

We apply the Eilenberg Moore Spectral sequence to it. The E^2-term writes

$$E^2 \cong (H_*((\Omega^n S^n)_0 \underset{W}{\square} Z_2) \otimes \mathrm{Cotor}^{H_*((\Omega^n E)_0)} (Z_2, Z_2)$$

$\square_W$ is the co-tensor product over

$W = \mathrm{Im}\, (\Omega^n x')_*$. The spectral sequence converges to $H_*(\Omega^n M)$.

$H_*((\Omega^n S^n)_0)$ is a co-module over W by the rule

$$\psi . H_*((\Omega^n S^n)_0 \overset{d_*}{\to} H_*((\Omega^n S^n)_0) \otimes H_* ((\Omega^n S^n)_0) \overset{1 \otimes (\Omega^n x')_*}{\to} H_*((\Omega^n S^n)_0) \otimes W .$$

d_* is induced by the diagonal.

Z_2 is a W-co-module by the natural map $\psi' \; Z_2 \to W \otimes Z_2$.
$H_*((\Omega^n S^n)_0) \underset{W}{\square} Z_2$ is the kernel of the map

$\psi \otimes 1 - 1 \otimes \psi' : H_*((\Omega^n S^n)_0) \to H_*((\Omega^n S^n)_0) \otimes W \otimes Z_2$.

Hence $u \otimes \varepsilon \in H_*((\Omega^n S^n)_0) \underset{W}{\square} Z_2$ if

$$\Sigma u' \otimes (\Omega^n x')_* u'' = u \otimes \varepsilon$$

(where $\psi(u) = \Sigma u' \otimes u''$)

$H_*((\Omega^n S^n)_0) \underset{W}{\square} Z_2$ survives to E^∞. To prove the assertion it suffices to show that

$$Q^I[1] * [-2^{l(I)}] \in H_*((\Omega^n S^n)_0) \underset{W}{\square} Z_2$$

Using the formulas

$$d_*(x * y) = d_*(x) * d_*(y)$$

$$d_*(Q^I(x)) = \sum_{I'+I''=I} Q^{I'}(x') \otimes Q^{I''}(x'')$$

(where $d_*(x) = \Sigma x' \otimes x''$ and the sum $I' + I''$ is defined to be the sum in each component. It is $l(I') = l(I'') = l(I)$.)

we have to show

$$\sum_{I'+I''=I} Q^{I'}[1] * [-2^{l(I')}] \otimes (\Omega^n x')_* (Q^{I''}[1] * [-2^{l(I)}])$$
$$= Q^I[1] * [-2^{l(I)}] \otimes [0]$$

If $l(I'') = l(I) \geq 2$ and $I'' \neq (0,\dots,0)$ then $(\Omega^n x')_*(Q^{I''}[1] * [-2^{l(I)}]) = 0$ by 2.1 and $Q^{(0,\dots,0)}[1] = [2^{l(I)}]$. Thus the equality holds and the proposition is proved.

To complete the proof of the Theorem we must use another description of M. M is the fiber production of x'' and j' in (7)

$$\begin{array}{ccc} M & \xrightarrow{h} & \Omega K' \\ \downarrow t & & \downarrow j' \\ S^n & \xrightarrow{x''} & E' \end{array}$$

We therefore have the following homotopy commutative diagram

$$\begin{array}{ccccccc} \Omega K' & \xrightarrow{j'} & E' & \rightarrow & E & \xrightarrow{\Phi(r)(p^*(\iota_n))} & K' = K_{n+q-1} \\ \uparrow h & & \uparrow x'' & & \uparrow \alpha'' & & \uparrow h' \\ M & \xrightarrow{t} & S^n & \rightarrow & C_t & \xrightarrow{v} & \Sigma M \\ \uparrow \alpha & & \| & & \uparrow \alpha' & & \uparrow \Sigma\alpha \\ \Sigma^n\Omega^n M & \xrightarrow{\beta} & S^n & \rightarrow & T(\xi) & \xrightarrow{\delta} & \Sigma^{n+1}\Omega^n M \end{array}$$

The lower two sequences are the Puppe-sequences of the maps t and β = adjoint of $\Omega^n t$. α is the adjoint of the identity on $\Omega^n M$. C_t is the mapping cone of t. h' is the adjoint map of h. α' and α'' are induced maps.

It follows that

$$\text{(9)} \qquad \begin{aligned} e(r)\cdot U &= [h' \circ \Sigma\alpha \circ \delta] \\ e(r)\cdot U &= \delta^* (\Sigma^n)^* (e(r)) \end{aligned}$$

(see [6] p. 127, remark 5.16)

Lemma 3.3:

If the homomorphism $(\Omega^n h)^* : H^*(K_{q-2}) \to H^*(\Omega^n M)$ satisfies $(\Omega^n h)^*(\iota_{q-2}) \neq 0$, then $e(r) \neq 0$.

We complete first the proof of the Theorem.

Consider

$$\begin{array}{ccc}
H_*(K_{q-2}) & \xrightarrow{(\Omega^n j')_*} & H_*((\Omega^n E')_0) \\
\uparrow (\Omega^n h)_* & & \uparrow (\Omega^n x'')_* \\
H_*(\Omega^n M) & \xrightarrow[(\Omega^n t)_*]{} & H_*((\Omega^n S^n)_0)
\end{array}$$

By 2.4 we have $(\Omega^n j')_*(C_{q-2}) = Q^{l_1-1} Q^{k_1-1} [1] * [-4]$
$= (\Omega^n x'')_*(Q^{l_1-1} Q^{k_1-1} [1] * [-4])$.

By 3.2 there exists an element $y \in H_{q-2}(\Omega^n M)$ such that
$(\Omega^n t)_*(y) = Q^{l_1-1} Q^{k_1-1} [1] * [-4]$.

Hence $(\Omega^n h)_* \neq 0$ and this implies $(\Omega^n h)^*(\iota_{q-2}) \neq 0$.

It remains to prove 3.3

This follows from the diagram

$$\begin{array}{ccccccc}
H^{q-2}(\Omega^n S^n) & \xrightarrow{(\Omega^n t)^*} & H^{q-2}(\Omega^n M) & \overset{(\Sigma^{n+1})^*}{\cong} & H^{n+q-1}(\Sigma^{n+1}\Omega^n M) & \xrightarrow{\delta^*} & H^{n+q-1}(T\xi) \\
\uparrow (\Omega^n x'')^* & & \uparrow (\Omega^n h)^* & & \uparrow (h' \circ \Sigma\alpha)^* & & \\
H^{q-2}(\Omega^n E') & \xrightarrow{(\Omega^n j)^*} & H^{q-2}(\Omega^{n+1} K_{n+q-1}) & \xleftarrow{(\sigma^*)^{n+1}} & H^{n+q-1}(K_{n+q-1}) & &
\end{array}$$

and (9), because

$e(r) \cdot U = \delta^* \circ (\Sigma^{n+1})^* (\Omega^n h)^* (\iota_{q-2})$.

Since δ^* is an isomorphism, the lemma is proved.

[1] J. F. Adams, On the non-existence of elements of Hopf invariant one, Ann. of Math. 72 (1960), 20-104.

[2] E. Dyer and R. Lashof, Homology of iterated loop spaces, Amer. J. Math. 84 (1962), 53-88.

[3] S. Gitler and J. Stasheff, The first exotic class of BF, Topology 4 (1965), 257-266.

[4] F. Hegenbarth, On a Cartan formula for exotic characteristic classes I, to appear in Math. Scand.

[5] P. May, A general algebraic approach to Steenrod operations, Proc. of the Conference of the Battelle Memorial Institute 1970, edited by F. P. Peterson, Springer Lecture Notes in Mathematics 168, 153-231.

[6] P. May, The homology of E_{∞} ring spaces, in F. R. Cohen, Th. J. Lada, J. P. May, The homology of iterated loop spaces, Springer Lecture Notes in Mathematics 533, 69-206.

[7] R. J. Milgram, The structure over the Steenrod algebra of some 2-stage Postnikov systems, Quart. J. Math. Oxford 20 (1969), 161-169.

[8] F. P. Peterson, Twisted cohomology operations and exotic characteristic classes, Advances in Math. 4 (1970), 81-90.

[9] D. C. Ravenel, A definition of exotic characteristic classes of spherical fibrations. Comm. Math. Helv. 47 (1972), 421-436.

Splittings of loop spaces, torsion in homotopy, and double suspension.

[after F. Cohen, J.C. Moore, and J. Neisendorfer,
J. Neisendorfer, and P. Selick]

by

D. Husemoller

This is a survey of the recent results on unstable homotopy by the above mentioned people. Their important progress is in many ways a natural extension of the early work by Serre, James, Moore, and Toda in the 1950's where the emphasis was on the loop space ΩS^m. The new techniques include a systematic use of mod p homotopy, its Bockstein spectral sequence, and Lie algebras. The decomposition theorems are based on properties of subalgebras of free Lie algebras and of differential Lie algebras.

The published material consists of Selick's thesis and his paper [1978] and one paper by Cohen, Moore, and Neisendorfer [1979] the first in a series of several papers. In prepublication form are several papers by the three authors and two by Neisendorfer himself on the exponent of Moore spaces and special features of the prime 3.

It is a pleasure to acknowledge many useful conversations with John Moore, Joseph Neisendorfer, Fred Cohen, and Frank Adams on this subject.

§1. Partial splitting of the double suspension map

Much of the early work on the global structure of the homotopy groups of spheres centered around the loop space ΩS^{m+1} and the suspension map $S^m \longrightarrow \Omega S^{m+1}$. There are two related methods for analysing the homotopy type of ΩS^{m+1}. The first uses the suspension map $S^m \longrightarrow \Omega S^{m+1}$ and various other maps of spheres into ΩS^{m+1} built out of the suspension map, and the second uses the decomposition and projections

$$S\Omega S^{m+1} \xrightarrow{\simeq} S^{m+1} \vee S^{2m+1} \vee \cdots \longrightarrow S^{km+1}.$$

The adjoint of the projection $h_k . \Omega S^{m+1} \longrightarrow \Omega S^{km+1}$ is made into a fibration, called the k-th Hopf invariant, and its fibre is compared with $S^m \longrightarrow \Omega S^{m+1}$ for $h_2 \quad \Omega S^{m+1} \longrightarrow \Omega S^{2m+1}$. The fibre of $h_p : \Omega S^{2n+1} \longrightarrow \Omega S^{2pn+1}$ is related to the double suspension $S^{2n-1} \longrightarrow \Omega^2 S^{2n+1}$.

The series of papers by the three authors Cohen, Moore, and Neisendorfer [1979,...] uses heavily the first method for the cases of Moore spaces and Neisendorfer's work on the exponent of Moore spaces also uses splittings of suspensions of spaces.

To illustrate the first type of construction, we recall that Serre in his thesis took a Lie bracket of $S^{2n-1} \longrightarrow \Omega S^{2n}$ with itself to obtain a map $S^{4n-2} \longrightarrow \Omega S^{2n}$ (called a Samelson or Whitehead product) which induces a map $\Omega S^{4n-1} \longrightarrow \Omega S^{2n}$. Multiplying this map with the suspension map, we obtain a map

$$S^{2n-1} \times \Omega S^{4n-1} \longrightarrow \Omega S^{2n}$$

which is a homology isomorphism with field coefficients for all characteristics except 2. Inverting the prime 2 on the spaces, we have a map

$$S^{2n-1}[1/2] \times \Omega S^{4n-1}[1/2] \longrightarrow \Omega S^{2n}[1/2]$$

which is a homotopy equivalence. In particular the suspension map $S^{2n-1}[1/2] \longrightarrow \Omega S^{2n}[1/2]$ has a left inverse. Further the homotopy inverse composed with the projection on the second factor

$$\begin{array}{ccc} \Omega S^{2n}[1/2] & \longrightarrow & S^{2n-1}[1/2] \times \Omega S^{4n-1}[1/2] \\ & \searrow & \downarrow \\ & & \Omega S^{4n-1}[1/2] \end{array}$$

is the 2nd Hopf invariant $h_2 : \Omega S^{2n} \longrightarrow \Omega S^{4n-1}$ localized at 1/2, and the fibre of the Hopf invariant is the suspension map $S^{2n-1} \longrightarrow \Omega S^{2n}$.

Hence for the odd primes the study of the homotopy of spheres reduces to the study of homotopy on the odd spheres because the induced homomorphism of the above map

$$\pi_*(S^{2n-1}) \times \pi_*(\Omega S^{4n-1}) \longrightarrow \pi_*(\Omega S^{2n})$$

has kernel and cokernel consisting of 2-torsion abelian groups. Now we concentrate only on the odd torsion and localize at an odd prime p. We are led to the consideration of the double suspension map for odd spheres localized at an odd prime.

$$S^{2n-1}_{(p)} \longrightarrow \Omega^2 S^{2n+1}_{(p)}$$

Toda [1956], using the James construction, factored the double suspension inclusion $S^{2n-1} \longrightarrow \Omega T \longrightarrow \Omega^2 S^{2n+1}$ such that the relative homotopy groups $\pi_*(\Omega^2 S^{2n+1}_{(p)}, \Omega T)$ and $\pi_*(\Omega T, S^{2n-1}_{(p)})$ are isomorphic to $\pi_*(\Omega^2 S^{2n+1}_{(p)})$ and $\pi_*(\Omega S^{2np-1}_{(p)})$ respectively.

In Husemoller [1975, Appendix 2] the Toda result was obtained by constructing fibrations of spaces localized at p

$$\begin{array}{ccc} S^{2n-1} & \longrightarrow & \Omega F_n(p) \\ & & \downarrow \\ & & \Omega S^{2np-1} \end{array} \qquad \begin{array}{ccc} F_n(p) & \longrightarrow & \Omega S^{2n+1} \\ & & \downarrow \\ & & \Omega S^{2np+1} \end{array}$$

where $F_n(p)$ is a version of the previous T. This approach depends on the analysis of a 2-cell complex $T_n(p) = S^{2n-1} \cup_\tau e^{2pn-2}$ where the attaching map τ generates $\pi_{2pn-3}(S^{2n-1}_{(p)}) = \mathbb{Z}/p$ and double suspends to zero. Then $S^2T_n(p)$ is the wedge of two spheres and $ST_n(p) \wedge \cdot\cdot^{(k)}\cdot\cdot \wedge T_n(p)$ is the wedge of 2^k spheres. It was pointed out to us by J.C. Moore that our claim that $T_n \wedge T_n$ was a wedge of 4 spheres was not true as asserted in the proof of Prop. 7.4, p. 308 of the Fibre Bundles (2nd Edition) but that the proposition was still true since $ST_n \wedge T_n$ is a wedge of 4 spheres. It is seen by looking at the attaching map of the top cell onto the bottom cell of the 4 cell complex. This suspends to zero in $S^2T_n \wedge T_n$ and is of odd torsion, but as observed above, the suspension map has a left inverse on an odd sphere so this attaching map is null homotopic. Included in the original work of Toda and this analysis is a homomorphism $\pi_*(\Omega^2 S^{2np+1}) \longrightarrow \pi_*(S^{2np-1})$ which when composed with the double suspension morphism $\pi_*(S^{2np-1}) \longrightarrow \pi_*(\Omega^2 S^{2np+1})$ is multiplication by p. The result anticipates the following far reaching result of Cohen, Moore, and Neisendorfer.

<u>Theorem 1</u>. For an odd prime the localized double suspension map $E^2 : S^{2n-1}_{(p)} \longrightarrow \Omega^2 S^{2n+1}_{(p)}$ has a related map $\pi : \Omega^2 S^{2n+1}_{(p)} \longrightarrow S^{2n-1}_{(p)}$ such that the composites

$$\pi \circ E^2 : S^{2n-1}_{(p)} \longrightarrow S^{2n-1}_{(p)} \quad \text{and} \quad E^2 \circ \pi \quad \Omega^2 S^{2n+1}_{(p)} \longrightarrow \Omega^2 S^{2n+1}_{(p)}$$

are each induced by multiplication by p on their respective H - spaces.

We call π a partial retraction of E^2. This theorem is proved in the second paper in the series by the three authors for $p \neq 3$ and by Neisendorfer for $p = 3$. It is a corollary of their study of the splitting of loop spaces using Lie algebra techniques.

§2. <u>Application of the partial retraction map to the order of torsion.</u>

In view of the partial retraction theorem for the double suspension we have the following implication for the homotopy groups localized at p.

$$p^s\pi_i(S^{2n-1})_{(p)} = 0 \quad \text{implies} \quad p^{s+1}\pi_i(\Omega^2 S^{2n+1})_{(p)} = 0.$$

For an odd sphere the torsion subgroup Tors $\pi_i(S^{2n-1})_{(p)} = \pi_i(S^{2n-1})_{(p)}$ for $i \neq 2n-1$. By considering the sequence of spheres

$$S^1,\ S^3,\ S^5,\ \ldots\ ,\ S^{2n-1},\ S^{2n+1},\ \ldots$$

and using the relation Tors $\pi_*(S^1) = 0$, we obtain the next theorem.

<u>Theorem 2</u>. $p^n \mathrm{Tors}\ \pi_*(S^{2n+1})_{(p)} = 0$ for all n where p is an odd prime.

In his thesis, appearing in [1978], Selick did the case of $n = 1$ for the 3-sphere namely $p\pi_{3+i}(S^3)_{(p)} = 0$ for $i > 0$. Previously, Toda in [1956], using his results on the double suspension, was able to show that p^{2n} Tors $\pi_*(S^{2n+1})_{(p)} = 0$ for $n \geq 0$ Theorem 2 was the conjecture for twenty years and was formulas by M. Barratt.

This result leads to the general question of exponents which is discussed in section 6 and in the Aarhus survey of Cohen, Moore, and Neisendorfer.

§3. <u>Sketch of the proof of theorem 1 from a loop space splitting.</u>

The proof of the theorem depends on a certain splitting of loop spaces, which we state below as theorem 3, associated with the mod p^r - sphere (or Moore space) $S^{m+1}(p^r) = S^m \cup_{p^r} e^{m+1}$ and the pinch map $S^{m+1}(p^r) \longrightarrow S^{m+1}$ defined by collasping S^m to a point We construct the map $\pi : \Omega^2 S^{2n+1}_{(p)} \longrightarrow S^{2n-1}_{(p)}$ by factoring $\Omega^2 p$ as $E^2 \circ \pi$ as maps defined $\Omega^2 S^{2n+1}_{(p)} \longrightarrow \Omega^2 S^{2n+1}_{(p)}$ using the following commutative diagram of fibre sequences where $F^{2n+1}\{p^r\} \longrightarrow S^{2n+1}(p^r)$ is the fibre of the pinch map $S^{2n+1}(p^r) \longrightarrow S^{2n+1}$

$$\begin{array}{ccccccc}
\Omega S^{2n+1} & \longrightarrow & F^{2n+1}\{p^r\} & \longrightarrow & S^{2n+1}(p^r) & \longrightarrow & S^{2n+1} \\
\downarrow \Omega p^r & & \downarrow & & \downarrow & & \downarrow p^r \\
\Omega S^{2n+1} & \xrightarrow{\sim} & \Omega S^{2n+1} & \longrightarrow & * & \longrightarrow & S^{2n+1}
\end{array}$$

This factors p^r through $F^{2n+1}\{p^r\}$ and hence $\Omega^2 p^r$ through the space $F^{2n+1}\{p^r\}$. Note, we have avoided the symbol $P^m(q)$ used by the above authors for the mod q sphere $S^m(q)$ since the symbol P^m usually refers to a projective space.

Now the pinch map $S^{2n+1}(p^r) \longrightarrow S^{2n+1}$ factors as a composite $S^{2n+1}(p^r) \longrightarrow S^{2n+1}\{p^r\}$ with $S^{2n+1}\{p^r\} \longrightarrow S^{2n+1}$ where the second map is the fibre of $S^{2n+1} \xrightarrow{p^r} S^{2n+1}$. The fibre of the second map $S^{2n+1}\{p^r\} \longrightarrow S^{2n+1}$ is ΩS^{2n+1}, and it is the base space and $F^{2n+1}\{p^r\}$ is the total space of a fibration represented vertically in the following commutative diagram.

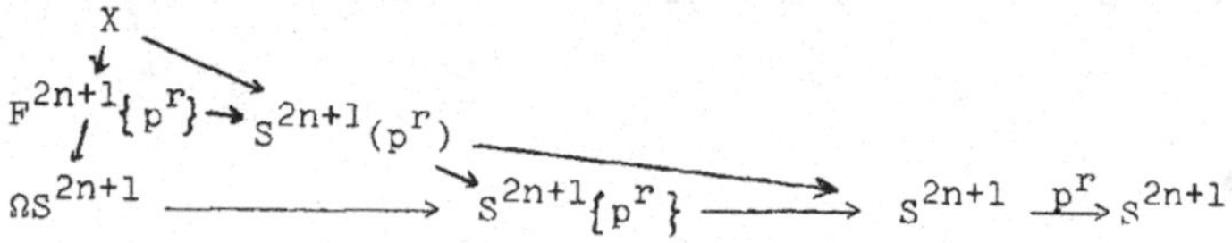
X
$F^{2n+1}\{p^r\}$
$S^{2n+1}(p^r)$
ΩS^{2n+1}
$S^{2n+1}\{p^r\}$
S^{2n+1}
p^r
S^{2n+1}

Here X is the fibre of both $S^{2n+1}(p^r) \longrightarrow S^{2n+1}\{p^r\}$ and $F^{2n+1}\{p^r\} \longrightarrow \Omega S^{2n+1}$.

Theorem 3. There is a space Y and a map $Y \longrightarrow \Omega X$ such that the following composite is a homotopy equivalence where the last map is loop space multiplication

$$S^{2n-1} \times Y \to \Omega F^{2n+1}\{p^r\} \times \Omega X \to \Omega F^{2n+1}\{p^r\} \times \Omega F^{2n+1}\{p^r\} \to \Omega F^{2n+1}\{p^r\}$$

Specializing to $r = 1$ and using the factorization of $\Omega^2 p$ in the triangle on the right, we have a commutative diagram where I is a homotopy inverse to the composite in theorem 3.

$$\begin{array}{ccccccc}
 & & \xleftarrow{\text{inverse } = \ I} & & & & \Omega^2 S^{2n+1}_{(p)} \\
 & & & & & \swarrow^{\lambda} & \\
S^{2n-1} \times Y & \longrightarrow & \Omega F^{2n+1}\{p\} \times \Omega F^{2n+1}\{p\} & \xrightarrow{\mu} & \Omega F^{2n+1}\{p\} & & \downarrow \Omega^2 p \\
\downarrow & & \downarrow & & & \searrow & \\
S^{2n-1} \times * & \longrightarrow & \Omega^2 S^{2n+1}_{(p)} \times \Omega^2 S^{2n+1}_{(p)} & & \xrightarrow{\qquad \mu \qquad} & & \Omega^2 S^{2n+1}_{(p)}
\end{array}$$

This leads to the factorization of $\Omega^2 p$

$$\Omega^2 S^{2n+1}_{(p)} \longrightarrow \Omega F^{2n+1}\{p\} \xrightarrow{pr \circ I} S^{2n-1}_{(p)} \xrightarrow{E^2} \Omega^2 S^{2n+1}_{(p)}$$

and $\pi \quad \Omega^2 S^{2n+1}_{(p)} \longrightarrow S^{2n-1}_{(p)}$ is defined to be the composite of the first two maps. Hence $E^2 \circ \pi$ is $\Omega^2 p$, and the verification that $\pi \circ E^2$ is multiplication by p is just an immediate check that this composite has degree p.

The space Y is constructed out of Moore spaces $S^m(p^r)$ and fibres $S^k\{p^{r+1}\}$ representing torsion homotopy elements in $\Omega S^{2n+1}(p^r)$ which are null homotopic when projected into $\Omega S^{2n+1}\{p^r\}$ and hence lift back into ΩX. An indication of where the jump in torsion from p^r to p^{r+1} comes from is given in section 5.

§4. Splitting of loops on a Moore space.

In this section all spaces are localized at an odd prime p. Observe that the Hopf algebra $H_*(\Omega S^{2n+2}(p^r), \mathbb{Z}/p^r)$ is the tensor algebra on two elements v_{2n+1}, u_{2n} of degrees 2n+1, 2n respectively. This tensor algebra $T(v_{2n+1}, u_{2n})$ is the universal enveloping algebra of the free Lie algebra (graded) $L = L(v_{2n+1}, u_{2n})$ on two generators. Since for the commutator subalgebra $[L,L]$ of L the following sequence is exact as Lie algebras

$$0 \longrightarrow [L,L] \longrightarrow L \longrightarrow L^{ab} \longrightarrow 0,$$

the universal enveloping algebra splits

$$T(v,u) = U(L) = U(L^{ab}) \otimes U[L,L]$$

as graded modules. In fact, it splits as right $U[L,L]$ modules and left $U(L^{ab})$ comodules.

Now under $S^{2n+1}\{p^r\} \longrightarrow \Omega S^{2n+2}(p^r)$ the homology injects onto the factor $U(L^{ab})$ in the above splitting. Hence with coefficients in $\mathbb{Z}/p^r$ we have an isomorphism $H_*(S^{2n+1}\{p^r\}) \otimes U[L,L] \longrightarrow U(L) = H_*(\Omega S^{2n+2}(p^r))$. Since $[L,L]$ is a subalgebra of a free Lie algebra L, it is free and its universal enveloping algebra $U[L,L]$ is the tensor algebra $T(W)$ on a free submodule $W \subset [L,L]$ where a basis of W is a free set of generators of $[L,L]$. The module of indecomposible elements $QU[L,L] = Q(T(W)) = W$.

As in the elementary case considered first by Serre, the elements $ad^k(u)[v,v]$ and $ad^k(u)[u,v]$ for $k \geq 0$ are examples of generators which are in the image of the mod p^r Hurewicz morphism. A Hopf algebra argument shows that they are linearly independent in $QU[L,L]$. A calculation with Poincaré series shows that they are a basis of $QU[L,L]$, hence a free set of generators of $[L,L]$.

Since the two generators are in the image of the mod p^r Hurewicz morphism, there is a map $S^{4n+2+2kn}(p^r) \longrightarrow \Omega S^{2n+2}(p^r)$ such that the image of $H_*(S^{4n+2+2kn}(p^r), \mathbb{Z}/p^r)$ is the free submodule generated by $ad^k(u)[v,v]$ and $ad^k(u)[u,v]$. The wedge product map $\bigvee_{k\geq 0} S^{4n+2+2kn}(p^r) \longrightarrow \Omega S^{2n+2}(p^r)$ has as image a loop space, so it extends by adjunction to loops on the suspension yielding a map

$$T = \Omega(\bigvee_{k\geq 0} S^{4n+3+2kn}(p^r)) \longrightarrow \Omega S^{2n+2}(p^r).$$

This map and the map $S^{2n+1}\{p^r\} \longrightarrow \Omega S^{2n+2}(p^r)$ multiply in the loop space to give a map

$$S^{2n+1}\{p^r\} \times T \longrightarrow \Omega S^{2n+2}(p^r)$$

which is a homology isomorphism since $H_*(T)$ maps isomorphically onto $U[L,L]$. Hence this map is a homotopy equivalence which gives the splitting of $\Omega S^{2n+2}(p^r)$.

This decomposition of $\Omega S^{2n+2}(p^r)$ is analogous to the Serre decomposition $\Omega S^{2n+2} \simeq S^{2n+1} \times \Omega S^{4n+3}$. Note that $S^{4n+3}(p^r)$ is the lowest dimensional term in the wedge of odd dimensional Moore spaces defining T. The question of the decomposition of $\Omega S^{2n+1}(p^r)$ is more difficult and is related to the decompositon of $\Omega F^{2n+1}(p^r)$ studied in the next section. Again $\Omega S^{2n+1}(p^r)$ is a product of loops on a wedge $\Omega(\bigvee_j S^{n(j)}(p^r))$ with a space $T^{2n+1}(p^r)$. The dimensions $n(j)$ of the Moore spaces in the wedge satisfy $n(j)$ converges to $+\infty$. The study of this space $T^{2n+1}(p^r)$ leads to the results on exponents given in section 6.

§5. Splitting of loops on the fibre of the pinch map.

All spaces are localized at an odd prime $p \neq 3$. The splitting of $F^{2n+1}\{p^r\}$ contains a new feature not present in the splitting of loops on Moore spaces. The spaces $S^{2m+1}\{p^{r+1}\}$ involving a higher power of p enter into the product decomposition. To see this, we look closer at the Bockstein spectral sequence.

Let k be a field of characteristic $p > 0$. In a commutative differential algebra A over k with differential β, the relation $\beta(x^{p^j}) = 0$ holds for any element x of A. In a general (associative) differential algebra $\beta(x^{p^j}) = ad^{p^j-1}(x)(\beta(x))$ so that

$$\beta[ad^{p^j-1}(x)(\beta(x))] = 0$$

where $ad(x)(y) = [x,y]$. In any differential Lie algebra $\beta[ad^{p^j-1}(x)(\beta(x))] = 0$, because it holds in the free Lie algebra on two generators x and $\beta(x)$ which is a subalgebra of the tensor differential algebra on x and $\beta(x)$

The map $S^{2n+1}(p^r) \longrightarrow S^{2n+1}$ onto the top cell induces a map $\Omega S^{2n+1}(p^r) \longrightarrow \Omega S^{2n+1}$ which with the Hurewicz morphism gives a commutative diagram of Lie algebras (recall $\pi_*(\Omega X, \mathbb{Z}/3)$ is not a Lie algebra since the Jacobi identity does not hold)

$$\begin{array}{ccccc} L(u_{2n}, v_{2n-1}) & \subset \pi_*(\Omega S^{2n+1}(p^r), \mathbb{Z}/p) & \longrightarrow & \pi_*(\Omega S^{2n+1}, \mathbb{Z}/p) \\ & \downarrow \phi & & \downarrow \phi \\ T(x_{2n}, y_{2n-1}) & = H_*(\Omega S^{2n+1}(p^r), \mathbb{F}_p) & \longrightarrow & H_*(\Omega S^{2n+1}, \mathbb{F}_p) = T(x_{2n}) \end{array}$$

where $L(u_{2n}, v_{2n-1})$ is the free Lie algebra on u_{2n} and v_{2n-1} and $T(x_{2n}, y_{2n-1})$ and $T(x_{2n})$ are tensor algebras. The elements u_{2n} and v_{2n-1} are chosen such that $\phi(u_{2n}) = x_{2n}$ and $\phi(v_{2n-1}) = y_{2n-1}$. The Bockstein differentials on $H_*(\Omega S^{2n+1}(p^r), \mathbb{F}_p)$ are determined by

$$\beta_s(x_{2n}) = 0 \quad \text{for } s < r \quad \text{and} \quad \beta_r(x_{2n}) = y_{2n-1}.$$

As remarked above, $\beta_r(\mathrm{ad}^{p^j-1}(u)(v)) = 0$ since $\beta_r(x_{2n}) = y_{2n-1}$ implies that $\beta_r(u) = v$. Now consider the possiblity that

$$\beta_r(w) = \mathrm{ad}^{p^j-1}(u)(v) \quad \text{for } j \geq 1 \text{ and } n > 1.$$

Then in the tensor algebra $H_*(\Omega S^{2n+1}(p^r), \mathbb{F}_p)$ we calculate

$$\beta_r(x^{p^j} - \phi(w)) = \mathrm{ad}(x)^{p^j-1}(y) - \phi(\mathrm{ad}^{p^j-1}(u)(v)) = 0$$

where ϕ is a morphism of Lie algebras. Since the homology of the Bockstein differential module $\bar{H}_*[H_*(\Omega S^{2n+1}(p^r), \mathbb{F}_p), \beta_r) = 0$, there exists z in $H_*(\Omega S^{2n+1}(p^r))$ in an odd degree with

$$\beta_r(z) = x^{p^j} - \phi(w).$$

This z maps to zero in $H_*(\Omega S^{2n+1})$ so that x^{p^j} and $\phi(w)$ map to the same element in $H_*(\Omega S^{2n+1})$, and this implies that

$$x^{p^j} \text{ is in } \mathrm{im}(\phi \quad \pi_*(\Omega S^{2n+1}, \mathbb{Z}/p) \longrightarrow H_*(\Omega S^{2n+1}(p^r), \mathbb{F}_p))$$

which is impossible by the nonexistence of the mod p Hopf invariant one elements. Hence no such w exists and $\mathrm{ad}^{p^j-1}(u)(v)$ defined an element in the r+1 terms of the Bockstein spectral sequence $E^{r+1}\pi_{2np^j-1}(\Omega S^{2n+1}(p^r))$ which projects to zero in $\pi_{2np^j-1}(\Omega S^{2n+1})$.

Using the fibration sequence

$$\Omega^2 S^{2n+1} \longrightarrow \Omega F^{2n+1}\{p^r\} \longrightarrow \Omega S^{2n+1}(p^r) \longrightarrow \Omega S^{2n+1},$$

we can lift this element in mod p^{r+1} homotopy of $\Omega S^{2n+1}(p^r)$ to $\Omega F^{2n+1}\{p^r\}$. The lifted element is used to define a map

$$S^{2np^j-1}\{p^{r+1}\} \longrightarrow \Omega F^{2n+1}\{p^r\}$$

as in the previous section, and we form a weak product of these maps for all j. The other factor in the splitting is the loop space on a wedge of a family of $S^{n(j)}(p^r)$ lifting back from the splitting of $\Omega S^{2n+1}(p^r)$ to that of $\Omega F^{2n+1}\{p^r\}$. The product of these spaces is $Y \longrightarrow \Omega F^{2n+1}\{p^r\}$ giving the splitting asserted in theorem 3. Observe that spaces with p^{r+1} torsion are needed to decompose $\Omega F^{2n+1}\{p^r\}$.

§6. Exponents

A simply connected space X is defined to have exponent p^r at a prime p provided $p^r \mathrm{Tors}\, \pi_*(X_{(p)}) = 0$. Theorem 2 says that S^{2n+1} has exponent p^n at p for an odd prime p. This is the best possible in the sense that S^{2n+1} does not have exponent p^{n-1} at p by results of B. Gray [1969].

An H - space X is defined to have a multiplicative exponent k provided the kth power map $k : X \longrightarrow X$ is null homotopic. Here $k : X \longrightarrow X$ is defined inductively by k = 0 the constant and k equals (k-1) times the identity $X \longrightarrow X$ with respect to the H - space structure.

If $X\{k\}$ denotes the fibre of the kth power map $k : X \longrightarrow X$ for an H - space X, then $\Omega^3(X\{k\})$ has the same homotopy type as $\mathrm{Map}(S^4(k),X)$ from the fibre sequence

$$\mathrm{Map}(S^4(k),X) \longrightarrow \mathrm{Map}(S^3,X) \xrightarrow{k} \mathrm{Map}(S^3,X) = \Omega^3 X.$$

For an H - space X, the H - space $\Omega^3(X\{k\})$ has exponent 2k, and in the case k is odd, it has exponent k. For k odd we have $k\pi_*(S^{2n+1}\{k\}) = 0$ since the localized sphere $S^{2n+1}[1/2]$ is an H - space and the previous considerations apply to $X = S^{2n+1}[1/2]$.

We list some of the known results on exponents which are proved in the papers of Cohen, Moore, and Neisendorfer, and of Neisendorfer.

(1) The connected component of $\Omega^{2n+1}S^{2n+1}$ has multiplicative exponent p^n.

This is a sharper version of the statement of theorem 2.

(2) The fibre of the double suspension $E^2 : S^{2n-1}_{(p)} \longrightarrow \Omega^2 S^{2n+1}_{(p)}$ has a natural H - space structure such that its multiplicative exponent is p.

Some of the modivation for the study of decompositions of loop spaces came from the problem of determining the exponent of a Moore space $S^m(p^r)$. The initial work of Cohen, Moore, and Neisendorfer gave an exponent of p^{2r+1} except possibly for $p = 3$, $r = 1$. Recent work of Neisendorfer, "Smaller exponents for Moore spaces," together with his work on the special technicalities at the prime 3 yield the following assertion

(3) The H - space $\Omega^2 S^m(p^r)$ has multiplicative exponent p^{r+2} for p odd and $m \geq 3$.

The most far reaching conjecture on the subject of exponents is:

Conjecture of Moore. Every simply connected finite complex with totally finite dimensional rational homotopy groups has a exponent at all primes p, that is, there exists n(p) for each prime p such that $p^{n(p)}\ \mathrm{Tors}\ \pi_*(X)_{(p)} = 0$.

BIBLIOGRAPHY

Adams, J F., The sphere, considered as an H - space mod p, Quart. J. Math. Oxford Ser (2), 12, (1961), 52-60.

Cohen, F.R., J.C. Moore, and J.A. Neisendorfer, Torsion in homotopy groups, Annals, 109, (1979), 121 - 168.

———, The double suspension and exponents of the homotopy groups of spheres, Annals (to appear)

———, Note on higher torsion in the homotopy groups of single suspensions, Illinois Journal, (to appear).

———, Decomposition of loop spaces and applications to exponents, Proceedings of the Aarhus Topology Symposium, 1978.

Husemoller, D., Fibre Bundles, 2nd Ed., Springer Verlag, 1966.

James, I., Reduced product spaces, Annals of Math., 62, (1955), 190 - 197.

———, On the suspension sequence, Annals of Math., 65, (1957), 74 - 107.

Moore, J.C., On homotopy groups of spaces with a single non-vanishing homology group, Annals of Math , 59, (1954), 549 - 557.

———, The double suspension and p - primary components of the homotopy groups of spheres, Boll. Soc. Mat. Mexicana, 1, (1956), 28 - 37.

Neisendorfer, J.A., Homotopy theory modulo an odd prime, Princeton University Thesis, 1972.

———, Primary homotopy theory, (to appear).

———, Smaller exponents for Moore spaces, (to appear).

———, 3 primary exponents, (to appear).

Selick, P.S , Odd primary torsion in the homotopy groups of spheres, Princeton University Thesis, 1977.

——, Odd primary torsion in $\pi_k(S^3)$, Topology, 17, (1978), 407 - 412.

Serre, J - P., Homologie singulière des espaces fibrés. Applications, Annals of Math., 54, (1951), 425 - 505.

——, Groupes d'homotopie et classes de groupes abeliens, Annals of Math., 58, (1953), 258 - 294.

Shimada, N and T Yamanoshita, On triviality of the mod p Hopf invariant, Japan J. of Math., 31, (1961), 1 - 25.

Toda, H , On the double suspension E^2, J. Inst. Polytech. Osaka City Univ. Ser A, 7, (1956), 103 - 145.

——, p - primary components of homotopy groups II, mod p Hopf invariant, Mem. Coll Sci Univ. of Kyoto Ser. A, 31, (1958), 143 - 160.

Haverford College
September 1979

Characteristic Classes of Representations over Imaginary Quadratic Fields

C. B. Thomas

In [7] there is a formula for the Newton polynomials in the Chern classes of an induced representation. This formula is obtained by combining a spectral sequence argument with estimates for the orders of the Chern classes of the representation of the symmetric group S_n by permutation matrices. More recent work on the integral cohomology of arithmetic groups suggests that the second part of this argument, originally used only as a tool, is as interesting as the original result, see also [3] and [4]. The purpose of this paper is to emphasise this aspect, and to show how a combination of Galois invariance (valid for infinite discrete groups) with calculations in the integral cohomology of S_n, proves the existence of an element in $H^{2k}(GL_{\bullet}(\mathbb{Z}), \mathbb{Z})_{(odd)}$ of order equal to $(B_k/k)_{(odd)}$. Since very little extra work is required, we shall work over the ring of integers in an imaginary quadratic field, rather than over $\mathbb{Z}$ itself. This also serves to show the delicacy of Galois invariance as a tool in estimating the order of torsion classes.

Let $\mathcal{O}$ denote the integral closure of $\mathbb{Z}$ in the imaginary quadratic field $\mathbb{Q}(\sqrt{-m})$. If $m \equiv 1$ or $2 \pmod 4$, $\mathcal{O} = \mathbb{Z}(\sqrt{-m})$; if $m \equiv 3 \pmod 4$, $\mathcal{O}$ is obtained by adjoining the complex number $\frac{1}{2}(1 + \sqrt{-m})$ to the rational integers. In both cases the elements of $\mathcal{O}$ define a lattice in the complex plane, and hence the arithmetic group $GL_n(\mathcal{O})$ embeds as a discrete subgroup in $GL_n(\mathbb{C})$. Consider the composition

$$S_n \xrightarrow[\tau_n]{} GL_n(\mathcal{O}) \xrightarrow[\pi_n]{} GL_n(\mathbb{C}),$$

in which S_n is the finite subgroup of permutation matrices. We propose to study the torsion classes $c_k(\pi_n)$ and $c_k(\pi_n\tau_n)$ in $H^{2k}(GL_n(\mathcal{O}), \mathbb{Z})$ and $H^{2k}(S_n, \mathbb{Z})$ respectively. As usual the characteristic class of a complex representation is defined to be the characteristic class of the associated flat vector bundle. For a fixed value of k both classes are independent of n, if $k << n$, see [2] and [6], so we may stabilise, and think of $c_k(\pi_\cdot)$ as a class in $H^{2k}(GL(\mathcal{O}), \mathbb{Z})$.

We adopt the following conventions. For an arbitrary discrete group Γ, $H^*(\Gamma, \mathbb{Z})_{(r)}$ denotes the r-torsion subgroup, and we define the kth. Bernoulli number B_k by the formula

$$\frac{e^t}{e^t - 1} + \frac{t}{2} = 1 + \sum_{k=2}^{\infty} \frac{B_k}{k!} t^k.$$

Thus, $B_{odd} = 0$ and $B_2 = \frac{1}{6}$, $B_4 = -\frac{1}{30}$, $B_6 = \frac{1}{42}, \ldots, B_{12} = -\frac{691}{2730}$,

THEOREM 1. Let k be even. (i) $c_k(\pi_\cdot)_{(odd)}$ has order equal to the maximal odd factor of den (B_k/k).

(ii) If $2^{N(k)}$ is the highest power of 2 dividing den $(B_k/4k)$, then $c_k(\pi_\cdot)_{(2)}$ has order equal either to

$2^{N(k)}$ or $2^{N(k)-1}$.

The proof is in two parts. We first show that den $(B_k/4k)$ is an upper bound for the order of $c_k(\pi_*)$, using Galois invariance, and then that half this number is a lower bound by calculating the order of $c_k(\pi_*)$ restricted to certain finite subgroups of $GL_n(\mathcal{O})$.

Galois invariance. If ρ is a representation of the arbitrary discrete group Γ in an algebraically closed field of characteristic prime to the order of the torsion subgroup of Γ, it is possible to define algebraic Chern classes $c_k(\rho) \in H^{2k}_{et}(\Gamma, \mathbb{T}_p^{\otimes k})$. Here μ_{p^n} denotes the group of p^n-th roots of unity in the field, and $\mathbb{T}_p^{\otimes k} = \varprojlim_n \mu_{p^n}$ is (non-canonically) isomorphic to $\hat{\mathbb{Z}}_p$, the ring of p-adic integers. One construction of these classes was sketched in [5], another which is perhaps clearer in the case of the complex numbers, is to start with the universal classes in $H^*_{\acute{e}t}(G_{r,s}, \hat{\mathbb{Z}}_p)$. As usual we approximate BU by the finite dimensional algebraic Grassmann varieties, and note that the generators in cohomology are dual to homology classes carried by subvarieties defined over the rational integers. The comparison theorem between classical and étale cohomology implies that the algebraic classes carry the p-torsion of the topological classes, informally defined above. Since the universal classes are defined over $\mathbb{Z}$, the algebraic Chern class $c_k(\rho)$ is invariant under the action of the Galois group $\mathcal{G} = \mathrm{Gal}\,(\bar{Q}, L)$ on the roots of unity. Here L is the field of definition for the representation ρ; in the case of π_n, $\mathcal{G}$ (abelianised) has index 2 in the group of units $\hat{\mathbb{Z}}^*$. From the definition it is quite easy to show, that if $\lambda(p, k)$ is the least power of p needed to kill $c_k(\pi_n)_{(p)}$, then

$$\lambda(p, k) = \min_{\alpha \in \mathcal{G}} \nu_p(\alpha^k - 1).$$

(Here ν_p denotes the usual p-adic valuation.)

Now $\mathcal{G} \subseteq \prod_p \mathbb{Z}_p^*$, where $\hat{\mathbb{Z}}^*{}_p = \mu_{p-1} + (1 + p\hat{\mathbb{Z}}_p)$ (respectively $\{\pm 1\} + (1 + 4\hat{\mathbb{Z}}_2)$), for p equal to an odd prime number (respectively p equals 2). The factor 4 enters the second decomposition, because of convergence problems with the 2-adic logarithm. There are two cases to consider; if p is odd,

$$\lambda(p, k) = \begin{cases} 0, & k \not\equiv 0 \pmod{(p-1)} \\ \min\limits_{\alpha \in p\hat{\mathbb{Z}}_p} (\nu_p(k\alpha)) = 1 + \nu_p(k), & k \equiv 0 \pmod{(p-1)}. \end{cases}$$

At the prime 2, because of the factor 4 above, and because of the index 2 property of $\mathcal{G}$, the estimate becomes

$$\lambda(2, k) = \min_{\alpha \in 8\hat{\mathbb{Z}}_2} (\nu_2(k\alpha)) = 3 + \nu_2(k), \quad k \text{ even}.$$

The statement about upper bounds now follows from the familiar properties of the Bernoulli numbers. Note in particular that

$$\text{den}(B_2/8) = 48, \text{ and hence } 2^{N(2)} = 16.$$

<u>Calculations for the symmetric group</u> S_n. In [4] L. Evens and D. Kahn show, that at odd primes p the upper bound, $\text{den}(B_k/2k)_{(odd)}$ for the order of $c_k(\pi\tau)_{(odd)}$ is actually attained, but that $c_k(\pi\tau)_{(2)}$ has exactly half the order predicted by the calculation above. We shall next give a variant of their argument, which is almost independent of spectral sequences, and perhaps easier to follow. Our starting point is the Riemann-Roch formula for group representations, proved in [7], and reproved below using an argument explained to me independently by I. Madsen and G. Segal (letter dated 19. March 1977).

If ρ is a representation of the subgroup Δ of finite index in Γ, the classical transfer map i_* is compatible under the flat bundle homomorphism with the transfer in K-theory. Furthermore there is a transfer, usually called "constriction", in ordinary cohomology. Write $s_k(\rho)$ for the kth. Newton polynomial in the Chern classes $c_j(\rho)$, for $j \leq k$. Then

$$s_k - c_1 s_{k-1} + \ldots + (-1)^k k c_k = 0.$$

Write $m(k) = \prod_p p^{[\frac{k}{p-1}]}$

PROPOSITION 2. With the notation above

$$\frac{m(k)}{k!}\left[s_k(\imath \times \rho) - \operatorname{cor}(s_k\rho)\right] = 0.$$

Outline proof. Let ch_{q+r} denote the (q+r)-component of the rational Chern character, and $BU(2q,\ldots,\infty)$ the representing space for connective K-theory in dimension 2q. In [1], J. F. Adams constructs elements

$$ch_{q,k} \in H^{2q+2k}(BU(2q,\ldots,\infty), \mathbb{Z}), \text{ such that}$$

(i) $\sigma^2(ch_{q,k}) = ch_{q-1,k}$, and

(ii) $ch_{q,k} = m(k)\, ch_{q+k}$ over $\mathbb{Q}$.

The spaces $BU(2q,\ldots,\infty)$ together with the Bott maps $\beta : S^2BU(2q-2,\ldots,\infty) \to BU(2q,\ldots,\infty)$ define a spectrum, and by (i) above, with k fixed we obtain a map of spectra

$$ch_{\bullet,k} : bu \times Z \to \Sigma^{2k} K(\mathbb{Z}).$$

In particular $ch_{0,k} : BU \times Z \to K(\mathbb{Z}, 2k)$ is an infinite loop map, and commutes with transfers. The formula now follows from (ii), once we note that

$$ch_{0,k} = m(k)\, ch_k = \frac{m(k)}{k!}\, s_k.$$

Before applying this formula we note, that if $k = \ell p^{m-1}(p-1)$

with $(\ell, p) = 1$, then the highest power of p dividing $\frac{m(k)}{k!}$ is p itself. Thus in the numerator p occurs to the (ℓp^{m-1})st. power, and in the denominator to the power

$$\ell(p-1)\{p^{m-2}+\ldots+p+1\}+(\ell-1)$$

$$= \ell(p-1)\left(\frac{p^{m-1}-1}{p-1}\right)+(\ell-1)$$

$$= \ell p^{m-1}-1.$$

In establishing a lower bound for the order of $c_k(\pi)_{(p)}$, it is enough to consider a subgroup of permutation matrices, for any bound attained on a finite subgroup will certainly hold for the larger infinite discrete group. First recall that the Wreath product $H \wr \mathbb{Z}/p$ denotes the semi-direct product

$$1 \longrightarrow \underbrace{H \times \ldots \times H}_{p} \longrightarrow G \longrightarrow \mathbb{Z}/p \longrightarrow 1,$$

where $\mathbb{Z}/p$ cyclically permutes the factors in the normal subgroup. Representative p-Sylow subgroups for the symmetric group S_{p^m} are then inductively defined by

P_1 = the subgroup generated by the cycle $(12\ldots p)$,
$P_m = P_{m-1} \wr P_1$.

The following result shows that the bound given by Galois invariance for $c_k(\pi_{,})$ is already attained, for infinitely many values of k, by the classes $c_k(\pi_n\tau_n)$. As usual in calculations involving the cohomology of groups it is enough to consider a Sylow subgroup for each prime p dividing the order. Furthermore, for a fixed odd value of p, the upper bound already obtained shows that the only interesting values of k are those divisible by $p-1$. Write $\sigma^{(p)}_{p^m}$ for the restriction of the representation $\sigma_{p^m} = \pi_{p^m}\tau_{p^m}$ to the subgroup P_m.

PROPOSITION 3. *If* p *is odd*, ν_p (order of $c_{p^{m-1}(p-1)}(\sigma_{p^m}))$ = $m = 1 + \nu_p(p^{m-1}(p-1))$.

Outline proof by induction on the exponent m.
When $m = 1$, $\sigma^{(p)}_p$ equals the regular representation of the cyclic group $P_1 \cong \mathbb{Z}/p$, and $c(\sigma^{(p)}_p) = 1 - \gamma^{p-1}$, where γ is the first Chern class of the usual faithful representation of P_1. Clearly the result holds in this case, so assume it for $r = 1, 2, \ldots, m-1$. Using the Wreath product decomposition above, it is easy to see that

$$\sigma_{p^m} = \sigma_{p^{m-1}} \wr 1 \; : P_m \to U_{p^m},$$

or in other words, that σ_{p^m} is the transfer of the representation $\sigma_{p^{m-1}}$ up from the subgroup P_{m-1} of index p.
Applying a little more than the formula in Proposition 2, we have that

$$s_{p^{m-1}(p-1)}(\sigma^{(p)}_{p^m}) = i_*(s_{p^{m-1}(p-1)}\sigma^{(p)}_{p^{m-1}}) + \text{correction terms.}$$

By the remark following Proposition 2 each of the correction terms has order at most p; the last one depends only on the quotient group P_m/P_{m-1} and involves the power

$\gamma^{p^{m-1}(p-1)}$ hence has order equal to p. Furthermore, a subsidiary induction argument, combined with the upper bound already obtained, shows that the Newton polynomial on the right also has order at most p. It follows that the left hand side

$$s_{p^{m-1}(p-1)}(\sigma^{(p)}_{p^m})$$

has order precisely equal to p. Using the recurrence formula for Newton polynomials it now follows that the order of the Chern class $c_{p^m(p-1)}(\sigma^{(p)}_{p^m})$ is p times the p-primary factor of $p^{m-1}(p-1)$, as claimed. [The characteristic classes of the permutation representation are in a certain sense universal for representations of finite groups, in particular independent, which implies that the coefficient of $c_{p^m(p-1)}$ in the recurrence formula is optimal]

Remarks. With a little more care one can handle values of k of the form $lp^{m-1}(p-1)$ with $l \geqslant 2$. Furthermore, one also has that

$$\nu_2(\text{order of } c_{2^{m-1}}(\sigma_{2^m})) = 2^m,$$

which is half the bound given by Galois invariance for the rational representation σ.

<u>Improving the lower bound at the prime 2</u>. Consider the binary dihedral group

$$D_8^* = \{A, B : A^4 = 1, A^2 = B^2, B^{-1}AB = A^{-1}\}.$$

There is a natural embedding $\xi : D_8^* \to GL_2(\mathbb{Z}(i))$, given by

$$\xi(A) = \begin{pmatrix} i & 0 \\ 0 & -i \end{pmatrix} \quad \text{and} \quad \xi(B) = \begin{pmatrix} 0 & -1 \\ 1 & 0 \end{pmatrix}, \text{ and}$$

the element $x = c_2(\xi) \in H^4(D_8^*, \mathbb{Z})$ generates a $\mathbb{Z}/8$-polynomial subalgebra in the cohomology ring. The n-fold direct sum of ξ with itself defines a faithful representation of D_8^* in $GL_{2n}(\mathbb{Z}(i))$, and

$$c(n\xi) = (1 + x)^n = (1 + nx + \frac{n(n-1)}{2} x^2 + \ldots + x^n).$$

If n is odd, $c_2(n\xi)$ has order 8, and it follows that $c_2(\pi)_{(2)}$ has order at least equal to 8. Referring to the statement of Theorem 1, we see that the lower bound given by the symmetric group i.e. $2^{N(k)-2}$, is too small at the prime 2.

Looking at the way in which the upper bound, den $(B_k/4k)$, was established, we see that one factor 2 enters because Gal $(\mathbb{Q}(i), \mathbb{Q}) = \mathbb{Z}/2$. Working from below this factor is detected by the representation $n\xi$ above. The second factor 2 enters because of problems with 2-adic convergence, and it is hardly surprising that it cannot be detected by restricting π to any <u>finite</u> subgroup. This also follows from [3]. The problem is therefore to find some infinite subgroup of $GL_n(\mathcal{O})$ with constructively computable characteristic classes (see the second part of the argument above). The first candidate

$$\Gamma = \underbrace{GL_2(\mathcal{O}) \oplus \ldots \oplus GL_2(\mathcal{O})}_{n}$$

is too small, since $GL_2(\mathcal{O})$ is a group of virtual cohomological dimension equal to 1, and $H^{2k}(GL_2(\mathcal{O}), \mathbb{Z})$ is detected by the lattice of finite subgroups. As we have already seen no characteristic class of order 16 can arise in this way. Similar arguments would seem to apply to $GL_3(\mathcal{O})$, although I have not written out the details.

References

1. J. F. Adams, On Chern characters and the structure of the unitary group, Proc. Camb. Phil. Soc. 57 (1961), 189 - 199.

2. R. Charney, Stability in the cohomology of general linear groups (to appear).

3. B. Eckmann and G. Mislin, Characteristic classes of representations and Galois symmetry (to appear).

4. L. Evens and D. Kahn, Chern classes of certain representations of symmetric groups (to appear in Trans. Amer. Math. Soc.

5. A. Grothendieck, Classes de Chern et représentations linéaires des groupes discrets, Dix exposés sur la cohomologie des schémas, North Holland, 1968.

6. M. Nakaoka, Decomposition theorem for homology groups of symmetric groups, Annals of Math., 71 (1960) 16 - 42.

7. C. B. Thomas, An integral Riemann-Roch formula for flat line bundles, Proc. London Math. Soc. 34 (1977), 87 - 101.

LOCALIZATION IN ALGEBRAIC L-THEORY

by Pierre VOGEL

Let $f : A \to B$ be a morphism of rings with involution. If B is a localization of A in the classical sense, Karoubi [1], Pardon [3], Ranicki [5] and Smith [6] have given exact sequences between the L-groups of A, the L-groups of B and relative groups which are defined in term of linking forms over torsion modules.

My purpose is to show that the localization exact sequence holds in a more general situation.

From f one can define a ring Λ endowed with a morphism $A \to \Lambda$ satisfying the following conditions

i) for any matrix α with entries in A such that $\alpha \otimes B$ is invertible, $\alpha \otimes \Lambda$ is invertible too,

ii) Λ is universal with respect to the property i).

We have a canonical homomorphism $\varepsilon : \Lambda \to B$. We will say that f is weakly locally epic if ε is epic and local if ε is an isomorphism. In this paper I will prove that the relative group $L_n(A \to \Lambda)$ depends only on the category $\mathcal{C}_B$ of finitely presented modules M with cohomology dimension 1 and satisfying $M \otimes B = \mathrm{Tor}_1(M,B) = 0$ if f is weakly locally epic and A doesn't contain any finitely generated submodule I which is B-perfect (i.e. $I \otimes B = 0$).

As a corollary we prove a Mayer-Vietoris exact sequence in L-theory for square of rings with involution .

$$\begin{array}{ccc} A & \to & B \\ \downarrow & & \downarrow \\ C & \to & D \end{array}$$

if $A \to B$ and $C \to D$ are local, A (resp. C) doesn't contain any finitely B-perfect (resp. D-perfect) submodule and the tensorization by C is an equivalence between the categories $\mathcal{C}_B$ dans $\mathcal{C}_D$.

For more simplicity I will consider the groups L_n^h only, but we have the same results with the groups L_n^α, α being any subgroup of $\tilde{K}_o$ or $\tilde{K}_1$ stable under involution ; we must just change a little the category $\mathcal{C}_B$.

§ 1 . QUADRATIC AND LINKING FORMS OVER COMPLEXES

Throughout this paper I will suppose that $f : A \to B$ is weakly locally epic and A doesn't contain any finitely generated B-perfect submodule.

Denote by $\mathcal{C}_B$ the class of "torsion modules" , i.e. the class of A-modules M having a resolution $0 \to C_1 \xrightarrow{d} C_o \to M \to 0$ by finitely generated free A-modules such that $d \otimes B$ is an isomorphism.

Let M be a torsion module and $0 \to C_1 \to C_o \to M \to 0$ be a resolution of M by finitely generated free A-modules. Since M is a finitely generated B-perfect module, $Hom(M,A)$ is zero and we have an exact sequence

$$0 \to Hom(C_o,A) \to Hom(C_1,A) \to Ext^1(M,A) \to 0 \quad .$$

Then $Ext^1(M,A)$ is a torsion module.

Denote by $\hat{M}$ the module $Ext^1(M,A)$. The correspondance $M \mapsto \hat{M}$ is a contravariant functor from $\mathcal{C}_B$ to itself and $\hat{\hat{M}}$ is canonically isomorphic to M .

If M is a torsion module a bilinear form over M is a map $M \to \hat{M}$. The set $B(M)$ of bilinear forms over M is endowed with an involution in the following way . if $\varphi : M \to \hat{M}$ is a bilinear form, $t\varphi$ is the composite map : $M \simeq \hat{\hat{M}} \xrightarrow{\hat{\varphi}} \hat{M}$.

Let us consider now bilinearforms over complexes

By definition a free (resp. torsion) complex will be a $\mathbb{Z}$-graded complex

$$\longrightarrow C_{n+1} \xrightarrow{d} C_n \xrightarrow{d} C_{n-1} \longrightarrow \cdots$$

where $\bigoplus_i C_i$ is a finitely generated free (resp. torsion) A-module. And a complex is a free or a torsion complex.

If C_* is a complex the dual complex $\hat{C}_*$ is the complex

$$\longleftarrow \hat{C}_{n+1} \xleftarrow{(-1)^{n+1}\hat{d}} \hat{C}_n \xleftarrow{(-1)^{n}\hat{d}} \hat{C}_{n-1} \longleftarrow$$

where $\wedge$ is the fonctor $\mathrm{Hom}(\ ,A)$ if C_* is free and $\mathrm{Ext}^1(\ ,A)$ if C_* is torsion, and $\hat{C}_n$ is of degree $-n$.

A bilinear form over C_* is a linear map $C_* \to \hat{C}_*$. The set $B_*(C_*) = B^{-*}(C_*) = \mathrm{Hom}(C_*,\hat{C}_*)$ of bilinear forms over C_* is a graded differential $\mathbb{Z}$-module endowed with an involution t by

$$\partial^\circ \varphi(u) = \partial^\circ\varphi + \partial^\circ u$$

$$d(\varphi(u)) = (d\varphi)u + (-1)^{\partial^\circ\varphi}\varphi(du)$$

$$(t\varphi)u = (-1)^{\partial^\circ u\, \partial^\circ \varphi(u)}\,\hat{\varphi}(u)$$

for any $u \in C_*$ and $\varphi \in B_*(C_*)$.

If $\varepsilon = \pm 1$, $B_*(C_*)^\varepsilon$ denote the complex $B_*(C_*)$ with the new involution $\varphi \mapsto \varepsilon t\varphi$, if C_* is free, and $\varphi \mapsto -\varepsilon t\varphi$ if C_* is torsion.

Definition 1.1

Let C_* be a free (resp. torsion) complex. A quadratic (resp. linking) n-form over C_* is an element of the group

$$Q^n(C_*) = H_{-n}(\mathbb{Z}/2, B_*(C_*)^\varepsilon) \qquad \varepsilon = (-1)^n .$$

If we take the standard resolution W_* of the $\mathbb{Z}[\mathbb{Z}/2]$-module $\mathbb{Z}$

$$\mathbb{Z}[\mathbb{Z}/2] \xleftarrow{1-t} \mathbb{Z}[\mathbb{Z}/2] \xleftarrow{1+t} \mathbb{Z}[\mathbb{Z}/2] \longleftarrow \cdots$$

any quadratic (resp. linking) n-form over C_* is represented by

$$e_0 \otimes \varphi_0 + e_1 \otimes \varphi_1 + \cdots \qquad \varphi_i \in B^{n+i}(C_*)$$

and we have

$$\forall\, i \geq 0 \qquad d\varphi_i = ((-1)^{i+1} + (-1)^n t)\,\varphi_{i+1}$$

$$(\text{resp. } d\varphi_i = ((-1)^{i+1} - (-1)^n t)\,\varphi_{i+1}) .$$

Definition 1.2

Let $\Sigma_* \to C_*$ be an epimorphism of free (resp. torsion) complexes. A quadratic (resp. linking) n-form over $\Sigma_* \to C_*$ is an element of the group

$$Q^n(\Sigma_* \to C_*) = H_{-n}(\mathbb{Z}/2, B_*(\Sigma_*)^{-\varepsilon}/B_*(C_*)^{-\varepsilon}) \qquad \varepsilon = (-1)^n .$$

Notations 1.3

Let C_* be a free (resp. torsion) complex and q be a quadratic (resp. linking) n-form over C_*. The image of q by the composite map

$$Q^n(C_*) \xrightarrow{\text{transfert}} H_{-n}(1,B_*(C_*)^{\varepsilon}) \xrightarrow{\sim} H_{-n}(B_*(C_*))$$

give a chain map from C_* to $\hat{C}_*$ of degree $-n$. This chain map, well defined up to homotopy, will be denoted by $\tilde{q}$.

Let $\Sigma_* \to C_*$ be an epimorphisme of free (resp. torsion) complexes and q be a quadratic (resp. linking) n-form over $\Sigma_* \to C_*$. If K_* is the kernel of $\Sigma_* \to C_*$, we get a chain map from K_* to $\hat{\Sigma}_*$, well defined up to homotopy, as the image of q by the composite map

$$Q^n(\Sigma_* \to C_*) \xrightarrow{\text{transfert}} H_{-n}(1,B_*(\Sigma_*)/B_*(C_*)) \longrightarrow H_{-n}(\text{Hom}_*(K_*,\hat{\Sigma}_*)) \quad .$$

This chain map will be denoted by $\tilde{q}$.

Definition 1.4

Let C_* (resp. $\Sigma_* \to C_*$) be a complex (resp. an epimorphism of complexes) and q be a quadratic or linking n-form over C_* (resp. $\Sigma_* \to C_*$). The form q is said non singular if $\tilde{q}$ is a homology equivalence. If C_* is free (resp. Σ_* and C_* are free), q is said B-non singular if $\tilde{q}$ is a B-homology equivalence.

Definition 1.5

Let $\mathscr{C}$ be the word free (resp. free, resp. B-acyclic free, resp. torsion) and $\mathscr{F}$ the words non singular quadratic (resp. B-non singular quadratic, resp. non singular quadratic, resp. non singular linking).

Let C_* be a $\mathscr{C}$ complex and q be a $\mathscr{F}$ n-form over C_*. The object (C_*,q) is called cobordant to zero if there exists an exact sequence of $\mathscr{C}$ complexes $0 \to K_* \to \Sigma_* \to C_* \to 0$ such that q is the boundary of a $\mathscr{F}$ n-1-form over $\Sigma_* \to C_*$.

Theorem 1.6 [7]

The group $L_n^h(A)$ (resp $L_n^h(\Lambda)$) is isomorphic to the group of free complexes together with non singular (resp. B-non singular) quadratic n-forms modulo the following relation (C_*,q) is cobordant to (C'_*,q') if $(C_*\oplus C'_*,q-q')$ is cobordant to zero.

Definition 1.7

The cobordism group of B-acyclic free (resp. torsion) complexes together with non singular quadratic (resp. linking) n-forms will be denoted by $L'_n(B,A)$ (resp. $L''_n(B,A)$).

Theorem 1.8

The group $L'_n(B,A)$ is isomorphic to $L^h_{n+1}(A \to A)$.

Proof

Since $L^h_n(A)$ is isomorphic to $\Gamma^h_n(A \to A)$ or $\Gamma^h_n(A \to B)$ [7], it suffices to prove that $L'_n(B,A)$ is isomorphic to the group $\Gamma^h_{n+1}\begin{pmatrix} A \to A \\ \downarrow \quad \downarrow \\ A \to B \end{pmatrix}$ and that is proved by Ranicki [5] and Smith [6] by using a dual point of view (a quadratic form over C_* in my sense is a quadratic form over C^* in the sense of Ranicki [4], [5]).

The main result of this paper is the following

Theorem 1.9

The group $L''_n(B,A)$ is isomorphic to $L^h_{n+2}(A \to A)$.

§ 2 . RELATIONS BETWEEN FREE COMPLEXES AND TORSION COMPLEXES

The first informations about B-acyclic free complexes and torsion complexes are the following

Lemma 2.1

Let $\ldots \to 0 \to C_p \xrightarrow{d} C_{p+1} \to 0 \ldots$ be a B-acyclic free complex of length two. Then d is monic and Coker d is a torsion module.

Proof

Since $d \otimes B$ is bijective, the complex $\ldots \to 0 \to \hat{C}_p \xrightarrow{\hat{d}} \hat{C}_{p+1} \to 0 \to \ldots$ is B-acyclic and Coker $\hat{d}$ is B-perfect. But A doesn't contain any finitely generated B-perfect submodule. Then the map $\mathrm{Hom}(\hat{C}_{p+1},A) \xrightarrow{\hat{d}^*} \mathrm{Hom}(\hat{C}_p,A)$ is monic and d is monic and Coker d is a torsion module.

Lemma 2.2

Let C_* be a B-acyclic free complex and $f : C_* \to K_*$ a morphism from C_* to a torsion complex K_* . Then there is a commutative diagram :

$$\begin{array}{ccc} & T_* & \\ {}^{g}\nearrow & & \searrow \\ C_* & \xrightarrow[f]{} & K_* \end{array}$$

such that T_* is a torsion complex and g is a homology equivalence.

Proof

Step 1

Suppose C_* is the complex $\ldots 0 \leftarrow C_p \leftarrow C_{p+1} \leftarrow 0 \leftarrow \ldots$. By setting

$$T_i = 0 \qquad \text{for } i \neq p, p+1$$

$$T_p = C_p \underset{C_{p+1}}{\oplus} K_{p+1} \quad \text{and} \quad T_{p+1} = K_{p+1}$$

we get a complex between C_* and K_* , and by lemma 2.1, T_* is a torsion complex, and $C_* \to T_*$ is a homology equivalence. Then the lemma is proved if C_* is of length two.

Step 2

Suppose we have an exact sequence of B-acyclic free complexes .

$$0 \to C_* \to C'_* \to C''_* \to 0$$

and suppose that C_* and C'_* satisfy the lemma. If $C''_* \to K_*$ is a morphism to a torsion complex, the composite map $C'_* \to K_*$ factorizes through a torsion complex T'_* and the map $C'_* \to T'_*$ is a homology equivalence. Up to homology equivalence we may suppose that $T'_* \overset{\alpha}{\to} K_*$ is epic. The map $C_* \to (\operatorname{Ker} \alpha)_*$ factorizes through a torsion complex T_* by a homology equivalence $C_* \to T_*$. Then it is not difficult to prove that the map $C''_* \to K_*$ factorizes by a homology equivalence through the maping cone of $T_* \to T'_*$.

$$\begin{array}{ccccccccc} 0 & \to & C_* & \to & C'_* & \to & C''_* & \to & 0 \\ & & \downarrow & & \downarrow & & \downarrow & & \\ & & T_* & \to & T'_* & \to & K_* & & \end{array}$$

The class of B-acyclic free complexes satisfying the lemma is stable under homotopy equivalence, suspension and quotient. Then it is stable under extension and contain all the complexes of length two. By [7] any B-acyclic complex is in this class and the lemma is proved.

Conversely we have

Lemma 2.3

Let T_* be a torsion complex. Then there exists a homology equivalence from a B-acyclic free complex to T_* .

Proof

This lemma will be proved by induction on the length of T_* . Let T_* be a n-dimensional torsion complex. By induction we have a homology equivalence f from a B-acyclic free complex C'_* to the complex $\to 0 \to T_{n-1} \to T_{n-2} \to \dots$. Take a finitely generated free resolution $0 \to F' \to F$ of T_n . Since the kernel of $C'_{n-1} \to C'_{n-2}$ maps onto the kernel of $T_{n-1} \to T_{n-2}$, the composite map $F \to T_n \to T_{n-1}$ factorizes through $\mathrm{Ker}(C'_{n-1} \to C'_{n-2})$. Moreover f is a homology equivalence and the composite map $F' \to F \to C'_{n-1}$ lift through C'_n . Then we get a morphism g from the complex $\to 0 \to F' \to F \to 0 \to \dots$ to C'_* and the induced map from the mapping cone of g to T_* is a homology equivalence

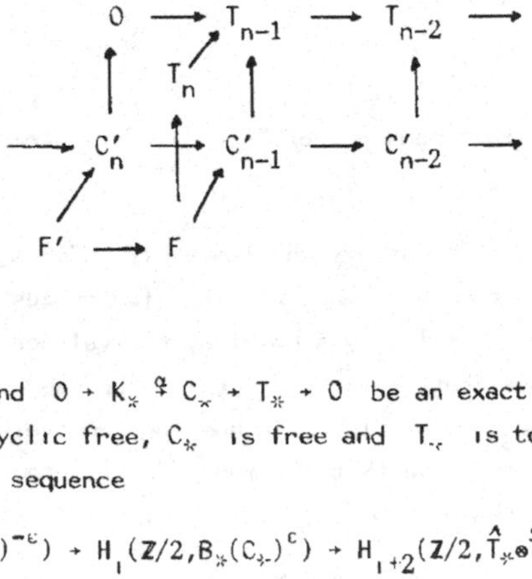

Lemma 2.4

Let $\varepsilon = \pm 1$ and $0 \to K_* \xrightarrow{q} C_* \to T_* \to 0$ be an exact sequence of complexes such that K_* is acyclic free, C_* is free and T_* is torsion. Then we have a canonical long exact sequence

$$\dots \to H_{i+1}(\mathbb{Z}/2, B_*(T_*)^{-\varepsilon}) \to H_i(\mathbb{Z}/2, B_*(C_*)^{\varepsilon}) \to H_{i+2}(\mathbb{Z}/2, \hat{T}_* \otimes \hat{T}_*) \to H_i(\mathbb{Z}/2, B_*(T_*)^{-\varepsilon}) \to .$$

where $\hat{T}_* \otimes \hat{T}_*$ is endowed with the involution $a \otimes b \to -\varepsilon(-1)^{\partial^\circ a \partial^\circ b}\, b \otimes a$.

Proof

The above exact sequence induces the following .

$$0 \leftarrow \hat{T}_* \overset{\partial}{\leftarrow} \hat{K}_* \overset{\hat{\alpha}}{\leftarrow} \hat{C}_* \leftarrow 0$$

and we get a complex of complexes

$$0 \leftarrow \hat{K}_* \otimes \hat{K}_* \overset{\mu}{\leftarrow} \hat{C}_* \otimes \hat{K}_* \oplus \hat{K}_* \otimes \hat{C}_* \overset{\lambda}{\leftarrow} \hat{C}_* \otimes \hat{C}_* \leftarrow 0$$

by setting
$$\lambda(u \otimes v) = u \otimes \hat{\alpha}(v) + \hat{\alpha}(u) \otimes v$$
$$\mu(u \otimes b + a \otimes v) = \hat{\alpha}(u) \otimes b - a \otimes \hat{\alpha}(v)$$

for any $a,b \in \hat{K}_*$ and $u,v \in \hat{C}_*$.

λ and μ are compatible with the differentials, and by setting

$$t(u \otimes v) = \varepsilon(-1)^{\partial^\circ u \partial^\circ v}\ v \otimes u$$
$$t(u \otimes b + a \otimes v) = \varepsilon(-1)^{\partial^\circ a \partial^\circ v}\ v \otimes a + \varepsilon(-1)^{\partial^\circ u \partial^\circ b}\ b \otimes u$$
$$t(a \otimes b) = -\ \varepsilon(-1)^{\partial^\circ a \partial^\circ b}\ b \otimes a$$

for any $a,b \in \hat{K}_*$ and $u,v \in \hat{C}_*$, the morphisms λ and μ are equivariant for this involution.

Now we get three exact sequences of differential $\mathbb{Z}[\mathbb{Z}/2]$-modules .

$$0 \leftarrow \hat{T}_* \otimes \hat{T}_* \leftarrow \hat{K}_* \otimes \hat{K}_* \leftarrow \mathrm{Im}\ \mu \leftarrow 0$$
$$0 \leftarrow \mathrm{Im}\ \mu \leftarrow \hat{C}_* \otimes \hat{K}_* \oplus \hat{K}_* \otimes \hat{C}_* \leftarrow \mathrm{Ker}\ \mu \leftarrow 0$$
$$0 \leftarrow B_*(T_*)^{-\varepsilon} \leftarrow \mathrm{Ker}\ \mu \leftarrow B_*(C_*)^{\varepsilon} \leftarrow 0 \quad .$$

The involution on $\hat{T}_* \otimes \hat{T}_*$ is defined by

$$t(a \otimes b) = -\ \varepsilon(-1)^{\partial^\circ a \partial^\circ b}\ b \otimes a \qquad \forall\ a,b \in \hat{T}_* \quad .$$

The third exact sequence comes from the isomorphism

$$\hat{C}_* \otimes \hat{C}_* = \mathrm{Hom}(C_*,\hat{C}_*) = B_*(C_*)$$
$$\mathrm{Ker}\ \mu/\mathrm{Im}\ \mu \simeq \mathrm{Tor}_1(\hat{T}_*,\hat{T}_*) \simeq \mathrm{Hom}(T_*,\hat{T}_*) \simeq B_*(T_*) \quad .$$

On the other hand $\hat{K}_*$ is contractible and $\hat{K}_* \otimes \hat{K}_*$ and $\hat{C}_* \otimes \hat{K}_* \oplus \hat{K}_* \otimes \hat{C}_*$ are acyclic. That implies the isomorphisms

$$H_{i+2}(\mathbb{Z}/2,\hat{T}_*\otimes\hat{T}_*) \overset{\sim}{\leftarrow} H_{i+1}(\mathbb{Z}/2,\mathrm{Im}\ \mu) \overset{\sim}{\leftarrow} H_i(\mathbb{Z}/2,\mathrm{Ker}\ \mu)$$

and by taking the homology exact sequence of the third above exact sequence, we prove the lemma.

Lemma 2.5

Let $C_* \xrightarrow{f} T_*$ be a homology equivalence from a free complex to a torsion complex. Then a linking n-1-form q over T_* induces a well defined quadratic n-form $f^*(q)$ on C_*. Furthermore q is non singular if and only if $f^*(q)$ is non singular.

Proof

Up to homotopy equivalence we may suppose that f is surjective with free kernel K_*. By 2.4 we get a map $Q^{n-1}(T_*) \xrightarrow{f^*} Q^n(C_*)$ and $f^*(q)$ is well defined.

Moreover f^* is induced by a boundary and the transfert commutes with the boundary. Then the cycle $\widetilde{f^*(q)}$ is the boundary of $\tilde{q}$ in the exact sequence $0 \to B_*(T_*)^{-\varepsilon} \to \operatorname{Ker} \mu \to B_*(C_*)^{\varepsilon} \to 0$ (see the proof of 2.4).

More precisely $\widetilde{f^*(q)}$ is the boundary of the composite map $C_* \to T_* \overset{\tilde{g}}{\to} \hat{T}_*$ in the exact sequence $0 \to \hat{C}_* \to \hat{K}_* \to \hat{T}_* \to 0$. But $\hat{K}_*$ is acyclic. Hence $\tilde{q}$ is a homology equivalence if and only if $\widetilde{f^*(q)}$ is a homotopy equivalence.

§ 3 . THE ISOMORPHISM $L''_{n-1}(B,A) \cong L'_n(B,A)$

Let T_* be a torsion complex and q a non singular linking n-1-form over T_*. By 2.3 there exists a homology equivalence f from a B-acyclic free complex C_* to T_*, and by 2.5 we get an element $F(T_*,q)$ in $L'_n(B,A)$ represented by $(C_*,f^*(q))$. If f' $C'_* \to T_*$ is an other choice there is a homotopy equivalence g. $C'_* \to C_*$ and $f \circ g$ is homotopic to f'. Then $g^*(f^*(q))$ is equal to $f'^*(q)$ and $(C_*,f^*(q))$ is cobordant to $(C'_*,f'^*(q))$. Hence $F(T_*,q)$ depends only on (T_*,q).

Lemma 3.1

The correspondance F induces a morphism from $L''_{n-1}(B,A)$ to $L'_n(B,A)$.

Proof

Clearly F is additive. Then the only thing to do is to prove that $F(T_*,q)$ vanishes if (T_*,q) is cobordant to zero.

Suppose we have an exact sequence of torsion complexes

$$0 \to R_* \to S_* \to T_* \to 0$$

and a non singular linking n-2-form u over $S_* \to T_*$ with boundary q.

By 2.3 there exist B-acyclic free complexes K_*, Σ_*, C_* and a commutative diagram

$$\begin{array}{ccccccccc} 0 & \to & K_* & \to & \Sigma_* & \to & C_* & \to & 0 \\ \downarrow & & \downarrow & & \downarrow f & & \downarrow f & & \downarrow \\ 0 & \to & R_* & \to & S_* & \to & T_* & \to & 0 \end{array}$$

such that the lines are exact and the vertical maps are homology equivalences.

With a relative version of 2.5 we get a commutative diagram

$$\begin{array}{ccc} Q^{n-2}(S_* \to T_*) & \xrightarrow{f^*} & Q^{n-1}(\Sigma_* \to C_*) \\ \partial \downarrow & & \downarrow \partial \\ Q^{n-1}(T_*) & \xrightarrow{f^*} & Q^{n}(C_*) \end{array} .$$

Then $f^*(u)$ is a quadratic n-1-form over $\Sigma_* \to C_*$ with boundary $f^*(q)$. Since u is non singular, $f^*(u)$ is non singular too and $(C_*, f^*(q))$ is cobordant to zero.

Lemma 3.2

Let f $C_* \to T_*$ be a homology equivalence from a free complex to a torsion complex. Let $\hat{T}_* \otimes \hat{T}_*$ be the graded differential module endowed with the involution $t(a \otimes b) = -\varepsilon(-1)^{\partial^\circ a \partial^\circ b}\, b \otimes a$ $(\varepsilon = \pm 1)$ for any $a,b \in \hat{T}_*$. Then for any element $u \in H_*(\mathbb{Z}/2, \hat{T}_* \otimes \hat{T}_*)$ there exist a torsion complex T'_* and a homology equivalence α $T'_* \to T_*$ such that f lifts through α and $\alpha^*(u)$ vanishes.

Proof

Any element in $H_*(\mathbb{Z}/2, \hat{T}_* \otimes \hat{T}_*)$ is represented by $\sum_{i,p,q} e_i \otimes u_{pq}$, $u_{pq} \in \hat{T}_p \otimes \hat{T}_q$. Then it suffices to prove that for any $v \in \hat{T}_p \otimes \hat{T}_q$ there exists a surjective homology equivalence α $T'_* \to T_*$ such that f lifts through α and v goes to zero in $\hat{T}'_p \otimes \hat{T}_q$.

By the canonical isomorphism $\hat{T}_p \otimes \hat{T}_q \simeq \mathrm{Ext}^1(T_p,\hat{T}_q)$ an element $v \in \hat{T}_p \otimes \hat{T}_q$ gives an extension $0 \to \hat{T}_q \to T'_p \to T_p \to 0$ and v goes to zero in $\hat{T}'_p \otimes \hat{T}_q$.

By setting

$$T'_i = \begin{cases} T_i & i \neq p,p+1 \\ T'_p \underset{T_p}{\times} T_{p+1} & i = p+1 \end{cases}$$

we get a torsion complex T'_* and a homology equivalence $\alpha\ T'_* \to T_*$ such that v vanishes in $\hat{T}'_p \otimes \hat{T}_q$. Moreover f is a homology equivalence and C_* is free, then f lifts through α and the lemma is proved.

<u>Theorem 3.3</u>

The morphism $F\ \ L''_{n-1}(B,A) \to L'_n(B,A)$ is an isomorphism.

<u>Proof</u>

<u>Surjectivity of F</u>

Let $w' \in L'_n(B,A)$ represented by a B-acyclic free complex C_* together with a non singular quadratic n-form q over C_*. By 2.2 there exists a homology equivalence f from C_* to a torsion complex T_*.

Consider the exact sequence (3.4)

$$Q^{n-1}(T_*) \xrightarrow{f^*} Q^n(C_*) \xrightarrow{\partial} H_{-n+2}(\mathbb{Z}/2,\hat{T}_* \otimes \hat{T}_*) \quad .$$

By 3.2 there exist a torsion complex T'_* and a homology equivalence $\alpha : T'_* \to T_*$ such that f lifts by f' through T'_* and ∂q vanishes in $H_{-n+2}(\mathbb{Z}/2,\hat{T}'_* \otimes \hat{T}'_*)$. Then there exists $q' \in Q^{n-1}(T'_*)$ such that $q = f'^*(q')$. By 2.5 q' is non singular and (T'_*,q') gives an element in $L''_{n-1}(B,A)$ which is going to w' by F.

<u>Injectivity of F</u>

Let T_* be a torsion complex and q be a non singular linking n-1-form over T_* such that $F(T_*,q)$ vanishes. Take a homology equivalence f from a B-acyclic free complex C_* to T_*. Since $(C_*,f^*(q))$ is cobordant to zero there exists an exact sequence of B-acyclic free complexes

$$0 \to K_* \to \Sigma_* \to C_* \to 0$$

together with a non singular quadratic n-1-form u over $\Sigma_* \to C_*$ with boundary $f^*(q)$.

By 2.2 we can construct a commutative diagram

$$\begin{array}{ccccccccc} 0 & \to & K_* & \to & \Sigma_* & \to & C_* & \to & 0 \\ \downarrow & & \downarrow & & \downarrow f & & \downarrow f & & \downarrow \\ 0 & \to & R_* & \to & S_* & \to & T_* & \to & 0 \end{array}$$

such that the lines are exact and the vertical maps are homology equivalences.

Consider the commutative diagram

$$\begin{array}{ccccccc} H_{-n+3}(\mathbb{Z}/2,\hat{S}_* \otimes \hat{S}_*) & & & & & & \\ \downarrow & & & & & & \\ Q^{n-1}(S_*) & \to & Q^{n-1}(T_*) & \xrightarrow{\partial} & Q^{n-2}(S_* \to T_*) & \to & H_{-n+2}(\mathbb{Z}/2,B_*(S_*)^{-\varepsilon}) \\ \downarrow & & \downarrow f^* & & \downarrow f^* & & \downarrow \\ Q^n(\Sigma_*) & \to & Q^n(C_*) & \xrightarrow{\partial} & Q^{n-1}(\Sigma_* \to C_*) & \to & H_{-n+1}(\mathbb{Z}/2,B_*(\Sigma_*)^{\varepsilon}) \\ & & & & & & \downarrow \\ & & & & & & H_{-n+3}(\mathbb{Z}/2,\hat{S}_* \otimes \hat{S}_*) \end{array}$$

All the lines and the columns of this diagram are exact.

Since $f^*(q)$ is the boundary of u, the image of q in $Q^{n-1}(S_*)$ comes from $H_{-n+3}(\mathbb{Z}/2,\hat{S}_* \otimes \hat{S}_*)$. By 3.2 we may as well suppose that q restricts to zero on S_* and is the boundary of an element $v' \in Q^{n-2}(S_* \to T_*)$. The obstruction to lift $f^*(v')-u$ in $Q^{n-2}(S_* \to T_*)$ is in $H_{-n+3}(\mathbb{Z}/2,\hat{S}_* \otimes \hat{S}_*)$. By 3.2 we may as well suppose that this obstruction vanishes and there exists an element $v \in Q^{n-2}(S_* \to T_*)$ such that $\partial v = q$ and $f^*v = u$.

Since u is non singular, v is non singular and (T_*,q) is cobordant to zero.

Corollary 3.4

The group $L''_{n-1}(B,A)$ is isomorphic to $L^h_{n+1}(A \to \Lambda)$.

Corollary 3.5

The group $L^h_{n+1}(A \to \Lambda)$ depends only on the category $\mathcal{C}_B$ of torsion modules.

Theorem 3.6

Let

$$\begin{array}{ccc} A & \to & B \\ \downarrow & & \downarrow \\ C & \to & D \end{array}$$

be a square of rings with involution such that $A \to B$ and $C \to D$ are local and A (resp. C) doesn't contain any finitely generated B-perfect (resp. D-perfect) submodule (that holds for example if $A \to B$ and $C \to D$ are monic). Suppose we have the following conditions

i) for any torsion module $M \in \mathcal{C}_B$ the map $M \to M \otimes C$ is an isomorphism and $\mathrm{Tor}_1^A(M,C) = 0$

ii) any torsion module $N \in \mathcal{C}_D$, considered as A-module, is in $\mathcal{C}_B$.

Then we have an long exact sequence

$$\dots \xrightarrow{\partial} L_n^h(A) \to L_n^h(B) \oplus L_n^h(C) \to L_n^h(D) \xrightarrow{\partial} L_{n-1}^h(A) \to \dots \quad .$$

Proof

The conditions i) and ii) imply that $\mathcal{C}_B$ and $\mathcal{C}_D$ are equivalent. Then the map $L_n^h(A \to B) \to L_n^h(C \to D)$ is an isomorphism and the Mayer-Vietoris exact sequence holds.

Remark 3.7

Actually it is possible to give an interpretation of the group $L_n^h(A \to \Lambda)$ in term of linking form over torsion modules as in [1], [3], [5], [6] for example. That will appear in a further paper.

BIBLIOGRAPHY

[1] M. KAROUBI . Localisation des formes quadratiques I . Ann. Scient. E.N.S. (1974) pp 359-404 .

[2] W. PARDON . The exact sequence of a localization of Witt groups . Springer Lecture Notes in Math. 551 (1976) pp 336-379 .

[3] W. PARDON . Local surgery and the exact sequence of a localization for Wall groups . A.M.S. Memoirs 196 (1977) .

[4] A. RANICKI . Algebraic L-theory I - Foundations . Proc. London Math. Soc. 27 (1973) pp 101-125 .

[5] A. RANICKI . The algebraic theory of surgery . Preprint .

[6] J. SMITH . Complements of codimension two submanifolds III - Cobordism theory . Preprint .

[7] P. VOGEL . On the homology surgery obstruction group . Preprint .

[8] C.T.C. WALL . Surgery on compact manifolds . Academic Press . New York and London (1970) .

Université de Nantes

Institut de Mathématiques
et d'Informatique

2, Chemin de la Houssinière

44072 NANTES Cedex (France)

Vol 640 J L Dupont, Curvature and Characteristic Classes X, 175 pages 1978

Vol 641 Seminaire d'Algèbre Paul Dubreil. Proceedings Paris 1976–1977 Edité par M P Malliavin IV, 367 pages 1978

Vol 642 Theory and Applications of Graphs, Proceedings, Michigan 1976 Edited by Y Alavi and D R Lick XIV, 635 pages 1978

Vol 643 M Davis, Multiaxial Actions on Manifolds VI, 141 pages 1978

Vol 644 Vector Space Measures and Applications I, Proceedings 1977 Edited by R M Aron and S Dineen VIII, 451 pages 1978

Vol 645 Vector Space Measures and Applications II, Proceedings 1977 Edited by R M Aron and S Dineen VIII, 218 pages 1978

Vol 646 O Tammi, Extremum Problems for Bounded Univalent Functions VIII, 313 pages 1978

Vol 647 L J Ratliff, Jr, Chain Conjectures in Ring Theory VIII, 133 pages 1978

Vol 648 Nonlinear Partial Differential Equations and Applications, Proceedings, Indiana 1976–1977 Edited by J M Chadam VI 206 pages 1978

Vol 649 Séminaire de Probabilités XII, Proceedings, Strasbourg, 1976–1977 Edité par C Dellacherie, P A Meyer et M Weil VIII, 805 pages 1978

Vol 650 C*-Algebras and Applications to Physics Proceedings 1977 Edited by H Araki and R V Kadison V, 192 pages 1978

Vol 651 P W Michor, Functors and Categories of Banach Spaces VI, 99 pages 1978

Vol 652 Differential Topology, Foliations and Gelfand-Fuks-Cohomology, Proceedings 1976 Edited by P A Schweitzer XIV, 252 pages 1978

Vol 653 Locally Interacting Systems and Their Application in Biology Proceedings, 1976 Edited by R L Dobrushin, V I Kryukov and A L Toom XI, 202 pages 1978

Vol 654 J P Buhler, Icosahedral Golois Representations III, 143 pages 1978

Vol 655 R Baeza, Quadratic Forms Over Semilocal Rings VI, 199 pages 1978

Vol 656 Probability Theory on Vector Spaces Proceedings, 1977 Edited by A Weron VIII, 274 pages 1978

Vol 657 Geometric Applications of Homotopy Theory I, Proceedings 1977 Edited by M G Barratt and M E Mahowald VIII, 459 pages 1978

Vol 658 Geometric Applications of Homotopy Theory II, Proceedings 1977 Edited by M G Barratt and M E Mahowald VIII, 487 pages 1978

Vol 659 Bruckner, Differentiation of Real Functions X, 247 pages 1978

Vol 660 Equations aux Derivee Partielles Proceedings, 1977 Edite par Pham The Lai VI, 216 pages 1978

Vol 661 P T Johnstone, R Pare, R D Rosebrugh, D Schumacher, R J Wood, and G C Wraith, Indexed Categories and Their Applications VII, 260 pages 1978

Vol 662 Akin, The Metric Theory of Banach Manifolds XIX, 306 pages 1978

Vol 663 J F Berglund, H D Junghenn, P Milnes, Compact Right Topological Semigroups and Generalizations of Almost Periodicity X, 243 pages 1978

Vol 664 Algebraic and Geometric Topology Proceedings, 1977 Edited by K C Millett XI, 240 pages 1978

Vol 665 Journees d'Analyse Non Lineaire Proceedings, 1977 Edite par P Benilan et J Robert VIII, 256 pages 1978

Vol 666 B Beauzamy, Espaces d'Interpolation Reels Topologie et Geometrie X, 104 pages 1978

Vol 667 J Gilewicz, Approximants de Padé XIV, 511 pages 1978

Vol 668 The Structure of Attractors in Dynamical Systems Proceedings, 1977 Edited by J C Martin, N G Markley and W Perrizo VI, 264 pages 1978

Vol 669 Higher Set Theory Proceedings, 1977 Edited by G H Muller and D S Scott XII, 476 pages 1978

Vol 670 Fonctions de Plusieurs Variables Complexes III, Proceedings, 1977 Edite par F Norguet XII, 394 pages 1978

Vol 671 R T Smythe and J C Wierman, First-Passage Perculation on the Square Lattice VIII, 196 pages 1978

Vol 672 R L Taylor, Stochastic Convergence of Weighted Sums of Random Elements in Linear Spaces VII, 216 pages 1978

Vol 673 Algebraic Topology, Proceedings 1977 Edited by P Hoffman, R Piccinini and D Sjerve VI, 278 pages 1978

Vol 674 Z Fiedorowicz and S Priddy, Homology of Classical Groups Over Finite Fields and Their Associated Infinite Loop Spaces VI, 434 pages 1978

Vol 675 J Galambos and S Kotz, Characterizations of Probability Distributions VIII, 169 pages 1978

Vol 676 Differential Geometrical Methods in Mathematical Physics II, Proceedings, 1977 Edited by K Bleuler, H R Petry and A Reetz VI, 626 pages 1978

Vol 677 Seminaire Bourbaki, vol 1976/77, Exposes 489–506 IV, 264 pages 1978

Vol 678 D Dacunha-Castelle, H Heyer et B Roynette Ecole d'Ete de Probabilites de Saint-Flour VII-1977 Edite par P L Hennequin IX, 379 pages 1978

Vol 679 Numerical Treatment of Differential Equations in Applications, Proceedings, 1977 Edited by R Ansorge and W Tornig IX, 163 pages 1978

Vol 680 Mathematical Control Theory, Proceedings, 1977 Edited by W A Coppel IX, 257 pages 1978

Vol 681 Seminaire de Theorie du Potentiel Paris, No 3, Directeurs M Brelot, G Choquet et J Deny Redacteurs F Hirsch et G Mokobodzki VII, 294 pages 1978

Vol 682 G D James, The Representation Theory of the Symmetric Groups V, 156 pages 1978

Vol 683 Varietes Analytiques Compactes, Proceedings, 1977 Edite par Y Hervier et A Hirschowitz V, 248 pages 1978

Vol 684 E E Rosinger, Distributions and Nonlinear Partial Differential Equations XI, 146 pages 1978

Vol 685 Knot Theory, Proceedings, 1977 Edited by J C Hausmann VII, 311 pages 1978

Vol 686 Combinatorial Mathematics, Proceedings, 1977 Edited by D A Holton and J Seberry IX, 353 pages 1978

Vol 687 Algebraic Geometry, Proceedings, 1977 Edited by L D Olson V, 244 pages 1978

Vol 688 J Dydak and J Segal, Shape Theory VI, 150 pages 1978

Vol 689 Cabal Seminar 76–77, Proceedings, 1976–77 Edited by A S Kechris and Y N Moschovakis V, 282 pages 1978

Vol 690 W J J Rey, Robust Statistical Methods VI, 128 pages 1978

Vol 691 G Viennot, Algebres de Lie Libres et Monoides Libres III, 124 pages 1978

Vol 692 T Husain and S M Khaleelulla, Barrelledness in Topological and Ordered Vector Spaces IX, 258 pages 1978

Vol 693 Hilbert Space Operators, Proceedings, 1977 Edited by J M Bachar Jr and D W Hadwin VIII, 184 pages 1978

Vol 694 Seminaire Pierre Lelong – Henri Skoda (Analyse) Annee 1976/77 VII, 334 pages 1978

Vol 695 Measure Theory Applications to Stochastic Analysis, Proceedings, 1977 Edited by G Kallianpur and D Kolzow XII, 261 pages 1978

Vol 696 P J Feinsilver, Special Functions, Probability Semigroups, and Hamiltonian Flows VI, 112 pages 1978

Vol 697 Topics in Algebra, Proceedings, 1978 Edited by M F Newman XI, 229 pages 1978

Vol 698 E Grosswald, Bessel Polynomials XIV, 182 pages 1978

Vol 699 R E Greene and H H Wu, Function Theory on Manifolds Which Possess a Pole III, 215 pages 1979

Vol 700 Module Theory, Proceedings, 1977 Edited by C Faith and S Wiegand X, 239 pages 1979

Vol 701 Functional Analysis Methods in Numerical Analysis, Proceedings, 1977 Edited by M Zuhair Nashed VII, 333 pages 1979

Vol 702 Yuri N Bibikov, Local Theory of Nonlinear Analytic Ordinary Differential Equations IX, 147 pages 1979

Vol 703 Equadiff IV, Proceedings, 1977 Edited by J Fábera XIX, 441 pages 1979

Vol 704 Computing Methods in Applied Sciences and Engineering, 1977, I Proceedings, 1977 Edited by R Glowinski and J L Lions VI, 391 pages 1979

Vol 705 O Forster und K Knorr, Konstruktion verseller Familien kompakter komplexer Raume VII, 141 Seiten 1979

Vol 706 Probability Measures on Groups, Proceedings 1978 Edited by H Heyer XIII 348 pages 1979

Vol 707 R Zielke, Discontinuous Čebyšev Systems VI, 111 pages 1979

Vol 708 J P Jouanolou, Equations de Pfaff algébriques V 255 pages 1979

Vol 709 Probability in Banach Spaces II Proceedings, 1978 Edited by A Beck V, 205 pages 1979

Vol 710 Séminaire Bourbaki vol 1977/78, Exposés 507–524 IV, 328 pages 1979

Vol 711 Asymptotic Analysis Edited by F Verhulst V, 240 pages 1979

Vol 712 Equations Différentielles et Systèmes de Pfaff dans le Champ Complexe Edité par R Gérard et J P Ramis V, 364 pages 1979

Vol 713 Séminaire de Théorie du Potentiel, Paris No 4 Edité par F Hirsch et G Mokobodzki VII, 281 pages 1979

Vol 714 J Jacod, Calcul Stochastique et Problèmes de Martingales X, 539 pages 1979

Vol 715 Inder Bir S Passi, Group Rings and Their Augmentation Ideals VI, 137 pages 1979

Vol 716 M A Scheunert, The Theory of Lie Superalgebras X, 271 pages 1979

Vol 717 Grosser, Bidualraume und Vervollstandigungen von Banachmoduln III, 209 pages 1979

Vol 718 J Ferrante and C W Rackoff, The Computational Complexity of Logical Theories X, 243 pages 1979

Vol 719 Categorial Topology, Proceedings, 1978 Edited by H Herrlich and G Preuß XII, 420 pages 1979

Vol 720 E Dubinsky, The Structure of Nuclear Fréchet Spaces V, 187 pages 1979

Vol 721 Séminaire de Probabilités XIII Proceedings Strasbourg, 1977/78 Edité par C Dellacherie, P A Meyer et M Weil VII, 647 pages 1979

Vol 722 Topology of Low-Dimensional Manifolds Proceedings, 1977 Edited by R Fenn VI, 154 pages 1979

Vol 723 W Brandal, Commutative Rings whose Finitely Generated Modules Decompose II, 116 pages 1979

Vol 724 D Griffeath, Additive and Cancellative Interacting Particle Systems V, 108 pages 1979

Vol 725 Algebres d'Opérateurs Proceedings, 1978 Edite par P de la Harpe VII, 309 pages 1979

Vol 726 Y-C Wong, Schwartz Spaces, Nuclear Spaces and Tensor Products VI, 418 pages 1979

Vol 727 Y Saito, Spectral Representations for Schrodinger Operators With Long-Range Potentials V, 149 pages 1979

Vol 728 Non Commutative Harmonic Analysis Proceedings, 1978 Edited by J Carmona and M Vergne V, 244 pages 1979

Vol 729 Ergodic Theory Proceedings, 1978 Edited by M Denker and K Jacobs XII, 209 pages 1979

Vol 730 Functional Differential Equations and Approximation of Fixed Points Proceedings, 1978 Edited by H O Peitgen and H-O Walther XV, 503 pages 1979

Vol 731 Y Nakagami and M Takesaki, Duality for Crossed Products of von Neumann Algebras IX, 139 pages 1979

Vol 732 Algebraic Geometry Proceedings, 1978 Edited by K Lønsted IV, 658 pages 1979

Vol 733 F Bloom, Modern Differential Geometric Techniques in the Theory of Continuous Distributions of Dislocations XII, 206 pages 1979

Vol 734 Ring Theory, Waterloo, 1978 Proceedings, 1978 Edited by D Handelman and J Lawrence XI, 352 pages 1979

Vol 735 B Aupetit, Propriétés Spectrales des Algebres de Banach XII, 192 pages 1979

Vol 736 E Behrends, M Structure and the Banach Stone Theorem X, 217 pages 1979

Vol 737 Volterra Equations Proceedings 1978 Edited by S-O Londen and O J Staffans VIII, 314 pages 1979

Vol 738 P E Conner, Differentiable Periodic Maps 2nd edition, IV, 181 pages 1979

Vol 739 Analyse Harmonique sur les Groupes de Lie II Proceedings, 1976–78 Edited by P Eymard et al VI, 646 pages 1979

Vol 740 Seminaire d'Algebre Paul Dubreil Proceedings, 1977–78 Edited by M-P Malliavin V, 456 pages 1979

Vol 741 Algebraic Topology, Waterloo 1978 Proceedings Edited by P Hoffman and V Snaith XI, 655 pages 1979

Vol 742 K Clancey, Seminormal Operators VII, 125 pages 1979

Vol 743 Romanian Finnish Seminar on Complex Analysis Proceedings, 1976 Edited by C Andreian Cazacu et al XVI, 713 pages 1979

Vol 744 I Reiner and K W Roggenkamp, Integral Representations VIII, 275 pages 1979

Vol 745 D K Haley, Equational Compactness in Rings III, 167 pages 1979

Vol 746 P Hoffman, τ-Rings and Wreath Product Representations V, 148 pages 1979

Vol 747 Complex Analysis, Joensuu 1978 Proceedings, 1978 Edited by I Laine, O Lehto and T Sorvali XV, 450 pages 1979

Vol 748 Combinatorial Mathematics VI Proceedings, 1978 Edited by A F Horadam and W D Wallis IX, 206 pages 1979

Vol 749 V Girault and P A Raviart, Finite Element Approximation of the Navier Stokes Equations VII, 200 pages 1979

Vol 750 J C Jantzen, Moduln mit einem hochsten Gewicht III, 195 Seiten 1979

Vol 751 Number Theory, Carbondale 1979 Proceedings Edited by M B Nathanson V, 342 pages 1979

Vol 752 M Barr, * Autonomous Categories VI, 140 pages 1979

Vol 753 Applications of Sheaves Proceedings, 1977 Edited by M Fourman, C Mulvey and D Scott XIV, 779 pages 1979

Vol 754 O A Laudal, Formal Moduli of Algebraic Structures III, 161 pages 1979

Vol 755 Global Analysis Proceedings, 1978 Edited by M Grmela and J E Marsden VII, 377 pages 1979

Vol 756 H O Cordes, Elliptic Pseudo Differential Operators – An Abstract Theory IX, 331 pages 1979

Vol 757 Smoothing Techniques for Curve Estimation Proceedings, 1979 Edited by Th Gasser and M Rosenblatt V, 245 pages 1979

Vol 758 C Năstăsescu and F Van Oystaeyen, Graded and Filtered Rings and Modules X, 148 pages 1979